Entropy in Dynamical Systems

This comprehensive text on entropy covers three major types of dynamics: measure-preserving transformations; continuous maps on compact spaces; and operators on function spaces.

Part I contains proofs of the Shannon–McMillan–Breiman Theorem, the Ornstein–Weiss Return Time Theorem, the Krieger Generator Theorem, the Sinai and Ornstein Theorems, and among the newest developments, the Ergodic Law of Series. In Part II, after an expanded exposition of classical topological entropy, the book addresses Symbolic Extension Entropy. It offers deep insight into the theory of entropy structure and explains the role of zero-dimensional dynamics as a bridge between measurable and topological dynamics. Part III explains how both measure-theoretic and topological entropy can be extended to operators on relevant function spaces.

Intuitive explanations, examples, exercises and open problems make this an ideal text for a graduate course on entropy theory. More experienced researchers can also find inspiration for further research.

TOMASZ DOWNAROWICZ is Full Professor in Mathematics at Wroclaw University of Technology, Poland.

NEW MATHEMATICAL MONOGRAPHS

All the titles listed below can be obtained from good booksellers or from Cambridge University Press. For a complete series listing visit www.cambridge.org/mathematics.

1 M. Cabanes and M. Enguehard *Representation Theory of Finite Reductive Groups*
2 J. B. Garnett and D. E. Marshall *Harmonic Measure*
3 P. Cohn *Free Ideal Rings and Localization in General Rings*
4 E. Bombieri and W. Gubler *Heights in Diophantine Geometry*
5 Y. J. Ionin and M. S. Shrikhande *Combinatorics of Symmetric Designs*
6 S. Berhanu, P. D. Cordaro and J. Hounie *An Introduction to Involutive Structures*
7 A. Shlapentokh *Hilbert's Tenth Problem*
8 G. Michler *Theory of Finite Simple Groups I*
9 A. Baker and G. Wüstholz *Logarithmic Forms and Diophantine Geometry*
10 P. Kronheimer and T. Mrowka *Monopoles and Three-Manifolds*
11 B. Bekka, P. de la Harpe and A. Valette *Kazhdan's Property (T)*
12 J. Neisendorfer *Algebraic Methods in Unstable Homotopy Theory*
13 M. Grandis *Directed Algebraic Topology*
14 G. Michler *Theory of Finite Simple Groups II*
15 R. Schertz *Complex Multiplication*
16 S. Bloch *Lectures on Algebraic Cycles (2nd Edition)*
17 B. Conrad, O. Gabber and G. Prasad *Pseudo-reductive Groups*

Entropy in Dynamical Systems

TOMASZ DOWNAROWICZ
Wroclaw University of Technology, Poland

CAMBRIDGE
UNIVERSITY PRESS

University Printing House, Cambridge CB2 8BS, United Kingdom

Published in the United States of America by Cambridge University Press, New York

Cambridge University Press is part of the University of Cambridge.

It furthers the University's mission by disseminating knowledge in the pursuit of education, learning and research at the highest international levels of excellence.

www.cambridge.org
Information on this title: www.cambridge.org/9780521888851

First published 2011

A catalogue record for this publication is available from the British Library

Library of Congress Cataloguing in Publication data

Downarowicz, Tomasz, 1956–
Entropy in Dynamical Systems / Tomasz Downarowicz.
p. cm. – (New Mathematical Monographs ; 18)
Includes bibliographical references and index.
ISBN 978-0-521-88885-1 (Hardback)
1. Topological entropy–Textbooks. 2. Topological dynamics–Textbooks. I. Title.
QA611.5.D685 2011
515'.39–dc22

2010050336

ISBN 978-0-521-88885-1 Hardback

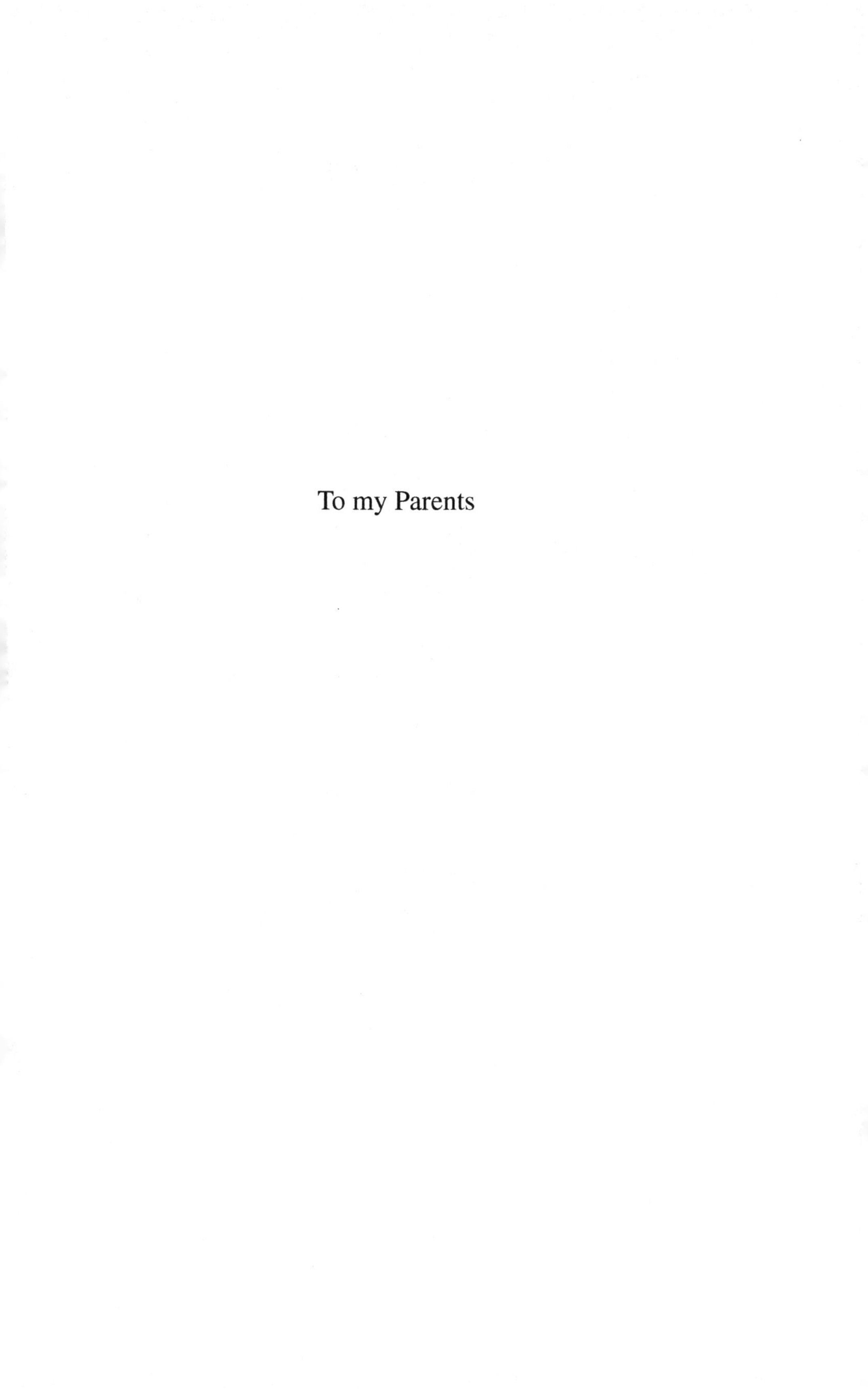

To my Parents

Contents

Preface

This book is designed as a comprehensive lecture on entropy in three major types of dynamics: measure-theoretic, topological and operator. In each case the study is restricted to the most classical case of the action of iterates of a single transformation (or operator) on either a standard probability space or on a compact metric space. We do not venture into studying actions of more general groups, dynamical systems on noncompact spaces or equipped with infinite measures. On the other hand, we do not restrict the generality by adding more structure to our spaces. The most structured systems addressed here in detail are smooth transformations of the compact interval. The primary intention is to create a self-contained course, from the basics through more advanced material to the newest developments. Very few theorems are quoted without a proof, mainly in the chapters or sections marked with an asterisk. These are treated as "nonmandatory" for the understanding of the rest of the book, and can be skipped if the reader chooses. Our facts are stated as generally as possible within the assumed scope, and wherever possible our proofs of classical theorems are different from those found in the most popular textbooks. Several chapters contain very recent results for which this is a textbook debut.

We assume familiarity of the reader with basics of ergodic theory, measure theory, topology and functional analysis. Nevertheless, the most useful facts are recalled either in the main text or in the appendix.

Some elementary statements and minor passages are left without a proof, as an exercise for the reader. Such statements are collected at the end of each chapter, together with other exercises of independent interest. It is planned that solutions to selected exercises will be made available shortly after the book has occurred in print, at the publisher's website www.cambridge.org/9780521888851.

Acknowledgments

First of all, I wish to express my gratitude to those who helped me with the mathematical issues, especially to Mike Boyle, Dan Rudolph, Jean-Paul Thouvenot and Benjy Weiss. I thank all those who read the preliminary version and helped me fix the mistakes. Large portions were proofread by Bartosz Frej, Jacek Serafin and David Burguet. I am grateful to the authorities of my Institute and Faculty for their understanding and a reduction on other university duties allowing me to better focus on writing the book. My warmest thanks are directed to Sylwia for her love and priceless help in organizing my work and everyday life during this busy time.

The research of the author was partially supported by the Polish Ministry of Education Grant Number N N201 394537.

Introduction

0.1 The leitmotiv

Nowadays, nearly every kind of information is turned into digital form. Digital cameras turn every image into a computer file. The same happens to musical recordings or movies. Even our mathematical work is registered mainly as computer files. Analog information is nearly extinct.

While studying dynamical systems (in any understanding of this term) sooner or later one is forced to face the following question: How can the information about the evolution of a given dynamical system be most precisely turned into a digital form? Researchers specializing in dynamical systems are responsible for providing the theoretical background for such a transition.

So suppose that we do observe a dynamical system, and that we indeed turn our observation into digital form. That means, from time to time, we produce a digital "report," a computer file, containing all our observations since the last report. Assume for simplicity that such reports are produced at equal time distances, say, at integer times. Of course, due to bounded capacity of our recording devices and limited time between the reports, our files have bounded size (in bits). Because the variety of digital files of bounded size is finite, we can say that at every integer moment of time we produce just one *symbol*, where the collection of all possible symbols, i.e. the *alphabet*, is finite.

An illustrative example is filming a scene using a digital camera. Every unit of time, the camera registers an image, which is in fact a bitmap of some fixed size (camera resolution). The camera turns the live scene into a sequence of bitmaps. We can treat every such bitmap as a single symbol in the alphabet of the "language" of the camera.

The sequence of symbols is produced as long as the observation is being conducted. We have no reason to restrict the global observation time, and we can agree that it goes on forever. Sometimes (but not always), we can imagine that the observation has been conducted since forever in the past as well. In this manner, the history of our recording takes on the form of a unilateral or bilateral sequence of symbols from some finite alphabet. Advancing in time by a unit corresponds, on one hand, to the unit-time evolution of the dynamical system, on the other, to shifting the enumeration of our sequence of symbols. This way we have come to the conclusion that the digital form of the observation is nothing else but an element of the space of all sequences of symbols, and the action on this space is the familiar shift transformation advancing the enumeration.

Now, in most situations, such a "digitalization" of the dynamical system will be *lossy*, i.e., it will capture only some aspects of the observed dynamical system, and much of the information will be lost. For example, the digital camera will not be able to register objects hidden behind other objects, moreover, it will not see objects smaller than one pixel or their movements until they pass from one pixel to another. However, it may happen that, after a while, each object will eventually become detectable, and we will be able to reconstruct its trajectory from the recorded information.

Of course, lossy digitalization is always possible and hence presents a lesser kind of challenge. We will be much more interested in *lossless* digitalization. When and how is it possible to digitalize a dynamical system so that no information is lost, i.e., in such a way that after viewing the entire sequence of symbols we can completely reconstruct the evolution of the system?

In this book the task of encoding a system with possibly smallest alphabet is refereed to as "data compression." The reader will find answers to the above question at two major levels: measure-theoretic, and topological. In the first case the digitalization is governed by the *Kolmogorov–Sinai entropy* of the dynamical system, the first major subject of this book. In the topological setup the situation is more complicated. Topological entropy, our second most important notion, turns out to be insufficient to decide about digitalization that respects the topological structure. Thus another parameter, called *symbolic extension entropy*, emerges as the third main object discussed in the book.

We also study entropy (both measure-theoretic and topological) for operators on function spaces, which generalize classical dynamical systems. The reference to data compression is not as clear here and we concentrate more on technical properties that carry over from dynamical systems, leaving the precise connection with information theory open for further investigation.

0.2 A few words about the history of entropy

Below we review very briefly the development of the notion of entropy focusing on the achievements crucial for the genesis of the basic concepts of entropy discussed in this book. For a more complete survey we refer to the expository article [Katok, 2007].

The term "entropy" was coined by a German physicist Rudolf Clausius from Greek "en-" = in + "trope" = a turning [Clausius, 1850]. The word reveals analogy to "energy" and was designed to mean the form of energy that any energy eventually and inevitably "turns into" – a useless heat. The idea was inspired by an earlier formulation by French physicist and mathematician Nicolas Léonard Sadi Carnot [Carnot, 1824] of what is now known as the *Second Law of Thermodynamics*: entropy represents the energy no longer capable to perform work, and in any isolated system it can only grow.

Austrian physicist Ludwig Boltzmann put entropy into the probabilistic setup of statistical mechanics [Boltzmann, 1877]. Entropy has also been generalized around 1932 to quantum mechanics by John von Neumann [see von Neumann, 1968].

Later this led to the invention of entropy as a term in probability and information theory by an American electronic engineer and mathematician Claude Elwood Shannon, now recognized as the father of information theory. Many of the notions have not changed much since they first occurred in Shannon's seminal paper *A Mathematical Theory of Communication* [Shannon, 1948]. Dynamical entropy in dynamical systems was created by one of the most influential mathematicians of modern times, Andrei Nikolaevich Kolmogorov, [Kolmogorov, 1958, 1959] and improved by his student Yakov Grigorevich Sinai who practically brought it to the contemporary form [Sinai, 1959].

The most important theorem about the dynamical entropy, so-called Shannon–McMillan–Breiman Theorem gives this notion a very deep meaning. The theorem was conceived by Shannon [Shannon, 1948], and proved in increasing strength by Brockway McMillan [McMillan, 1953] (L^1-convergence), Leo Breiman [Breiman, 1957] (almost everywhere convergence), and Kai Lai Chung [Chung, 1961] (for countable partitions). In 1970 Wolfgang Krieger obtained one of the most important results, from the point of view of data compression, about the existence (and cardinality) of finite generators for automorphisms with finite entropy [Krieger, 1970].

In 1970 Donald Ornstein proved that Kolmogorov–Sinai entropy was a *a complete invariant* in the class of *Bernoulli systems*, a fact considered one of the most important features of entropy (alternatively of Bernoulli systems) [Ornstein, 1970a].

In 1965, Roy L. Adler, Alan G. Konheim and M. Harry McAndrew carried the concept of dynamical entropy over to topological dynamics [Adler *et al.*, 1965] and in 1970 Efim I. Dinaburg and (independently) in 1971 Rufus Bowen redefined it in the language of metric spaces [Dinaburg, 1970; Bowen, 1971]. With regard to entropy in topological systems, probably the most important theorem is the Variational Principle proved by L. Wayne Goodwyn (the "easy" direction) and Timothy Goodman (the "hard" direction), which connects the notions of topological and Kolmogorov–Sinai entropy [Goodwyn, 1971; Goodman, 1971] (earlier Dinaburg proved both directions for finite-dimensional spaces [Dinaburg, 1970]).

The theory of symbolic extensions of topological systems was initiated by Mike Boyle around 1990 [Boyle, 1991]. The outcome of this early work is published in [Boyle *et al.*, 2002]. The author of this book contributed to establishing that invariant measures and their entropies play a crucial role in computing the so-called symbolic extension entropy [Downarowicz, 2001; Boyle and Downarowicz, 2004; Downarowicz, 2005a].

Dynamical entropy generalizing the Kolmogorov–Sinai dynamical entropy to noncommutative dynamics occurred as an adaptation of von Neumann's quantum entropy in a work of Robert Alicki, Johan Andries, Mark Fannes and Pim Tuyls [Alicki *et al.*, 1996] and then was applied to doubly stochastic operators by Igor I. Makarov [Makarov, 2000]. The axiomatic approach to entropy of doubly stochastic operators, as well as topological entropy of Markov operators have been developed in [Downarowicz and Frej, 2005].

The term "entropy" is used in many other branches of science, sometimes distant from physics or mathematics (such as sociology), where it no longer maintains its rigorous quantitative character. Usually, it roughly means "disorder," "chaos," "decay of diversity" or "tendency toward uniform distribution of kinds."

0.3 Multiple meanings of entropy

In the following paragraphs we review some of the various meanings of the word "entropy" and try to explain how they are connected. We devote a few pages to explain how dynamical entropy corresponds to data compression rate; this interpretation plays a central role in the approach to entropy in dynamical systems presented in the book. The notation used in this section is temporary.

0.3.1 Entropy in physics

In classical physics, a physical system is a collection of objects (bodies) whose *state* is parametrized by several characteristics such as the distribution of density, pressure, temperature, velocity, chemical potential, etc. The change of entropy of a physical system, as it passes from one state to another, is

$$\Delta S = \int \frac{dQ}{T},$$

where dQ denotes an element of heat being absorbed (or emitted; then it has the negative sign) by a body, T is the absolute temperature of that body at that moment, and the integration is over all elements of heat active in the passage. The above formula allows us to compare entropies of different states of a system, or to compute entropy of each state up to an additive constant (this is satisfactory in most cases). Notice that when an element dQ of heat is transmitted from a warmer body of temperature T_1 to a cooler one of temperature T_2 then the entropy of the first body changes by $-dQ/T_1$, while that of the other rises by dQ/T_2. Since $T_2 < T_1$, the absolute value of the latter fraction is larger and jointly the entropy of the two-body system increases (while the global energy remains the same).

A system is *isolated* if it does not exchange energy or matter (or even information) with its surroundings. By virtue of the First Law of Thermodynamics, the conservation of energy principle, an isolated system can pass only between states of the same global energy. The Second Law of Thermodynamics introduces irreversibility of the evolution: an isolated system cannot pass from a state of higher entropy to a state of lower entropy. Equivalently, it says that it is impossible to perform a process whose only final effect is the transmission of heat from a cooler medium to a warmer one. Any such transmission must involve an outside work, the elements participating in the work will also change their states and the overall entropy will rise.

The first and second laws of thermodynamics together imply that an isolated system will tend to the state of maximal entropy among all states of the same energy. The energy distributed in this state is incapable of any further activity. The state of maximal entropy is often called the "thermodynamical death" of the system.

Ludwig Boltzmann gave another, probabilistic meaning to entropy. For each state A the (negative) difference between the entropy of A and the entropy of the "maximal state" B is nearly proportional to the logarithm of the probability that the system spontaneously assumes state A,

$$S(A) - S_{max} \approx k \log_2(\mathsf{Prob}(A)).$$

The proportionality factor k is known as the Boltzmann constant. In this approach the probability of the maximal state is almost equal to 1, while the probabilities of states of lower entropy are exponentially small. This provides another interpretation of the Second Law of Thermodynamics: the system spontaneously assumes the state of maximal entropy simply because all other states are extremely unlikely.

Example Consider a physical system consisting of an ideal gas enclosed in a cylindrical container of volume 1. The state B of maximal entropy is clearly the

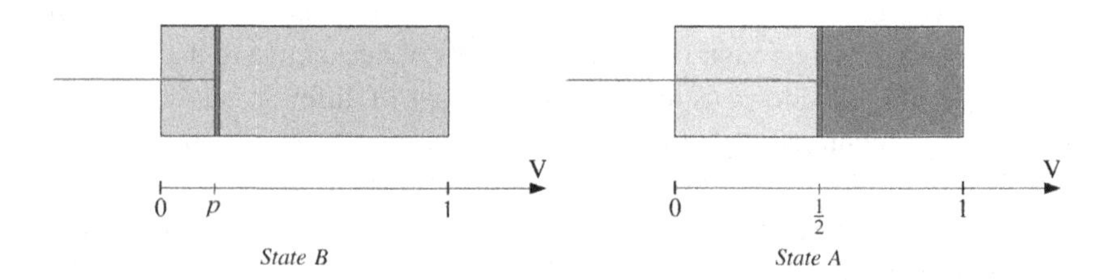

one where both pressure and temperature are constant (P_0 and T_0, respectively) throughout the container. Any other state can be achieved only with help from outside. Suppose one places a piston at a position $p < \frac{1}{2}$ in the cylinder (the left figure; thermodynamically, this is still the state B) and then slowly moves the piston to the center of the cylinder (position $\frac{1}{2}$), allowing the heat to flow between the cylinder and its environment, where the temperature is T_0, which stabilizes the temperature at T_0 all the time. Let A be the final state (the right figure). Note that both states A and B have the same energy level inside the system.

To compute the jump of entropy one needs to examine what exactly happens during the passage. The force acting on the piston at position x is proportional to the difference between the pressures:

$$F = c\left(P_0 \frac{1-p}{1-x} - P_0 \frac{p}{x}\right).$$

Thus, the work done while moving the piston equals:

$$W = \int_p^{\frac{1}{2}} F\,dx = cP_0\big((1-p)\ln(1-p) + p\ln p + \ln 2\big).$$

The function

$$p \mapsto (1-p)\ln(1-p) + p\ln p$$

is negative and assumes its minimal value $-\ln 2$ at $p = \frac{1}{2}$.

Thus the above work W is positive and represents the amount of energy delivered to the system from outside. During the process the compressed gas on the right emits heat, while the depressed gas on the left absorbs heat. By conservation of energy (applied to the enhanced system including the outside world), the gas altogether will emit heat to the environment equivalent to the delivered work

$\Delta Q = -W$. Since the temperature is constant all the time, the change in entropy between states B and A of the gas is simply $1/T_0$ times ΔQ, i.e.,

$$\Delta S = \frac{1}{T_0} \cdot cP_0\big(-(1-p)\ln(1-p) - p\ln p - \ln 2\big).$$

Clearly ΔS is negative. This confirms, what was already expected, that the outside intervention has lowered the entropy of the gas.

This example illustrates very clearly Boltzmann's interpretation of entropy. Assume that there are N particles of the gas independently wandering inside the container. For each particle the probability of falling in the left or right half of the container is 1/2. The state A of the gas occurs spontaneously if pN and $(1-p)N$ particles fall in the left and right halves of the container, respectively. By elementary combinatorics formulae, the probability of such an event equals

$$\mathsf{Prob}(A) = \frac{N!}{(pN)!((1-p)N)!}2^{-N}.$$

By Stirling's formula ($\ln n! \approx n\ln n - n$ for large n), the logarithm of $\mathsf{Prob}(A)$ equals approximately

$$N\big(-(1-p)\ln(1-p) - p\ln p - \ln 2\big),$$

which is indeed proportional to the drop ΔS of entropy between the states B and A (see above).

0.3.2 Shannon entropy

In probability theory, a *probability vector* $\mathbf{p}$ is a sequence of finitely many non-negative numbers $\{p_1, p_2, \dots, p_n\}$ whose sum equals 1. The Shannon entropy of a probability vector $\mathbf{p}$ is defined as

$$H(\mathbf{p}) = -\sum_{i=1}^{n} p_i \log_2 p_i$$

(where $0\log_2 0 = 0$). Probability vectors occur naturally in connection with finite partitions of a probability space. Consider an abstract space Ω equipped with a probability measure μ assigning probabilities to measurable subsets of Ω. A finite partition $\mathcal{P}$ of Ω is a collection of pairwise disjoint measurable sets $\{A_1, A_2, \dots, A_n\}$ whose union is Ω. Then the probabilities $p_i = \mu(A_i)$ form a probability vector $\mathbf{p}_{\mathcal{P}}$. One associates the entropy of this vector with the (ordered) partition $\mathcal{P}$:

$$H_\mu(\mathcal{P}) = H(\mathbf{p}_{\mathcal{P}}).$$

In this setup entropy can be linked with *information*. Given a measurable set A, the information $I(A)$ associated with A is defined as $-\log_2(\mu(A))$. The *information function* $I_{\mathcal{P}}$ associated with a partition $\mathcal{P} = \{A_1, A_2, \dots, A_n\}$ is defined on the space Ω and it assumes the constant value $I(A_i)$ at all points ω belonging to the set A_i. Formally,

$$I_{\mathcal{P}}(\omega) = \sum_{i=1}^{n} -\log_2(\mu(A_i))\mathbb{I}_{A_i}(\omega),$$

where $\mathbb{I}_{A_i}$ is the characteristic function of A_i. One easily verifies that the expected value of this function with respect to μ coincides with the entropy $H_\mu(\mathcal{P})$.

We shall now give an interpretation of the information function and entropy, the key notions in entropy theory. The partition $\mathcal{P}$ of the space Ω associates with each element $\omega \in \Omega$ the "information" that gives an answer to the question "in which A_i are you?". That is the best knowledge we can acquire about the points, based solely on the partition. One bit of information is equivalent to acquiring an answer to a binary question, i.e., a question of a choice between two possibilities. Unless the partition has two elements, the question "in which A_i are you?" is not binary. But it can be replaced by a series of binary questions and one is free to use any arrangement (tree) of such questions. In such an arrangement, the number of questions $N(\omega)$ (i.e., the amount of information in bits) needed to determine the location of the point ω within the partition may vary from point to point (see the example below). The smaller the expected value of $N(\omega)$ the better the arrangement. It turns out that the best arrangement satisfies $I_{\mathcal{P}}(\omega) \leq N(\omega) \leq I_{\mathcal{P}}(\omega) + 1$ for μ-almost every ω. The difference between $I_{\mathcal{P}}(\omega)$ and $N(\omega)$ follows from the crudeness of the measurement of information by counting binary questions; the outcome is always a positive integer. The real number $I_{\mathcal{P}}(\omega)$ can be interpreted as the precise value. Entropy is the expected amount of information needed to locate a point in the partition.

Example Consider the unit square representing the space Ω, where the probability is the Lebesgue measure (i.e., the surface area), and the partition $\mathcal{P}$ of Ω into four sets A_i of probabilities $\frac{1}{8}, \frac{1}{4}, \frac{1}{8}, \frac{1}{2}$, respectively, as shown in the figure.

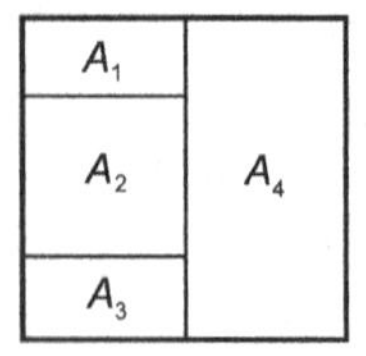

The information function equals $-\log_2(\frac{1}{8}) = 3$ on A_1 and A_3, $-\log_2(\frac{1}{4}) = 2$ on A_2 and $-\log_2(\frac{1}{2}) = 1$ on A_4. The entropy of $\mathcal{P}$ equals

$$H(\mathcal{P}) = \frac{1}{8} \cdot 3 + \frac{1}{4} \cdot 2 + \frac{1}{8} \cdot 3 + \frac{1}{2} \cdot 1 = \frac{7}{4}.$$

The arrangement of questions that optimizes the expected value of the number of questions asked is the following:

1. *Are you in the left half?*

The answer "no", locates ω in A_4 using one bit. Otherwise the next question is:

2. *Are you in the central square of the left half?*

The "yes" answer locates ω in A_2 using two bits. If not, the last question is:

3. *Are you in the top half of the whole square?*

Now "yes" or "no" locate ω in A_1 or A_3, respectively. This takes three bits.

$$\text{Question 1} \begin{cases} yes \to \text{Question 2} \begin{cases} yes \to A_2 \text{ (2 bits)} \\ no \to \text{Question 3} \begin{cases} yes \to A_1 \text{ (3 bits)} \\ no \to A_3 \text{ (3 bits)} \end{cases} \end{cases} \\ no \to A_4 \text{ (1 bit)} \end{cases}$$

In this example the number of questions equals exactly the information function at every point and the expected number of question equals the entropy $\frac{7}{4}$. There does not exist a better arrangement of questions. Of course, such an accuracy is possible only when the probabilities of the sets A_i are integer powers of 2; in general the information is not integer valued.

Another interpretation of Shannon entropy deals with the notion of *uncertainty*. Let X be a random variable defined on the probability space Ω and assuming values in a finite set $\{x_1, x_2, \ldots, x_n\}$. The variable X generates a partition $\mathcal{P}$ of Ω into the sets $A_i = \{\omega \in \Omega : \mathrm{X}(\omega) = x_i\}$ (called the preimage partition). The probabilities $p_i = \mu(A_i) = \mathsf{Prob}\{\mathrm{X} = x_i\}$ form a probability vector called the *distribution* of X. Suppose an experimenter knows the distribution of X and tries to guess the outcome of X before performing the experiment, i.e., before picking some $\omega \in \Omega$ and reading the value $\mathrm{X}(\omega)$. His/her *uncertainty* about the outcome is the expected value of the information he/she is *missing* to be certain. As explained above that is exactly the entropy $H_\mu(\mathcal{P})$.

0.3.3 Connection between Shannon and Boltzmann entropy

Both notions in the title of this subsection refer to probability and there is an evident similarity in the formulae. But the analogy fails to be obvious. In the literature many different attempts toward understanding the relation can be found. In simple words, the interpretation relies on the distinction between the macroscopic state considered in classical thermodynamics and the microscopic states of statistical mechanics. A thermodynamical state A (a distribution of pressure, temperature, etc.) can be realized in many different ways ω at the microscopic level, where one distinguishes all individual particles, their positions and velocity vectors. As explained above, the difference of Boltzmann entropies $S(A) - S_{max}$ is proportional to $\log_2(\mathsf{Prob}(A))$, the logarithm of the probability of the macroscopic state A in the probability space Ω of all microscopic states ω. This leads to the equation

$$S_{max} - S(A) = k \cdot I(A), \tag{0.3.1}$$

where $I(A)$ is the probabilistic information associated with the set $A \subset \Omega$. So, Boltzmann entropy seems to be closer to Shannon information rather than Shannon entropy. This interpretation causes additional confusion, because $S(A)$ appears in this equation with negative sign, which reverses the direction of monotonicity; the more information is "associated" with a macrostate A the smaller its Boltzmann entropy. This is usually explained by interpreting what it means to "associate" information with a state. Namely, the information about the state of the system is an information available to an outside observer. Thus it is reasonable to assume that this information acually "escapes" from the system, and hence it should receive the negative sign. Indeed, it is the knowledge about the system possessed by an outside observer that increases the usefulness of the energy contained in that system to do physical work, i.e., it decreases the system's entropy.

The interpretation goes further: each microstate in a system appearing to the observer as being in macrostate A still "hides" the information about its "identity." Let $I_h(A)$ denote the joint information still hiding in the system if its state is identified as A. This entropy is clearly maximal at the maximal state, and then it equals S_{max}/k. In a state A it is diminished by $I(A)$, the information already "stolen" by the observer. So, one has

$$I_h(A) = \frac{S_{max}}{k} - I(A).$$

This, together with (0.3.1), yields

$$S(A) = k \cdot I_h(A),$$

which provides a new interpretation to the Boltzmann entropy: it is proportional to the information still "hiding" in the system provided the macrostate A has been detected.

So far the entropy was determined up to an additive constant. We can compute the *change* of entropy when the system passes from one state to another. It is very hard to determine the proper additive constant of the Boltzmann entropy, because the entropy of the maximal state depends on the level of precision of identifying the microstates. Without a quantum approach, the space

Ω is infinite and so is the maximal entropy. However, if the space of states is assumed finite, the absolute entropy obtains a new interpretation, already in terms of the Shannon entropy (not just of the information function). Namely, in such case, the highest possible Shannon entropy $H_\mu(\mathcal{P})$ is achieved when $\mathcal{P} = \xi$ is the partition of the space Ω into single states ω and μ is the uniform measure on Ω, i.e., such that each state has probability $(\#\Omega)^{-1}$. It is thus natural to set

$$S_{max} = k \cdot H_\mu(\xi) = k \log_2 \#\Omega.$$

The detection that the system is in state A is equivalent to acquiring the information $I(A) = -\log_2(\mu(A)) = -\log_2(\frac{\#A}{\#\Omega})$. By Equation (0.3.1) we get

$$S(A) = k(-\log_2 \#\Omega + \log_2(\tfrac{\#A}{\#\Omega})) = k \log_2 \#A.$$

The latter equals (k times) the Shannon entropy of μ_A, the normalized uniform measure restricted to A. In this manner we have compared the Boltzmann entropy directly with the Shannon entropy and we have gotten rid of the unknown additive constant.

The whole interpretation above is a subject of much discussion, as it makes entropy of a system depend on the seemingly nonphysical notion of "knowledge" of a mysterious observer. The classical *Maxwell's paradox* [Maxwell, 1871] is based on the assumption that it is possible to acquire information about the parameters of individual particles without any expense of heat or work. To avoid such paradoxes, one must agree that every bit of acquired information has its physical entropy equivalent (equal to the Boltzmann constant k), by which the entropy of the memory of the observer increases. In consequence, erasing one bit of information from a memory (say, of a computer) at temperature T, results in the emission of heat in amount kT to the environment. Such calculations set limits on the theoretical maximal speed of computers, because the heat can be driven away with a limited speed only.

0.3.4 Dynamical entropy

This is the key entropy notion in ergodic theory; a version of the Kolmogorov–Sinai entropy for one partition. It refers to Shannon entropy, but it differs significantly as it makes sense only in the context of a measure-preserving transformation. Let T be a measurable transformation of the space Ω, which preserves the probability measure μ, i.e., such that $\mu(T^{-1}(A)) = \mu(A)$ for every measurable set $A \subset \Omega$. Let $\mathcal{P}$ be a finite measurable partition of Ω and let $\mathcal{P}^n$ denote the partition $\mathcal{P} \vee T^{-1}(\mathcal{P}) \vee \cdots \vee T^{-n+1}(\mathcal{P})$ (the least common refinement of n preimages of $\mathcal{P}$). By a subadditivity argument, the sequence of Shannon entropies $\frac{1}{n}H_\mu(\mathcal{P}^n)$ converges to its infimum. The limit

$$h_\mu(T, \mathcal{P}) = \lim_n \frac{1}{n} H_\mu(\mathcal{P}^n) \tag{0.3.2}$$

is called *the dynamical entropy of the process generated by* $\mathcal{P}$ *under the action of* T. This notion has a very important physical interpretation, which we now try to capture.

First of all, one should understand that in the passage from a physical system to its mathematical model (a dynamical system) (Ω, μ, T), the points $\omega \in \Omega$ should not be interpreted as particles nor the transformation T as the way the particles move around the system. Such an interpretation is sometimes possible, but has a rather restricted range of applications. Usually a point ω (later we will use the letter x) represents the physical state of the entire physical system. The space Ω is hence called the *phase space*. The transformation T is interpreted as the *set of physical rules* causing the system that is currently at some state ω to assume in the following instant of time (for simplicity we consider models with discrete time) the state $T\omega$. Such a model is *deterministic* in the sense that the initial state has "imprinted" the entire future evolution. Usually, however, the observer cannot fully determine the "identity" of the initial state. The observer knows only the values of a few measurements, which give only a rough information, and the future of the system is, from his/her standpoint, random. In particular, the values of future measurements are random variables. As time passes, the observer learns more and more about the evolution (by repeating his measurements) through which, in fact, he/she learns about the initial state ω. A finite-valued random variable X imposes a finite partition $\mathcal{P}$ of the phase space Ω. After time n, the observer has learned the values $\mathrm{X}(\omega), \mathrm{X}(T\omega), \ldots, \mathrm{X}(T^n\omega)$ i.e., he/she has learned which element of the partition $\mathcal{P}^n$ contains ω. His/her acquired *information* about the "identity" of ω equals $I_{\mathcal{P}^n}(\omega)$, the expected value of which is $H_\mu(\mathcal{P}^n)$. It is now seen directly from the definition that:

- *The dynamical entropy equals the average (over time and the phase space) gain in one step of information about the initial state.*

Notice that it does not matter whether in the end (at time infinity) the observer determines the initial state completely, or not. What matters is the "gain of information in one step."

If the transformation T is invertible, we can also assume that the evolution of the system runs from time $-\infty$, i.e., it has an infinite past. In such case ω

should be called the *current state* rather than initial state (in a process that runs from time $-\infty$, there is no initial state). Then the entropy $h_\mu(T, \mathcal{P})$ can be computed alternatively using conditional entropy:

$$h_\mu(T, \mathcal{P}) = \lim_n H(\mathcal{P}|T(\mathcal{P}) \vee T^2(\mathcal{P}) \cdots \vee T^{n-1}(\mathcal{P})) = H(\mathcal{P}|\mathcal{P}^-),$$

where $\mathcal{P}^-$ is the sigma-algebra generated by all partitions $T^n(\mathcal{P})$ $(n \geq 0)$ and is called *the past.* This formula provides another interpretation:

- *The dynamical entropy equals the expected amount of information about the current state ω acquired, in addition to was already known from the infinite past, by learning the element of the partition $\mathcal{P}$ to which ω belongs.*

Notice that in this last formulation the averaging over time is absent.

0.3.5 Dynamical entropy as data compression rate

The interpretation of entropy given in this subsection is going to be fundamental for our understanding of dynamical entropy, in fact, we will also refer to a similar interpretation when discussing topological dynamics.

We will distinguish two kinds of data compression: "horizontal" and "vertical." In horizontal data compression we are interested in replacing computer files by other files, as short as possible. We want to "shrink them horizontally." Vertical data compression concerns infinite sequences of symbols interpreted as *signals*. Such signals occur for instance in any "everlasting" data transmission, such as television or radio broadcasting. Vertical data compression attempts to losslessly translate the signal maintaining the same speed of transmission (average lengths of incoming files) but using a smaller alphabet. We call it "vertical" simply by contrast to "horizontal." One can imagine that the symbols of a large alphabet, say of cardinality 2^k, are binary columns of k zeros or ones, and then the vertical data compression will reduce not the length but the "height" of the signal. This kind of compression is useful for data transmission "in real time"; a compression device translates the incoming signal into the optimized alphabet and sends it out at the same speed as the signal arrives (perhaps with some delay).

First we discuss the connection between entropy and the horizontal data compression. Consider a collection of computer files, each in form of a long string B (we will call it a *block*) of symbols belonging to some finite alphabet Λ. For simplicity let us assume that all files are binary, i.e., that $\Lambda = \{0, 1\}$.

Suppose we want to compress them to save the disk space. To do it, we must establish a coding algorithm ϕ which replaces our files B by some other (preferably shorter) files $\phi(B)$ so that no information is lost, i.e., we must also have a decoding algorithm ϕ^{-1} allowing us to reconstruct the original files when needed. Of course, we assume that our algorithm is efficient, that is, it compresses the files as much as possible. Such an algorithm allows us to measure the effective information content of every file: a file carries s bits of information (regardless of its original size) if it can be compressed to a binary file of length $s(B) = s$. This complies with our previous interpretation of information: each symbol in the compressed file is an answer to a binary question, and $s(B)$ is the optimized number of answers needed to identify the original file B.

Somewhat surprisingly, the amount of information $s(B)$ depends not only on the initial size $m = m(B)$ of the original file B but also on subtle properties of its structure. Evidently $s(B)$ is not the simple-minded Shannon information function. There are 2^m binary blocks of a given length m, all of them are "equally likely" so that each has "probability" 2^{-m}, and hence each should carry the same "amount of information" equal to $m \log_2 2 = m$. But $s(B)$ does not behave that simply!

Example Consider the two bitmaps shown in this figure. They have the same

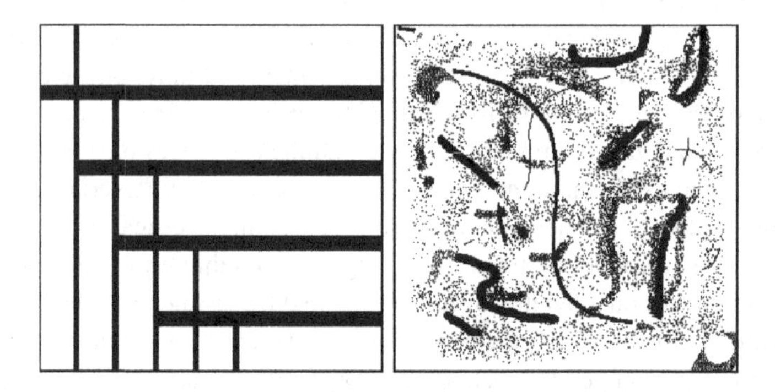

dimensions and the same "density," i.e., the same amount of black pixels. As uncompressed computer files, they occupy exactly the same amount of disk space. However, if we compress them, using nearly any available "zipping" program, the sizes of the zipped files will differ significantly. The left-hand side picture will shrink nearly 40 times, while the right-hand side one only 8 times. Why? To quickly get an intuitive understanding of this phenomenon imagine that you try to pass these pictures over the phone to another person, so that he/she can literally copy it based on your verbal description. The left picture can be precisely described in a few sentences containing the precise coordinates of only two points, while the second picture, if we want it precisely copied, requires tediously dictating the coordinates of nearly all black pixels. Evidently, the right-hand side picture carries more information. A file can be strongly compressed if it reveals some regularity or predictability, which can be used to shorten its description. The more random it looks, the more information must be passed over to the recipient, and the less it can be compressed no matter how intelligent a zipping algorithm is used.

How can we *a priori*, i.e., without experimenting with compression algorithms, just by looking at the file's internal structure, predict the *compression rate* $s(B)/m(B)$ of a given block B? Here is an idea: The compression rate should be interpreted as the *average information content per symbol.* Recall that the dynamical entropy was interpreted similarly, as the expected gain of information per step. If we treat our long block as a portion of the orbit of some point ω representing a shift-invariant measure μ on the symbolic space $\Lambda^{\mathbb{N}\cup\{0\}}$ of all sequences over Λ, then the global information carried by this block should be approximately equal to its length (number of steps in the shift map) times the dynamical entropy of μ. It will be only an approximation, but it should work. The alphabet Λ plays the role of the finite partition $\mathcal{P}$ of the symbolic space, and the partition $\mathcal{P}^n$ used in the definition of the dynamical entropy can be identified with Λ^n – the collection of all blocks over Λ of length n. Any shift-invariant measure on $\Lambda^{\mathbb{N}\cup\{0\}}$ assigns values to all blocks $A \in \Lambda^n$ ($n \in \mathbb{N}$) following some rules of consistency; we skip discussing them now. It is enough to say that a long block B (of a very large length m) nearly determines a shift-invariant measure: for subblocks A of lengths n much smaller than m (but still very large) it determines their *frequencies*:

$$\mu_{(B)}(A) = \frac{\#\{1 \le i \le m-n+1 : B[i, i+n-1] = A\}}{m-n+1},$$

i.e., it associates with A the probability of seeing A in B at a randomly chosen "window" of length n. Of course, this measure is not completely defined (values on longer blocks are not determined), so we cannot perform the full computation of the dynamical entropy. But instead, we can use the approximate value $\frac{1}{n}H_{\mu_{(B)}}(\Lambda^n)$ (see (0.3.2)), which is defined and practically computable for some reasonable length n. We call it *the combinatorial entropy of the block B*. In other words, we decide that the compression rate should be approximately

$$\frac{s(B)}{m(B)} \approx \frac{1}{n}H_{\mu_{(B)}}(\Lambda^n). \tag{0.3.3}$$

As we will prove later, this idea works perfectly well; in most cases the combinatorial entropy estimates the compression rate very accurately. For now we replace a rigorous proof with a simple example.

Example We will construct a lossless compression algorithm and apply it to a file B of a finite length m. The compressed file will consist of a *decoding instruction* followed by the coded image $\phi(B)$ of B. To save on the output length, the decoding instruction must be relatively short compared to m. This is easily achieved in codes which refer to relatively short components of the block B. For example, the instruction of the code may consist of the complete list of subblocks A (appearing in B) of some carefully chosen length n followed by the list of their images $\Phi(A)$. The images may have different lengths (as short as possible). The assignment $A \mapsto \Phi(A)$ will depend on B, therefore it must be included in the output file. The coded image $\phi(B)$ is obtained by cutting B into subblocks $B = A_1A_2 \ldots A_k$ of length n and concatenating the images of these subblocks: $\phi(B) = \Phi(A_1)\Phi(A_2)\cdots\Phi(A_k)$. There are additional issues here: in order for such a code to be invertible, the images $\Phi(A)$ must form a *prefix free* family (i.e., no block in this family is a prefix of another). Then there is always a unique way of cutting $\phi(B)$ back into the images $\Phi(A_i)$. But this does not affect essentially the computations. For best compression results, it is reasonable to assign shortest images to the subblocks appearing in B with highest frequencies. For instance, consider a long binary block

$$B = 010001111001111...110 = 010, 001, 111, 001, 111, ..., 110$$

On the right, B is shown divided into subblocks of length $n = 3$. Suppose that the frequencies of the subblocks in this division are:

$$\begin{array}{llll} 000 - 0\% & 001 - 40\% & 010 - 10\% & 011 - 10\% \\ 100 - 0\% & 101 - 0\% & 110 - 10\% & 111 - 30\% \end{array}$$

The theoretical value of the compression rate (obtained using the formula (0.3.3) for $n = 3$) is

$$(-0.4\log_2(0.4) - 0.3\log_2(0.3) - 3\cdot 0.1\log_2(0.1))/3 \approx 68.2\%.$$

A binary prefix free code giving shortest images to most frequent subblocks is

$$\begin{aligned} 001 &\mapsto 0, \\ 111 &\mapsto 10, \\ 010 &\mapsto 110, \\ 011 &\mapsto 1110, \\ 110 &\mapsto 1111. \end{aligned}$$

The compression rate achieved on B using this code equals

$$(0.4 \times 1 + 0.3 \times 2 + 0.1 \times 3 + 0.1 \times 4 + 0.1 \times 4)/3 = 70\%$$

(ignoring the finite length of the decoding instruction, which is simply a recording of the above code). This code is nearly optimal (at least for this file).

We now focus on the vertical data compression. Its connection with the dynamical entropy is easier to describe but requires a more advanced apparatus. Since we are dealing with an infinite sequence (the signal), we can assume it represents some genuine (not only approximate as it was for a long but finite block) shift-invariant probability measure μ on the symbolic space $\Lambda^{\mathbb{Z}}$. Recall that the dynamical entropy $h = h_\mu(\sigma, \Lambda)$ (where σ denotes the shift map) is the expected amount of new information per step (i.e., per incoming symbol of the signal). We intend to replace the alphabet by a possibly small one. It is obvious that if we manage to losslessly replace the alphabet by another, say Λ_0, then the entropy h cannot exceed $\log_2 \#\Lambda_0$. Conversely, it turns out that any alphabet of cardinality $\#\Lambda_0 > 2^h$ is sufficient to encode the signal. This is a consequence of the famous Krieger Generator Theorem (in this book it is Theorem 4.2.3). Thus we have the following connection:

$$\log_2(\#\Lambda_0 - 1) \leq h \leq \log_2 \#\Lambda_0,$$

where Λ_0 is the smallest alphabet allowing to encode the signal. In this manner the cardinality of the optimal alphabet is completely determined by the entropy. If 2^h happens to be an integer we seem to have two choices, but there is an easy way to decide which one to choose (see Theorem 4.2.3).

0.3.6 Entropy as disorder

The interplay between Shannon and Boltzmann entropy has led to associating with the word "entropy" some colloquial understanding. In all its strict meanings (described above), entropy can be viewed as a measure of disorder and chaos, as long as by "order" one understands that "things are segregated by their kind" (e.g. by similar properties or parameter values). Chaos is the state of a system (physical or dynamical) in which elements of all "kinds" are mixed evenly throughout the space. For example, a container with gas is in its state of maximal entropy when the temperature and pressure are constant. That means there is approximately the same amount of particles in every unit of the volume, and the proportion between slow and fast particles is everywhere the same. States of lower entropy occur when particles are "organized": slower ones in one area, faster ones in another. A signal (an infinite sequence of symbols) has large entropy (i.e., compression rate) when all subblocks of a given length n appear with equal frequencies in all sufficiently long blocks. Any trace of "organization" and "logic" in the structure of the file allows for its compression and hence lowers its entropy. These observations generated a colloquial meaning of entropy. To have order in the house, means to have food separated from utensils and plates, clothing arranged in the closet by type, trash segregated and deposited in appropriate recycling containers, etc. When these things get mixed together "entropy" increases causing disorder and chaos. Entropy is a term in social sciences, too. In a social system, order is associated with classification of the individuals by some criteria (stratification, education, skills, etc.) and assigning to them appropriate positions and roles in the system. Law and other mechanisms are enforced to keep such order. When this classification and assignment fails, the system falls into chaos called "entropy." Entropy equals lack of diversity.

0.4 Conventions

In the main body of the book (Parts I – III) we are using a consistent notational system. Every symbol has an assigned fixed meaning throughout the book. If a letter is multiply used, the meanings are distinguished by font types. The complete list of symbols is provided at the end.

The main conventions include:

- The capital letters X, Y, Z (sometimes with primes or subscripts) are reserved to denote phase spaces of dynamical systems, lowercase x, y, z are their elements. The lowercase Greek letters μ, ν, ξ denote probability measures, while Gothic capitals $\mathfrak{A}, \mathfrak{B}$, etc. stand for sigma-algebras. The letters T, S, R are used for transformations of the phase space that govern the dynamical system. Boldface $\boldsymbol{T}$ represents an operator on a function space. Factor maps and other auxiliary maps between spaces are π, ϕ, ψ. Dual maps on relevant spaces of measures are denoted by the same letter as the map on points (exception: $\boldsymbol{T}^*$ denotes the dual to a Markov operator). The images by major maps of elements of their domains are written (whenever possible) without parentheses, for example Tx, $T\mu$, $\pi\mu$, $\boldsymbol{T}f$.
- The script capitals $\mathcal{P}, \mathcal{Q}, \mathcal{R}$ stand for measurable partitions with elements (cells) denoted A, B, C, etc. The letters B and C are also used to denote finite blocks and their associated cylinders (which in fact are cells of certain partitions of appropriate symbolic spaces). The alphabet in a symbolic system is Λ (rarely Δ). If we need to distinguish between the alphabet and the associated zero-coordinate partition of the symbolic space, we use $\mathcal{P}_\Lambda$ for the latter. A special meaning is reserved to the Gothic capital $\mathfrak{P}$ (with subscripts); it is used for various spaces whose elements are partitions.
- The letters $\mathcal{U}, \mathcal{V}$ represent open covers and their cells are U, V, while $\mathcal{F}, \mathcal{G}, \mathcal{H}$ represent finite families of functions (measurable or continuous) on X.
- The symbols $\mathbb{Z}$, $\mathbb{N}$, $\mathbb{N}_0$ and $\mathbb{R}$ denote the sets of all integers, positive integers (natural numbers), nonnegative integers and real numbers, respectively. The letter $\mathbb{S}$ is used as either $\mathbb{Z}$ or $\mathbb{N}_0$. We try to consistently reserve n for integers representing the time; whereas k indexes refining sequences of partitions or covers, while i, j, l, m (sometimes also p, q, r, s, t) are integer indices of all kinds.
- The letters H and $\mathbf{H}$ are reserved to denote various notions of static entropy, with the boldface version used for topological notions. Similarly, h and $\mathbf{h}$ will be used for dynamical entropy, respectively, measure-theoretic and topological. Calligraphic $\mathcal{H}$ is used for a net or sequence of functions such as an entropy structure.

Some other conventions:

- From now on we choose to use only logarithms to base 2. We write just log.
- A sequence will be written as $(a_i)_{i \geq 1}$ or (a_i), or just "the sequence a_i," when this is not ambiguous.
- Throughout this book, in order to avoid confusingly sounding words we use "decreasing" and "increasing" in the meaning of "nonincreasing" and "nondecreasing," with the adverb "strictly" when the monotonicity is sharp.

Part I

Entropy in ergodic theory

1

Shannon information and entropy

1.1 Information and entropy of probability vectors

We agree (applying the continuous extension) that the real function

$$\eta(t) = -t \log t \tag{1.1.1}$$

assumes the value 0 at $t = 0$. It is strictly concave, i.e., $\eta(pt + qs) > p\eta(t) + q\eta(s)$ for every $t, s \in [0, 1]$, where $p \in (0, 1)$, $q = 1 - p$. Like every concave nonnegative function on $[0, 1]$, η satisfies the *subadditivity condition*

$$\eta(t + s) \leq \eta(t) + \eta(s),$$

whenever $t, s, t + s \in [0, 1]$ (Exercise 1.1). By iterating and by continuity, we also obtain *countable subadditivity*

$$\eta\Big(\sum_{i=1}^{\infty} t_i\Big) \leq \sum_{i=1}^{\infty} \eta(t_i),$$

whenever all above arguments of η belong to $[0, 1]$.

Let $\mathbf{P}$ and $\mathbf{S}$ denote the set of all countable probability vectors (i.e., nonnegative, with sum equal to 1) and subprobability vectors (likewise, but with sum in $[0, 1]$), respectively. Both sets are contained in the space ℓ^1 of all absolutely summable sequences, and we will regard them with the ℓ^1 topology. It is an elementary exercise to check that relatively on $\mathbf{P}$ this topology coincides with the topology of the pointwise convergence (Exercise 1.2), but on $\mathbf{S}$ this is no longer true. For instance $\mathbf{P}$ is closed in ℓ^1, while it is dense in $\mathbf{S}$ in the topology of the pointwise convergence. Of course, we are mainly interested in probability vectors. Subprobabilistic vectors will be technically useful in one place in the proof of Fact 1.1.11, so until then we are forced to check all statements for them as well.

Below, we define the key notions of entropy theory.

Definition 1.1.2 If $\mathbf{p} = (p_i)_{i\in\mathbb{N}}$ is a probability vector, its associated *information function* $I_{\mathbf{p}} : \mathbb{N} \to [0,\infty]$ is defined by

$$I_{\mathbf{p}}(i) = -\log p_i.$$

The *entropy* of $\mathbf{p}$ is defined as

$$H(\mathbf{p}) = \sum_{i=1}^{\infty} p_i I_{\mathbf{p}}(i) = -\sum_{i=1}^{\infty} p_i \log(p_i) = \sum_{i=1}^{\infty} \eta(p_i).$$

This nonnegative value can be infinite but it is certainly finite for vectors with at most finitely many nonzero terms and vectors tending to zero sufficiently fast (see Fact 1.1.4 below). The function H can be applied to any countable sequence with values in $[0,1]$ (in particular to subprobabilistic vectors) and here it satisfies the following:

Fact 1.1.3 *The function H is concave and on the set where H is finite the concavity is strict.*

Proof Let $\mathbf{p} = (p_1, p_2, \dots)$, $\mathbf{q} = (q_1, q_2, \dots)$ and $\mathbf{r} = (r_1, r_2, \dots)$ belong to $[0,1]^{\mathbb{N}}$, and suppose that $\mathbf{r} = p\mathbf{p} + q\mathbf{q}$ where $p \in (0,1)$, $q = 1 - p$. Then by concavity of the function η

$$H(\mathbf{r}) = \sum_{i=1}^{\infty} \eta(pp_i + qq_i) \geq \sum_{i=1}^{\infty} (p\eta(p_i) + q\eta(q_i)) = pH(\mathbf{p}) + qH(\mathbf{q}),$$

and since η is strictly concave and all terms of the above sums are nonnegative, equality holds when either $p_i = q_i$ for all i, or both sides are infinite. □

We note the following criterion for finiteness of the function H on probability vectors:

Fact 1.1.4 *If a probability vector $\mathbf{p} = (p_i)$ satisfies $\sum_{i=1}^{\infty} ip_i < \infty$, then $H(\mathbf{p}) < \infty$.*

Proof Because the function $-\log t$ is decreasing, while $-t\log t$ is increasing (certainly for values below $1/4$), we have

$$H(\mathbf{p}) = -\sum_i p_i \log(p_i) =$$

$$p_1 \log p_1 + \sum_{i\geq 2: p_i > 2^{-i}} p_i(-\log(p_i)) + \sum_{i\geq 2: p_i \leq 2^{-i}} (-p_i \log(p_i)) \leq$$

$$p_1 \log p_1 + \sum_{i\geq 2: p_i > 2^{-i}} p_i(-\log(2^{-i})) + \sum_{i\geq 2: p_i \leq 2^{-i}} (-2^{-i} \log(2^{-i})) \leq$$

$$p_1 \log p_1 + \sum_i i p_i + \sum_i i 2^{-i} < \infty.$$

□

Moreover, for vectors as above the following holds: If we let $p = 1/\sum_i ip_i$ (clearly, $p \in (0, 1]$), then $H(\mathbf{p}) \leq \frac{1}{p} H(p, 1-p)$, and equality is attained if and only if $\mathbf{p}$ is the geometric distribution $p_i = p(1-p)^{i-1}$. Although this fact can be proved using analysis (constrained maximum), we will prove it using dynamical methods much later, in Section 4.3 (Fact 4.3.7).

Let $\mathbf{P}_m$ (respectively, $\mathbf{S}_m$) denote the subset of $\mathbf{P}$ (respectively, of $\mathbf{S}$) consisting of all m-dimensional probability (respectively, subprobability) vectors, i.e., satisfying $p_i = 0$ for all $i > m$. Obviously, $\mathbf{P}_m$ (and $\mathbf{S}_m$) are compact, and the function H is continuous (hence uniformly continuous) on these sets, and assumes the maximal value equal to $\log m$ at the probability vector $\mathbf{p} = (\frac{1}{m}, \frac{1}{m}, \dots, \frac{1}{m}, 0, 0, 0, \dots)$.

Below we provide a tool very useful for handling countable vectors (and later countable partitions):

Definition 1.1.5 For $\mathbf{p} \in \mathbf{P}$ we let $\mathbf{p}_{(m)} \in \mathbf{P}_m$ denote the vector obtained from $\mathbf{p}$ by taking its $m-1$ largest terms and, as the mth term, the sum of the rest, and ordering the resulting m terms decreasingly. For $\mathbf{p} \in \mathbf{S}$, $\mathbf{p}_{(m)}$ is defined identically, and it belongs to $\mathbf{S}_m$.

It is not hard to see that the map $\mathbf{p} \mapsto \mathbf{p}_{(m)}$ is uniformly continuous in ℓ^1. Moreover, we have

Fact 1.1.6

$$H(\mathbf{p}) = \lim_m \uparrow H(\mathbf{p}_{(m)}).$$

Proof By the finite and countable subadditivity of η we have

$$H(\mathbf{p}_{(m)}) \leq H(\mathbf{p}_{(m+1)}) \text{ and } H(\mathbf{p}_{(m)}) \leq H(\mathbf{p}).$$

On the other hand, ordering the terms p_i of $\mathbf{p}$ decreasingly, we can write

$$H(\mathbf{p}) = \lim_m \sum_{i=1}^{m-1} \eta(p_i) \leq \lim_m H(\mathbf{p}_{(m)}). \tag{1.1.7}$$

□

Combining the above fact with the uniform continuity of the map $\mathbf{p} \mapsto \mathbf{p}_{(m)}$ and that of H on $\mathbf{P}_m$ (and on $\mathbf{S}_m$), we conclude the following

Fact 1.1.8 *The functions $\mathbf{p} \mapsto H(\mathbf{p}_{(m)})$ are ℓ^1-uniformly continuous and $\mathbf{p} \mapsto H(\mathbf{p})$ is ℓ^1-lower semicontinuous on $\mathbf{P}$ (and on $\mathbf{S}$) (see Appendix A.1.4 for the definition of lower semicontinuity).* □

We shall be needing another observation:

Fact 1.1.9 *For each $0 \leq M < \infty$ the set of all decreasingly ordered countable probability vectors $\mathbf{p}$ with $H(\mathbf{p}) \leq M$ is compact in ℓ^1. The same holds for subprobability vectors.*

Before the proof we note that the statement does not hold without the ordering. Indeed, if $\mathbf{p}_n$ is the probability vector whose all terms are 0 except the nth term which is 1, then $H(\mathbf{p}_n) = 0$, and the set $\{\mathbf{p}_n : n \geq 1\}$ is 2-separated in ℓ^1.

Proof of Fact 1.1.9 Let $\mathbf{p}$ be a decreasingly ordered probability vector. If $H(\mathbf{p}) \leq M$, then for every $\varepsilon > 0$ the joint mass of the terms p_i smaller than $2^{-\frac{M}{\varepsilon}}$ is at most ε, for otherwise already the sum of $-p_i \log p_i$ over these terms would exceed $\varepsilon \cdot \frac{M}{\varepsilon} = M$. The cardinality of the terms larger than or equal to $2^{-\frac{M}{\varepsilon}}$ is clearly bounded by $K(\varepsilon) = 2^{\frac{M}{\varepsilon}}$. Thus, $\mathbf{p}$ has the following property:

- For every $\varepsilon > 0$ the sum of the terms above index $K(\varepsilon)$ is at most ε.

The set of all probability vectors with this property is totally bounded in ℓ^1. Indeed, every such vector can be, up to ε, approximated by its restriction to the initial $K(\varepsilon)$ terms, while the set of all subprobability vectors of dimension $K(\varepsilon)$ obviously has a finite ε-net. This net becomes a 2ε-net in the set in question. On the other hand, by lower semicontinuity of H, the set of probability vectors with $H(\mathbf{p}) \leq M$ is closed in ℓ^1, and its subset of decreasing vectors is also closed. We have shown that the set of decrasingly ordered probability vectors $\mathbf{p}$ with $H(\mathbf{p}) \leq M$ is closed in ℓ^1 and contained in a totally bounded set. By completeness of the space ℓ^1, such a set is compact. The proof for subprobability vectors is identical. □

Before we continue we need some more notation. Let ξ be a probability distribution on $[0,1]^{\mathbb{N}}$. The *barycenter* of ξ is the sequence $x^\xi = (x_1^\xi, x_2^\xi, \dots)$ such that for each natural i, $x_i^\xi = \int x_i \, d\xi(x)$ (here $x = (x_1, x_2, \dots)$). This notion generalizes convex combinations of vectors, which correspond to barycenters of finitely supported probability distributions ξ. Let $\mathbf{p}^\xi = (p_1^\xi, p_2^\xi, \dots)$ be the barycenter of a probability distribution ξ supported on $\mathbf{P}$. We claim that then $\mathbf{p}^\xi \in \mathbf{P}$. Indeed,

$$\sum_{i=1}^{\infty} p_i^\xi = \sum_{i=1}^{\infty} \int p_i \, d\xi = \int \sum_{i=1}^{\infty} p_i \, d\xi = 1,$$

where the central equality follows from monotone convergence of the finite sums to the infinite sum and linearity of the integral. By the same argument, the barycenter of a distribution supported by $\mathbf{S}$ belongs to $\mathbf{S}$.

A real function f on $\mathbf{P}$ (respectively on $\mathbf{S}$) is *supharmonic* if for every probability distribution ξ on $\mathbf{P}$ (respectively on $\mathbf{S}$), we have $f(\mathbf{p}^\xi) \geq \int f(\mathbf{p}) \, d\xi$. (The notions of barycenter and of supharmonic function are discussed in a more general context in Appendix A.2.3.) The following holds.

Fact 1.1.10 *As a concave lower semicontinuous function, the entropy H is supharmonic on $\mathbf{P}$ and on $\mathbf{S}$ (see Fact A.2.10).* □

The next fact will become important in Section 3.1. It says that on the set of probability vectors $\mathbf{p}$ such that $H(\mathbf{p}) \leq M$, the supharmonic property of H is *ℓ^1-uniformly strict*, in the following sense:

Fact 1.1.11 *Fix some positive number M. For every $\varepsilon > 0$ there exists $\delta > 0$ such that whenever ξ is a probability distribution on $\mathbf{P}$ with barycenter $\mathbf{p}^\xi$ such that $H(\mathbf{p}^\xi) \leq M$ and $\int H(\mathbf{p}) \, d\xi > H(\mathbf{p}^\xi) - \delta$, then*

$$\int \|\mathbf{p}^\xi - \mathbf{p}\|_1 \, d\xi < \varepsilon,$$

where $\|\cdot\|_1$ denotes the norm in ℓ^1.

Proof The ℓ^1-uniform strictness of the concavity of H is obvious on the interval $[0,1]$ because this set is compact, as is the set of all probability measures supported by this set, and H (which is equal to η) is uniformly continuous and strictly concave. This property easily passes to any finite-dimensional cube $[0,1]^m$ ($m \in \mathbb{N}$) and thus to $\mathbf{S}_m$.

Let us proceed to countable probability vectors, as in the assertion. We can change the order of coordinates so that $\mathbf{p}^\xi$ becomes decreasingly ordered.

Let m be such that $\sum_{i \geq m} p_i^\xi < \frac{\varepsilon}{4}$. Let δ be such that on $\mathbf{S}_m$ we have

$$\int H(x)\,d\xi > H(x^\xi) - \delta \implies \int \|x^\xi - x\|_1\,d\xi < \tfrac{\varepsilon}{4}.$$

Suppose that $\int H(\mathbf{p})\,d\xi > H(\mathbf{p}^\xi) - \delta$. Then, letting $\pi(x) = (x_1, x_2, \ldots, x_m) \in \mathbf{S}_m$ and $\phi(x) = (x_{m+1}, x_{m+2}, \ldots) \in \mathbf{S}$, we can write:

$$H(\pi(\mathbf{p}^\xi)) + H(\phi(\mathbf{p}^\xi)) - \delta = H(\mathbf{p}^\xi) - \delta < \int H(\mathbf{p})\,d\xi =$$

$$\int H(\pi(\mathbf{p}))\,d\xi + \int H(\phi(\mathbf{p}))\,d\xi \leq \int H(\pi(\mathbf{p}))\,d\xi + H(\phi(\mathbf{p}^\xi)),$$

where the last inequality follows by the supharmonic property of H on $\mathbf{S}$ and the fact that $\phi(\mathbf{p}^\xi)$ is the barycenter of the measure ξ composed with ϕ (this measure is supported on $\mathbf{S}$; it is here that we actually need the set of subprobabilistic vectors and the properties of H on this set). Cancelling $H(\phi(\mathbf{p}^\xi))$ on both ends we obtain $H(\pi(\mathbf{p}^\xi)) - \delta < \int H(\pi(\mathbf{p}))\,d\xi$, which implies $\int \|\pi(\mathbf{p}^\xi) - \pi(\mathbf{p})\|_1\,d\xi < \frac{\varepsilon}{4}$. In particular, since $\sum_{i=1}^m p_i^\xi \geq 1 - \frac{\varepsilon}{4}$, we have $\int \sum_{i=1}^m p_i\,d\xi \geq 1 - \frac{\varepsilon}{2}$. This implies that $\int \sum_{i=m+1}^\infty p_i\,d\xi \leq \frac{\varepsilon}{2}$, and finally

$$\int \|\mathbf{p}^\xi - \mathbf{p}\|_1\,d\xi \leq \int \|\pi(\mathbf{p}^\xi) - \pi(\mathbf{p})\|_1\,d\xi + \int \|\phi(\mathbf{p}^\xi) - \phi(\mathbf{p})\|_1\,d\xi <$$

$$\tfrac{\varepsilon}{4} + \sum_{i=m+1}^\infty p_i^\xi + \int \sum_{i=m+1}^\infty p_i\,d\xi \leq \tfrac{\varepsilon}{4} + \tfrac{\varepsilon}{2} + \tfrac{\varepsilon}{4} = \varepsilon.$$

So far, the choice of δ depended on m hence also on $\mathbf{p}^\xi$. Because the inequality defining the parameter m is sharp, the choice of m is stable under small ℓ^1-perturbations of $\mathbf{p}^\xi$, i.e., m is good for all vectors in an open set around $\mathbf{p}^\xi$. The same applies to the parameter δ, which depends only on m. Now, by compactness of the considered set of vectors $\mathbf{p}^\xi$ ordered decreasingly (Fact 1.1.9), there is a universal choice of δ on this whole set. Finally, notice that the terms appearing in the assertion of the lemma are insensitive to the ordering of the coordinates, so the lemma holds as it is stated. □

Now we introduce a somewhat exotic notion of *information and entropy with respect to variable lengths*. Anticipating a bit, this notion can be interpreted as corresponding to a partition of the symbolic space into cylinder sets of different lengths. This notion and the following lemma will be used in the proof of one of the key theorems in classical entropy theory – the Shannon–McMillan–Breiman Theorem, where the above interpretation will become clear.

Definition 1.1.12 Consider a pair of vectors: a probability vector $\mathbf{p} = (p_i)_{i\in\mathbb{N}}$ and an arbitrary vector of positive integers (called *lengths*), $\mathbf{n} = (n_i)_{i\in\mathbb{N}}$. With this pair we associate its *length-information function* $I_{\mathbf{p},\mathbf{n}} : \mathbb{N} \to [0,\infty]$ defined by

$$I_{\mathbf{p},\mathbf{n}}(i) = -\frac{\log p_i}{n_i}.$$

The expected value of $I_{\mathbf{p},\mathbf{n}}$ will be called the *length-entropy* of the pair $(\mathbf{p},\mathbf{n})$:

$$H(\mathbf{p},\mathbf{n}) = \sum_{i=1}^{\infty} p_i I_{\mathbf{p},\mathbf{n}}(i) = \sum_{i=1}^{\infty} \frac{\eta(p_i)}{n_i}.$$

Let $\mathbf{n}_\mathbf{p} = \sum_i p_i n_i$ denote the "average length" (which may be infinite, but not for vectors $\mathbf{p}$ with finitely many nonzero terms). We have the following:

Lemma 1.1.13 *Let* $\mathbf{n} = (n_1, n_2, \dots)$ *be a vector of natural numbers. Let* $\mathbf{p}$ *be a finite-dimensional probability vector (i.e., with finitely many nonzero terms), and* $\mathbf{p}'$ *an arbitrary (countable) probability vector. Then*

$$\frac{H(\mathbf{p})}{\mathbf{n}_\mathbf{p}} \le \max_i I_{\mathbf{p}',\mathbf{n}}(i) + \max_i \tfrac{1}{n_i}. \tag{1.1.14}$$

If c *is such that* $\sum_i 2^{-cn_i} \le 1$, *then for any countable probability vector* $\mathbf{p}$,

$$H(\mathbf{p},\mathbf{n}) \le c + \max_i \tfrac{1}{n_i}. \tag{1.1.15}$$

Proof By straightforward computations of partial derivatives it is clear that with $\mathbf{n}$ fixed, the maximal length-entropy among all m-dimensional probability vectors $\mathbf{p}$ is assumed for the unique $\mathbf{p}_0$ such that the function $I_{\mathbf{p}_0,\mathbf{n}}(i) - \frac{1}{n_i}$ is constant on $i \in \{1,2,\dots,m\}$. Any other m-dimensional probability vector $\mathbf{p}$ has, on one hand, a smaller length-entropy, on the other, a larger maximal value of $I_{\mathbf{p},\mathbf{n}}(i) - \frac{1}{n_i}$, thus

$$\begin{aligned} &H(\mathbf{p},\mathbf{n}) \le H(\mathbf{p}_0,\mathbf{n}) \le \max_{i\le m} I_{\mathbf{p}_0,\mathbf{n}}(i) \le \\ &\max_{i\le m}(I_{\mathbf{p}_0,\mathbf{n}}(i) - \tfrac{1}{n_i}) + \max_{i\le m}\tfrac{1}{n_i} \le \max_{i\le m}(I_{\mathbf{p}'_{(m)},\mathbf{n}}(i) - \tfrac{1}{n_i}) + \max_{i\le m}\tfrac{1}{n_i} \le \\ &\quad \max_{i\le\infty}(I_{\mathbf{p}',\mathbf{n}}(i) - \tfrac{1}{n_i}) + \max_{i\le\infty}\tfrac{1}{n_i} \le \max_i I_{\mathbf{p}',\mathbf{n}}(i) + \max_i \tfrac{1}{n_i}. \end{aligned} \tag{1.1.16}$$

(There is a subtlety in comparing $I_{\mathbf{p}'_{(m)},\mathbf{n}}(i)$ with $I_{\mathbf{p}',\mathbf{n}}(i)$ for $i = m$: on the left-hand side we have $-\frac{\log(\sum_{i\ge m} p'_i)}{n_m}$, while on the right-hand side we have $-\frac{\log p'_m}{n_m}$. Clearly, the right-hand side is larger.) Applying this to the probability vector $\mathbf{p}'' = \left(\frac{p_i n_i}{\sum_i p_i n_i}\right)_{i \le m}$ and by convexity of the function $-\log t$, we obtain

$$\max_i I_{\mathbf{p}',\mathbf{n}}(i) + \max_i \tfrac{1}{n_i} \ge H(\mathbf{p}'', \mathbf{n}) =$$

$$\frac{1}{\sum_i p_i n_i}\left(-\sum_i p_i \log p_i - \sum_i p_i \log n_i + \log\Big(\sum_i p_i n_i\Big)\right) \ge \frac{H(\mathbf{p})}{\mathbf{n}_{\mathbf{p}}}.$$

This proves (1.1.14).

For (1.1.15) note that there is $c' \le c$ such that $\mathbf{p}' = (2^{-c' n_i})_i$ is a probability vector and then $I_{\mathbf{p}',\mathbf{n}}(i) = c'$ for each i. Now (1.1.16) yields $H(\mathbf{p}, \mathbf{n}) \le c' + \max_i \frac{1}{n_i} \le c + \max_i \frac{1}{n_i}$ for any finite-dimensional $\mathbf{p}$. Approximating an arbitrary probability vector $\mathbf{p}$ by the finite-dimensional vectors $\mathbf{p}_{(m)}$ and because

$$H(\mathbf{p}, \mathbf{n}) = \sup_m \Big\{-\sum_{i=1}^{m-1} \frac{\log p_i}{n_i}\Big\} \le \sup_m H(\mathbf{p}_{(m)}, \mathbf{n}),$$

we extend the inequality to all probability vectors. □

1.2 Partitions and sigma-algebras

Let $(X, \mathfrak{A}, \mu)$ be a *standard probability space*, i.e., a probability space isomorphic to a compact metric space with the Borel sigma-algebra and a Borel probability measure (also called a *Lebesgue space*). The sigma-algebra $\mathfrak{A}$ is necessarily *completed* with respect to μ, i.e., every subset of a measurable set of measure zero is agreed to be measurable. From now on, if not specified otherwise, by a *partition* we will mean an at most countable partition of X into measurable sets $\mathcal{P} = \{A_i : i \in \mathbb{N}\}$, $\mu(\bigcup_{i=1}^{\infty} A_i) = 1$, and $i \ne j \implies A_i \cap A_j = \emptyset$. In this section we will view all finite partitions as countable by attaching countably many copies of the empty set. Still, if all but finitely many elements A_i have measure zero we will call $\mathcal{P}$ a *finite partition*. While the "master" measure μ is fixed, we will identify partitions equal modulo sets of measure μ zero and write $\mathcal{P} = \mathcal{Q}$ instead of $\mathcal{P} = \mathcal{Q}$ (mod μ). The elements A_i of a partition $\mathcal{P}$ will be referred to as *cells*.

A partition $\mathcal{P}$ is *finer* than (or is a *refinement* of) $\mathcal{Q}$, which we write as $\mathcal{P} \succcurlyeq \mathcal{Q}$, when each cell of $\mathcal{P}$ is (up to measure) contained in a cell of $\mathcal{Q}$. By disjointness, each cell of $\mathcal{Q}$ is then a union of some cells of $\mathcal{P}$; we will also say that $\mathcal{Q}$ is $\mathcal{P}$-*measurable*.

By the *join* (sometimes called the *least common refinement*) of two partitions $\mathcal{P}$ and $\mathcal{Q}$ we shall mean the partition

$$\mathcal{P} \vee \mathcal{Q} = \{A \cap B : A \in \mathcal{P}, B \in \mathcal{Q}\}. \tag{1.2.1}$$

It is easy to see that $\mathcal{P} \succcurlyeq \mathcal{Q} \iff \mathcal{P} \vee \mathcal{Q} = \mathcal{P}$.

We will also consider sub-sigma-algebras $\mathfrak{B}$ of $\mathfrak{A}$, and call them simply *sigma-algebras*, always assuming that $\mathfrak{B}$ are completed. The fact stated below, concerning standard probability spaces, will be of crucial importance in many arguments throughout the book. We refer the reader to [Rokhlin, 1952] for the proof and for more background on standard spaces.

Fact 1.2.2 *Let $(X, \mathfrak{A}, \mu)$ be a standard probability space.*

- *If $\mathfrak{B} \subset \mathfrak{A}$ is a sigma-algebra, then there exists a standard probability space $(Y, \mathfrak{B}', \nu)$ and a measurable map $\pi : X \to Y$ such that $\pi^{-1}(\mathfrak{B}') = \mathfrak{B}$, and which sends the measure μ to ν "by preimage", i.e., $\nu(B) = \mu(\pi^{-1}(B))$ $(B \in \mathfrak{B}')$. We will write $\nu = \pi\mu$.*
- *If $\pi : X \to Y$ is a map as described above and, moreover, it is injective, then π is an isomorphism of measure spaces (i.e., it is, up to measure, a bimeasurable bijection).*

By *atoms* of $\mathfrak{B}$ we will understand the preimages $\pi^{-1}(y)$ of points $y \in Y$. The above fact implies that any $\mathfrak{A}$-measurable set which is a union of atoms of $\mathfrak{B}$, is $\mathfrak{B}$-measurable. It is so, because sets of the above kind form a sigma-algebra contained in $\mathfrak{A}$ and containing $\mathfrak{B}$, with the same atoms as $\mathfrak{B}$, and hence determining the same (up to isomorphism) space $(Y, \mathfrak{B}, \nu)$ (from now on we can identify $\mathfrak{B}'$ with $\mathfrak{B}$) and the same map π. We remark that the above fails without assuming that $\mathfrak{A}$ is standard, for example on the interval the sigma-algebra $\mathfrak{A}$ of all sets is essentially larger than the Borel sigma-algebra, while any of its elements is a union of the atoms of the latter.

Recall that a sequence of partitions $(\mathcal{Q}_k)_{k \geq 1}$ is said to *generate* the sigma-algebra $\mathfrak{B}$ if $\mathfrak{B}$ is the smallest (complete) sigma-algebra which contains all elements of each $\mathcal{Q}_k$. If additionally the sequence $\mathcal{Q}_k$ is *refining*, i.e., such that $\mathcal{Q}_{k+1} \succcurlyeq \mathcal{Q}_k$ for each k, then for every $\mathfrak{B}$-measurable set B and every $\varepsilon > 0$ there is a k such that B can be approximated up to ε (in terms of the measure of the symmetric difference) by a union of some cells of $\mathcal{Q}_k$; (see also (1.7.3) below).

In a standard probability space every sigma-algebra $\mathfrak{B}$ admits a generating sequence of finite partitions. It is convenient to denote the sigma-algebra generated by a sequence of partitions $\mathcal{Q}_k$ by

$$\mathfrak{B} = \bigvee_{k=1}^{\infty} \mathcal{Q}_k.$$

Taking $\mathcal{Q}'_k = \bigvee_{i=1}^{k} \mathcal{Q}_i$, we can always replace an arbitrary sequence of partitions by a refining one, which generates the same sigma-algebra. For sigma-algebras $\mathfrak{B}$, $\mathfrak{C}$, the join $\mathfrak{B} \vee \mathfrak{C}$ will denote the (completed) sigma-algebra generated by the union of $\mathfrak{B}$ and $\mathfrak{C}$, while $\mathfrak{B} \succcurlyeq \mathfrak{C}$ is synonymous with $\mathfrak{B} \supset \mathfrak{C}$. This notation is consistent with that for partitions if partitions are replaced by their generated sigma-algebras. In this spirit, we can (and will) also join partitions with sigma-algebras.

With each ordered countable partition $\mathcal{P} = \{A_i, i \in \mathbb{N}\}$ we associate a probability vector $\mathbf{p}(\mu, \mathcal{P}) = (p_i)_{i \in \mathbb{N}}$ called *the distribution vector of* $\mathcal{P}$, defined by $p_i = \mu(A_i)$. Recall that the set of all probability vectors is convex. The following decomposition rule, involving another countable partition $\mathcal{Q}$, holds, by the Law of Total Probability:

$$\mathbf{p}(\mu, \mathcal{P}) = \sum_{B \in \mathcal{Q}} \mu(B) \mathbf{p}(\mu_B, \mathcal{P}), \tag{1.2.3}$$

where $\mu_B(\cdot) = \mu(\cdot \cap B)/\mu(B)$ is the *conditional measure* on B, for B of positive measure, and μ_B is an arbitrarily chosen probability measure supported by B if B has measure zero.

1.3 Information and static entropy of a partition

For a partition $\mathcal{P}$ of a probability space $(X, \mathfrak{A}, \mu)$ we define its associated *information function* $I_{\mu,\mathcal{P}} : X \to [0, \infty]$ by

$$I_{\mu,\mathcal{P}}(x) = I_{\mathcal{P}}(x) = -\log \mu(A_x),$$

where A_x is the cell of $\mathcal{P}$ containing x. Whenever μ is fixed, we will use the simplified notation $I_{\mathcal{P}}(x)$.

Definition 1.3.1 The *static* (or *Shannon*) *entropy* of $\mathcal{P}$ with respect to μ, $H(\mu, \mathcal{P})$ is defined as the expected value of the information function:

$$H(\mu, \mathcal{P}) = \int I_{\mathcal{P}}(x)\, d\mu(x) = -\sum_{A \in \mathcal{P}} \mu(A) \log \mu(A) = H(\mathbf{p}(\mu, \mathcal{P})).$$

In considerations involving one "master measure" μ we will abbreviate $H(\mu, \mathcal{P})$ as $H(\mathcal{P})$, while $H_B(\mathcal{P})$ will stand for $H(\mu_B, \mathcal{P})$ (for $B \in \mathfrak{A}$). One of the fundamental estimates of the entropy of finite partitions is

$$H(\mathcal{P}) \leq \log \#\mathcal{P}. \tag{1.3.2}$$

We note that convexity of the function $t \mapsto -\log(t)$ immediately implies that the information function of a given partition at a given point is a convex function of the measure μ, contrary to the entropy, which is concave (this follows immediately from the concavity of H on probability vectors, see Fact 1.1.3). We gather these two facts below:

Fact 1.3.3 *If μ is a convex combination of two measures, $\mu = p\mu_1 + q\mu_2$, where $p \in [0, 1], q = 1 - p$, then, for every countable partition $\mathcal{P}$ and every point $x \in X$ we have*

$$I_{\mu,\mathcal{P}}(x) \leq pI_{\mu_1,\mathcal{P}}(x) + qI_{\mu_2,\mathcal{P}}(x), \tag{1.3.4}$$

$$H(\mu, \mathcal{P}) \geq pH(\mu_1, \mathcal{P}) + qH(\mu_2, \mathcal{P}). \tag{1.3.5}$$

□

1.4 Conditional static entropy

For given two partitions $\mathcal{P}$ and $\mathcal{Q}$ one defines the *conditional information function*

$$I_{\mu,\mathcal{P}|\mathcal{Q}}(x) = I_{\mathcal{P}|\mathcal{Q}}(x) = I_{\mathcal{P}\vee\mathcal{Q}}(x) - I_{\mathcal{Q}}(x) = - \log \frac{\mu(A_x \cap B_x)}{\mu(B_x)} = - \log \mu_{B_x}(A_x) \tag{1.4.1}$$

(A_x and B_x denote the cells of $\mathcal{P}$ and $\mathcal{Q}$ containing x, respectively). This function can be interpreted as the "gain of precision" with which one can locate x when, already knowing its position with respect to $\mathcal{Q}$, one learns its position with respect to $\mathcal{P}$.

Definition 1.4.2 We define the *conditional static entropy of $\mathcal{P}$ given $\mathcal{Q}$* as

$$H(\mu, \mathcal{P}|\mathcal{Q}) = \int I_{\mathcal{P}|\mathcal{Q}}(x)\, d\mu(x).$$

As before, $H(\mathcal{P}|\mathcal{Q})$ will stand for $H(\mu, \mathcal{P}|\mathcal{Q})$ when this causes no confusion. Clearly, if $H(\mathcal{Q})$ is finite, by (1.4.1) we have $H(\mathcal{P}|\mathcal{Q}) = H(\mathcal{P} \vee \mathcal{Q}) - H(\mathcal{Q})$.

In order to avoid the undefined term $\infty - \infty$ in the infinite case, it is always safe to write

$$H(\mathcal{P} \vee \mathcal{Q}) = H(\mathcal{P}|\mathcal{Q}) + H(\mathcal{Q}). \tag{1.4.3}$$

In any case we have

$$H(\mathcal{P}|\mathcal{Q}) = -\sum_{B\in\mathcal{Q},A\in\mathcal{P}} \mu(A \cap B) \log \mu_B(A) =$$
$$-\sum_{B\in\mathcal{Q}} \mu(B) \sum_{A\in\mathcal{P}} \mu_B(A) \log \mu_B(A) = \sum_{B\in\mathcal{Q}} \mu(B) H_B(\mathcal{P}).$$

The resulting formula

$$H(\mathcal{P}|\mathcal{Q}) = \sum_{B\in\mathcal{Q}} \mu(B) H_B(\mathcal{P}) \tag{1.4.4}$$

(or $H(\mu, \mathcal{P}|\mathcal{Q}) = \sum_{B\in\mathcal{Q}} \mu(B) H(\mu_B, \mathcal{P})$, in the expanded version), is often used as an alternative definition of the conditional entropy.

We will now introduce the conditional entropy given a sigma-algebra, which generalizes the preceding notion. Let $\mathcal{P}$ be a countable partition and let $\mathfrak{B} \preccurlyeq \mathfrak{A}$ be a sigma-algebra.

Definition 1.4.5 The *conditional entropy of $\mathcal{P}$ given* $\mathfrak{B}$ is defined as

$$H(\mu, \mathcal{P}|\mathfrak{B}) = H(\mathcal{P}|\mathfrak{B}) = \inf\{H(\mathcal{P}|\mathcal{Q}) : \mathcal{Q} \preccurlyeq \mathfrak{B}\},$$

where $\mathcal{Q}$ ranges over all countable partitions measurable with respect to $\mathfrak{B}$.

For a fixed partition, conditioning partition, and a point, the conditional information as a function of the measure, being a difference between two convex functions, is in general, neither convex nor concave. Simple examples on the four-point space show this. The conditional entropy, however, maintains the concavity property:

Fact 1.4.6 *Let μ be a convex combination of two measures, $\mu = p\mu_1 + q\mu_2$, where $p \in [0, 1], q = 1 - p$. Then, for any pair of countable partitions $\mathcal{P}, \mathcal{Q}$, alternatively, for any $\mathcal{P}$ and a sigma-algebra $\mathfrak{B}$, it holds that*

$$H(\mu, \mathcal{P}|\mathcal{Q}) \geq pH(\mu_1, \mathcal{P}|\mathcal{Q}) + qH(\mu_2, \mathcal{P}|\mathcal{Q}), \tag{1.4.7}$$
$$H(\mu, \mathcal{P}|\mathfrak{B}) \geq pH(\mu_1, \mathcal{P}|\mathfrak{B}) + qH(\mu_2, \mathcal{P}|\mathfrak{B}). \tag{1.4.8}$$

Proof The concavity (1.4.7) follows from (1.3.5) via (1.4.4). Then (1.4.8) follows from Definition 1.4.5 and the mere fact that infimum of concave functions is concave. □

1.5 Conditional entropy via probabilistic tools*

In this section we review some alternative definitions of the conditional entropy given a sigma-algebra in terms of the conditional expectation and of the disintegration of the measure. The proofs that these definitions coincide with the definition given at the beginning of this section rely on the Martingale Convergence Theorem. We refer for example to the book by Karl Petersen [Petersen, 1983] for a more detailed exposition.

The definition of the conditional information function of $\mathcal{P}$ given a countable partition $\mathcal{Q}$, involves the term $\mu_{B_x}(A_x)$, which can be viewed as $\mathsf{E}_\mu(\mathbb{I}_{A_x}|\mathcal{Q})(x)$, the value at x of the conditional expectation of the characteristic function of A_x given the sigma-algebra generated by the partition $\mathcal{Q}$. By pure analogy we define

$$I_{\mu,\mathcal{P}|\mathfrak{B}}(x) = I_{\mathcal{P}|\mathfrak{B}}(x) = -\log \mathsf{E}_\mu(\mathbb{I}_{A_x}|\mathfrak{B})(x), \tag{1.5.1}$$

and call it the *conditional information function of* $\mathcal{P}$ *given the sigma-algebra* $\mathfrak{B}$. The alternative definition of the conditional entropy of $\mathcal{P}$ given $\mathfrak{B}$ is via the integral

$$H(\mathcal{P}|\mathfrak{B}) = H(\mu,\mathcal{P}|\mathfrak{B}) = \int I_{\mathcal{P}|\mathfrak{B}}(x)\, d\mu(x).$$

We briefly recall the notion of disintegration of the measure μ with respect to a sigma-algebra $\mathfrak{B} \preccurlyeq \mathfrak{A}$. Fact 1.2.2 says that in a standard probability space $(X,\mathfrak{A},\mu)$ any sub-sigma-algebra $\mathfrak{B}$ determines a standard probability space $(Y,\mathfrak{B},\nu)$ and a projection $\pi : X \to Y$ where the elements $y \in Y$ are the *atoms* of $\mathfrak{B}$ and π is determined by the inclusion $x \in \pi x$. In this setup we identify $\mathfrak{B}$ with $\pi^{-1}(\mathfrak{B})$ and the measure ν is the restriction of μ to $\mathfrak{B}$. The *disintegration* of μ is the ν-almost everywhere defined assignment $y \mapsto \mu_y$, where μ_y is a probability measure on $\mathfrak{A}$ supported by the atom y, such that for every $A \in \mathfrak{A}$ we have

$$\mu(A) = \int \mu_y(A)\, d\nu(y), \tag{1.5.2}$$

which we can write, using the Petis integral, as

$$\mu = \int_Y \mu_y\, d\nu.$$

The disintegration provides alternative formulae for the conditional information function and the conditional entropy, completely analogous to (1.4.1) and to (1.4.4):

$$I_{\mu,\mathcal{P}|\mathfrak{B}}(x) = I_{\mathcal{P}|\mathfrak{B}}(x) = -\log \mu_{\pi x}(A_x) \tag{1.5.3}$$

$$H(\mu, \mathcal{P}|\mathfrak{B}) = \int_Y H(\mu_y, \mathcal{P})\, d\nu(y). \tag{1.5.4}$$

1.6 Basic properties of static entropy

In this section we fix one measure μ and drop it in the denotation, while we treat the partition as a variable. We gather all basic properties of the static and conditional static entropy which can be classified as "monotonicity" and "subadditivity." Continuity properties are gathered in the next section. The first fact gives two very useful equalities that generalize (1.4.4) and (1.4.3).

Fact 1.6.1 *Let $\mathcal{P}$, $\mathcal{Q}$ and $\mathcal{R}$ be countable partitions. Then*

$$H(\mathcal{P}|\mathcal{Q} \vee \mathcal{R}) = \sum_{C \in \mathcal{R}} \mu(C) H_C(\mathcal{P}|\mathcal{Q}), \tag{1.6.2}$$

$$H(\mathcal{P} \vee \mathcal{Q}|\mathcal{R}) = H(\mathcal{P}|\mathcal{Q} \vee \mathcal{R}) + H(\mathcal{Q}|\mathcal{R}). \tag{1.6.3}$$

Proof By (1.4.4) we can write:

$$H(\mathcal{P}|\mathcal{Q} \vee \mathcal{R}) = \sum_{B \cap C \in \mathcal{Q} \vee \mathcal{R}} \mu(B \cap C) H_{B \cap C}(\mathcal{P}) =$$

$$\sum_{C \in \mathcal{R}} \mu(C) \sum_{B \in \mathcal{Q}} \mu_C(B) H_{B \cap C}(\mathcal{P}) = \sum_{C \in \mathcal{R}} \mu(C) H_C(\mathcal{P}|\mathcal{Q}),$$

and then

$$H(\mathcal{P}|\mathcal{Q} \vee \mathcal{R}) + H(\mathcal{Q}|\mathcal{R}) = \sum_{C \in \mathcal{R}} \mu(C) H_C(\mathcal{P}|\mathcal{Q}) + \sum_{C \in \mathcal{R}} \mu(C) H_C(\mathcal{Q}) =$$

$$\sum_{C \in \mathcal{R}} \mu(C) H_C(\mathcal{P} \vee \mathcal{Q}) = H(\mathcal{P} \vee \mathcal{Q}|\mathcal{R}).$$

□

If $H(\mathcal{R})$ is finite, (1.6.3) can be gotten faster, from (1.4.3), see Exercise 1.3. The next fact contains monotonicity properties.

Fact 1.6.4 *For countable partitions $\mathcal{P}$, $\mathcal{Q}$ and $\mathcal{R}$, the following holds*

$$\mathcal{P} \succcurlyeq \mathcal{Q} \iff H(\mathcal{Q}|\mathcal{P}) = 0, \tag{1.6.5}$$

$$\mathcal{P} \succcurlyeq \mathcal{Q} \implies H(\mathcal{P}|\mathcal{R}) \geq H(\mathcal{Q}|\mathcal{R}), \;\; H(\mathcal{P}) \geq H(\mathcal{Q}), \tag{1.6.6}$$

$$\mathcal{Q} \succcurlyeq \mathcal{R} \implies H(\mathcal{P}|\mathcal{Q}) \leq H(\mathcal{P}|\mathcal{R}). \tag{1.6.7}$$

Before the proof we list some consequences of (1.6.3) and the monotonicities (1.6.6) and (1.6.7). Their easy derivation is left to the reader (Exercise 1.3). The first four of them can be viewed as various kinds of subadditivity. The last three will be useful in the context of the Rokhlin metric later.

Corollary 1.6.8 *For countable partitions $\mathcal{P}, \mathcal{P}', \mathcal{Q}, \mathcal{Q}'$, and $\mathcal{R}$ we have*

$$H(\mathcal{P} \vee \mathcal{Q}|\mathcal{R}) \le H(\mathcal{P}|\mathcal{R}) + H(\mathcal{Q}|\mathcal{R}), \tag{1.6.9}$$

$$H(\mathcal{P} \vee \mathcal{Q}) \le H(\mathcal{P}) + H(\mathcal{Q}), \tag{1.6.10}$$

$$H(\mathcal{P} \vee \mathcal{P}'|\mathcal{Q} \vee \mathcal{Q}') \le H(\mathcal{P}|\mathcal{Q}) + H(\mathcal{P}'|\mathcal{Q}'), \tag{1.6.11}$$

$$H(\mathcal{P}|\mathcal{R}) \le H(\mathcal{P}|\mathcal{Q}) + H(\mathcal{Q}|\mathcal{R}), \tag{1.6.12}$$

$$|H(\mathcal{P}|\mathcal{R}) - H(\mathcal{Q}|\mathcal{R})| \le \max\{H(\mathcal{P}|\mathcal{Q}), H(\mathcal{Q}|\mathcal{P})\}, \tag{1.6.13}$$

$$|H(\mathcal{P}|\mathcal{Q}) - H(\mathcal{P}|\mathcal{R})| \le \max\{H(\mathcal{Q}|\mathcal{R}), H(\mathcal{R}|\mathcal{Q})\}, \tag{1.6.14}$$

$$|H(\mathcal{P}) - H(\mathcal{Q})| \le \max\{H(\mathcal{P}|\mathcal{Q}), H(\mathcal{Q}|\mathcal{P})\}. \tag{1.6.15}$$

(in each of the last three statements we assume that at least one of the terms on the left is finite). □

Proof of Fact 1.6.4 Both directions of (1.6.5) are immediate, by the formula (1.4.4) and the fact that $H(\mu, \mathcal{P}) = 0 \iff \mathcal{P}$ is the trivial partition.

We continue by proving (1.6.7) for the trivial partition $\mathcal{R}$. By (1.4.4), the left-hand side of (1.6.7) equals the countable convex combination (with coefficients $\mu(B)$) of the values the function H assumes at the probability vectors $\mathbf{p}(\mu_B, \mathcal{P})$. The right-hand side of (1.6.7) equals H applied to $\mathbf{p}(\mu, \mathcal{P})$, the convex combination (with the same coefficients) of the vectors $\mathbf{p}(\mu_B, \mathcal{P})$ (see (1.2.3)). By the supharmonic property of the entropy on probability vectors (see Fact 1.1.10), we obtain (1.6.7). For the full version of (1.6.7) assume $\mathcal{Q} \succcurlyeq \mathcal{R}$ and, using (1.6.2) and (1.6.7) already proved for trivial $\mathcal{R}$, write:

$$H(\mathcal{P}|\mathcal{Q}) = H(\mathcal{P}|\mathcal{Q} \vee \mathcal{R}) = \sum_{C \in \mathcal{R}} \mu(C) H_C(\mathcal{P}|\mathcal{Q}) \le \sum_{C \in \mathcal{R}} \mu(C) H_C(\mathcal{P}) = H(\mathcal{P}|\mathcal{R}).$$

We move on to (1.6.6). Suppose $\mathcal{P} \succcurlyeq \mathcal{Q}$. By (1.6.3), we have $H(\mathcal{P}|\mathcal{R}) = H(\mathcal{P} \vee \mathcal{Q}|\mathcal{R}) = H(\mathcal{P}|\mathcal{Q} \vee \mathcal{R}) + H(\mathcal{Q}|\mathcal{R}) \ge H(\mathcal{Q}|\mathcal{R})$. For trivial $\mathcal{R}$ we get the second inequality. □

Recall that two partitions are said to be *stochastically independent* (we will write $\mathcal{P} \perp \mathcal{Q}$) if $\mu(A \cap B) = \mu(A)\mu(B)$ for all $A \in \mathcal{P}, B \in \mathcal{Q}$. The following is an immediate consequence of Fact 1.1.3:

Fact 1.6.16 $\mathcal{P}\perp\mathcal{Q} \implies H(\mathcal{P}|\mathcal{Q}) = H(\mathcal{P})$. *If $H(\mathcal{P}) < \infty$, then the converse implication holds.* □

Analogously as for probability vectors, for a partition $\mathcal{P}$ and $m \in \mathbb{N}$, by $\mathcal{P}_{(m)}$ we will denote the finite m-element partition obtained from $\mathcal{P}$ by uniting all but the largest $m-1$ cells. (In case several cells have equal measures we fix some order among them.) We note the following convergence:

Fact 1.6.17

$$H(\mathcal{P}) = \lim_m \uparrow H(\mathcal{P}_{(m)}), \tag{1.6.18}$$

$$H(\mathcal{P}|\mathcal{Q}) = \lim_m \uparrow H(\mathcal{P}_{(m)}|\mathcal{Q}). \tag{1.6.19}$$

Proof (1.6.18) is literally Fact 1.1.6, while (1.6.19) follows from (1.4.4), then (1.6.18) applied to each μ_B, and the mere fact that since the series in (1.1.7) is in fact an increasing limit, the order of limits can be reversed. □

Lemma 1.6.20 *Suppose $H(\mathcal{P}) < \infty$. Then $\mu(C)H_C(\mathcal{P})$ tends to 0 as $\mu(C)$ tends to 0.*

Proof Let $\mathcal{R} = \{C, C^c\}$. Because

$$H(\mathcal{P}) \geq H(\mathcal{P}|\mathcal{R}) = \mu(C)H_C(\mathcal{P}) + \mu(C^c)H_{C^c}(\mathcal{P}),$$

it suffices to show that $H_{C^c}(\mathcal{P})$ is larger than $H(\mathcal{P}) - \varepsilon$ whenever $\mu(C^c)$ is sufficiently close to 1. This, in turn, follows from the ℓ^1 lower semicontinuity of the entropy on probability vectors (Fact 1.1.8) and the fact that the probability vectors $\mathbf{p}(\mu_{C^c}, \mathcal{P})$ converge in ℓ^1 to $\mathbf{p}(\mu, \mathcal{P})$ as the measure of $\mu(C^c)$ tends to 1. □

Fact 1.6.21 *Suppose $H(\mathcal{P}) < \infty$. Then*

$$H(\mathcal{P}|\mathcal{Q}) = \lim_m \downarrow H(\mathcal{P}|\mathcal{Q}_{(m)}).$$

Proof By (1.6.7), the sequence $H(\mathcal{P}|\mathcal{Q}_{(m)})$ decreases. We order $\mathcal{Q} = \{B_1, B_2, \dots\}$ decreasingly and let C_m denote the union of B_i over $i \geq m$. Then

$$H(\mathcal{P}|\mathcal{Q}) = H(\mathcal{P}|\mathcal{Q}_{(m)}) - \mu(C_m)H_{C_m}(\mathcal{P}) + \sum_{i\geq m} \mu(B_i)H_{B_i}(\mathcal{P}).$$

Since $H(\mathcal{P}|\mathcal{Q}) \leq H(\mathcal{P})$ is finite, the last sum is the tail of a convergent series. Thus, and by the preceding lemma, we obtain $H(\mathcal{P}|\mathcal{Q}) = \lim_m H(\mathcal{P}|\mathcal{Q}_{(m)})$. □

Remark 1.6.22 The above may fail if $\mathcal{P}$ has infinite entropy. For example, if $\mathcal{Q} = \mathcal{P}$, then $H(\mathcal{P}|\mathcal{Q}) = 0$, while the conditional entropy of $\mathcal{P}$ given any finite partition is infinite.

Corollary 1.6.23 *Fact 1.6.21 implies that in Definition 1.4.5 (of $H(\mathcal{P}|\mathfrak{B})$), for $\mathcal{P}$ with finite entropy, it suffices to take the infimum over* finite *$\mathfrak{B}$-measurable partitions $\mathcal{Q}$ only.*

Next, we discuss properties analogous to the ones above, but involving conditional entropy given a sigma-algebra.

Fact 1.6.24 *For any countable partitions $\mathcal{P}$ and $\mathcal{Q}$ and any sigma-algebra $\mathfrak{B}$ it holds that*

$$H(\mathcal{P}|\mathcal{Q}\vee\mathfrak{B}) = \sum_{B\in\mathcal{Q}} \mu(B)H_B(\mathcal{P}|\mathfrak{B}), \tag{1.6.25}$$

$$H(\mathcal{P}\vee\mathcal{Q}|\mathfrak{B}) = H(\mathcal{P}|\mathcal{Q}\vee\mathfrak{B}) + H(\mathcal{Q}|\mathfrak{B}). \tag{1.6.26}$$

Proof We have, by (1.6.2),

$$H(\mathcal{P}|\mathcal{Q}\vee\mathfrak{B}) = \inf_{\mathcal{R}} H(\mathcal{P}|\mathcal{Q}\vee\mathcal{R}) = \inf_{\mathcal{R}} \sum_{B\in\mathcal{Q}} \mu(B)H_B(\mathcal{P}|\mathcal{R}) \geq$$
$$\sum_{B\in\mathcal{Q}} \mu(B)\inf_{\mathcal{R}} H_B(\mathcal{P}|\mathcal{R}) = \sum_{B\in\mathcal{Q}} \mu(B)H_B(\mathcal{P}|\mathfrak{B}),$$

where $\mathcal{R}$ range over all $\mathfrak{B}$-measurable partitions.

We proceed with the reversed inequality. It holds trivially if the right-hand side is infinite. In the finite case, for each $B\in\mathcal{Q}$ there is a $\mathfrak{B}$-measurable partition $\mathcal{R}_B$ which realizes the infimum in the definition of $H_B(\mathcal{P}|\mathfrak{B})$ up to ε. We let $\mathcal{R}$ be the countable partition obtained as a refinement of $\mathcal{Q}$ by applying $\mathcal{R}_B$ relatively to each $B\in\mathcal{Q}$. This partition is measurable with respect to $\mathcal{Q}\vee\mathfrak{B}$. We have, by (1.6.2) again,

$$\sum_{B\in\mathcal{Q}} \mu(B)H_B(\mathcal{P}|\mathfrak{B})+\varepsilon \geq \sum_{B\in\mathcal{Q}} \mu(B)H_B(\mathcal{P}|\mathcal{R}_B) = \sum_{B\in\mathcal{Q}} \mu(B)H_B(\mathcal{P}|\mathcal{R}) =$$
$$H(\mathcal{P}|\mathcal{Q}\vee\mathcal{R}) \geq H(\mathcal{P}|\mathcal{Q}\vee\mathfrak{B}).$$

Elementary properties of infima (see Appendix A.1.5) and (1.6.3) imply the inequality "$\geq$" in (1.6.26). For the converse, take two $\mathfrak{B}$-measurable partitions which nearly realize the infima defining the two terms on the right. By monotonicity with respect to the conditioning partition, their join realizes both. For this join apply (1.6.3), then apply infimum on the left. □

Fact 1.6.27

$$\mathcal{P} \preccurlyeq \mathfrak{B} \iff H(\mathcal{P}|\mathfrak{B}) = 0, \tag{1.6.28}$$

$$\mathcal{P} \succcurlyeq \mathcal{Q} \implies H(\mathcal{P}|\mathfrak{B}) \geq H(\mathcal{Q}|\mathfrak{B}), \tag{1.6.29}$$

$$\mathfrak{B} \succcurlyeq \mathfrak{C} \implies H(\mathcal{P}|\mathfrak{B}) \leq H(\mathcal{P}|\mathfrak{C}). \tag{1.6.30}$$

As before, prior to the proof, we list a number of easy consequences of (1.6.26) and the last two monotonicities.

Corollary 1.6.31

$$H(\mathcal{P} \vee \mathcal{Q}|\mathfrak{B}) \leq H(\mathcal{P}|\mathfrak{B}) + H(\mathcal{Q}|\mathfrak{B}), \tag{1.6.32}$$

$$H(\mathcal{P} \vee \mathcal{P}'|\mathcal{Q} \vee \mathcal{Q}' \vee \mathfrak{B}) \leq H(\mathcal{P}|\mathcal{Q} \vee \mathfrak{B}) + H(\mathcal{P}'|\mathcal{Q}' \vee \mathfrak{B}), \tag{1.6.33}$$

$$H(\mathcal{P}|\mathfrak{B}) \leq H(\mathcal{P}|\mathcal{Q}) + H(\mathcal{Q}|\mathfrak{B}), \tag{1.6.34}$$

$$H(\mathcal{P}|\mathcal{R} \vee \mathfrak{B}) \leq H(\mathcal{P}|\mathcal{Q} \vee \mathfrak{B}) + H(\mathcal{Q}|\mathcal{R} \vee \mathfrak{B}), \tag{1.6.35}$$

$$|H(\mathcal{P}|\mathfrak{B}) - H(\mathcal{Q}|\mathfrak{B})| \leq \max\{H(\mathcal{P}|\mathcal{Q}), H(\mathcal{Q}|\mathcal{P})\}, \tag{1.6.36}$$

$$|H(\mathcal{P}|\mathcal{Q} \vee \mathfrak{B}) - H(\mathcal{P}|\mathcal{R} \vee \mathfrak{B})| \leq \max\{H(\mathcal{Q}|\mathcal{R}), H(\mathcal{R}|\mathcal{Q})\} \tag{1.6.37}$$

(each of the last two statements requires at least one of the terms on the left to be finite). □

Proof of Fact 1.6.27 If $\mathcal{P}$ is $\mathfrak{B}$-measurable, then $H(\mathcal{P}|\mathfrak{B}) \leq H(\mathcal{P}|\mathcal{P}) = 0$. Conversely, if $H(\mathcal{P}|\mathfrak{B}) = 0$, then for every $\delta > 0$ there exists an $\mathfrak{B}$-measurable partition $\mathcal{Q}$ such that

$$H(\mathcal{P}|\mathcal{Q}) = \sum_{B \in \mathcal{Q}} \mu(B) H_B(\mathcal{P}) < \delta.$$

This means (by the elementary "rectangle rule," Fact A.1.1 in the Appendix A.1) that at least $1 - \sqrt{\delta}$ of the space is covered by cells $B \in \mathcal{Q}$ for which $H_B(\mathcal{P}) \leq \sqrt{\delta}$. For small δ such a B is partitioned by $\mathcal{P}$ nearly trivially, i.e., with one dominating cell A with $\mu_B(A) > 1 - \varepsilon$. Every cell A of $\mathcal{P}$ is now approximated (up to $\varepsilon + \sqrt{\delta}$) by the union of those cells B of $\mathcal{Q}$ in which A dominates. Since $\mathfrak{B}$ is completed, every set which can be arbitrarily well approximated by $\mathfrak{B}$-measurable sets is $\mathfrak{B}$-measurable.

The inequality (1.6.29) follows from (1.6.6), while (1.6.30) requires nothing but Definition 1.4.5. □

Fact 1.6.38 *Assume that $H(\mathcal{P}) < \infty$. Then $\mathcal{P} \perp \mathfrak{B} \iff H(\mathcal{P}|\mathfrak{B}) = H(\mathcal{P})$.*

Proof If $\mathcal{P} \perp \mathfrak{B}$, then $\mathcal{P} \perp \mathcal{R}$ for any $\mathfrak{B}$-measurable partition $\mathcal{R}$, and Fact 1.6.16 implies the entropy condition. Suppose the entropy condition holds. For a $\mathfrak{B}$-measurable set A let $\mathcal{R} = \{A, A^c\}$. Then $H(\mathcal{P}) \geq H(\mathcal{P}|\mathcal{R}) \geq H(\mathcal{P}|\mathfrak{B}) = H(\mathcal{P})$, so, by Fact 1.6.16, $\mathcal{P} \perp \mathcal{R}$, in particular $\mathcal{P} \perp A$. □

Fact 1.6.39 *For a sigma-algebra $\mathfrak{B}$ and a countable partition $\mathcal{P}$ we have*

$$H(\mathcal{P}|\mathfrak{B}) = \lim_m \uparrow H(\mathcal{P}_{(m)}|\mathfrak{B}).$$

Proof Monotonicity and the inequality "$\geq$" are obvious. We order the cells of $\mathcal{P}$ decreasingly in measure, $\mathcal{P} = \{A_1, A_2, \dots\}$. For every $\varepsilon > 0$ there exists a subsequence m_k of the natural numbers, such that the partition

$$\mathcal{P}' = \{A'_1, A'_2, \dots\} = \{A_1 \cup A_2 \cup \dots \cup A_{m_1}, \\ A_{m_1+1} \cup A_{m_1+2} \cup \dots \cup A_{m_2}, \dots\}$$

has entropy smaller than ε (see Exercise 1.5). By (1.6.26), and since $\mathcal{P} \succcurlyeq \mathcal{P}'$, we have

$$H(\mathcal{P}|\mathfrak{B}) = H(\mathcal{P} \vee \mathcal{P}'|\mathfrak{B}) = \\ H(\mathcal{P}|\mathcal{P}' \vee \mathfrak{B}) + H(\mathcal{P}'|\mathfrak{B}) < H(\mathcal{P}|\mathcal{P}' \vee \mathfrak{B}) + \varepsilon. \quad (1.6.40)$$

For each k, on each of the sets A'_j, where $j \leq k$, the restrictions of $\mathcal{P}_{(m_k)}$ and of $\mathcal{P}$ coincide. On the other hand, for $j > k$, the restriction of $\mathcal{P}_{(m_k)}$ to A'_j is trivial. So,

$$H(\mathcal{P}_{(m_k)}|\mathfrak{B}) \geq H(\mathcal{P}_{(m_k)}|\mathcal{P}' \vee \mathfrak{B}) = \sum_{j=1}^{\infty} \mu(A'_j) H_{A'_j}(\mathcal{P}_{(m_k)}|\mathfrak{B}) = \\ \sum_{j=1}^{k} \mu(A'_j) H_{A'_j}(\mathcal{P}|\mathfrak{B}).$$

The last terms increase with k to

$$\sum_{j=1}^{\infty} \mu(A'_j) H_{A'_j}(\mathcal{P}|\mathfrak{B}) = H(\mathcal{P}|\mathcal{P}' \vee \mathfrak{B}),$$

which was shown in (1.6.40) to be larger than $H(\mathcal{P}|\mathfrak{B}) - \varepsilon$ (or infinite). We have shown that $H(\mathcal{P}|\mathfrak{B}) \leq \lim_k H(\mathcal{P}_{(m_k)}|\mathfrak{B})$. By monotonicity, this ends the proof. □

Fact 1.6.41 *Suppose $H(\mathcal{P}|\mathfrak{B}) < \infty$. Then*

$$H(\mathcal{P}|\mathcal{Q} \vee \mathfrak{B}) = \lim_m \downarrow H(\mathcal{P}|\mathcal{Q}_{(m)} \vee \mathfrak{B}).$$

We will prove this fact near the end of the next section, as a particular case of a more general Fact 1.7.10.

1.7 Metrics on the space of partitions

There is a natural pseudometric between countable partitions:

$$d_1(\mathcal{P},\mathcal{Q}) = \frac{1}{2}\inf\Big\{\sum_{i=1}^{\infty}\mu(A_i \ominus B_{\pi(i)})\Big\} =$$

$$1 - \sup\Big\{\sum_{i=1}^{\infty}\mu(A_i \cap B_{\pi(i)})\Big\}, \qquad (1.7.1)$$

where $\mathcal{P} = \{A_i, i \in \mathbb{N}\}, \mathcal{Q} = \{B_i, i \in \mathbb{N}\}$, $\ominus$ denotes the symmetric difference and the infimum (and supremum) runs through all permutations π of the natural numbers. This pseudometric becomes a metric once factored to classes of partitions modulo measure zero.

It is elementary to see that

$$d_1(\mathcal{P}_1 \vee \mathcal{P}_2, \mathcal{Q}_1 \vee \mathcal{Q}_2) \le d_1(\mathcal{P}_1,\mathcal{Q}_1) + d_1(\mathcal{P}_2,\mathcal{Q}_2). \qquad (1.7.2)$$

One of the important features of this metric is the possibility of approximating a partition measurable with respect to $\bigvee_k \mathcal{Q}_k$ by $\mathcal{Q}_k$-measurable partitions. We skip the elementary measure-theoretic proof:

Fact 1.7.3 *Let $\mathfrak{B} = \bigvee_{k=1}^{\infty}\mathcal{Q}_k$, where $\mathcal{Q}_{k+1} \succcurlyeq \mathcal{Q}_k$ for all k and let $\mathcal{P}$ be a partition measurable with respect to $\mathfrak{B}$. Then for every ε there exists a k and a $\mathcal{Q}_k$-measurable partition $\mathcal{P}_k$ with $d_1(\mathcal{P},\mathcal{P}_k) < \varepsilon$. If $\mathcal{P}$ is finite, then each $\mathcal{P}_k$ can be made finite with the same cardinality as $\mathcal{P}$.* □

It is a well-known fact [see e.g. Rokhlin, 1952] that in standard spaces this metric on the set of all (classes of) countable partitions is complete and separable. The same is true for every $m \in \mathbb{N}$ and the set of all partitions with at most m cells of positive measure. The corresponding Polish spaces of partitions will be denoted $\mathfrak{P}_{\aleph_0}$ and $\mathfrak{P}_m$, respectively.

The following obvious fact will play a crucial role in this section.

Fact 1.7.4 *The assignment $\mathcal{P} \mapsto \mathcal{P}_{(m)}$ from $\mathfrak{P}_{\aleph_0} \to \mathfrak{P}_m$ is continuous in d_1 at every partition $\mathcal{P}$ whose $(m–1)$st and mth largest cells differ in measure.* □

Note that if $\mathcal{P}$ is infinite, the assumption is satisfied for infinitely many m's.

Conditional entropy gives rise to an alternative metric among partitions with finite entropy, called the *Rokhlin metric* :

Definition 1.7.5

$$d_{\mathrm{R}}(\mathcal{P},\mathcal{Q}) = \max\{H(\mathcal{P}|\mathcal{Q}), H(\mathcal{Q}|\mathcal{P})\}.$$

Indeed, $H(\mathcal{P}|\mathcal{Q}) = H(\mathcal{Q}|\mathcal{P}) = 0$ if and only if both $\mathcal{Q} \succcurlyeq \mathcal{P}$ and $\mathcal{P} \succcurlyeq \mathcal{Q}$, i.e., $\mathcal{P} = \mathcal{Q}$. The triangle inequality is (1.6.12).

Remark 1.7.6 In most texts the Rokhlin metric is defined as $H(\mathcal{P}|\mathcal{Q}) + H(\mathcal{Q}|\mathcal{P})$. Ours is, of course, a uniformly equivalent version.

Fact 1.7.7 *The metrics d_{R} and d_1 are uniformly equivalent on the spaces $\mathfrak{P}_m$.*

Proof If

$$H(\mathcal{P}|\mathcal{Q}) = \sum_{B \in \mathcal{Q}} \mu(B) H_B(\mathcal{P}) \leq \delta,$$

where $\delta < 1$, then, by the rectangle rule (Fact A.1.1), $H_B(\mathcal{P}) < \sqrt{\delta}$ on sets B of joint measure at least $1 - \sqrt{\delta}$. In such a B there is a dominating $A \in \mathcal{P}$ (we have already used this argument in the proof of (1.6.28)). This time we will estimate the value $t = \mu_B(A)$ of the dominating cell more accurately. Since $\sqrt{\delta} < 1$ it must be that $t > \frac{1}{2}$. Then

$$\sqrt{\delta} \geq H_B(\mathcal{P}) \geq -t \log t - (1-t)\log(1-t) \geq 2t(1-t) \geq 1 - t.$$

Thus $t \geq 1 - \sqrt{\delta}$ (we have used twice the inequality $-\log t \geq 1 - t$). This easily implies that $d_1(\mathcal{Q}, \mathcal{P} \vee \mathcal{Q}) \leq 2\sqrt{\delta}$ (see (1.7.1)). If the same holds with the roles of $\mathcal{P}$ and $\mathcal{Q}$ reversed, then $d_1(\mathcal{P}, \mathcal{Q})$ does not exceed $4\sqrt{\delta}$. In case $\delta \geq 1$ we have $d_1 < 4\sqrt{\delta}$ since d_1 never exceeds 1. We have proved that

$$d_1 \leq 4\sqrt{d_{\mathrm{R}}} \tag{1.7.8}$$

(we did not strive to get the best estimate here).

Conversely, let $\varepsilon = d_1(\mathcal{P}, \mathcal{Q})$. Let $C^c = \bigcup_{n=1}^{m} A_n \cap B_{\pi(n)}$ where π is such that $\mu(C) = \varepsilon$ (see (1.7.1)). Let $\mathcal{R} = \{C, C^c\}$. Then, using (1.6.3),

$$H(\mathcal{P}|\mathcal{Q}) \leq H(\mathcal{P} \vee \mathcal{R}|\mathcal{Q}) = H(\mathcal{P}|\mathcal{R} \vee \mathcal{Q}) + H(\mathcal{R}|\mathcal{Q}) \leq$$
$$\mu(C^c) H_{C^c}(\mathcal{P}|\mathcal{Q}) + \mu(C) H_C(\mathcal{P}|\mathcal{Q}) + H(\mathcal{R}) \leq 0 + \varepsilon H_C(\mathcal{P}) + H(\varepsilon, 1-\varepsilon),$$

and on $\mathfrak{P}_m$ the term $H_C(\mathcal{P})$ is estimated by $\log m$. Thus the right-hand side decreases to zero with ε. The same estimate applies to $H(\mathcal{Q}|\mathcal{P})$. □

Note that the latter direction of the equivalence between the metrics fails for countable partitions, even if we assume that their entropies are bounded. On nonatomic measure spaces, the trivial partition $\mathcal{P}$ can be approximated in d_1 by countable partitions $\mathcal{P}_n$ with arbitrary entropy, and then $d_{\mathrm{R}}(\mathcal{P}_n, \mathcal{P}) = H(\mathcal{P}_n)$ need not converge to 0.

We pass to investigating continuity properties of the entropy with respect to varying partitions.

Fact 1.7.9 *The functions $H(\cdot)$, $H(\cdot|\mathcal{Q})$ and $H(\cdot|\mathfrak{B})$ (for an arbitrary fixed countable partition $\mathcal{Q}$, and an arbitrary fixed sigma-algebra $\mathfrak{B}$) are uniformly continuous on $\mathfrak{P}_m$ and lower semicontinuous on $\mathfrak{P}_{\aleph_0}$.*

Proof On $\mathfrak{P}_m$ we can use the Rokhlin metric, and then (1.6.15), (1.6.13) and (1.6.36) yield the desired uniform continuities. To prove lower semicontinuity (in d_1) of each of the above functions at any $\mathcal{P} \in \mathfrak{P}_{\aleph_0}$ we first choose a subsequence m_k of integers for which the assumption of Fact 1.7.4 is satisfied for $\mathcal{P}$, and then the considered three functions become increasing limits (along m_k) of functions continuous at $\mathcal{P}$ (see (1.6.18), (1.6.19) and Fact 1.6.39). This suffices for lower semicountinuity at $\mathcal{P}$ (see Fact A.1.11 in Appendix A.1). □

Fact 1.7.10 *Assume $H(\mathcal{P}|\mathfrak{B}) < \infty$ where $\mathfrak{B}$ is a sigma-algebra. Then the function $H(\mathcal{P}|\cdot \vee \mathfrak{B})$ is uniformly continuous on $\mathfrak{P}_{\aleph_0}$. For trivial $\mathfrak{B}$ this reads: if $H(\mathcal{P}) < \infty$, then the function $H(\mathcal{P}|\cdot)$ is uniformly continuous on $\mathfrak{P}_{\aleph_0}$.*

Proof Assume that $H(\mathcal{P}|\mathfrak{B})$ is finite. Then, by (1.6.26) and (1.6.30),

$$H(\mathcal{P}|\mathcal{Q}\vee\mathfrak{B})-H(\mathcal{P}_{(m)}|\mathcal{Q}\vee\mathfrak{B}) = H(\mathcal{P}|\mathcal{P}_{(m)}\vee\mathcal{Q}\vee\mathfrak{B}) \le H(\mathcal{P}|\mathcal{P}_{(m)}\vee\mathfrak{B}) = H(\mathcal{P}|\mathfrak{B}) - H(\mathcal{P}_{(m)}|\mathfrak{B}),$$

so, by Fact 1.6.39, the function $\mathcal{Q} \mapsto H(\mathcal{P}|\mathcal{Q} \vee \mathfrak{B})$ is a uniform limit of the functions $\mathcal{Q} \mapsto H(\mathcal{P}_{(m)}|\mathcal{Q} \vee \mathfrak{B})$. It remains to prove that the considered function is uniformly continuous for a finite partition $\mathcal{P}$. Regard two partitions $\mathcal{Q}$ and $\mathcal{Q}'$ and set $\delta = d_1(\mathcal{Q}, \mathcal{Q}')$. Let $C^c = \bigcup_{n=1}^m B_n \cap B'_{\pi(n)}$ where π is such that $\mu(C) = \delta$ (see (1.7.1)) and let $\mathcal{R} = \{C, C^c\}$. Then, by (1.6.26) (applied twice to $H(\mathcal{P} \vee \mathcal{R}|\mathcal{Q} \vee \mathfrak{B})$), we get

$$H(\mathcal{P}|\mathcal{Q} \vee \mathfrak{B}) = H(\mathcal{P}|\mathcal{R} \vee \mathcal{Q} \vee \mathfrak{B}) + H(\mathcal{R}|\mathcal{Q} \vee \mathfrak{B}) - H(\mathcal{R}|\mathcal{P} \vee \mathcal{Q} \vee \mathfrak{B}),$$

and the same for $\mathcal{Q}'$. Since $\mathcal{R}$ has small entropy depending only on δ, we can ignore the last two terms at a cost of a uniform error ε (if δ is small enough). Thus

$$\begin{aligned}\left|H(\mathcal{P}|\mathcal{Q}\vee\mathfrak{B})-H(\mathcal{P}|\mathcal{Q}'\vee\mathfrak{B})\right| &\le \left|H(\mathcal{P}|\mathcal{R}\vee\mathcal{Q}\vee\mathfrak{B})-H(\mathcal{P}|\mathcal{R}\vee\mathcal{Q}'\vee\mathfrak{B})\right|+2\varepsilon \le \\ &\mu(C^c)\left|H_{C^c}(\mathcal{P}|\mathcal{Q} \vee \mathfrak{B}) - H_{C^c}(\mathcal{P}|\mathcal{Q}' \vee \mathfrak{B})\right|+ \\ &+ \mu(C)\left|H_C(\mathcal{P}|\mathcal{Q} \vee \mathfrak{B}) - H_C(\mathcal{P}|\mathcal{Q}' \vee \mathfrak{B})\right| + 2\varepsilon.\end{aligned}$$

On C^c the partitions $\mathcal{Q}$ and $\mathcal{Q}'$ coincide, so the first term in the last line is zero. The next term is, by (1.3.2), at most $\delta \log \#\mathcal{P}$. This ends the proof. □

We can now provide the missing proof of Fact 1.6.41.

Proof of Fact 1.6.41 Notice that $\mathcal{Q}_{(m)}$ converges to $\mathcal{Q}$ in $\mathfrak{P}_{\aleph_0}$ and apply Fact 1.7.10. □

Fact 1.6.41 is a particular case of a more general property:

Lemma 1.7.11 *Assume that $H(\mathcal{P}) < \infty$. If $\mathcal{Q}_k$ is a refining sequence of partitions, then*

$$H(\mathcal{P}|\bigvee_k \mathcal{Q}_k) = \lim_k \downarrow H(\mathcal{P}|\mathcal{Q}_k). \tag{1.7.12}$$

More generally, for any monotone sequence of sigma-algebras $\mathfrak{B}_k$, we have

$$\forall_k\, \mathfrak{B}_k \preccurlyeq \mathfrak{B}_{k+1} \implies H(\mathcal{P}|\bigvee_k \mathfrak{B}_k) = \lim_k \downarrow H(\mathcal{P}|\mathfrak{B}_k), \tag{1.7.13}$$

$$\forall_k\, \mathfrak{B}_k \succcurlyeq \mathfrak{B}_{k+1} \implies H(\mathcal{P}|\bigcap_k \mathfrak{B}_k) = \lim_k \uparrow H(\mathcal{P}|\mathfrak{B}_k). \tag{1.7.14}$$

Proof For (1.7.12), by Corollary 1.6.23, it suffices to consider finite partitions $\mathcal{Q}$ measurable with respect to $\bigvee_k \mathcal{Q}_k$. Every such partition can be approximated in d_1 by $\mathcal{Q}_k$-measurable partitions, say $\mathcal{R}_k$, of the same cardinality. By the continuity stated in Fact 1.7.10, we can have $H(\mathcal{P}|\mathcal{R}_k)$ arbitrarily close to $H(\mathcal{P}|\mathcal{Q})$, which can be chosen close to $H(\mathcal{P}|\bigvee_k \mathcal{Q}_k)$. More precisely, we can arrange that the following inequalities hold:

$$H(\mathcal{P}|\mathcal{Q}_k) \le H(\mathcal{P}|\mathcal{R}_k) \le H(\mathcal{P}|\mathcal{Q}) + \varepsilon \le H(\mathcal{P}|\bigvee_k \mathcal{Q}_k) + 2\varepsilon.$$

This implies that $\inf_k H(\mathcal{P}|\mathcal{Q}_k) \le H(\mathcal{P}|\bigvee_k \mathcal{Q}_k)$. The other inequality is obvious, by the monotonicity (1.6.30).

To prove (1.7.13), for each k let $(\mathcal{R}_{k,i})_{i\in\mathbb{N}}$ be a refining sequence of partitions generating $\mathfrak{B}_k$. We can arrange that for fixed i, $\mathcal{R}_{k+1,i} \succcurlyeq \mathcal{R}_{k,i}$ for all k. Consider the expressions $H(\mathcal{P}|\mathcal{R}_{k,i})$. The assertion follows by applying, on one hand the infimum over the pairs (k,i), (and (1.7.12)), on the other, the iterated infimum, first over i (and (1.7.12) again), then over k.

The statement (1.7.14) will never be used in this book, so we can afford to sketch an argument using the tools from the "asterisk sections." In fact one gets both statements (1.7.13) and (1.7.14) essentially strengthened. For $\mathcal{P}$ with finite entropy, since the function $t \mapsto -\log t$ is convex, the corresponding sequence of conditional information functions $I_{\mathcal{P}|\mathfrak{B}_k}$ is a (forward or backward) *submartingale*. Using the martingale theory, it can be proved that in both cases, this submartingale converges almost everywhere and in $L^1(\mu)$ to the conditional information function given $\mathfrak{B}$, of which the convergence of the conditional entropies is an immediate consequence. □

We conclude this section with the following topological statement:

Fact 1.7.15 *The space $\mathfrak{P}_R$ of all countable partitions with finite static entropy is complete and separable in d_R.*

Proof By (1.7.8), any d_R-Cauchy sequence $\mathcal{P}_k$ is d_1-Cauchy and by (1.6.15) has bounded entropy. By completeness of $\mathfrak{P}_{\aleph_0}$, there exists a d_1-limit, $\mathcal{P}$, which, by lower semicontinuity of $H(\cdot)$ (Fact 1.7.9), has finite entropy. Now, $H(\mathcal{P}|\mathcal{P}_k)$ converges to zero, by the continuity of $H(\mathcal{P}|\cdot)$ (Fact 1.7.10). It remains to prove that also $H(\mathcal{P}_k|\mathcal{P}) \to 0$. By the last mentioned continuity, for each k,

$$H(\mathcal{P}_k|\mathcal{P}) = \lim_{k'} H(\mathcal{P}_k|\mathcal{P}_{k'}) \le \lim_{k'} d_R(\mathcal{P}_k, \mathcal{P}_{k'}),$$

which is small for large k, by the Cauchy condition.

Separability follows from the fact that every partition with finite entropy can be approximated in d_R by a finite partition and then, by separability of the sigma-algebra $\mathfrak{A}$ in the standard metric "measure of the symmetric difference," this finite partition can be approximated in d_1 (equivalently in d_R), by a finite partition with elements belonging to a preselected countable family of sets. □

1.8 Mutual information*

The term addressed in the title is a notion popularly used in information theory. Its name is a bit misleading, because it is much closer to entropy than to an information function.

Definition 1.8.1 For two partitions $\mathcal{P}$ and $\mathcal{Q}$, their *mutual information* is defined as

$$I(\mathcal{P};\mathcal{Q}) = \sum_{A\in\mathcal{P},B\in\mathcal{Q}} \mu(A\cap B)\log\frac{\mu(A\cap B)}{\mu(A)\mu(B)}.$$

Mutual information is the "missing value" when comparing the entropy of the join with the sum of entropies: $H(\mathcal{P})+H(\mathcal{Q}) = H(\mathcal{P}\vee\mathcal{Q})+I(\mathcal{P};\mathcal{Q})$, which can be rewritten as $H(\mathcal{P}) = H(\mathcal{P}|\mathcal{Q})+I(\mathcal{P};\mathcal{Q})$ or $H(\mathcal{Q}) = H(\mathcal{Q}|\mathcal{P})+I(\mathcal{P};\mathcal{Q})$. Mutual information is indispensable only when both $H(\mathcal{P}|\mathcal{Q})$ and $H(\mathcal{Q}|\mathcal{P})$ are infinite, otherwise it can be expressed as a difference of entropies (conditional and unconditional). In particular, mutual information allows us to determine stochastic independence without any finiteness restrictions (compare Fact 1.6.16):

Fact 1.8.2 $\mathcal{P}\perp\mathcal{Q} \iff I(\mathcal{P};\mathcal{Q}) = 0.$

Proof By elementary operations, directly from the definition one derives

$$I(\mathcal{P};\mathcal{Q}) = \sum_{A\in\mathcal{P}}\Big[-\mu(A)\log(\mu(A)) + \sum_{B\in\mathcal{Q}}\mu(B)\mu_B(A)\log(\mu_B(A))\Big] =$$
$$\sum_{A\in\mathcal{P}}\Big[\eta(\mu(A)) - \sum_{B\in\mathcal{Q}}\mu(B)\eta(\mu_B(A))\Big],$$

where η is the strictly concave function (1.1.1). Since $\mu(A) = \sum_{B\in\mathcal{Q}}\mu(B)\mu_B(A)$, for each $A\in\mathcal{P}$ the corresponding term in the last displayed sum is non-negative, and it equals zero if and only if $\mu_B(A) = \mu(A)$ for every $B\in\mathcal{Q}$. □

In this book mutual information will not be used. Its usefulness as a measure of correlation between two partitions is illustrated in the anecdote below.[1]

Example 1.8.3 Imagine a place on the Earth, where it rains precisely one day per year, yet one never knows which day it is. Suppose three weather stations cover this place. The first one uses an extremely primitive algorithm, in fact no algorithm at all: they predict "no rain" for each day of the year. Notice that this station is wrong only one day per year. The second station runs a simulation, which copies precisely the strategy of the nature: they draw randomly one of 365 numbers and they predict rain for the corresponding day. This station is wrong a little below 2 times per year (with very small probability they may be accurate, otherwise they miss the rainy day and they predict rain on a sunny day). There is also a third station. These guys conduct very complicated research, study the patterns from the past, use advanced simulations etc. Each year they obtain three equal peaks of probability for the rainy day to occur. Their method is so good that the true rainy day always occurs in one of the three peak days. Their official prediction is rain for each of these three days. Notice that they are wrong full two times per year.

Judging the "reliability" by the number of errors per year, the first station is the best, the last one is the worse. However, it is intuitively clear that only the last station provides us with valuable information. This "valuable information" is precisely the mutual information of two partitions: the partition of the year (365 days) into "rain" and "no rain" predicted by the station, and similar partition occurring in reality. The higher the mutual information, the better job done by the station.

And so, the entropy of the partition $\mathcal{P}$ provided by nature equals $H(\frac{1}{365},\frac{364}{365})$ (we do not really need to calculate this value). The partition $\mathcal{Q}_1$ provided by the first station is the trivial partition, has entropy zero, and so is the mutual information. The partition $\mathcal{Q}_2$ given by the second station is already nontrivial and has entropy $H(\frac{1}{365},\frac{364}{365})$, the same as $\mathcal{P}$, still it is independent of $\mathcal{P}$ and the mutual information is again zero. The partition $\mathcal{Q}_3$ provided by the third station has entropy $H(\frac{3}{365},\frac{362}{365})$. The joined partition $\mathcal{P}\vee\mathcal{Q}_3$ divides the year into three sets of days: "no rain predicted and no rain" (362 days), "rain predicted but no rain" (2 days), and "rain predicted and rain" (1 day). The mutual information is the difference

$$H(\tfrac{1}{365},\tfrac{364}{365}) + H(\tfrac{3}{365},\tfrac{362}{365}) - H(\tfrac{1}{365},\tfrac{2}{365},\tfrac{362}{365}) =$$
$$\tfrac{1}{365}\big((365\log 365 - 364\log 364) - (3\log 3 - 2\log 2)\big).$$

[1] This example was told to the author over lunch by Mike Keane, who attributed it to Jack van Lint. In van Lint's story the place was Death Valley in California and there were just two weather stations. The modifications are due to the author.

By elementary calculus, this is approximately (and not less than) $\frac{\log 121}{365} > 0$.

1.9 Non-Shannon inequalities*

This section is slightly off our main course, hence we will not provide detailed proofs of all facts quoted here. It can be treated as a mention of a kind of curiosity in entropy theory, of which it is good to be aware.

For two or three partitions with finite entropy (or just finite, for simplicity), the inequalities of Fact 1.6.4 (combined with the equality (1.4.3)) bound the set of possible vectors

$$\mathbf{v}_{(\mathcal{P},\mathcal{Q})} = \langle H(\mathcal{P}), H(\mathcal{Q}), H(\mathcal{P} \vee \mathcal{Q}) \rangle$$

(for two partitions) or

$$\mathbf{v}_{(\mathcal{P},\mathcal{Q},\mathcal{R})} = \\ \langle H(\mathcal{P}), H(\mathcal{Q}), H(\mathcal{R}), H(\mathcal{P} \vee \mathcal{Q}), H(\mathcal{P} \vee \mathcal{R}), H(\mathcal{Q} \vee \mathcal{R}), H(\mathcal{P} \vee \mathcal{Q} \vee \mathcal{R}) \rangle$$

(for three partitions). Of course, these bounds (called *Shannon inequalities*) are applicable also to collections of four or more partitions; any two or three joins composed from such a collection must obey them. Thus, the following problem arises: Are these the only restrictions? Surprisingly, the answer is positive only for two partitions; any nonnegative vector $\mathbf{v} = \langle a, b, c \rangle$ equals $\mathbf{v}_{(\mathcal{P},\mathcal{Q})}$ for some partitions, if and only if $\max\{a, b\} \le c \le a + b$ (see Exercise 1.7). We will show that already for $k = 3$ the set of corresponding vectors cannot be described by homogeneous linear inequalities (i.e., without additive constants), because it is not invariant under positive scaling (i.e., it is not a *cone*).

Fact 1.9.1 *The vector $\langle 1, 1, 1, 2, 2, 2, 2 \rangle$ can be obtained as $\mathbf{v}_{(\mathcal{P},\mathcal{Q},\mathcal{R})}$, while its half, $\langle \frac{1}{2}, \frac{1}{2}, \frac{1}{2}, 1, 1, 1, 1 \rangle$, cannot.*

Proof The first statement is seen by taking three two-element partitions of the standard space modeled by the unit square with the Lebesgue measure: $\mathcal{P}$ is the partition into the vertical halves, $\mathcal{Q}$ is the partition into the horizontal halves, and $\mathcal{R} \preccurlyeq \mathcal{P} \vee \mathcal{Q}$ is the "chessboard partition" consisting of two sets: the union of the top-left and bottom-right corner squares, and its complement.

The second statement needs some computation. Suppose the "half vector" equals $\mathbf{v}_{(\mathcal{P},\mathcal{Q},\mathcal{R})}$ for some partitions. Because the entropies of each pair behave additively under joining, each pair is stochastically independent (see Fact 1.6.16). On the other hand, since joining the third partition does not increase the entropy of the join of the other two, it follows that $\mathcal{R} \preccurlyeq \mathcal{P} \vee \mathcal{Q}$ (see

Fact 1.6.16). Notice that $H(\mathcal{P}) \leq 1/2$ forces $\mathcal{P}$ to contain a set of measure larger than $1/\sqrt{2}$. Clearly this set is the largest in this partition. Analogously for $\mathcal{Q}$ and $\mathcal{R}$. Let a, b and c denote the measures of the largest sets A, B and C in $\mathcal{P}$, $\mathcal{Q}$ and $\mathcal{R}$, respectively. Permuting, if necessary, the names of the partitions we can assume that $c \leq a$ and $c \leq b$. By independence, $\mu(A \cap B) = ab > 1/2$, and since C is a union of such intersections, $A \cap B$ must be a part of C. On the other hand $\mu(A \cap C) = ac \leq ab$ implies that $A \cap C = A \cap B$ and similarly, $B \cap C = A \cap B$. This yields $ac = bc = ab$, hence $a = b = c$. This also implies that $C \setminus A \cap B \subset A^c \cap B^c$, i.e., that $a - a^2 \leq (1-a)^2$. This quadratic inequality solves as $a \leq 1/2$ or $a \geq 1$. In the first case the entropies of the three partitions are larger than or equal to 1, in the other case they all equal zero. □

We are facing the following general problem: Consider the standard probability space $(X, \mathfrak{A}, \mu)$, and a collection of $k \in \mathbb{N}$ measurable partitions $\mathcal{P}_i$ of X $(i = 1, \ldots, k)$. Associated with these is the $(2^k - 1)$-dimensional vector of Shannon entropies

$$\mathbf{v}_{(\mathcal{P}_1,\ldots,\mathcal{P}_k)} = \langle H_\mu(\mathcal{P}_F) : \emptyset \neq F \subset \{1, \ldots, k\} \rangle,$$

where, $\mathcal{P}_F$ abbreviates the join $\bigvee_{i \in F} \mathcal{P}_i$, and the indexing sets F are ordered increasingly by cardinality (so that the first k coordinates are just the entropies of the $\mathcal{P}_i$'s, the last coordinate is the entropy of the join of all of them). Describe the set

$$\Gamma_k = \{\mathbf{v}_{(\mathcal{P}_1,\ldots,\mathcal{P}_k)} : \mathcal{P}_i, \ldots, \mathcal{P}_k \text{ are countable partitions with finite entropy}\}.$$

Question 1.9.2 It is clear that the set Γ_k^* obtained by admitting, in its definition, only finite partitions (but without bounding their cardinality) is dense in Γ_k. Is it the same set?

Another pathology of the set Γ_k, for $k \geq 3$ (in spite of not being a cone) is that it is not even closed. This will follow immediately from the example above and the statement proved below:

Fact 1.9.3 *For each $k \geq 1$, the closure $\overline{\Gamma_k}$ is a cone.*

Proof If $\mathbf{v} = \mathbf{v}_{(\mathcal{P}_1,\ldots,\mathcal{P}_k)}$ and $m \in \mathbb{N}$, then $m\mathbf{v}$ also belongs to Γ_k. This is seen by regarding the product space X^m with the product measure μ^m and taking partitions $\mathcal{P}'_i = \bigvee_{j=1}^m \mathcal{P}_i^{(j)}$, where $\mathcal{P}_i^{(j)}$ is the partition $\mathcal{P}_i$ applied to the jth copy of X (and crossed with the trivial partitions at other coordinates). Now, for each $\emptyset \neq F \subset \{1, \ldots, k\}$, $\mathcal{P}'_F$ is the join of independent copies of $\mathcal{P}_F$, hence its entropy is precisely m times larger.

Because every multiplier can be approximated by a rational one with very large denominator, in order to prove Fact 1.9.3 it now suffices, for each $\mathbf{v} \in \Gamma_k$, to approximate $\frac{1}{n}\mathbf{v}$ by an element of Γ_k up to an error small for large n.

So, again, assume that $\mathbf{v} = \mathbf{v}_{(\mathcal{P}_1,\dots,\mathcal{P}_k)}$. Let X' be the product of X with the set $\{1,\dots,n\}$, equipped with the product sigma-algebra $\mathfrak{A}' = \mathfrak{A}\times\{$all subsets$\}$ and the product measure $\mu' = \mu \times \mathsf{Prob}$, where Prob is the normalized counting measure on $\{1,\dots,n\}$. Now, $(X', \mathfrak{A}', \mu')$ is again a standard probability space. In this space, the set $A = X \times \{1\}$ has measure $1/n$ and the complement $B = X \times \{2, ..., n\}$ has measure $(n-1)/n$. Let $\mathcal{Q} = \{A, B\}$ be the associated partition of X'.

Now define partitions $\mathcal{P}'_i$ $(i = 1,\dots,k)$ of X' as follows: each of them refines the partition $\mathcal{Q}$; on the small part A they are copies of $\mathcal{P}_i$, respectively (formally they are $\mathcal{P}_i \times \{1\}$), and on B they are trivial, i.e., they all contain the large set B in one piece. We fix a nonempty set $F \subset \{1,\dots,k\}$ and we calculate the Shannon entropy of the join $\mathcal{P}'_F$. Because $\mathcal{Q}$ is refined by each $\mathcal{P}'_i$, it is also refined by $\mathcal{P}'_F$, thus we have

$$H(\mathcal{P}'_F) = H(\mathcal{P}'_F \vee \mathcal{Q}) = H(\mathcal{P}'_F|\mathcal{Q}) + H(\mathcal{Q}) =$$
$$\mu'(B)H_B(\mathcal{P}'_F) + \mu'(A)H_A(\mathcal{P}'_F) + H(\mathcal{Q}) = \tfrac{n-1}{n}\cdot 0 + \tfrac{1}{n}H_\mu(\mathcal{P}_F) + H(\mathcal{Q}) =$$
$$\tfrac{1}{n}H_\mu(\mathcal{P}_F) + H(\mathcal{Q}).$$

The error term $H(\mathcal{Q})$ of this approximation depends only on n and converges to zero as $n \to \infty$. This concludes the proof. □

The closure $\overline{\Gamma_k}$ remains hard to describe; only for $k \le 3$ it is determined by the Shannon inequalities. For $k \ge 4$ this set is not even a polyhedral cone, i.e., it cannot be described by a system of linear inequalities. The known constraints (inequalities) embracing the set $\overline{\Gamma_k}$, and not following from the Shannon inequalities are called "non-Shannon inequalities" and are usually highly nontrivial. The list of such inequalities is not exhausted and every now and then new non-Shannon inequalities are being discovered [see Makarychev *et al.*, 2002, and references therein]. Below we replicate (without a proof) from [Zhang and Yeung, 1997] an example of such an inequality valid for $k = 4$:

$$H(\mathcal{P}_1\vee\mathcal{P}_2)+H(\mathcal{P}_1\vee\mathcal{P}_3)+3(H(\mathcal{P}_2\vee\mathcal{P}_3)+H(\mathcal{P}_3\vee\mathcal{P}_4)+H(\mathcal{P}_2\vee\mathcal{P}_4)) \ge$$
$$2H(\mathcal{P}_2)+2H(\mathcal{P}_3)+H(\mathcal{P}_4)+H(\mathcal{P}_1\vee\mathcal{P}_4)+H(\mathcal{P}_1\vee\mathcal{P}_2\vee\mathcal{P}_3)+4H(\mathcal{P}_2\vee\mathcal{P}_3\vee\mathcal{P}_4).$$

Question 1.9.4 There is another, similar set of vectors, say $\widetilde{\Gamma}_k$, arising as dynamical entropies of all possible joins composed out of k partitions, under an action of measure-preserving transformation of the standard probability space (see the next chapter). The definition of $\widetilde{\Gamma}_k$ involves all possible systems of k partitions as well as all possible transformations. Directly from the definition of the dynamical entropy (which is the limit of expressions of the form $\frac{1}{n}H(\mathcal{P}^n)$, where $\mathcal{P}^n$ is a partition), and from scalability of the set $\overline{\Gamma}_k$, it is immediately seen that this new set is contained in $\overline{\Gamma}_k$. Whether it is closed (hence $\widetilde{\Gamma}_k = \overline{\Gamma}_k$) seems to be, at the moment this book was written, an open problem.

Exercises

1.1 Show that every nonnegative concave function f on $[0, a]$ or on $[0, \infty)$ is subadditive, i.e., $f(x + y) \leq f(x) + f(y)$ whenever $x, y, x + y$ belong to the domain.

1.2 Prove that a sequence of countable probability vectors converges in ℓ^1 to a probability vector if and only if the convergence is coordinatewise.

1.3 Consider an abstract function $H : \mathfrak{X} \to [0, \infty)$, where $\mathfrak{X}$ is some set equipped with an associative and commutative operation $\vee$. Assume that for all $a, b \in \mathfrak{X}$, $H(a \vee b) \geq H(a)$ and define $H(a|b)$ as $H(a \vee b) - H(b)$. Derive

$$H(a \vee b|c) = H(a|b \vee c) + H(b|c),$$
$$H(a \vee b|c) \geq H(a|c).$$

Assuming additionally that $H(a|b) \geq H(a|b \vee c)$ (for all $a, b, c \in \mathfrak{X}$) and that there exists $e \in \mathfrak{X}$ such that $a \vee e = a$ for all $a \in \mathfrak{X}$ (e is a "unity" for $\vee$), derive also

$$\begin{aligned}
H(a \vee b|c) &\leq H(a|c) + H(b|c),\\
H(a \vee b) &\leq H(a) + H(b),\\
H(a \vee a'|b \vee b') &\leq H(a|b) + H(a'|b'),\\
H(a|c) &\leq H(a|b) + H(b|c),\\
|H(a|c) - H(b|c)| &\leq \max\{H(a|b), H(b|a)\},\\
|H(a|b) - H(a|c)| &\leq \max\{H(b|c), H(c|b)\},\\
|H(a) - H(b)| &\leq \max\{H(a|b), H(b|a)\}.
\end{aligned}$$

1.4 Prove that for every finite-dimensional probability vector $\mathbf{p} = \{p_1, \ldots, p_l\}$ we have

$$H(\mathbf{p}) \leq \max_{1 \leq i \leq l} (1 - p_i) \log l + 1.$$

1.5 Let $\mathbf{p} = (p_i)_{i \geq 1}$ be a probability vector. Show that for every $\varepsilon > 0$ there exists a subsequence m_k of the natural numbers, such that the vector

$\mathbf{p}' = (p'_i)$ where $p'_i = \sum_{i=m_{i-1}+1}^{m_i} p_i$ (we set $m_0 = 0$) has entropy smaller than ε.

1.6 Give an example showing that mutual information is not subadditive, i.e., that the inequality $I(\mathcal{P} \vee \mathcal{Q}; \mathcal{R}) \leq I(\mathcal{P}; \mathcal{R}) + I(\mathcal{Q}; \mathcal{R})$ may fail.

1.7 Show that for any triple $a, b, c \in [0, \infty)$ such that $\max\{a, b\} \leq c \leq a+b$ there exist two partitions $\mathcal{P}, \mathcal{Q}$ (say, of the unit square with the Lebesgue measure) such that $H(\mathcal{P}) = a$, $H(\mathcal{Q}) = b$ and $H(\mathcal{P} \vee Q) = c$.

2

Dynamical entropy of a process

2.1 Subadditivity

Subadditivity is a property of real functions (we have already mentioned it for the function η), sequences, and also sequences of functions (so-called cocycles) defined on a dynamical system. Because it plays a very important role in the theory of entropy, we isolate this short section.

We will distinguish four classes of sequences $(a_n)_{n\geq 1}$ of nonnegative numbers:

(A) the sequence (a_n) has *decreasing increments* if $(a_n - a_{n-1})$ (with a_0 defined as 0) is decreasing;
(B) the sequence (a_n) has *decreasing nths* if $(\frac{1}{n}a_n)$ is decreasing;
(C) the sequence (a_n) is *subadditive* if $a_{m+n} \leq a_n + a_m$, for any $m, n \in \mathbb{N}$;
(D) the sequence (a_n) has *descending nths* if $(\frac{1}{n}a_n)$ converges to its infimum.

Fact 2.1.1 (A) $\implies$ (B) $\implies$ (C) $\implies$ (D). *None of the implications can be reversed. If (A) holds, then* (a_n) *increases and* $\lim_n(a_{n+1} - a_n) = \lim_n \frac{1}{n}a_n$.

Proof The term $\frac{1}{n}a_n$ is the arithmetic average of the first n increments. Averages of a decreasing sequence decrease to the same limit, so (A) implies (B) and equality of the appropriate limits. If (a_n) was not increasing, there would be a negative increment and all following increments would be even smaller, leading eventually to negative values of a_n (assumed to be nonnegative).

Further, (B) implies $\frac{1}{m+n}a_{m+n} \leq \min\{\frac{1}{m}a_m, \frac{1}{n}a_n\}$, hence

$$\frac{1}{m+n}a_{m+n} \leq \frac{m}{m+n}\frac{1}{m}a_m + \frac{n}{m+n}\frac{1}{n}a_n,$$

which, after cancellation, yields (C). Assume (C). If $m = kn + r$ with $0 \leq r < n$, then

$$\frac{a_m}{m} \leq \frac{ka_n}{m} + \frac{a_r}{m} \leq \frac{a_n}{n} + \frac{na_1}{m},$$

hence, for every n, $\limsup_m \frac{a_m}{m} \leq \frac{a_n}{n}$. This implies (D). Examples justifying the last statement are elementary. □

The following lemma will be used to show that some sequences have decreasing $n$ths.

Lemma 2.1.2 *Let* $\overline{a} = \frac{1}{n}(a_1+a_2+\cdots+a_n)$, $\overline{b} = \frac{1}{n-1}(b_1+b_2+\cdots+b_{n-1})$. *If* $b_i \geq a_i$ *and* $b_i \geq a_{i+1}$, *for each* $i = 1, 2, \ldots, n-1$, *then* $\overline{b} \geq \overline{a}$.

Proof

$$\begin{aligned}
&n(b_1 + b_2 + \cdots + b_{n-1}) = \\
&(n-1)b_1 + [b_1 + (n-2)b_2] + \cdots + [(n-2)b_{n-2} + b_{n-1}] + (n-1)b_{n-1} \geq \\
&(n-1)a_1 + \quad (n-1)a_2 \quad + \cdots + \quad (n-1)a_{n-1} \quad + \; (n-1)a_n = \\
&\qquad\qquad\qquad\qquad (n-1)(a_1 + a_2 + \cdots + a_n).
\end{aligned}$$

□

For certain sequences related to dynamical entropy, subadditivity is usually the property most easy to verify, and sufficient for the existence of appropriate limits. Nevertheless, in most cases, either (A) or at least (B) also holds. We will comment on that in due course.

Now we pass to discussing subadditive cocycles.

Definition 2.1.3 Let X be any space and $T : X \to X$ any transformation. By a *subadditive cocycle* on the system (X, T) we will mean a sequence of nonnegative functions $(f_n)_{n\geq 1}$ such that for every $x \in X$ and every natural m and n, it holds that

$$f_{m+n}(x) \leq f_n(x) + f_m(T^n x).$$

If (f_n) is a subadditive cocycle such that f_1 is bounded from above, then the sequence (a_n) defined by $a_n = \sup_{x\in X} f_n(x)$ is subadditive. Moreover, if the cocycle is defined on a measure space, T is measure-preserving and each f_n is integrable, then the sequence (b_n), where $b_n = \int f_n \, d\mu$, is also subadditive.

Although the next theorem does not refer directly to entropy we give here a relatively elementary proof. For a more general version see [Krengel, 1985, Theorem 5.3]. In the formulation, we anticipate a bit our notation. It is explained at the beginning of the next section.

Theorem 2.1.4 (Subadditive Ergodic Theorem) *Let $(X, \mathfrak{A}, \mu, T, \mathbb{S})$ be an ergodic measure-preserving transformation and let $(f_n)_{n \in \mathbb{N}}$ be a measurable subadditive cocycle with f_1 bounded from above. Then, μ-almost surely*

$$\lim_n \frac{1}{n} f_n(x) = \lim_n \frac{1}{n} \int f_n \, d\mu. \tag{2.1.5}$$

The limit on the right-hand side can be replaced by the infimum.

Proof Denote by C the limit on the right, whose existence (and the last statement) follows from the subadditivity of the sequence of integrals. Assume that f_1 is bounded from above by a constant a. Then, for each $n \in \mathbb{N}$, $\frac{1}{n} f_n$ is also bounded from above by a. Fix two natural numbers, $m > n$ and write $m = kn + l$ $(l < n)$. By subadditivity of f_n we have

$$f_m(x) \leq \sum_{i=0}^{k-1} f_n(T^{ni} x) + la.$$

Substituting x by $Tx, \ldots, T^{n-1}x$ and adding on both sides we obtain

$$\sum_{i=0}^{n-1} f_m(T^i x) \leq \sum_{i=0}^{m-1} f_n(T^i x) + nla. \tag{2.1.6}$$

On the other hand, applying subadditivity again, for each $0 \leq i < n$ we have

$$f_{m+n}(x) \leq f_i(x) + f_m(T^i x) + f_{n-i}(T^{m+i} x),$$

where the sum of the first and last terms does not exceed na. Averaging over $0 \leq i < n$ and applying the previous estimate we get

$$f_{m+n}(x) \leq \frac{1}{n} \sum_{i=0}^{n-1} f_m(T^i x) + na \leq \frac{1}{n} \sum_{i=0}^{m-1} f_n(T^i x) + 2na. \tag{2.1.7}$$

Dividing by m and letting $m \to \infty$ we obtain, by the ergodic theorem,

$$\limsup_{m \to \infty} \frac{1}{m} f_m(x) \leq \frac{1}{n} \int f_n \, d\mu,$$

for μ-almost every x. Since n is arbitrary, we have, almost surely,

$$\limsup_{m \to \infty} \frac{1}{m} f_m(x) \leq C.$$

It remains to prove the reversed inequality with $\liminf$. Recall that a measurable function g is called *subinvariant* (or *supinvariant*) if, for almost all $x \in X$, $g(Tx) \leq g(x)$ (or $g(Tx) \geq g(x)$), and that in ergodic systems only constants are subinvariant (supinvariant). So, suppose for contrary that for some positive ε,

$$\liminf_{n\to\infty} \tfrac{1}{n} f_n(x) < C - 3a\varepsilon \tag{2.1.8}$$

on a positive measure set. Denote $c = C - 3a\varepsilon$. By subadditivity, $f_{n+1}(x) \le a + f_n(Tx)$, hence the function $\liminf_n \frac{1}{n} f_n$ is supinvariant, and thus constant μ-almost everywhere. This implies that the inequality (2.1.8) holds in fact on a full measure set. In particular,

$$\mu\Big(\bigcup_{n=1}^{\infty} \{x : \tfrac{1}{n} f_n(x) < c\}\Big) = 1.$$

Then there exists n_0 such that the set

$$E = \bigcup_{n=1}^{n_0} \{x : f_n(x) < nc\} \tag{2.1.9}$$

has measure larger than $1 - \varepsilon$. By the Ergodic Theorem applied to the characteristic function of E there exists a positive integer m_0 larger than n_0/ε such that the set

$$F = \Big\{x : \tfrac{1}{m_0} \#\{i : 0 \le i < m_0, T^i x \in E\} > 1 - \varepsilon\Big\} \tag{2.1.10}$$

has measure also larger than $1 - \varepsilon$. By the definition of C in (2.1.5) we have

$$m_0 C \le \int f_{m_0}\, d\mu = \int_F f_{m_0}\, d\mu + \int_{X\setminus F} f_{m_0}\, d\mu.$$

The second integral is smaller than $m_0 a\varepsilon$. We will arrive at a contradiction with (2.1.8) by estimating the first integral by $m_0(c + 2a\varepsilon)$. This will be done by showing that $f_{m_0}(x) \le m_0(c + 2a\varepsilon)$ for every $x \in F$. We fix such an x and proceed as follows:

We denote by i_1 the smallest nonnegative integer with $T^{i_1}x \in E$ and we choose an $n_1 \le n_0$ with $f_{n_1}(T^{i_1}x) < n_1 c$ (see the definition of E in (2.1.9)). Inductively, for each $k > 1$, we let i_k be the smallest integer satisfying $i_k \ge i_{k-1} + n_{k-1}$ and $T^{i_k}x \in E$, and we choose an $n_k \le n_0$ with $f_{n_k}(T^{i_k}x) < n_k c$. We call $[i_k, i_k + n_k)$ a *good* interval. The number of positive integers smaller than m_0 not contained in good intervals is at most $m_0\varepsilon$ (see the definition of F in (2.1.10)). The length of the last incomplete part of a good interval intersecting $[0, m_0)$ (if such exists) is at most n_0, also smaller than $m_0\varepsilon$. The sum $n_1 + n_2 + \cdots + n_{k_0}$ representing the joint length of good intervals fully contained in $[0, m_0)$ is thus larger than $m_0(1 - 2\varepsilon)$. A final application of subadditivity allows us to write

$$f_{m_0}(x) \leq \sum_{k=1}^{k_0} f_{n_k}(T^{i_k}x) + (m_0 - \sum_{k=1}^{k_0} n_k)a,$$

where the first sum comes from the good intervals and the second one estimates the rest. Replacing each $f_{n_k}(T^{i_k}x)$ by $n_k c$ we obtain

$$f_{m_0}(x) \leq m_0 c + 2m_0 \varepsilon a < m_0(c + 2a\varepsilon),$$

as claimed. □

2.2 Preliminaries on dynamical systems

We assume familiarity of the reader with basics of ergodic theory, nonetheless we recall the notation. Let $(X, \mathfrak{A}, \mu)$ be a standard (completed) probability space and let $T : X \to X$ be a measurable measure-preserving map (an *endomorphism*), i.e., such that $T^{-1}(A) \in \mathfrak{A}$ and $\mu(T^{-1}(A)) = \mu(A)$, for every $A \in \mathfrak{A}$. The semigroup of nonnegative integers acts on X by iterates of T, with the convention that T^0 is the identity map. We call T an *automorphism* when, after discarding a set of measure zero, T becomes injective. Then T^{-1} can be defined almost everywhere, and in standard spaces it is automatically measurable and preserves μ. In such case we can also consider the action of the group which includes the iterates of T^{-1}. In some aspects it is very important to remember which action one has in mind. In case of an automorphism we can still have two different actions and the corresponding (measure-theoretic) *dynamical systems* will be denoted by $(X, \mathfrak{A}, \mu, T, \mathbb{N}_0)$ and $(X, \mathfrak{A}, \mu, T, \mathbb{Z})$, respectively, or by $(X, \mathfrak{A}, \mu, T, \mathbb{S})$ ($\mathbb{S} \in \{\mathbb{N}_0, \mathbb{Z}\}$) if we want to include both choices. The set $\mathbb{S}$ will be referred to as *the acting semigroup*.

Example 2.2.1 Let X be a compact metric space and let $T : X \to X$ be a continuous map (or homeomorphism). The triple $(X, T, \mathbb{S})$ is called a *topological dynamical system*. It is known that the collection of T-invariant probability measures μ on the Borel sigma-algebra $\mathfrak{A}_X$ in X is nonempty. Every such measure produces a dynamical system $(X, \mathfrak{A}_\mu, \mu, T, \mathbb{S})$ ($\mathfrak{A}_\mu$ is the Borel sigma-algebra completed with respect to μ).

A dynamical system $(Y, \mathfrak{B}, \nu, S, \mathbb{S})$ is a *factor* of $(X, \mathfrak{A}, \mu, T, \mathbb{S})$ (equivalently, $(X, \mathfrak{A}, \mu, T, \mathbb{S})$ is an *extension* of $(Y, \mathfrak{B}, \nu, S, \mathbb{S})$) (notice that we require the acting semigroup to be the same) if there is a measurable map (called a *factor map*) $\pi : X \to Y$ which is *equivariant*, i.e., such that $\pi \circ T = S \circ \pi$ and $\pi\mu = \nu$. Every factor $(Y, \mathfrak{B}, \nu, S, \mathbb{S})$ determines a sigma-algebra in X, $\pi^{-1}(\mathfrak{B}) \preccurlyeq \mathfrak{A}$, whenever possible also denoted by $\mathfrak{B}$. This sigma-algebra is *subinvariant*, i.e., $T^{-1}(\mathfrak{B}) \preccurlyeq \mathfrak{B}$ or, in case S is invertible (for instance, when $\mathbb{S} = \mathbb{Z}$), it is *invariant*, i.e., $T^{-1}(\mathfrak{B})$ after completing equals $\mathfrak{B}$. Notice that by preservation of the measure, after discarding a set of measure zero, T is a surjection, hence the last condition also implies $T(\mathfrak{B}) = \mathfrak{B}$. Conversely, it is well known in ergodic theory (compare Fact 1.2.2) that every subinvariant sigma-algebra $\mathfrak{B}$ (invariant if $\mathbb{S} = \mathbb{Z}$) defines a factor of the system $(X, \mathfrak{A}, \mu, T, \mathbb{S})$. The space Y of that factor corresponds to the collection of all atoms of the sigma-algebra $\mathfrak{B}$. The action on this factor is invertible if and only if $\mathfrak{B}$ is invariant.

Two systems $(X, \mathfrak{A}, \mu, T, \mathbb{S})$ and $(Y, \mathfrak{B}, \nu, S, \mathbb{S})$ are *isomorphic* if there exists a factor map $\pi : X \to Y$ which is invertible. It is important to realize that a factor of a system associated with a (sub)invariant proper sub-sigma-algebra can be isomorphic to the whole system via another map (see Exercise 2.1).

An extremely important class of systems are *symbolic dynamical systems*, in which X is the space $\Lambda^{\mathbb{S}'}$ ($\mathbb{S}' \in \{\mathbb{N}_0, \mathbb{Z}\}$) of unilateral (i.e., with $\mathbb{S}' = \mathbb{N}_0$) or bilateral (i.e., with $\mathbb{S}' = \mathbb{Z}$) sequences $(x_i)_{i\in\mathbb{S}'}$ over a countable set Λ called the *alphabet*. In such spaces we will always regard only one "master" sigma-algebra, namely the completed product sigma-algebra in $\Lambda^{\mathbb{S}'}$, where each copy of Λ is equipped with the sigma-algebra of all subsets. The transformation T will be typically the *shift transformation* σ defined by $\sigma(x_i)_{i\in\mathbb{S}'} = (x_{i+1})_{i\in\mathbb{S}'}$. The measure μ can be chosen as any shift-invariant measure. In full generality we assume that $\mathbb{S} \subset \mathbb{S}'$. The remaining configuration, $\mathbb{S}' = \mathbb{N}_0$ and $\mathbb{S} = \mathbb{Z}$, is possible only when the shift map is invertible on unilateral sequences, which has very strong consequences (see Fact 2.3.12). We will denote symbolic systems by $(\Lambda^{\mathbb{S}'}, \mu, \sigma, \mathbb{S})$, skipping the obvious sigma-algebra.

Let $\mathcal{P}$ be a finite or countable partition of X and let Λ be a set of labels assigned bijectively to the elements of $\mathcal{P}$. The map

$$x \mapsto \pi x = (x_n)_{n\in\mathbb{S}} \in \Lambda^{\mathbb{S}}$$

defined by the rule $x_n = a \iff T^n x \in A$, where $a \in \Lambda$ is the label of $A \in \mathcal{P}$, is a factor map from $(X, \mathfrak{A}, \mu, T, \mathbb{S})$ to the symbolic system over the alphabet Λ with the shift-invariant measure $\pi\mu$. This symbolic factor system will be called the *process* generated by $\mathcal{P}$ and denoted by $(X, \mathcal{P}, \mu, T, \mathbb{S})$. The reader will easily distinguish between a dynamical system denoted by $(X, \mathfrak{A}, \mu, T, \mathbb{S})$ (with the sigma-algebra in Gothic) and a process $(X, \mathcal{P}, \mu, T, \mathbb{S})$, where the emphasis is on one selected partition. The sequence $(x_n)_{n\in\mathbb{S}}$ corresponding to a point $x \in X$ will be referred to as the $\mathcal{P}$-*name* of x.

Every symbolic system $(\Lambda^{\mathbb{S}'}, \mu, \sigma, \mathbb{S})$ is a process in the above sense, as long as $\mathbb{S}' = \mathbb{S}$. It suffices to take $X = \Lambda^{\mathbb{S}}$ and consider the *zero-coordinate partition* $\mathcal{P}_\Lambda = \{A_a : a \in \Lambda\}$, where $A_a = \{x \in \Lambda^{\mathbb{S}} : x_0 = a\}$. Every x is now identical with its own $\mathcal{P}_\Lambda$-name, so the process generated by $\mathcal{P}_\Lambda$ is equal to the original system. From now on the term "process" becomes almost synonymous with "symbolic dynamical system," except that a process requires $\mathbb{S}' = \mathbb{S}$ and we want to maintain the reference to the "master" dynamical system and the selected partition.

In a symbolic system, for $(x_n) \in \Lambda^{\mathbb{S}}$ and $i \leq j \in \mathbb{S}$, by $x[i,j]$ we will denote the *block* $(x_i, x_{i+1}, \ldots, x_j) \in \Lambda^{j-i+1}$. With each block $B \in \Lambda^n$ we associate its *cylinder set*, $U_B = \{x : x[0, n-1] = B\}$. In a process $(X, \mathcal{P}, \mu, T, \mathbb{S})$ generated by a partition $\mathcal{P}$ labeled by an alphabet Λ the cylinder is formally the preimage of U_B by π. For $n \geq 1$ we will denote:

$$\mathcal{P}^n = \bigvee_{i=0}^{n-1} T^{-i}(\mathcal{P}).$$

By convention, $\mathcal{P}^0$ equals the trivial partition. It is easy to see that $\mathcal{P}^n$ equals the partition into the cylinder sets $\pi^{-1}(U_B)$ over all blocks $B \in \Lambda^n$. From now on we will identify the blocks B with their cylinders U_B (in symbolic systems) or with $\pi^{-1}(U_B)$ (in processes) and denote by B both the block and the cylinder, depending on the context. We shall not use U_B or $\pi^{-1}(U_B)$ again. In other words, we will identify the Cartesian product Λ^n with the above defined join $\mathcal{P}^n$.

We will also use the following notation: if $\mathbb{D}$ is a finite subset of $\mathbb{S}$, then

$$\mathcal{P}^{\mathbb{D}} = \bigvee_{i \in \mathbb{D}} T^{-i}(\mathcal{P}).$$

If $\mathbb{D}$ is infinite, $\mathcal{P}^{\mathbb{D}}$ will be used to denote the smallest completed sigma-algebra containing all the partitions $T^{-i}(\mathcal{P})$, where $i \in \mathbb{D}$. Intuitively, $\mathcal{P}^{\mathbb{D}}$ is the partition (or sigma-algebra) with atoms determined by the entries the $\mathcal{P}$-names assume at the coordinates $i \in \mathbb{D}$. In addition to the above introduced convention $\mathcal{P}^{[0,n-1]} = \mathcal{P}^n$, we will also abbreviate $\mathcal{P}^{-n} = \mathcal{P}^{[-n,-1]}$, $\mathcal{P}^- = \mathcal{P}^{(-\infty,-1]}$ and $\mathcal{P}^+ = \mathcal{P}^{[1,\infty)}$. The partitions $\mathcal{P}^{-n}$, $\mathcal{P}^n$ and the sigma-algebras $\mathcal{P}^-$, $\mathcal{P}^+$ will be called the nth *past*, the nth *present*, the *full past* and the *full future* of the process, respectively. Of course, the notions involving the past apply only to the invertible case with $\mathbb{S} = \mathbb{Z}$. By the *full history* of the process we shall mean $\mathcal{P}^{\mathbb{S}}$, the full completed sigma-algebra generated by $\mathcal{P}$ via the dynamics.

Remark 2.2.2 We apologize for not sticking to the most commonly used notation $\mathcal{P}_m^n$ in place of our $\mathcal{P}^{[m,n]}$. There are two major reasons for that: (1) we will often use partitions with a subscript, which would collide with the "lower bound" m, (2) we will use more complicated sets $\mathbb{D}$ than intervals, for which the traditional denotation is insufficient.

2.3 Dynamical entropy of a process

Dynamical entropy of a process depends only on its future, so it is not important whether we deal with the action of $\mathbb{N}_0$ or $\mathbb{Z}$. Nevertheless, since some properties related to entropy of the process do depend on the acting semigroup, we will keep it in mind (and $\mathbb{S}$ in the denotation). In order to avoid repetitions of very similar constructions and statements, we will discuss dynamical entropy and conditional dynamical entropy at the same time.

Let $(X, \mathfrak{A}, \mu, T, \mathbb{S})$ be a dynamical system. As long as the transformation and measure are fixed, we will skip them in the denotation of the entropy notions. We will consider two countable partitions $\mathcal{P}$, $\mathcal{Q}$ and a sigma-algebra $\mathfrak{B}$ which represents a factor (recall that $\mathfrak{B}$ is subinvariant, i.e., $T^{-1}(\mathfrak{B}) \preccurlyeq \mathfrak{B}$, in case $\mathbb{S} = \mathbb{N}_0$, or invariant, i.e., $T^{-1}(\mathfrak{B}) = \mathfrak{B}$, in case $\mathbb{S} = \mathbb{Z}$). We will be interested in the following four sequences:

$$H(\mathcal{P}^n), \quad H(\mathcal{P}^n|\mathfrak{B}), \quad H(\mathcal{P}^n|\mathcal{Q}^n), \quad H(\mathcal{P}^n|\mathcal{Q}^n \vee \mathfrak{B}).$$

Subadditivity of these sequences can be immediately derived from (1.6.33), subinvariance of $\mathfrak{B}$ and T-invariance of the measure, and in most cases this subadditivity would be enough for us. But in fact these sequences satisfy stronger conditions. Since the proof in full generality seems not to occur in other textbooks, we have decided to give it.

Fact 2.3.1 *The sequences $H(\mathcal{P}^n), H(\mathcal{P}^n|\mathfrak{B}), H(\mathcal{P}^n|\mathcal{Q}^n), H(\mathcal{P}^n|\mathcal{Q}^n \vee \mathfrak{B})$ have decreasing $n$ths. Moreover, the sequences $H(\mathcal{P}^n)$ and $H(\mathcal{P}^n|\mathfrak{B})$ have decreasing increments.*

Proof Of course, it suffices to prove the decreasing $n$ths for $H(\mathcal{P}^n|\mathcal{Q}^n \vee \mathfrak{B})$. This term decomposes, by (1.6.26), into n terms

$$H(\mathcal{P}|\mathcal{Q}^n \vee \mathfrak{B}) + H(T^{-1}(\mathcal{P})|\mathcal{P} \vee \mathcal{Q}^n \vee \mathfrak{B}) + H(T^{-2}(\mathcal{P})|\mathcal{P}^2 \vee \mathcal{Q}^n \vee \mathfrak{B}) + \cdots$$
$$\cdots + H(T^{-(n-1)}(\mathcal{P})|\mathcal{P}^{n-1} \vee \mathcal{Q}^n \vee \mathfrak{B}) = \sum_{i=0}^{n-1} H(T^{-i}(\mathcal{P})|\mathcal{P}^i \vee \mathcal{Q}^n \vee \mathfrak{B}),$$

and $H(\mathcal{P}^{n-1}|\mathcal{Q}^{n-1} \vee \mathfrak{B})$ decomposes analogously, into $n-1$ terms

$$\sum_{i=0}^{n-2} H(T^{-i}(\mathcal{P})|\mathcal{P}^i \vee \mathcal{Q}^{n-1} \vee \mathfrak{B}).$$

Clearly, the ith term below is not smaller than the ith term above. But it is also not smaller than the $(i+1)$st term above:

$$H(T^{-i}(\mathcal{P})|\mathcal{P}^i \vee \mathcal{Q}^{n-1} \vee \mathfrak{B}) = H(T^{-(i+1)}(\mathcal{P})|\mathcal{P}^{[1,i]} \vee \mathcal{Q}^{[1,n-1]} \vee T^{-1}(\mathfrak{B})) \geq H(T^{-(i+1)}\mathcal{P}|\mathcal{P}^{i+1} \vee \mathcal{Q}^n \vee \mathfrak{B}).$$

It is now seen that $\overline{a} = \frac{1}{n}H(\mathcal{P}^n|\mathcal{Q}^n \vee \mathfrak{B})$ and $\overline{b} = \frac{1}{n-1}H(\mathcal{P}^{n-1}|\mathcal{Q}^{n-1} \vee \mathfrak{B})$ satisfy the requirements of Lemma 2.1.2, so $\overline{b} \geq \overline{a}$.

The second statement is much easier (though stronger for the sequences to which it applies). By (1.6.26), the nth increment is

$$H(\mathcal{P}^n|\mathfrak{B}) - H(\mathcal{P}^{n-1}|\mathfrak{B}) = H(T^{-n+1}(\mathcal{P})|\mathcal{P}^{n-1} \vee \mathfrak{B}). \tag{2.3.2}$$

Similarly, the $(n+1)$st increment is $H(T^{-n}(\mathcal{P})|\mathcal{P}^n \vee \mathfrak{B})$. Applying T^{-1}, the nth increment can be rewritten as $H(T^{-n}(\mathcal{P})|\mathcal{P}^{[1,n-1]} \vee T^{-1}(\mathfrak{B}))$. Because $\mathcal{P}^n \vee \mathfrak{B} \succcurlyeq \mathcal{P}^{[1,n-1]} \vee T^{-1}(\mathfrak{B})$, the $(n+1)$st increment is not larger than the nth. □

We can now define the major notions of this section (and, perhaps, of the entire book): the dynamical entropy of a process interpreted as the average gain of information per iterate.

Definition 2.3.3 Assume $H(\mathcal{P}) < \infty$. The value

$$h(\mu, T, \mathcal{P}) = h(\mathcal{P}) = \lim_n \downarrow \tfrac{1}{n}H(\mathcal{P}^n)$$

will be called the *dynamical entropy* of (the process generated by) $\mathcal{P}$. The simplified notation will be used when the measure and the transformation are fixed. The values

$$h(\mu, T, \mathcal{P}|\mathfrak{B}) = h(\mathcal{P}|\mathfrak{B}) = \lim_n \downarrow \tfrac{1}{n}H(\mathcal{P}^n|\mathfrak{B}),$$

$$h(\mu, T, \mathcal{P}|\mathcal{Q}) = h(\mathcal{P}|\mathcal{Q}) = \lim_n \downarrow \tfrac{1}{n}H(\mathcal{P}^n|\mathcal{Q}^n), \text{ and}$$

$$h(\mu, T, \mathcal{P}|\mathcal{Q}, \mathfrak{B}) = h(\mathcal{P}|\mathcal{Q}, \mathfrak{B}) = \lim_n \downarrow \tfrac{1}{n}H(\mathcal{P}^n|\mathcal{Q}^n \vee \mathfrak{B})$$

(assuming $H(\mathcal{P}|\mathfrak{B}) < \infty$, or $H(\mathcal{P}|\mathcal{Q}) < \infty$, or $H(\mathcal{P}|\mathcal{Q} \vee \mathfrak{B}) < \infty$, respectively) are called the *conditional dynamical entropy* of $\mathcal{P}$ *given* $\mathfrak{B}$, *given* $\mathcal{Q}$, and *given* both $\mathcal{Q}$ and $\mathfrak{B}$.

Let us remark that the finiteness assumption is completely natural and necessary. Without it, the corresponding dynamical entropy is infinite, even for the action of the identity transformation, i.e., the infinite value is completely useless for describing the "complexity" of the dynamical system.

We can derive another formula which is often used as an alternative definition of the dynamical entropy. It allows for another interpretation of $h(\mathcal{P})$, as the expected gain of information in one step given all the information from the future (or the past, depending on how we interpret the direction of the time) of the process.

Fact 2.3.4 *Let $\mathfrak{B}$ denote an invariant sigma-algebra (in particular, trivial) such that $H(\mathcal{P}|\mathfrak{B}) < \infty$. Then*

$$h(\mathcal{P}|\mathfrak{B}) = H(\mathcal{P}|\mathcal{P}^+ \vee \mathfrak{B}), \tag{2.3.5}$$

$$h(\mathcal{P}) = H(\mathcal{P}|\mathcal{P}^+). \tag{2.3.6}$$

Proof Since now $T^{-1}(\mathfrak{B}) = \mathfrak{B}$, we can rework the increment in the sequence $H(\mathcal{P}^n|\mathfrak{B})$ differently:

$$H(\mathcal{P}^{n+1}|\mathfrak{B}) - H(\mathcal{P}^n|\mathfrak{B}) = H(\mathcal{P}^{n+1}|\mathfrak{B}) - H(T^{-1}(\mathcal{P}^n)|\mathfrak{B}) = H(\mathcal{P}|\mathcal{P}^{[1,n]} \vee \mathfrak{B}).$$

By Fact 2.1.1, these increments converge to the same limit as $\frac{1}{n}H(\mathcal{P}^n|\mathfrak{B})$, i.e., to $h(\mu, T, \mathcal{P}|\mathfrak{B})$. On the other hand, by (1.7.12), their limit equals $H(\mathcal{P}|\mathcal{P}^+ \vee \mathfrak{B})$. □

We shall now explain that the last two notions in Definition 2.3.3 (conditional dynamical entropy given $\mathcal{Q}$ and given $\mathcal{Q}, \mathfrak{B}$) reduce to the preceding notion of conditional dynamical entropy given a subinvariant sigma-algebra.

Fact 2.3.7

$$H(\mathcal{P}|\mathcal{Q}) < \infty \implies h(\mathcal{P}|\mathcal{Q}) = h(\mathcal{P}|\mathcal{Q}^{\mathbb{S}}), \tag{2.3.8}$$

$$H(\mathcal{P}|\mathcal{Q} \vee \mathfrak{B}) < \infty \implies h(\mathcal{P}|\mathcal{Q}, \mathfrak{B}) = h(\mathcal{P}|\mathcal{Q}^{\mathbb{S}} \vee \mathfrak{B}). \tag{2.3.9}$$

Notice that the terms on the right are more universal, as they require weaker finiteness assumptions ($H(\mathcal{P}|\mathcal{Q}^{\mathbb{S}}) < \infty$ or $H(\mathcal{P}|\mathcal{Q}^{\mathbb{S}} \vee \mathfrak{B}) < \infty$). Nevertheless, the limits involving $\mathcal{Q}^n$ rather than $\mathcal{Q}^{\mathbb{S}}$ are often more convenient to use.

Proof of Fact 2.3.7 It suffices to prove (2.3.9). It follows directly from the definition that $h(\mathcal{P}|\mathcal{Q}, \mathfrak{B}) \geq h(\mathcal{P}|\mathcal{Q}^{\mathbb{S}} \vee \mathfrak{B})$. To get the reversed inequality for $\mathbb{S} = \mathbb{N}_0$ we write

$$\begin{aligned} H(\mathcal{P}^{m+n}|\mathcal{Q}^{m+n} \vee \mathfrak{B}) &\leq H(\mathcal{P}^m|\mathcal{Q}^{m+n} \vee \mathfrak{B}) + H(T^{-m}(\mathcal{P}^n)|\mathcal{Q}^{m+n} \vee \mathfrak{B}) \leq \\ H(\mathcal{P}^m|\mathcal{Q}^{m+n} \vee \mathfrak{B}) &+ H(T^{-m}(\mathcal{P}^n)|T^{-m}(\mathcal{Q}^n) \vee T^{-m}(\mathfrak{B})) = \\ &H(\mathcal{P}^m|\mathcal{Q}^{m+n} \vee \mathfrak{B}) + H(\mathcal{P}^n|\mathcal{Q}^n \vee \mathfrak{B}). \end{aligned}$$

If $\mathbb{S} = \mathbb{Z}$, $\mathfrak{B}$ is necessarily invariant and, then we write

$$\begin{aligned} &H(\mathcal{P}^{m+2n}|\mathcal{Q}^{m+2n} \vee \mathfrak{B}) \leq \\ &H(\mathcal{P}^n|\mathcal{Q}^{m+2n} \vee \mathfrak{B}) + H(T^{-n}(\mathcal{P}^m)|\mathcal{Q}^{m+2n} \vee \mathfrak{B}) + H(T^{-n-m}(\mathcal{P}^n)|\mathcal{Q}^{m+2n} \vee \mathfrak{B}) \\ &\leq 2H(\mathcal{P}^n|\mathcal{Q}^n \vee \mathfrak{B}) + H(\mathcal{P}^m|\mathcal{Q}^{[-n,m+n-1]} \vee \mathfrak{B}). \end{aligned}$$

Dividing by m and taking infimum over m we obtain,

$$h(\mathcal{P}|\mathcal{Q},\mathfrak{B}) \le \inf_m \tfrac{1}{m} H(\mathcal{P}^m|\mathcal{Q}^{m+n} \vee \mathfrak{B}) \quad \text{or}$$
$$h(\mathcal{P}|\mathcal{Q},\mathfrak{B}) \le \inf_m \tfrac{1}{m} H(\mathcal{P}^m|\mathcal{Q}^{[-n,m+n-1]} \vee \mathfrak{B}),$$

depending on the choice of $\mathbb{S}$. Because this is true for every n, we can apply infimum over n on the right. We can also reverse the order of the infima:

$$h(\mathcal{P}|\mathcal{Q},\mathfrak{B}) \le \inf_m \tfrac{1}{m} \inf_n H(\mathcal{P}^m|\mathcal{Q}^{m+n} \vee \mathfrak{B}) = \inf_m \tfrac{1}{m} H(\mathcal{P}^m|\mathcal{Q}^{\mathbb{N}_0} \vee \mathfrak{B}) \quad \text{or}$$
$$h(\mathcal{P}|\mathcal{Q},\mathfrak{B}) \le \inf_m \tfrac{1}{m} \inf_n H(\mathcal{P}^m|\mathcal{Q}^{[-n,m+n-1]} \vee \mathfrak{B}) = \inf_m \tfrac{1}{m} H(\mathcal{P}^m|\mathcal{Q}^{\mathbb{Z}} \vee \mathfrak{B}).$$

□

Remark 2.3.10 For actions of $\mathbb{S} = \mathbb{Z}$ the sigma-algebra $\mathcal{Q}^{\mathbb{S}}$ is invariant (and so is $\mathfrak{B}$), hence, by (2.3.5), we also have

$$h(\mathcal{P}|\mathcal{Q},\mathfrak{B}) = H(\mathcal{P}|\mathcal{P}^+ \vee \mathcal{Q}^{\mathbb{S}} \vee \mathfrak{B}). \tag{2.3.11}$$

There is now an interesting interpretation of dynamical entropy zero related to invertibility.

Fact 2.3.12 *Consider a process $(X,\mathcal{P},\mu,T,\mathbb{S})$. The three conditions below are equivalent:*

(a) $h(\mu,T,\mathcal{P}) = 0$;
(b) $\mathcal{P}$ is $\mathcal{P}^+$-measurable;
(c) the shift map σ on the unilateral symbolic space $(\Lambda^{\mathbb{N}_0},\pi\mu)$ is invertible,

where Λ is a set of labels bijectively assigned to the elements of the partition $\mathcal{P}$, π is the factor map sending each x to its unilateral $\mathcal{P}$-name and $\pi\mu$ is μ transported via π (by preimage).

Proof The condition $h(\mu,T,\mathcal{P}) = 0$ reads $H(\mathcal{P}|\mathcal{P}^+) = 0$. The statement (1.6.28) now establishes the equivalence between (a) and (b). If (b) holds, then for μ-almost every $x \in X$ the unilateral $\mathcal{P}$-name of Tx determines the element of $\mathcal{P}$ containing x. In other words, the unilateral $\mathcal{P}$-name of Tx determines the unilateral $\mathcal{P}$-name of x, and this is exactly (c). In the unilateral symbolic system $(\Lambda^{\mathbb{N}_0},\pi\mu,\sigma,\mathbb{N}_0)$, $\mathcal{P}^{\mathbb{N}_0}$ is, by definition, the full (product) sigma-algebra. If (c) holds, then $T(\mathcal{P})$ is measurable (i.e., $\mathcal{P}^{\mathbb{N}_0}$-measurable). Applying T^{-1} we get that $\mathcal{P}$ is $\mathcal{P}^+$-measurable, i.e., (b). □

The above Fact 2.3.12 can be expressed as follows:

- Every zero-entropy process is in fact invertible.
- An invertible process has entropy zero if and only if $\mathcal{P}^{\mathbb{Z}} = \mathcal{P}^+ (= \mathcal{P}^-)$.

In yet other words, if a process has entropy zero, then regardless of whether it is an action of $\mathbb{N}_0$ or $\mathbb{Z}$, almost every point has a well-defined *backward* $\mathcal{P}$-*name* or *past* $(\ldots, x_{-2}, x_{-1})$ and this past determines the future of x. Such processes are often called *deterministic*. See also Section 3.2 devoted to the Pinsker factor.

Example 2.3.13 Let $\mathbb{T}$ denote the unit circle on the complex plane and let $R: \mathbb{T} \to \mathbb{T}$ be the *rotation* map given by $R(z) = \varrho z$, where $\varrho \in \mathbb{T}$. We have a topological dynamical system $(\mathbb{T}, R, \mathbb{S})$. The normalized Lebesgue measure λ is preserved by R. So, we have a dynamical system $(\mathbb{T}, \mathfrak{A}_\lambda, \lambda, R, \mathbb{S})$. It is well known, that this system is ergodic if and only if ϱ is not a root of unity (then we deal with an *irrational rotation*). Every partition $\mathcal{P}$ into two complementary arcs A_0, A_1 produces a so-called Sturmian process[1] $(\mathbb{T}, \mathcal{P}, \lambda, R, \mathbb{S})$. The past of every point determines this point, so the process has entropy zero.

Remark 2.3.14 Since, for an invertible map, $h(\mu, T, \mathcal{P}) = h(\mu, T^{-1}, \mathcal{P})$, it follows from Fact 2.3.12 (b) that $\mathcal{P}$ is measurable with respect to $\mathcal{P}^+$ if and only if it is measurable with respect to $\mathcal{P}^-$. This fact refers only to sigma-algebras and measurability, in particular it is expressed without using the entropy. It would be very interesting to have a proof based exclusively on manipulating sigma-algebras. Such a proof, however, is not known. This is a good example showing how powerful a tool the entropy theory is. See also Question 3.2.3.

The opposite class to processes of entropy zero are independent processes.

Definition 2.3.15 A process $(X, \mathcal{P}, \mu, T, \mathbb{S})$ is *independent* if $h(\mathcal{P}) = H(\mathcal{P})$, equivalently, $H(\mathcal{P}|\mathcal{P}^+) = H(\mathcal{P})$, i.e, $\mathcal{P}$ is independent of the future (see Fact 1.6.38).

Independent processes have only one possible symbolic realization – as *Bernoulli shifts*.[2]

Let $\mathbf{p} = (p_1, p_2, \ldots)$ be a probability distribution on a countable (or finite) set of symbols Λ, satisfying $H(\mathbf{p}) < \infty$. On $X = \Lambda^{\mathbb{S}}$ the product measure $\mu = \mathbf{p}^{\mathbb{S}}$ is shift invariant. It is easy to see that, for each n, $H(\mathcal{P}_\Lambda^n) = nH(\mathbf{p})$, so $h(\mathcal{P}_\Lambda) = H(\mathcal{P}_\Lambda) = H(\mathbf{p})$, so the process generated by $\mathcal{P}_\Lambda$ on the symbolic system $(\Lambda^{\mathbb{S}}, \mu, \sigma, \mathbb{S})$ is an independent process.

[1] Some authors restict the name "Sturmian" only to the process generated by the partition with cuts at 1 and ϱ.

[2] Bernoulli shifts should not be confused with Bernoulli processes (or Bernoulli systems), by which we understand any process (system) isomorphic to a Bernoulli shift. See also Section 4.5.

2.4 Properties of dynamical entropy

We start with a useful statement:

Fact 2.4.1 *Suppose $H(\mathcal{Q}) < \infty$ and $\mathcal{Q} \preccurlyeq \mathcal{P}^{\mathbb{N}_0}$. Then*

$$h(\mathcal{Q}) \le h(\mathcal{P}).$$

Proof Note that $\mathcal{Q} \preccurlyeq \mathcal{P}^{\mathbb{N}_0}$ implies $\mathcal{Q}^+ \preccurlyeq \mathcal{P}^+$. By (1.6.26) and (1.6.28) we have

$$h(\mathcal{Q}) \le h(\mathcal{P} \vee \mathcal{Q}) = H(\mathcal{P} \vee \mathcal{Q}|\mathcal{P}^+ \vee \mathcal{Q}^+) = H(\mathcal{P} \vee \mathcal{Q}|\mathcal{P}^+) =$$
$$H(\mathcal{Q}|\mathcal{P} \vee \mathcal{P}^+) + H(\mathcal{P}|\mathcal{P}^+) = 0 + h(\mathcal{P}).$$

□

Applying the decreasing limit of Definition 2.3.3 to appropriate rules for static entropy ((1.6.26), Fact 1.6.27 and Corollary 1.6.31) we derive the list of already familiar monotonicity and subadditivity properties, this time for dynamical entropy. The last two statements are consequences of the preceding two and the inequality $h(\mathcal{P}|\mathcal{Q}) \le H(\mathcal{P}|\mathcal{Q})$. Of course, all the statements are valid with $\mathfrak{B}$ replaced by a partition, or the trivial sigma-algebra. We skip rewriting these versions, except the first one which is most frequently used.

Fact 2.4.2 *Let $\mathcal{P}, \mathcal{Q}, \mathcal{R}$ be any countable partitions, and $\mathfrak{B}$ a subinvariant sigma-algebra. Then*

$$h(\mathcal{P} \vee \mathcal{Q}|\mathfrak{B}) = h(\mathcal{P}|\mathcal{Q}, \mathfrak{B}) + h(\mathcal{Q}|\mathfrak{B}), \tag{2.4.3}$$
$$h(\mathcal{P} \vee \mathcal{Q}) = h(\mathcal{P}|\mathcal{Q}) + h(\mathcal{Q}), \tag{2.4.4}$$
$$\mathcal{P} \succcurlyeq \mathcal{Q} \implies h(\mathcal{P}|\mathfrak{B}) \ge h(\mathcal{Q}|\mathfrak{B}), \tag{2.4.5}$$
$$\mathfrak{B} \succcurlyeq \mathfrak{C} \implies h(\mathcal{P}|\mathfrak{B}) \le h(\mathcal{P}|\mathfrak{C}), \tag{2.4.6}$$
$$h(\mathcal{P} \vee \mathcal{Q}|\mathfrak{B}) \le h(\mathcal{P}|\mathfrak{B}) + h(\mathcal{Q}|\mathfrak{B}), \tag{2.4.7}$$
$$h(\mathcal{P}|\mathfrak{B}) \le h(\mathcal{P}|\mathcal{Q}) + h(\mathcal{Q}|\mathfrak{B}), \tag{2.4.8}$$
$$h(\mathcal{P}|\mathcal{R}, \mathfrak{B}) \le h(\mathcal{P}|\mathcal{Q}, \mathfrak{B}) + h(\mathcal{Q}|\mathcal{R}, \mathfrak{B}), \tag{2.4.9}$$
$$|h(\mathcal{P}|\mathfrak{B}) - h(\mathcal{Q}|\mathfrak{B})| \le \max\{H(\mathcal{P}|\mathcal{Q}), H(\mathcal{Q}|\mathcal{P})\}, \tag{2.4.10}$$
$$|h(\mathcal{P}|\mathcal{Q}, \mathfrak{B}) - h(\mathcal{P}|\mathcal{R}, \mathfrak{B})| \le \max\{H(\mathcal{Q}|\mathcal{R}), H(\mathcal{R}|\mathcal{Q})\}. \tag{2.4.11}$$

□

Recall that if $\mathcal{P}$ is countable, then $\mathcal{P}_{(m)}$ denotes the m-element partition obtained by uniting all but the largest $m-1$ cells. Because $\mathcal{P}_{(m)} \preccurlyeq \mathcal{P}_{(m+1)}$, the entropies $h(\mathcal{P}_{(m)}|\mathcal{Q})$ increase.

Fact 2.4.12 *Assume $H(\mathcal{P}|\mathfrak{B}) < \infty$ (with $\mathfrak{B}$ subinvariant). Then*

$$h(\mathcal{P}|\mathfrak{B}) = \lim_m \uparrow h(\mathcal{P}_{(m)}|\mathfrak{B}).$$

In particular, replacing $\mathfrak{B}$ by $\mathcal{Q}^{\mathbb{S}}$ or the trivial partition we get

$$h(\mathcal{P}|\mathcal{Q}) = \lim_m \uparrow h(\mathcal{P}_{(m)}|\mathcal{Q}),$$

$$h(\mathcal{P}) = \lim_m \uparrow h(\mathcal{P}_{(m)}).$$

Proof By (2.4.10), and since $\mathcal{P} \succcurlyeq \mathcal{P}_{(m)}$, we have $|h(\mathcal{P}|\mathfrak{B}) - h(\mathcal{P}_{(m)}|\mathfrak{B})| \leq H(\mathcal{P}|\mathcal{P}_{(m)}) = H(\mathcal{P}) - H(\mathcal{P}_{(m)})$, which by (1.6.18) decreases to zero. Monotonicity is obvious by (2.4.5). □

We remark that the finiteness assumption cannot be skipped. For instance, let T be the identity map on a nonatomic space. There exist partitions with infinite static (hence dynamical) entropy, while every finite partition has dynamical entropy zero.

We pass to the continuity properties.

Fact 2.4.13 *The functions $h(\cdot|\mathfrak{B})$, where $\mathfrak{B}$ is a fixed subinvariant sigma-algebra, in particular $h(\cdot|\mathcal{Q})$ for any fixed partition $\mathcal{Q}$, and $h(\cdot)$, are uniformly continuous on $\mathfrak{P}_{\mathrm{R}}$ (with the Rokhlin metric d_{R}), while they are only lower semicontinuous on the same space with respect to d_1. If $\mathcal{P}$ satisfies $H(\mathcal{P}|\mathfrak{B}) < \infty$, then the function $h(\mathcal{P}|\cdot, \mathfrak{B})$, (in particular $h(\mathcal{P}|\cdot)$ where $H(\mathcal{P}) < \infty$) is uniformly continuous on $\mathfrak{P}_{\mathrm{R}}$ and upper semicontinuous on $\mathfrak{P}_{\aleph_0}$ (with respect to d_1).*

Proof Uniform continuity of $h(\cdot|\mathfrak{B})$ and $h(\mathcal{P}|\cdot, \mathfrak{B})$ in d_{R} is literally (2.4.10) and (2.4.11), respectively.

Lower semicontinuity of $h(\cdot|\mathfrak{B})$ in d_1 is proved at each $\mathcal{P}$ separately. Since on finite partitions the metrics d_{R} and d_1 are equivalent, $h(\mathcal{P}_{(m)}|\mathfrak{B})$ is continuous in d_1 at $\mathcal{P}_{(m)}$ for each m. Recall Fact 1.7.4, saying that the map $\mathcal{P} \mapsto \mathcal{P}_{(m)}$ is continuous in d_1 for infinitely many indices m. For these indices $\mathcal{P} \mapsto h(\mathcal{P}_{(m)}|\mathfrak{B})$ is continuous at $\mathcal{P}$. The increasing convergence of Fact 2.4.12 implies lower semicontinuity of $h(\cdot|\mathfrak{B})$ at $\mathcal{P}$ and thus globally.

Since $h(\mathcal{P}|\mathcal{Q}, \mathfrak{B}) = \lim_n \downarrow \frac{1}{n} H(\mathcal{P}^n|\mathcal{Q}^n \vee \mathfrak{B})$, Fact 1.7.10 ($d_1$-continuity of $H(\mathcal{P}| \cdot \vee \mathfrak{B})$), implies that $h(\mathcal{P}|\cdot, \mathfrak{B})$ is upper semicontinuous in d_1. □

Corollary 2.4.14 *If $\mathcal{P}$ satisfies $H(\mathcal{P}|\mathfrak{B}) < \infty$, then for any countable partition $\mathcal{Q}$ we have*

$$h(\mathcal{P}|\mathcal{Q} \vee \mathfrak{B}) = \lim_m \downarrow h(\mathcal{P}|\mathcal{Q}_{(m)} \vee \mathfrak{B}).$$

Proof Since $\mathcal{Q}_{(m)}$ clearly converge to $\mathcal{Q}$ in d_1, we can apply the above upper semicontinuity to obtain $h(\mathcal{P}|\mathcal{Q}\vee\mathfrak{B}) \ge \lim_m h(\mathcal{P}|\mathcal{Q}_{(m)}\vee\mathfrak{B})$. The other inequality is trivial by the monotonicity (2.4.6). □

Lack of continuity of $h(\cdot)$ in d_1, even for partitions with bounded static entropy (i.e., on $\mathfrak{P}_R$), is illustrated below. The same example shows lack of continuity in d_1 of $h(\cdot|\mathcal{Q})$ and $h(\mathcal{P}|\cdot)$.

Example 2.4.15 This example works for both cases of $\mathbb{S}$. Consider the Bernoulli shift on $X = \{0,1\}^{\mathbb{S}}$ (with the product Borel sigma-algebra), and the product measure $\{\frac{1}{2},\frac{1}{2}\}^{\mathbb{S}}$. Let $\mathcal{P}$ denote the zero-coordinate partition $\mathcal{P}_\Lambda$ (into the cylinders 0 and 1). Let $\mathcal{Q}_k$ be the following partition: the cylinder associated with the block $0^k = (000\dots0)$ is partitioned into 2^{2^k} cylinders (blocks) of the form 0^kB, where $B \in \mathcal{P}^{2^k}$, the rest of X is left in one piece (of measure $1-2^{-k}$), labeled by $*$. Notice that the static entropy of $\mathcal{Q}_k$ exceeds $2^{-k}\cdot\log 2^{2^k} = 1$. Clearly, the partitions $\mathcal{Q}_k$ converge in d_1 to the trivial partition $\mathcal{Q} = \{X\}$. The process generated by $\mathcal{Q}_k$ is the factor of the full shift obtained by the following code: wherever we find a block 0^k in the $\mathcal{P}$-name of x, we maintain the zeros and the following 2^k symbols. The rest is replaced by the stars. Since in almost every x the gaps between the occurrences of the block 0^k have nearly the exponential distribution with the expected value 2^k, the fraction of coordinates copied by the code has a value v_k stabilizing for large k at a positive v. It is obvious that the dynamical entropy of the factor process generated by $\mathcal{Q}_k$ is at least v_k times $h(\mathcal{P})$, hence $\limsup_k h(\mathcal{Q}_k) \ge v > 0$. On the other hand, $\mathcal{Q}_k \to \mathcal{Q}$ in d_1, and $h(\mathcal{Q}) = 0$.

Now observe the conditional entropies $h(\mathcal{P}|\mathcal{Q}_k)$. Clearly $\mathcal{Q}_k \preccurlyeq \mathcal{P}^{\mathbb{N}_0}$ which implies $h(\mathcal{P}|\mathcal{Q}_k) = h(\mathcal{P}\vee\mathcal{Q}_k) - h(\mathcal{Q}_k) \le h(\mathcal{P}) - h(\mathcal{Q}_k)$ (see Fact 2.4.1) and thus their lim inf is at most $1-v$. On the other hand, $h(\mathcal{P}|\mathcal{Q}) = h(\mathcal{P}) = 1$.

A convergence as in (1.7.13) holds also for dynamical conditional entropies, generalizing Corollary 2.4.14:

Fact 2.4.16 *Let $\mathcal{P}$ be a partition with finite static entropy. If $\mathfrak{B}_k$ is an increasing sequence of subinvariant sigma-algebras and $\mathfrak{B} = \bigvee_k \mathfrak{B}_k$, then*

$$h(\mathcal{P}|\mathfrak{B}) = \lim_k \downarrow h(\mathcal{P}|\mathfrak{B}_k). \tag{2.4.17}$$

If $\mathfrak{B}_k$ is a decreasing sequence of subinvariant sigma-algebras, and $\mathfrak{B} = \bigcap_k \mathfrak{B}_k$, then

$$h(\mathcal{P}|\mathfrak{B}) \ge \lim_k \uparrow h(\mathcal{P}|\mathfrak{B}_k). \tag{2.4.18}$$

Proof Since the limit in Definition (2.3.3) is an infimum, we can apply (1.7.13) (where the limit is also an infimum) and exchange the order of infima. The inequality in (2.4.18) is obvious by monotonicity. □

The inequality (2.4.18) cannot be reversed even for $\mathbb{S} = \mathbb{Z}$ when all the sigma-algebras are invariant. See Fact 3.2.8.

We conclude this section with what we call the *power rule*, the calculation of entropy in a process in which T is replaced by its iterate. There is a slight inconvenience in the notation when regarding several transformations: the terms of the form $\mathcal{P}^n$ require verbal explanation, to the action of which transformation does the "power" n refer. Each time it refers to an action different than that of T, we will explicitly say it, managing to avoid the nasty notation of the kind $\mathcal{P}^{T,n}$.

Given a process $(X, \mathcal{P}, \mu, T, \mathbb{S})$, we fix some $n \in \mathbb{S}$ and let $\bar{\mathcal{P}} = \mathcal{P}^n$. Next, we consider the *power process* $(X, \bar{\mathcal{P}}, \mu, T^n, \mathbb{S})$. We have

Fact 2.4.19 (The Power Rule)

$$h(\mu, T^n, \mathcal{P}^{|n|}) = |n| h(\mu, T, \mathcal{P}),$$
$$h(\mu, T^n, \mathcal{P}^{|n|} | \mathfrak{B}) = |n| h(\mu, T, \mathcal{P} | \mathfrak{B}),$$

where negative values of n require $\mathfrak{B}$ to be invariant (otherwise it is assumed subinvariant).

Proof For $n \geq 0$ the partition $\bar{\mathcal{P}}^m$, where m refers to the action of T^n, coincides with $\mathcal{P}^{nm}$ (now in the original process). This easily implies the assertion for such n. For actions of $\mathbb{Z}$, by invariance of $\mathfrak{B}$, we have $H(\mathcal{P}^n | \mathfrak{B}) = H(\mathcal{P}^{-n} | \mathfrak{B})$, hence $h(\mu, T, \mathcal{P} | \mathfrak{B}) = h(\mu, T^{-1}, \mathcal{P} | \mathfrak{B})$. □

Remark 2.4.20 By the way, for actions of $\mathbb{Z}$ we also have $H(\mathcal{P} | \mathcal{P}^+) = H(\mathcal{P} | \mathcal{P}^-)$, which explains why entropy can be interpreted as the one step information given the future as well as given the past.

2.5 Affinity of dynamical entropy

So far, we have been concerned with a fixed measure space, a fixed transformation, and we have been varying the partition. In this section we will change this point of view. Given a measurable space $(X, \mathfrak{A})$ and a measurable transformation T there may exist many probability measures on $\mathfrak{A}$ preserved by T. So, we can study the behavior of $h(\mu, T, \mathcal{P})$ as a function of μ, with the other parameters fixed.

Theorem 2.5.1 *Dynamical entropy and conditional dynamical entropy of a fixed partition are affine functions of the invariant measure. That is, for a partition $\mathcal{P}$ of a measurable space $(X, \mathfrak{A})$, a measurable transformation T of X, and two probability measures μ and ν preserved by T, for any $p \in [0, 1]$ and*

$q = 1 - p$, the measure $p\mu + q\nu$ is preserved by T and

$$h(p\mu + q\nu, T, \mathcal{P}) = ph(\mu, T, \mathcal{P}) + qh(\nu, T, \mathcal{P}).$$

If $\mathfrak{B}$ is a T-subinvariant sigma-algebra, then also

$$h(p\mu + q\nu, T, \mathcal{P}|\mathfrak{B}) = ph(\mu, T, \mathcal{P}|\mathfrak{B}) + qh(\nu, T, \mathcal{P}|\mathfrak{B}).$$

Proof It is obvious that $p\mu + q\nu$ is T-invariant. Concavity of dynamical entropy follows from concavity of static entropy (see (1.3.5)) via the limit passage defining dynamical entropy. For convexity, consider a model in which X is replaced by disjoint union of two copies X_1, X_2 of the original space X. The transformation T on this union is defined naturally, as T inside each of the copies. The partition $\mathcal{P}$ of the united space is defined by uniting the pairs of corresponding sets $A \in \mathcal{P}$ in both copies. The measure μ is regarded as concentrated on X_1, and ν analogously on X_2. Also let $\mathcal{Q}$ denote the partition into the two copies. In this model, the measure $p\mu + q\nu$ assigns to each cell A of $\mathcal{P}$ the convex combination $p\mu(A) + q\nu(A)$, the same as the combination of measures does in the original model. The same holds for $\mathcal{P}^n$, so the dynamical entropy of $p\mu + q\nu$ in this model is the same as in the original system. In this model, for each n, we use the monotonicity (1.6.6), the formulae (1.4.3) and (1.4.4), and we get

$$\begin{aligned}
&H(p\mu + q\nu, \mathcal{P}^n) \le \\
&\quad H(p\mu + q\nu, \mathcal{P}^n \vee \mathcal{Q}) = H(p\mu + q\nu, \mathcal{P}^n|\mathcal{Q}) + H(p\mu + q\nu, \mathcal{Q}) \le \\
&\quad \mu(X_1)H_{X_1}(p\mu + q\nu, \mathcal{P}^n) + \mu(X_2)H_{X_2}(p\mu + q\nu, \mathcal{P}^n) + \log 2 = \\
&\qquad\qquad\qquad pH(\mu, \mathcal{P}^n) + qH(\nu, \mathcal{P}^n) + \log 2.
\end{aligned}$$

Dividing by n and passing to the limit we arrive at the desired convexity condition. The proof of the conditional version is identical, except that now we need to use concavity of the conditional entropy as stated in (1.4.8), monotonicity (1.6.29), then (1.6.26) and (1.6.25). □

2.6 Conditional dynamical entropy via disintegration*

In this section we present a different approach to conditional dynamical entropy given a subinvariant sigma-algebra, based on disintegration. This approach will not be used in the main stream of argumentation, and we include it for completeness purposes. Recall the formula (1.5.2) defining the disintegration of μ with respect to a sigma-algebra $\mathfrak{B}$. If π is a factor map between dynamical systems $(X, \mathfrak{A}, \mu, T, \mathbb{S})$ and $(Y, \mathfrak{B}, \nu, S, \mathbb{S})$, and $\mathfrak{B} = \pi^{-1}(\mathfrak{B})$, then a desirable property of the disintegration is the following

Definition 2.6.1 The disintegration $y \mapsto \mu_y$ is *equivariant* if

$$\mu_{Sy} = T\mu_y$$

for ν-almost every $y \in Y$.

In general, the disintegration with respect to a subinvariant sigma-algebra is not guaranteed to be equivariant. It is so, however, when the sigma-algebra is invariant [see e.g. Furstenberg, 1981, Proposition 5.9], or when $(X, \mathfrak{A}, \mu, T, \mathbb{S})$ is a *skew product extension* of $(Y, \mathfrak{B}, \nu, S, \mathbb{S})$ [see e.g. Petersen, 1983, for the definition of a skew product].

The following definition is a direct generalization of a function appearing in [Abramov and Rokhlin, 1962] in the definition of the fiber entropy in skew products:

Definition 2.6.2 Let $\mathcal{P}$ be a partition of X with $H(\mathcal{P}) < \infty$. By the *fiber entropy* of $\mathcal{P}$ with respect to the measure ν we shall mean the function $y \mapsto h(\mathcal{P}|y)$ defined ν-almost everywhere on Y by the formula

$$h(\mathcal{P}|y) = \lim_n \downarrow H(\mu_y, \mathcal{P}|\mathcal{P}^{[1,n)}).$$

Notice that (1.5.4) implies that $H(\mu_y, \mathcal{P})$ (hence also $h(\mathcal{P}|y)$) is finite for ν-almost every y. It is interesting to note that fiber entropy as a function of y need not be invariant, (nor, in the ergodic case, constant). We leave finding an example to the reader, as Exercise 2.7.

Translating to our notation, Abramov and Rokhlin have proved that

$$\sup_{\mathcal{P}} \int h(\mathcal{P}|y) d\nu(y) = \sup_{\mathcal{P}} h(\mathcal{P}|\mathfrak{B}).$$

The precise meaning of the last expression will be discussed in Section 4.1. Now we shall strengthen the Abramov–Rokhlin Theorem by showing that it holds also for a fixed partition (i.e., that the supremum can be dropped on both sides).

Theorem 2.6.3 *If $(Y, \mathfrak{B}, \nu, S, \mathbb{S})$ is a factor of $(X, \mathfrak{A}, \mu, T, \mathbb{S})$ via a map π such that the disintegration of μ with respect to $\mathfrak{B} = \pi^{-1}(\mathfrak{B})$ is equivariant, then, for every partition $\mathcal{P}$ of X with $H(\mathcal{P}) < \infty$, we have*

$$h(\mathcal{P}|\mathfrak{B}) = \int h(\mathcal{P}|y) d\nu(y).$$

Proof We have

$$\int h(\mathcal{P}|y)d\nu(y) = \int \lim_n H(\mu_y, \mathcal{P}|\mathcal{P}^{[1,n]})d\nu(y) =$$

$$\lim_n \int \left[H(\mu_y, \mathcal{P}^{[0,n]}) - H(\mu_y, \mathcal{P}^{[1,n]})\right] d\nu(y) =$$

$$\lim_n \left[\int H(\mu_y, \mathcal{P}^{n+1})d\nu(y) - \int H(\mu_y, \mathcal{P}^n)d\nu(y)\right],$$

using the Lebesgue Monotone Theorem, equivariance of the disintegration and invariance of the measure ν. Note that all these integrals are finite for almost every y. The limit of the last sequence is the same as the limit of its averages, which, after cancellation, reads:

$$\lim_n \frac{1}{n} \int H(\mu_y, \mathcal{P}^n)d\nu(y).$$

The integral, by the formula (1.5.4), equals $H(\mathcal{P}^n|\mathfrak{B})$, so the last limit is exactly $h(\mathcal{P}|\mathfrak{B})$ (see Definition 2.3.3). □

A particular case of Theorem 2.6.3 occurs when $\mathfrak{B}$ is the sigma-algebra of invariant sets. Clearly, this is an invariant sigma-algebra and the corresponding disintegration is trivially equivariant. The disintegration formula $\mu = \int \mu_y \, d\nu$ then corresponds precisely to the ergodic decomposition of μ; the measures μ_y are the *ergodic components* of μ. Since T acts on the factor corresponding to $\mathfrak{B}$ by identity, its dynamical entropy is zero. Thus, using appropriate monotonicities and (2.4.4) we get, for any fixed partition $\mathcal{P}$ of finite static entropy and any finite $\mathfrak{B}$-measurable partition $\mathcal{Q}$, $h(\mathcal{P}) \geq h(\mathcal{P}|\mathcal{Q}) \geq h(\mathcal{P}) - h(\mathcal{Q}) = h(\mathcal{P})$. Using a sequence of partitions $\mathcal{Q}$ which generate $\mathfrak{B}$, by (2.4.17), we obtain $h(\mathcal{P}|\mathfrak{B}) = h(\mathcal{P})$. In this case Theorem 2.6.3 takes on the following form

Theorem 2.6.4 *Let $\mu = \int \mu_y \, d\nu$ be the ergodic decomposition of μ. Then, for any partition $\mathcal{P}$ with $H(\mu, \mathcal{P})$ finite, the following holds*

$$h(\mathcal{P}) = \int h(\mathcal{P}|y)d\nu(y). \quad \square$$

2.7 Summary of the properties of entropy

In the following tables we gather the major properties of static (top table) and dynamical (bottom table) unconditional and conditional entropies treated as functions of $\mathcal{P}$, $\mathcal{Q}$, μ (and, in the boxes, of $\mathfrak{B}$). All abbreviations and unclear terms are explained below the table. In the second table $\mathfrak{B}$ is subinvariant.

	$H(\mu, \mathcal{P})$	$H(\mu, \mathcal{P}\|\mathcal{Q})$	$H(\mu, \mathcal{P}\|\mathfrak{B})$	$H(\mu, \mathcal{P}\|\mathcal{Q} \vee \mathfrak{B})$
$\mathcal{P}$	increasing subadditive limit in (m) $\mathfrak{P}_m$ – u. cont. $\mathfrak{P}_{\aleph_0}$ – l.s.c.	increasing subadditive limit in (m) $\mathfrak{P}_m$ – u. cont. $\mathfrak{P}_{\aleph_0}$ – l.s.c.	increasing subadditive limit in (m) $\mathfrak{P}_m$ – u. cont. $\mathfrak{P}_{\aleph_0}$ – l.s.c.	increasing subadditive limit in (m) $\mathfrak{P}_m$ – u. cont. $\mathfrak{P}_{\aleph_0}$ – l.s.c.
$\mathcal{Q}$		decreasing limit in $(m)^*$ $\mathfrak{P}_{\aleph_0}$ – u. cont.*	$\mathfrak{B}$: decreasing incr. lim eq* decr. lim eq*	decreasing limit in $(m)^{***}$ $\mathfrak{P}_{\aleph_0}$ – u. cont.***
μ	concave	concave	concave	concave

	$h(\mu, T, \mathcal{P})^*$	$h(\mu, T, \mathcal{P}\|\mathcal{Q})^{**}$	$h(\mu, T, \mathcal{P}\|\mathfrak{B})^{***}$	$h(\mu, T, \mathcal{P}\|\mathcal{Q} \vee \mathfrak{B})^{****}$
$\mathcal{P}$	increasing subadditive power limit in (m) $\mathfrak{P}_R$ – u. cont. d_1 – l.s.c.	increasing subadditive power limit in (m) $\mathfrak{P}_R$ – u. cont. d_1 – l.s.c.	increasing subadditive power limit in (m) $\mathfrak{P}_R$ – u. cont. d_1 – l.s.c.	increasing subadditive power limit in (m) $\mathfrak{P}_R$ – u. cont. d_1 – l.s.c.
$\mathcal{Q}$		decreasing limit in $(m)^*$ $\mathfrak{P}_m$ – u. cont.* $\mathfrak{P}_{\aleph_0}$ – u.s.c*	$\mathfrak{B}$: decreasing incr. lim eq* decr. lim ineq*	decreasing limit in $(m)^{***}$ $\mathfrak{P}_m$ – u. cont.*** $\mathfrak{P}_{\aleph_0}$ – u.s.c.***
μ	affine	affine	affine	affine

The meaning of terms:

increasing (decreasing) = increasing (decreasing) as the partition (sigma-algebra) refines

subadditive = subadditive under the join of partitions

limit in (m) = equal to the limit over the partitions $\mathcal{P}_{(m)}$ (or $\mathcal{Q}_{(m)}$)

$\mathfrak{P}_m$ = the space of all m-element partitions with either d_1 or d_R

$\mathfrak{P}_{\aleph_0}$ = the space of all countable partitions with d_1

$\mathfrak{P}_\mathrm{R}$ = the space of all finite entropy partitions with d_R

u. cont. = uniformly continuous

l.s.c. (u.s.c.) = lower (upper) semicontinuous

incr. (decr.) lim eq = limit equality for an increasing (decreasing) sequence of sigma-algebras

decr. lim ineq = limit inequality for a decreasing sequence of sigma-algebras

power = for $\mathcal{P}^n$ (and $\mathcal{Q}^n$) under the action of T^n the function grows $|n|$ times

concave = concave under convex combinations of any Borel probability measures

affine = affine under convex combinations of invariant probability measures

* = requires the assumption $H(\mathcal{P}) < \infty$

** = requires the assumption $H(\mathcal{P}|\mathcal{Q}^\mathbb{S}) < \infty$

*** = requires the assumption $H(\mathcal{P}|\mathfrak{B}) < \infty$

**** = requires the assumption $H(\mathcal{P}|\mathcal{Q}^\mathbb{S} \vee \mathfrak{B}) < \infty$

2.8 Combinatorial entropy

Combinatorial entropy is a notion that mimics the idea of dynamical entropy, but assigns entropy directly to finite blocks. It can be applied universally to any sufficiently long block in the symbolic space, without fixing any shift-invariant measure. The block itself provides a substitute for such a measure. In this section the alphabet Λ is assumed finite and its cardinality is denoted by l.

Pick $n \in \mathbb{N}$ and $m \in \mathbb{N}$ much larger than n. Let $B = B[0, m-1] \in \Lambda^m$ be a block over Λ. With each block A of length n we can associate its *frequency* in B defined as

$$\mathsf{fr}_B(A) = \frac{\#\{0 \le i \le m-n : B[i, i+n-1] = A\}}{m-n+1}.$$

Notice that the frequencies of all blocks of length n form a probability vector

$$\mathbf{p}_{n,B} = \{\mathsf{fr}_B(A) : A \in \Lambda^n\}.$$

With the block B we can thus associate its nth *combinatorial entropy* defined as one nth of the entropy of $\mathbf{p}_{n,B}$:

$$H_n(B) = \tfrac{1}{n} H(\mathbf{p}_{n,B}).$$

Another possibility, which gives slightly different values (with the difference vanishing for large m) is as follows. For a block B consider the periodic point $\dots BBB \dots$ in the shift space $\Lambda^{\mathbb{Z}}$ (the infinite concatenation of copies of B). Its orbit under the shift is (at most) m-periodic and it carries exactly one ergodic measure, which we denote by $\mu_{(B)}$. We can thus define the nth *periodic combinatorial entropy* of B as

$$H_{(n)}(B) = \tfrac{1}{n} H(\mu_{(B)}, \mathcal{P}_\Lambda^n).$$

Given a block A of length n, the value $\mu_{(B)}(A)$ is the frequency with which A occurs in $...BBB...$, and, for $m \gg n$, it differs from that in B by the frequency of occurrences at the contact places between the concatenated copies of B. This difference is not larger than n/m. Thus, for fixed n and with m growing to infinity the vectors of frequencies $\mathbf{p}_{n,B}$ and the vectors $\mathbf{p}_{n,(B)} = \{\mu_{(B)}(A) : A \in \Lambda^n\}$ are close, uniformly for all blocks B of length m. By the uniform continuity of the entropy on l^n-dimensional vectors this implies that $H_{(n)}(B)$ and $H_n(B)$ get together uniformly for all blocks B of length m as m grows to infinity. This is why we can use either definition of combinatorial entropy, depending on the convenience. The advantage of $H_{(n)}$ over H_n is that the former is computed with respect to a genuine shift-invariant measure defined on the symbolic space, while the latter is computed with respect to a probability vector which does not represent a measure on the symbolic space – it assigns values to short cylinders only.

As was stated in the Introduction, combinatorial entropy can be used to estimate the compression rate. In order to make sure that there is no mistake we are interested in counting the blocks B over an alphabet Λ, of some fixed length m, whose compression rate does not exceed some fixed value $c < 1$. The count must not exceed 2^{mc+1} (at least for large m) because these blocks are supposed to be encoded (by the data compression code) in a injective way, by binary blocks of lengths not exceeding mc, and there are 2^{mc+1} such binary blocks. Let us see ...

Definition 2.8.1 By $\mathbf{C}[n, m, c]$ we will denote the cardinality of the collection of blocks B of length m with $H_{(n)}(B) \leq c$.

Lemma 2.8.2 *Fix some $n \in \mathbb{N}$ and $c > 0$. Then*

$$\limsup_{m\to\infty} \frac{\log(\mathbf{C}[n, m, c])}{m} \leq c. \tag{2.8.3}$$

Proof We first prove the lemma for $n = 1$. In this case there is no difference between $H_{(n)}(B)$ and $H_n(B)$. Let $\Lambda = \{a_1, a_2, \dots, a_l\}$. The frequencies of the symbols a_i in B form a probability vector $\mathbf{p}_{1,B} = (p_1, p_2, \dots, p_l)$ with $p_i = k_i/m$ where k_i is the number of times a_i occurs in B. The number of blocks B of length m producing the same vector $\mathbf{p} = \mathbf{p}_{1,B}$ is

$$\mathbf{C}_\mathbf{p} = \frac{m!}{k_1!\, k_2! \,\cdots\, k_l!}.$$

By a direct application of Stirling's formula: $\log(n!) \approx n \log n - n$ [see e.g. Feller, 1968] we can write

$$\log(\mathbf{C}_\mathbf{p}) = -m\Big(\sum_{i=1}^{l} p_i \log(p_i) \pm \delta_m\Big) = m(H(\mathbf{p}) \pm \delta_m), \tag{2.8.4}$$

where $\delta_m \to 0$ as m grows. Thus, the cardinality of all blocks B of length m with $H(\mathbf{p}_{1,B}) \leq c$ is not larger than $m(H(\mathbf{p}) \pm \delta_m)$ times the cardinality of all l-dimensional probability vectors $\mathbf{p}$ with rational entries with denominator m. Now, this latter cardinality is simply m^l, the logarithm of which is $l \log m$. Eventually, we have obtained

$$\log(\mathbf{C}[1, m, c]) \leq m(c \pm \delta_m) + l \log m. \tag{2.8.5}$$

The assertion follows by dividing by m and letting m grow to infinity.

Now consider $n > 1$. Assume for a while that m is a multiple of n. By grouping the symbols in n-tuples, the periodic sequence $...BBB...$ can be viewed as a periodic sequence ... $B^o B^o B^o$ ..., where B^o is over Λ^n and has length m/n. There are n possible such representations depending on the positioning of the first cut. We denote the blocks B^o so obtained by $B^{(1)}, B^{(2)}, \ldots, B^{(n)}$. It is easy to see that the vector of probabilities $\mathbf{p}_{n,(B)} = \{\mu_{(B)}(A) : A \in \Lambda^n\}$ equals the arithmetic average (over $i = 1, \ldots, n$) of the vectors $\mathbf{p}_{1,B^{(i)}} = \{\mu_{B^{(i)}}(A) : A \in \Lambda^n\}$, where now the blocks A are treated as single symbols. By concavity of the entropy on probability vectors,

$$nc \geq nH_{(n)}(B) = H(\mathbf{p}_{n,(B)}) \geq \min_i H(\mathbf{p}_{1,B^{(i)}}) = \min_i H_1(B^{(i)}).$$

Consider the index i which realizes this minimum. The wanted cardinality $\mathbf{C}[n, m, c]$ of all blocks B with $H_{(n)}(B) \leq c$ is not larger than n times (due to the choice of the cutting position) the estimate of the cardinality of all blocks which can play the role of $B^{(i)}$. Such blocks are over the alphabet Λ^n, have length m/n, and have the 1st combinatorial entropy bounded by nc, so their cardinality is $\mathbf{C}[1, \frac{m}{n}, nc]$. Applying (2.8.3) already proved for $n = 1$, we thus have

$$\frac{\log(\mathbf{C}[n, m, c])}{m} \leq \frac{\log(n\mathbf{C}[1, \frac{m}{n}, c])}{m} \leq \frac{\log n}{m} + \frac{nc\frac{m}{n}}{m}.$$

The right-hand side converges to c as m grows to infinity. This concludes the proof for multiples of n. For other lengths m we estimate the desired cardinality by one obtained for the nearest $m' > m$ divisible by n. Since the ratio m'/m tends to 1, we will obtain the same estimate for lim sup. □

We shall also need a conditional version of the above facts. Let $\Lambda = \Lambda_1 \times \Lambda_2$ be a product of two finite sets having l_1 and l_2 elements, respectively. Let $B \in \Lambda^m$. Such B can be viewed as a two-row block having some $B_1 \in \Lambda_1^m$ in the first row and some $B_2 \in \Lambda_2^m$ in the second. We define the *nth conditional periodic combinatorial entropy* of B (given the first row) as

$$H_{(n)}(B|B_1) = \tfrac{1}{n} H(\mu_{(B)}, \mathcal{P}_\Lambda^n | \mathcal{P}_{\Lambda_1}^n).$$

Since $\mathcal{P}_\Lambda^n$ is a refinement of $\mathcal{P}_{\Lambda_1}^n$, it follows that

$$H_{(n)}(B|B_1) = H_{(n)}(B) - H_{(n)}(B_1).$$

By analogy, we can also define the *nth conditional combinatorial entropy* of B as

$$H_n(B|B_1) = H_n(B) - H_n(B_1).$$

(We will not use this term, except for $n = 1$, when $H_1 = H_{(1)}$.)

We introduce the notation for cardinalities, analogous to the one used before:

Definition 2.8.6 For $D \in \Lambda_1^m$ let $\mathbf{C}_D[n,m,c]$ denote the cardinality of the collection of all blocks B of length m over Λ such that $B_1 = D$ and $H_{(n)}(B|B_1) \leq c$. Next, define

$$\mathbf{C}_{\text{cond}}[n,m,c] = \sup_{D \in \Lambda_1^m} \mathbf{C}_D[n,m,c].$$

Lemma 2.8.7 *Fix some $n \in \mathbb{N}$, $c > 0$. Then*

$$\limsup_{m \to \infty} \frac{\log(\mathbf{C}_{\text{cond}}[n,m,c])}{m} \leq c. \tag{2.8.8}$$

Proof We first prove the lemma for $n = 1$. Fix some m, $D \in \Lambda_1^m$ and $\varepsilon > 0$. We can assume that $c < \log l_2$, otherwise the statement holds trivially for every m. Denote by $\mathcal{E}$ the family of blocks whose cardinality we want to estimate:

$$\mathcal{E} = \{B \in \Lambda^m : B_1 = D, H_1(B|B_1) < c\}.$$

It is straightforward to see that

$$H_1(B|B_1) = \sum_{i=1}^{l_1} \frac{k_i}{m} H_1(B_2^{(i)}),$$

where $B_2^{(i)}$ is the block over Λ_2 of a certain length k_i, obtained by collecting all positions in B_2 where in the first row there appears the ith symbol of Λ_1. The numbers k_i are determined by D. We fix an integer s and we divide the interval $[0, \log l_2)$ into s subintervals of equal lengths. With every block $B \in \mathcal{E}$ we associate the formal sequence $\mathcal{I} = \{[a_i, b_i) : i = 1, \ldots, l_1\}$ of the above subintervals determined by the inclusions $\frac{k_i}{m} H_1(B_2^{(i)}) \in [a_i, b_i)$. Clearly, the number of such sequences is limited by s^{l_1} and only such $\mathcal{I}$ will appear (for some B) for which

$$\sum_{i=1}^{l_1} b_i \leq c + l_1 \frac{\log l_2}{s}.$$

For every $\mathcal{I}$ we can estimate the logarithm of the number of the associated blocks B by the sum over i of the logarithms of the quantities Q_i of blocks $B_2^{(i)}$ for which $\frac{k_i}{m} H_1(B_2^{(i)})$ falls in the subinterval $[a_i, b_i)$. For indices i for which $k_i \leq \sqrt{m}$ we use the trivial estimate

$$\log Q_i \leq \sqrt{m} \log l_2.$$

For the remaining indices i we apply the preceding lemma, and get

$$\log Q_i \leq k_i(\tfrac{m}{k_i} b_i + \varepsilon),$$

where the error term ε is small for large m. The sum of these logarithms amounts to at most

$$\sqrt{m}\,l_1 \log l_2 + m(\sum_i b_i + \varepsilon) \leq \sqrt{m}\,l_1 \log l_2 + m(c + \frac{l_1 \log l_2}{s} + \varepsilon) \leq m(c + \frac{l_1 \log l_2}{s} + 2\varepsilon),$$

for large enough m. The cardinality of $\mathcal{E}$ is at most s^{l_1} times larger than the product of the Q_i's, which adds $l_1 \log s$ to the logarithm, and vanishes after dividing by m and passing with m to infinity. Since ε is arbitrarily small and s arbitrarily large, we arrive at the hypothesis.

The proof for $n > 1$ is identical to the corresponding part of the proof of Lemma 2.8.2. Suppose, for simplicity, that m is a multiple of n. After grouping the symbols in n-tuples, and denoting the blocks over Λ^n so obtained by $B^{(1)}, B^{(2)}, \ldots, B^{(n)}$ (depending on the first cutting place) one shows that for at least one index i, $H_1(B^{(i)}|B_1^{(i)}) \leq nc$. Applying (2.8.8) already proved for $n = 1$, we thus have

$$\frac{\log(\mathbf{C}_{\text{cond}}[n,m,c])}{m} \leq \frac{\log(n\mathbf{C}_{\text{cond}}[1,\frac{m}{n},c])}{m} \leq \frac{\log n}{m} + \frac{nc\frac{m}{n}}{m},$$

which converges to c as m grows to infinity. □

Also the inequality converse to (2.8.3) is valid for lim inf, hence the limit exists (see Corollary 2.8.10 below). We will prove a slightly stronger statement. (Usually, a slightly weaker statement is derived with the help of the Shannon–McMillan–Breiman Theorem.)

Theorem 2.8.9 *Let $(X, \mathcal{P}, \mu, T, \mathbb{S})$ be an ergodic process with $\mathcal{P}$ finite. Denote $h = h(\mu, T, \mathcal{P})$. Choose $a \in (0, 1]$, $\varepsilon > 0$, and let n be so large that $H(\mu, \mathcal{P}^n) < n(h + a\varepsilon)$. Then, for m large enough, any set A with $\mu(A) \geq a$ intersects at least $2^{m(h-2\varepsilon)}$ blocks of length m whose nth combinatorial entropy is smaller than $h + \varepsilon$.*

Corollary 2.8.10 *Because, for $c \leq \log l$ (recall that $l = \#\Lambda$), there exists an independent process of any entropy $h < c$ over the alphabet Λ, (so that $H(\mu, \mathcal{P}^n_\Lambda) = nh$ for every n), the above lemma (applied to $A = X$) together with Lemma 2.8.2 implies that*

$$\lim_m \frac{\log(\mathbf{C}[n,m,c])}{m} = c. \qquad \square$$

Proof of Theorem 2.8.9 By the Ergodic Theorem, a subset X' of measure $1 - \delta$ is covered by blocks B of length m generating frequency vectors $\mathbf{p}_{n,B}$

so close to the distribution of μ on $\mathcal{P}^n$ that their nth combinatorial entropies are smaller than $h + a\varepsilon$ (see Exercise 2.8). By Lemma 2.8.2, there are no more than $2^{m(h+2a\varepsilon)}$ such blocks (if m is large enough), hence the conditional entropy of $\mathcal{P}^m$ on any subset of X' cannot exceed $m(h + 2a\varepsilon)$ (we will use this for $X' \setminus A$). Let $\mathcal{Q}$ be the partition by three sets: $A \cap X'$ (of measure at most $\mu(A)$), $X' \setminus A$ (of measure at most $1 - \mu(A)$) and $X \setminus X'$ (of measure at most δ). Then

$$\begin{aligned} &mh \le H(\mathcal{P}^m) \le H(\mathcal{P}^m|\mathcal{Q}) + H(\mathcal{Q}) \le \\ &\mu(A)H_{A\cap X'}(\mathcal{P}^m) + (1-\mu(A))H_{X'\setminus A}(\mathcal{P}^m) + \delta H_{X\setminus X'}(\mathcal{P}^m) + \log 3 \le \\ &\qquad \mu(A)H_{A\cap X'}(\mathcal{P}^m) + (1-\mu(A))m(h+2a\varepsilon) + \delta m \log \#\mathcal{P} + \log 3. \end{aligned}$$

This implies that the conditional entropy of $\mathcal{P}^m$ on $A \cap X'$ is at least

$$\begin{aligned} &\frac{mh - (1-\mu(A))m(h+2a\varepsilon) - \delta m \log \#\mathcal{P} - \log 3}{\mu(A)} \ge \\ &\qquad m\Big(h - \frac{2a\varepsilon}{\mu(A)} + 2\varepsilon - \frac{\delta \log \#\mathcal{P}}{a} - \frac{\log 3}{ma}\Big) \ge m(h - 2\varepsilon), \end{aligned}$$

if δ is chosen sufficiently small and m is large enough. Such entropy cannot be achieved on fewer than $2^{m(h-2\varepsilon)}$ blocks of length m. All of them intersect A and, since they are part of X', have the nth combinatorial entropy smaller than $h + a\varepsilon \le h + \varepsilon$. □

Exercises

2.1 Consider the Bernoulli shift (unilateral or bilateral) on two symbols 0 and 1 where each symbol has measure $1/2$. Let π be the factor map given by the code $(\pi x)_n = x_n + x_{n+1}$ mod 2. Show that although the map is far from being invertible, the factor process is isomorphic to the original.

2.2 Use the power rule to show that in a bilateral process for every $n \in \mathbb{N}$ we have the equality $H(\mathcal{P}^n|\mathcal{P}^-) = nh(\mathcal{P})$.

2.3 Provide an example showing that the sequence $H(\mathcal{P}^n|\mathcal{Q}^n)$ need not have decreasing increments; moreover, may fail to be increasing.

2.4 For subinvariant $\mathfrak{B}$ show that $h(\mathcal{P}|\mathcal{Q}, \mathfrak{B}) \le H(\mathcal{P}|\mathcal{P}^+ \vee \mathcal{Q}^{\mathbb{N}_0} \vee \mathfrak{B})$.

2.5 For subinvariant $\mathfrak{B}$ the formula (2.3.5) need not hold. In particular, the inequality in the preceding exercise cannot be reversed even for trivial (hence invariant) $\mathfrak{B}$. Provide an appropriate example.

2.6 Prove Fact 2.4.1, i.e., $h(\mathcal{Q}) \le h(\mathcal{P})$ in case T is invertible and $\mathcal{Q} \preccurlyeq \mathcal{P}^{\mathbb{Z}}$. Hint: apply (2.3.11).

2.7 Provide an example of an ergodic process $(X, \mathcal{P}, \mu, T, \mathbb{S})$ and its factor, such that the fiber entropy $h(\mathcal{P}|y)$ is not constant on the factor space Y. Hint: The factor can be periodic on a two-element space.

2.8 Consider an ergodic process $(X, \mathcal{P}, \mu, T, \mathbb{S})$. For $x \in X$ denote $B_m(x) = x[0, m-1]$. Prove that for every n the combinatorial entropies $H_n(B_m(x))$ converge almost surely to $\frac{1}{n}H(\mu, \mathcal{P}^n)$.

3

Entropy theorems in processes

3.1 Independence and ε-independence

We begin by introducing the notion of ε-independence for partitions. The goal is to extend this notion to processes.

Recall (Fact 1.6.16) that stochastic independence between $\mathcal{P}$, with $H(\mathcal{P}) < \infty$, and any countable $\mathcal{Q}$ is equivalent to the equality $H(\mathcal{P}|\mathcal{Q}) = H(\mathcal{P})$.

Definition 3.1.1 Two countable partitions $\mathcal{P}$ and $\mathcal{Q}$ are *ε-independent* (we write $\mathcal{P}\perp^{\varepsilon}\mathcal{Q}$) if

$$\sum_{A\in\mathcal{P},B\in\mathcal{Q}} |\mu(A\cap B) - \mu(A)\cdot\mu(B)| \leq \varepsilon. \tag{3.1.2}$$

Note that ε-independence for all $\varepsilon > 0$ is equivalent to independence of the partitions. The connection between entropy and ε-independence is captured by the following fact.

Fact 3.1.3 *For every $M > 0$ and $\varepsilon > 0$ there is $\delta > 0$ such that for any two countable partitions $\mathcal{P}$ and $\mathcal{Q}$, with $H(\mathcal{P}) < M$ the following implications hold:*

$$\mathcal{P}\perp^{\delta}\mathcal{Q} \implies H(\mathcal{P}|\mathcal{Q}) \geq H(\mathcal{P}) - \varepsilon,$$

and

$$H(\mathcal{P}|\mathcal{Q}) \geq H(\mathcal{P}) - \delta \implies \mathcal{P}\perp^{\varepsilon}\mathcal{Q}.$$

Proof Assume δ-independence. The probability vector $\mathbf{p}(\mu,\mathcal{P})$ associated with the partition $\mathcal{P}$ can be ordered decreasingly. Reformulation of (3.1.2) with δ replacing ε reads

$$\sum_{B\in\mathcal{Q}} \mu(B)\,\|\mathbf{p}(\mu_B,\mathcal{P}) - \mathbf{p}(\mu,\mathcal{P})\|_1 \leq \delta, \tag{3.1.4}$$

By the rectangle rule (Fact A.1.2), for a collection of B's of joint measure at least $1-\sqrt{\delta}$ it holds that $\|\mathbf{p}(\mu_B, \mathcal{P})-\mathbf{p}(\mu, \mathcal{P})\|_1 < \sqrt{\delta}$. By lower semicontinuity of $H(\cdot)$, and compactness stated in Fact 1.1.9, for such B's, $H(\mathbf{p}(\mu_B, \mathcal{P})) > H(\mathbf{p}(\mu, \mathcal{P})) - \varepsilon/2$, if δ is *a priori* (independently of $\mathcal{P}$, depending only on M) chosen small enough. Finally,

$$H(\mathcal{P}|\mathcal{Q}) = \sum_{B\in\mathcal{Q}} \mu(B)\, H(\mathbf{p}(\mu_B, \mathcal{P})) > (1-\sqrt{\delta})\big(H(\mathbf{p}(\mu, \mathcal{P})) - \tfrac{\varepsilon}{2}\big) \geq H(\mathcal{P}) - \varepsilon,$$

again, for a well chosen δ.

The second implication follows directly from Lemma 1.1.11 and the reformulation (3.1.4) (with ε put back in place of δ) of the ε-independence. □

Fact 3.1.3 allows one to define an alternative notion of ε-independence for a partition $\mathcal{P}$ of finite entropy.

Definition 3.1.5 We say that $\mathcal{P}$ is *ε-entropy independent of* $\mathcal{Q}$ if

$$H(\mathcal{P}|\mathcal{Q}) \geq H(\mathcal{P}) - \varepsilon.$$

Among partitions of finite entropy this is a symmetric relation (use (1.4.3)).

Definition 3.1.6 We say that a partition $\mathcal{P}$ with finite entropy is *ε-independent (ε-entropy independent) of a sigma-algebra* $\mathfrak{B}$ if it is ε-independent (ε-entropy independent) of any countable $\mathfrak{B}$-measurable partition $\mathcal{Q}$.

Now we pass to ε-independence for processes. Recall (Definition 2.3.15) that independent processes are characterized by the property that $\mathcal{P}$ is independent of the future $\mathcal{P}^+$ or, equivalently, by the equality $h(\mathcal{P}) = H(\mathcal{P})$. By analogy we will consider two notions:

Definition 3.1.7 The process $(X, \mathcal{P}, \mu, T, \mathbb{S})$ is called ε-independent (ε-entropy independent) if $\mathcal{P}$ is ε-independent (ε-entropy independent) of $\mathcal{P}^+$.

In the finite entropy case ε-entropy independence of a process can be written as

$$h(\mathcal{P}) \geq H(\mathcal{P}) - \varepsilon.$$

Remark 3.1.8 Clearly, if $H(\mathcal{P})$ is smaller than ε, then the generated process is (trivially) ε-entropy independent. It is thus natural to require, for nontriviality of the notion, that ε is much smaller than $H(\mathcal{P})$.

Remark 3.1.9 If a process is ε-entropy independent for every $\varepsilon > 0$, then it is an independent process.

We will now provide examples of ε-entropy independent processes, appearing naturally in processes of positive entropy. Before we continue, we recall the notion of an induced system. Induced systems are a good source of ε-independent processes.

Definition 3.1.10 Let $(X, \mathfrak{A}, \mu, T, \mathbb{S})$ be a dynamical system and let $B \in \mathfrak{A}$ be a set of positive measure. For $x \in B$ we define the *return time to* B as

$$\mathrm{R}_B(x) = \min\{n > 0 : T^n x \in B\} \tag{3.1.11}$$

(this is well defined for μ_B almost every x by the Poincaré Recurrence Theorem). Then the map

$$T_B(x) = T^{\mathrm{R}_B(x)}(x) \tag{3.1.12}$$

is defined μ_B-almost everywhere, it is measurable and preserves the measure μ_B (we skip the argument here). It is called the *induced map* while the system $(X, \mathfrak{A}, \mu_B, T_B, \mathbb{S})$ (in fact X can be replaced by B) is called the *induced system*.

It is clear that if μ is ergodic under T so is μ_B under T_B.

Theorem 3.1.13 *Let $(X, \mathcal{P}, \mu, T, \mathbb{S})$ be a process with finite positive entropy and let $\varepsilon > 0$ be given. Then for n sufficiently large there is a set $X_n \subset X$ of measure at least $1 - \varepsilon$ being a union of cylinders $B \in \mathcal{P}^{[1,n]}$ with the property that the process $(B, \mathcal{P}, \mu_B, T_B, \mathbb{S})$ is ε-entropy independent.*

Proof Because $H(\mathcal{P}|\mathcal{P}^{[1,n]}) \searrow H(\mathcal{P}|\mathcal{P}^+)$, for large n, $H(\mathcal{P}|\mathcal{P}^{[1,n]}) - \varepsilon^2 < H(\mathcal{P}|\mathcal{P}^+)$. Now

$$\sum_{B \in \mathcal{P}^{[1,n]}} \mu(B) H_B(\mathcal{P}) - \varepsilon^2 = H(\mathcal{P}|\mathcal{P}^{[1,n]}) - \varepsilon^2 < H(\mathcal{P}|\mathcal{P}^+) =$$

$$H(\mathcal{P}|\mathcal{P}^{[1,n]} \vee \mathcal{P}^+) = \sum_{B \in \mathcal{P}^{[1,n]}} \mu(B) H_B(\mathcal{P}|\mathcal{P}^+)$$

(we have used (1.4.4) and (1.6.25)). Clearly, for every B, the term $H_B(\mathcal{P})$ dominates $H_B(\mathcal{P}|\mathcal{P}^+)$. However, the above calculation shows that the weighted average of the first terms exceeds the weighted average of the latter terms by no more than ε^2. Thus, by the rectangle rule (Fact A.1.2),

$$H_B(\mathcal{P}) \le H_B(\mathcal{P}|\mathcal{P}^+) + \varepsilon,$$

except for sets B of joint measure at most ε. We let X_n be the union of the cylinders B satisfying the above. Because the future of the induced process is contained (as a sigma-algebra) in the full future $\mathcal{P}^+$, we have

$$H_B(\mathcal{P}|\mathcal{P}^+) \le h(\mu_B, T_B, \mathcal{P})$$

for every B. Thus for every B contained in X_n, $H_B(\mathcal{P}) \le h(\mu_B, T_B, \mathcal{P}) + \varepsilon$, i.e., the process generated by $\mathcal{P}$ for the induced map is ε-entropy independent, as claimed. □

The process $(B, \mathcal{P}, \mu_B, T_B, \mathbb{Z})$ for $B \in \mathcal{P}^{[1,n]}$ is illustrated in Figure 3.1; its $\mathcal{P}$-name $(\cdots a_{-1}a_0a_1a_2a_3\cdots)$ consists of concatenated single symbols that precede the subsequent repetitions of the block B observed in a $\mathcal{P}$-name of the master process $(X, \mathcal{P}, \mu, T, \mathbb{Z})$. (For $\mathbb{S} = \mathbb{N}_0$ the picture starts at coordinate zero and the above $\mathcal{P}$-name is just $(a_0a_1a_2a_3\cdots)$.)

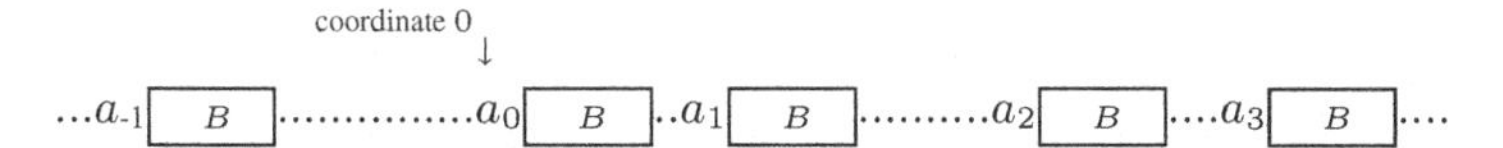

Figure 3.1 The process generated by $\mathcal{P}$ for the map induced on B.

If $(X, \mathcal{P}, \mu, T, \mathbb{S})$ has entropy zero, then the ε-entropy independence as stated in Theorem 3.1.13 is trivial (see Remark 3.1.8). This means that for the majority of blocks $B \in \mathcal{P}^{[1,n]}$, the partition $\mathcal{P}$ restricted to B is nearly the one-element partition, hence the process $(B, \mathcal{P}, \mu_B, T_B, \mathbb{S})$ is nearly trivial; every name is dominated by repetitions of one symbol.

We now turn to the case of two partitions $\mathcal{P}$ and $\mathcal{Q}$. Assume $H(\mathcal{P}) < \infty$.

Definition 3.1.14 We will say that the process $(X, \mathcal{P}, \mu, T, \mathbb{S})$ generated by a partition $\mathcal{P}$ with finite entropy is *ε-entropy limit-independent of the process* generated by $\mathcal{Q}$ if

$$h(\mathcal{P}|\mathcal{Q}) \ge h(\mathcal{P}) - \varepsilon, \tag{3.1.15}$$

equivalently,

$$\lim_n \tfrac{1}{n}(H(\mathcal{P}^n) - H(\mathcal{P}^n|\mathcal{Q}^n)) \le \varepsilon. \tag{3.1.16}$$

If $\mathcal{Q}$ also has finite entropy, then (3.1.15) can be rewritten as

$$h(\mathcal{P}) + h(\mathcal{Q}) - h(\mathcal{P} \vee \mathcal{Q}) \le \varepsilon,$$

which proves that in such case the relation is symmetric.

Notice that ε-entropy limit-independence between two processes for every $\varepsilon > 0$ does not imply the usual (stochastic) independence. For instance, a process of entropy zero is, for every $\varepsilon > 0$, (trivially) ε-entropy limit-independent of any process, even of itself. This is why we introduce a stronger notion.

Definition 3.1.17 The process generated by $\mathcal{P}$ is *ε-entropy independent of the process* generated by $\mathcal{Q}$ if

$$\tfrac{1}{n}H(\mathcal{P}^n|\mathcal{Q}^n) \geq \tfrac{1}{n}H(\mathcal{P}^n) - \varepsilon,$$

for every n.

This notion is also symmetric among finite entropy partitions. Now, ε-entropy independence for every $\varepsilon > 0$ does imply stochastic independence between the processes. As before, to avoid triviality, ε should be taken smaller than both $H(\mathcal{P})$ and $H(\mathcal{Q})$.

The notions of ε-entropy independence and ε-entropy limit-independence coincide (via change of the parameter) if one of the processes is itself an ε-entropy independent process:

Fact 3.1.18 *Suppose $(X, \mathcal{P}, \mu, T, \mathbb{S})$ is an ε-entropy independent process and that it is ε-entropy limit-independent of the process generated by another partition $\mathcal{Q}$. Then the first process is 2ε-entropy independent of the latter.*

Proof

$$\tfrac{1}{n}H(\mathcal{P}^n|\mathcal{Q}^n) \geq h(\mathcal{P}|\mathcal{Q}) \geq h(\mathcal{P}) - \varepsilon \geq H(\mathcal{P}) - 2\varepsilon \geq \tfrac{1}{n}H(\mathcal{P}^n) - 2\varepsilon.$$

□

Question 3.1.19 Let B be a set of positive measure and let $\mathcal{Q} = \{Q_n\}$ be the countable partition of B given by the first return time, $Q_n = \{x \in B : \mathrm{R}_B(x) = n\}$. The following question is open: Is it true that for the majority of sufficiently long cylinders B, the ε-entropy independent process $(B, \mathcal{P}, \mu_B, T_B, \mathbb{S})$ of Theorem 3.1.13 is also ε-entropy limit-independent of the process of return times $(B, \mathcal{Q}, \mu_B, T_B, \mathbb{S})$? (If yes, then by Fact 3.1.18, these processes are 2ε-entropy independent of one another.)

Later we will show a partial result in this direction (see Lemma 5.3.11).

A rich source of examples of pairs of mutually ε-independent processes will be provided in the section concerning joinings, where we prove the following representation theorem: Given a process and its factor-process, both with the action of $\mathbb{Z}$, then the larger process is a joining of the given factor-process with another factor-process such that the two factors are ε-entropy limit-independent of one another (see Theorem 4.4.6).

3.2 The Pinsker sigma-algebra in a process

Definition 3.2.1 Let $(X, \mathcal{P}, \mu, T, \mathbb{S})$ be a process with finite entropy. The *Pinsker sigma-algebra* of this process is $\Pi_{\mathcal{P}} = \bigcap_{n=1}^{\infty} \mathcal{P}^{[n,\infty)}$. It is sometimes referred to as the *remote future*.

It is clear that $\Pi_{\mathcal{P}}$ is an invariant sigma-algebra. Its meaning is explained in the following theorem originating from [Rokhlin and Sinai, 1961].

Theorem 3.2.2 (Rokhlin–Sinai) *Assume $H(\mathcal{P}) < \infty$. A countable, measurable with respect to $\mathcal{P}^{\mathbb{S}}$, partition $\mathcal{R}$ with $H(\mathcal{R}) < \infty$ is $\Pi_{\mathcal{P}}$-measurable if and only if $h(\mathcal{R}) = 0$. For actions of $\mathbb{Z}$ we also have $\Pi_{\mathcal{P}} = \bigcap_{n=1}^{\infty} \mathcal{P}^{(-\infty,-n]}$.*

Proof Assume that $\mathcal{R}$ is $\Pi_{\mathcal{P}}$-measurable. By invariance of $\Pi_{\mathcal{P}}$, the transformation on the associated factor is invertible, so $\mathcal{R}^{\mathbb{Z}}$ is a well-defined invariant sigma-algebra contained in $\Pi_{\mathcal{P}}$ and hence also in $\mathcal{P}^{+}$. Then, using (2.3.11), we get $h(\mathcal{P}|\mathcal{R}) = h(\mathcal{P})$. On the other hand, since $H(\mathcal{R}) < \infty$ and $\mathcal{P} \vee \mathcal{R} \preccurlyeq \mathcal{P}^{\mathbb{N}_0}$, by Fact 2.4.1 we get $h(\mathcal{P} \vee \mathcal{R}) = h(\mathcal{P})$ which by (2.4.4) implies $h(\mathcal{R}) = 0$.

The proof of the other implication is different for the actions of $\mathbb{N}_0$ and $\mathbb{Z}$.

For $\mathbb{S} = \mathbb{N}_0$ consider a $\mathcal{P}^{\mathbb{N}_0}$-measurable partition $\mathcal{R}$ such that $h(\mathcal{R}) = 0$. Then, by Fact 2.3.12, $\mathcal{R}$ is measurable with respect to $\mathcal{R}^{+}$ and, by an easy induction, also with respect to $\Pi_{\mathcal{R}}$. Since $\mathcal{R}$ is measurable with respect to $\mathcal{P}^{\mathbb{N}_0}$, $\mathcal{R}^{[n,\infty)}$ is contained in $(\mathcal{P}^{\mathbb{N}_0})^{[n,\infty)} = \mathcal{P}^{[n,\infty)}$ and hence $\Pi_{\mathcal{R}} \subset \Pi_{\mathcal{P}}$.

If $\mathbb{S} = \mathbb{Z}$, let $\mathcal{R}$ be a $\mathcal{P}^{\mathbb{Z}}$-measurable partition with $h(\mathcal{R}) = 0$. If $\mathcal{R}$ is measurable with respect to $\mathcal{P}^{\mathbb{N}_0}$, then the preceding argument applies. So, suppose it is not. Then, by (1.6.28), $H(\mathcal{R}|\mathcal{P}^{\mathbb{N}_0}) > c > 0$. By Fact 1.6.39, $H(\mathcal{R}_{(m)}|\mathcal{P}^{\mathbb{N}_0}) > c$ for some m. Of course $H(\mathcal{R}_{(m)}|\mathcal{R}) = 0$. By Fact 1.7.3, for every ε there exists a $k \in \mathbb{N}$ and a $\mathcal{P}^{[-k,k]}$-measurable partition $\mathcal{R}' \in \mathfrak{P}_m$ which approximates $\mathcal{R}_{(m)}$ as accurately as we need. By uniform continuity of conditional entropy in $\mathfrak{P}_m$, we can obtain $\mathcal{R}'$ for which $H(\mathcal{R}'|\mathcal{P}^{\mathbb{N}_0}) > c$ and $H(\mathcal{R}'|\mathcal{R}) < c$. Since $T^{-nk}(\mathcal{R}')$ is $\mathcal{P}^{\mathbb{N}_0}$-measurable for every $n \geq 1$, we have

$$c < H(\mathcal{R}'|\mathcal{P}^{\mathbb{N}_0}) \leq H(\mathcal{R}'|\bigvee_{n=1}^{\infty} T^{-nk}(\mathcal{R}')).$$

The last term equals $h(\mu, T^k, \mathcal{R}')$, the dynamical entropy of $\mathcal{R}'$ under the action of T^k. On the other hand (no matter what action we consider), by Fact 2.3.1 and (2.4.4), we have

$$H(\mathcal{R}'|\mathcal{R}) \geq h(\mu, T^k, \mathcal{R}'|\mathcal{R}) = h(\mu, T^k, \mathcal{R}' \vee R) - h(\mu, T^k, \mathcal{R}),$$

i.e., $h(\mu, T^k, \mathcal{R}) \geq h(\mu, T^k, \mathcal{R}') - H(\mathcal{R}'|\mathcal{R})$ which, in this case, is positive (recall, we have $h(\mu, T^k, \mathcal{R}') > c$ and $H(\mathcal{R}'|\mathcal{R}) < c$). This implies that under the action of T^k, $\mathcal{R}$ has positive entropy. A contradiction, because, by the power rule, Fact 2.4.19,

$$h(\mu, T^k, \mathcal{R}) \le h(\mu, T^k, \mathcal{R}^k) = kh(\mu, T, \mathcal{R}) = 0.$$

We have proved that $\mathcal{R}$ is $\Pi_{\mathcal{P}}$-measurable.

For the last statement, note that since a partition has dynamical entropy zero for an invertible T if and only if it does for T^{-1}, $\Pi_{\mathcal{P}}$ coincides with the identical sigma-algebra defined for T^{-1}. □

Question 3.2.3 It is not known whether and how the last statement of the theorem (equality between the "remote future" and "remote past" sigma-algebras) can be proved without using entropy.

Remark 3.2.4 A process whose Pinsker sigma-algebra is trivial is called a *K-system.* Such systems are extremely important in classical ergodic theory and are subject of an extensive study. Whether there exist K-systems other than independent processes (or isomorphic to such) has been a long-standing open problem first solved positively by Donald Ornstein [Ornstein, 1973] and refined in a work with Paul Shields [Ornstein and Shields, 1973]. Later a much more explicit example was constructed by Steve Kalikow [Kalikow, 1982]. His example involves a so-called skew product transformation and is popularly known as the "T-T-inverse" transformation. In this book K-systems will play only a marginal role, hence we skip any detailed presentation of this class. More detailed information on K-systems can be found in the book by Paul Shields [Shields, 1996].

Although in $\mathbb{Z}$-actions both sequences of sigma-algebras $\mathcal{P}^{(-\infty,-n]}$ and $\mathcal{P}^{[n,\infty)}$ decrease to $\Pi_{\mathcal{P}}$, their joins do not have to. This shows that the operation "join" does not commute with countable intersections of sigma-algebras. There are for example so-called *bilaterally deterministic* processes of positive entropy, in which every join $\mathcal{P}^{(-\infty,-n]} \vee \mathcal{P}^{[n,\infty)}$ equals $\mathcal{P}^{\mathbb{Z}}$:

Example 3.2.5 We construct a bilaterally deterministic process of positive entropy. Let Λ be a finite alphabet disjoint of the set of natural numbers. Let Λ_1 be the collection of blocks of length $n_1 = 2$ over $\Lambda \cup \mathbb{N}$ of the form $a1$, where $a \in \Lambda, 1 \in \mathbb{N}$. Suppose, for some $k \ge 1$, that Λ_k has been defined as a collection of r_k blocks over $\Lambda \cup \mathbb{N}$, so that each member of Λ_k has length n_k and ends with the symbol $k \in \mathbb{N}$. Now we define Λ_{k+1} as the collection of all *permutation concatenations* over Λ_k, i.e., blocks of length $n_{k+1} = r_k n_k$, each being a concatenation of all blocks from Λ_k using each of them exactly once. In every such block we replace the terminal symbol k by $k+1$. The cardinality of Λ_{k+1} is $r_{k+1} = r_k!$. Let X be the (closed) set of all doubly infinite sequences over the alphabet $\Lambda \cup \mathbb{N} \cup \{\infty\}$ which, for every k, are concatenations of the blocks from Λ_k perhaps with the last symbol changed to some $k' > k$. Let $\mathcal{P}$ be the zero-coordinate partition of X. Let

μ be some shift-invariant ergodic measure μ supported by X. Notice that the symbol ∞ may appear in a sequence x only once. Clearly, the set of points x which contain the infinite symbol has measure zero. We claim that the symbolic system $(\Lambda^{\mathbb{Z}}, \mu, \sigma, \mathbb{Z})$ is bilaterally deterministic, i.e., that for μ-almost every x and every $n \in \mathbb{N}$, the block $x[-n+1, n-1]$ is completely determined by the pair $x(-\infty, -n]$ and $x[n, \infty)$. Indeed, assume x does not contain the infinite symbol and suppose we see all entries of the $\mathcal{P}$-name of x except on the interval $[-n+1, n-1]$. By periodicity of the symbols from $\mathbb{N}$, we can determine all natural symbols in x. Let $k-1$ be the largest natural symbol in $x[-n+1, n-1]$. By examining the entries of x far enough to the left and right we will see completely all but one (the one covering the coordinate zero) blocks from the family Λ_k which constitute the block C from Λ_{k+1} covering the considered interval. Because every block from Λ_k is used in C exactly once, by elimination, we will be able to determine the missing block from Λ_k and hence all symbols in $x[-n+1, n-1]$. So, the process is bilaterally deterministic. We leave the verification of positive entropy to the reader (see Exercise 3.4).

Remark 3.2.6 The "bilaterally deterministic" property of a process is far from being exceptional. Ornstein and Weiss have proved that every dynamical system $(X, \mathfrak{A}, \mu, T, \mathbb{Z})$ has a bilaterally deterministic generator [Ornstein and Weiss, 1975].

The finiteness of $H(\mathcal{P})$ in Theorem 3.2.2 is essential:

Example 3.2.7 For a partition $\mathcal{P}$ of infinite entropy the remote future $\Pi_{\mathcal{P}}$ defined as $\bigcap_{n=1}^{\infty} \mathcal{P}^{[n,\infty)}$ may admit a partition with positive dynamical entropy.

We begin with an arbitrary ergodic process $(X, \mathcal{R}, \mu, T, \mathbb{Z})$ of positive dynamical entropy. The zero-coordinate partition is either finite or countable with finite static entropy, and is (exceptionally) denoted by $\mathcal{R}$ (and will play the role of the partition measurable with respect to the remote future), while $\mathcal{P}$ will denote another partition, of infinite entropy, and finer than $T(\mathcal{R})$.

To construct $\mathcal{P}$ we divide X into infinitely (and of course countably) many cylinders $B_i \in \mathcal{R}^{[-n_i,-1]}$ of positive measure (such partition exists except in periodic processes; the sequence of lengths n_i of B_i is then unbounded). For each i we let τ_i denote the random variable defined on X as the waiting time for the first visit to the cylinder B_i after time i: $\tau_i(x) = \min\{n \geq i : T^n x \in B_i\}$. By ergodicity this variable is almost surely finite and by definition not smaller than i. So, there exist integers $N_i \geq i$ such that the sets $C_i = \{x : \tau_i(x) \leq N_i\}$ have measures converging to 1. This implies, of course, that almost every x belongs to C_i for infinitely many indices i. We can also choose $N_i \geq n_i$.

Now, we subdivide each B_i into cylinders $A_{i,1}, \ldots, A_{i,m_i} \in \mathcal{R}^{[-N_i,-1]}$ (each block $A_{i,k}$ is obtained from B_i by extending it on the left) and the partition of X so obtained we denote by $\mathcal{P}$. Any time $x \in C_i$, we know that $T^n x \in B_i$ for some $n \in [i, N_i]$. If we know the $\mathcal{P}$-name of x at positions $i, i+1, i+2, \ldots$, then we also know the value of n and the nth $\mathcal{P}$-symbol, i.e., the set $A_{i,k}$ containing $T^n x$. But the length of $A_{i,k}$ is N_i, so that this block extends in the $\mathcal{R}$-name of x from the position $n-1$ on the right to $n - N_i \leq 0$ on the left. This implies that the coordinate 0 (with respect to $\mathcal{R}$) in x is then determined, i.e., $\mathcal{R}$ is (conditionally on C_i) measurable with respect to $\mathcal{P}^{[i,\infty)}$. But C_i also belongs to $\mathcal{P}^{[i,\infty)}$. Since x belongs to infinitely many sets C_i, the partition $\mathcal{R}$ is measurable with respect to $\Pi_{\mathcal{P}}$.

Of course, $\mathcal{R}$ generates a process with positive dynamical entropy. We conclude that $\mathcal{P}$ must have infinite static entropy, otherwise Theorem 3.2.2 would apply implying $h(\mathcal{R}) = 0$.

We apply the Pinsker sigma-algebra to demonstrate that conditional dynamical entropy does not pass via countable intersections of invariant sigma-algebras (compare (2.4.18)).

Fact 3.2.8 *There exists a process* $(X, \mathcal{P}, \mu, T, \mathbb{Z})$ *and a decreasing sequence of invariant sigma-algebras* $\mathfrak{B}_k$ *such that* $h(\mathcal{P}|\mathfrak{B}) > \lim_k \uparrow h(\mathcal{P}|\mathfrak{B}_k)$, *where* $\mathfrak{B} = \bigcap_k \mathfrak{B}_k$.[1]

Proof Consider the bilateral Bernoulli shift on two symbols $0, 1$ with equal probabilities $1/2, 1/2$. Denote $\mathfrak{B}_1 = \mathcal{P}^{\mathbb{Z}}$ and $\mathcal{P}_1 = \mathcal{P}$. Consider the map π sending each point $x = (x_n)$ to the point $\pi x = (y_n)$, where

$$y_n = x_{n-1} + x_n + x_{n+1} \text{ mod } 2.$$

This map is a factor map to the same Bernoulli shift (compare Exrecise 2.1), hence it defines an invariant sigma-algebra $\mathfrak{B}_2 = \pi^{-1}(\mathfrak{B}_1)$. Now, $\mathfrak{B}_2$ has a two-element generator

$$\mathcal{P}_2 = \pi^{-1}(\mathcal{P}_1) = \{[000]\cup[011]\cup[101]\cup[110]\,,\ [001]\cup[010]\cup[100]\cup[111]\},$$

and, according to the formulae (2.3.8) and (2.4.4),

$$h(\mathcal{P}_1|\mathfrak{B}_2) = h(\mathcal{P}_1|\mathcal{P}_2) = h(\mathcal{P}_1 \vee \mathcal{P}_2) - h(\mathcal{P}_2) = h(\mathcal{P}_1) - h(\mathcal{P}_2) = 0,$$

where the last but one equality follows from the fact that $\mathcal{P}_2 \preccurlyeq \mathcal{P}_1^{[-1,1]}$ (and $h(\mathcal{P}_1^{[-1,1]}) = h(\mathcal{P}_1^3) = h(\mathcal{P}_1)$) and the last equality follows from the fact that the process generated by $\mathcal{P}_2$ is isomorphic to that generated by $\mathcal{P}_1$ (of course, π is not the isomorphism here). Now we apply the same map π again to obtain $\mathfrak{B}_3 = \pi^{-1}(\mathfrak{B}_2) = \pi^{-2}(\mathfrak{B}_1)$. The conditional entropy of the process given $\mathfrak{B}_3$ is zero, because the factor by π^2 is again isomorphic to the original process. And so on. We construct a decreasing sequence of invariant sigma-algebras $\mathfrak{B}_k$ such that the conditional entropies $h(\mathcal{P}_1|\mathfrak{B}_k)$ are all zeros.

Finally consider the intersection $\mathfrak{B} = \bigcap_k \mathfrak{B}_k$. We claim that this sigma-algebra is trivial, hence $h(\mathcal{P}_1|\mathfrak{B}) = h(\mathcal{P}_1) = 1$, so that the limit passage fails as desired. To prove triviality of $\mathfrak{B}$ we need the following observation. Consider the partition $\mathcal{Q} = \mathcal{P}^2 = \{[00], [01], [10], [11]\}$. By an elementary verification, $\pi^{-1}(\mathcal{Q})$ is seen to be a partition into four sets, each of measure 1/4 and independent of $\mathcal{Q}$. By induction, one verifies that the process generated by $\mathcal{Q}$ under the $\mathbb{N}_0$-action of the iterates of π is in fact an independent Bernoulli

[1] This example is a slight modification of one suggested to the author by B. Weiss

shift on four symbols of equal measures (for the same measure). Next, notice that the unilateral $\mathcal{Q}$-name of a point x in this process completely determines this point (i.e., determines its bilateral $\mathcal{P}$-name under the action of the shift). Indeed, the element of $\mathcal{Q}$ containing x determines both x_0 and x_1. Then, the element of $\pi^{-1}(\mathcal{Q})$ to which x belongs, tells us the values of two sums mod 2: $x_{-1}+x_0+x_1$ and $x_0+x_1+x_2$, which, combined with the knowledge of x_0 and x_1 determines both x_{-1} and x_2. And so on: knowing the initial k entries of the $\mathcal{Q}$-name of x we know its (original) coordinates from $-k+1$ to k. We have proved that $\mathcal{Q}$ generates the full sigma-algebra $\mathfrak{B}_1$ under the $\mathbb{N}_0$-action of π, which we can write as $\mathfrak{B}_1 = \mathcal{Q}^{[0,\infty)}$ (here the exponent refers to the action of π). It is now completely obvious that $\mathfrak{B}_2 = \pi^{-1}(\mathfrak{B}_1) = \mathcal{Q}^{[1,\infty)}$ and generally $\mathfrak{B}_k = \mathcal{Q}^{[k-1,\infty)}$. We obtain that $\mathfrak{B}$ (the intersection of the $\mathfrak{B}^k$'s) coincides with the Pinsker sigma-algebra for the process generated by $\mathcal{Q}$ under the $\mathbb{N}_0$-action of π. But we already know that this process is a unilateral Bernoulli shift, and in any Bernoulli shift the Pinsker sigma-algebra is trivial (see Exercise 3.3). □

3.3 The Shannon–McMillan–Breiman Theorem

In this section we present one of the most important entropy theorems in measurable dynamics. The traditional proof relies on the *maximal inequality* and then the *Martingale Convergence Theorem* [see e.g. Petersen, 1983]. We present a quite different approach using only the notion of length-entropy and Lemma 1.1.13. Although this proof is not shorter than the traditional one, in our opinion, it provides a new intuition about the mechanisms behind this result.

Theorem 3.3.1 (Shannon–McMillan–Breiman) *Let $(X, \mathcal{P}, \mu, T, \mathbb{S})$ be an ergodic process on finitely or countably many states. Assume that $H(\mathcal{P}) < \infty$. For $x \in X$ and $n \in \mathbb{N}$ denote $I_{\mathcal{P}^n}(x) = -\log \mu(A_x^n)$, where A_x^n is the unique cell of $\mathcal{P}^n$ which contains x (the cylinder $x[0, n-1]$). Then*

$$\lim_{n\to\infty} \tfrac{1}{n} I_{\mathcal{P}^n}(x) = h(\mu, T, \mathcal{P}) \ \ \mu\text{-a.e.}$$

Proof Notice that, since $A_x^n \subset T^{-1}(A_{Tx}^{n-1})$, the functions

$$x \mapsto \liminf_{n\to\infty} \tfrac{1}{n} I_{\mathcal{P}^n}(x) \ \text{ and } \ x \mapsto \limsup_{n\to\infty} \tfrac{1}{n} I_{\mathcal{P}^n}(x)$$

are subinvariant, and hence, by the Ergodic Theorem, equal almost everywhere to some constants c and C, respectively. We need to show that $C \le h(\mathcal{P}) \le c$.

We begin by proving $h(\mathcal{P}) \le c$ for a process generated by a finite l-element partition $\mathcal{P}$.

By the definition of c, given $\varepsilon > 0$, for every n_0 and almost every x there is some $n_x \geq n_0$ such that the cylinder $W_x = A_x^{n_x}$ (of length n_x and containing x) has measure larger than $2^{-n_x(c+\varepsilon)}$. If, for each x, we choose the smallest possible n_x, then the collection $\{W_x\}$ becomes a countable partition $\{W_i : i = 1, 2, \dots\}$ of cylinders of various lengths $n_i \geq n_0$, and each cell W_i has measure larger than $2^{-n_i(c+\varepsilon)}$. For a fixed $\delta > 0$ there is a set Z of measure larger than $1 - \delta$ covered by a finite collection $\mathcal{W}$ of such cylinders W_i. We denote by N_0 the maximal length n_i in this finite collection. We restrict our attention to the set Z, the conditional measure μ_Z and the finite partition of Z by the sets W_i. Notice that μ_Z assigns to the sets W_i at least as large values as μ. The length-information function associated with the pair $(\mathbf{p}', \mathbf{n})$, where $\mathbf{p}' = (\mu_Z(W_i))$ and $\mathbf{n} = (n_i)$ satisfies

$$\max_i I_{\mathbf{p}',\mathbf{n}}(i) + \max_i \tfrac{1}{n_i} \leq c + \varepsilon + \tfrac{1}{n_0} \leq c + 2\varepsilon, \tag{3.3.2}$$

(for an appropriately *a priori* chosen n_0).

By the Ergodic Theorem, for m sufficiently large, all points x from a set $X' \subset X$ of measure $1 - \delta$ visit the complement of Z no more than $m\delta$ times within the first m iterates. Assume also that $N_0/m \leq \delta$. Then X' can be covered by some finite number $\mathbf{C}$ of cylinders B of length m, such that each of them can be represented (perhaps in several ways) as a concatenation of the blocks from $\mathcal{W}$ and no more than $2m\delta$ other entries (at most $m\delta$ visits in the complement of Z and, at the end, a possible prefix of an incomplete block W_i of length at most N_0). The structure of such a block B is shown in Figure 3.2 below.

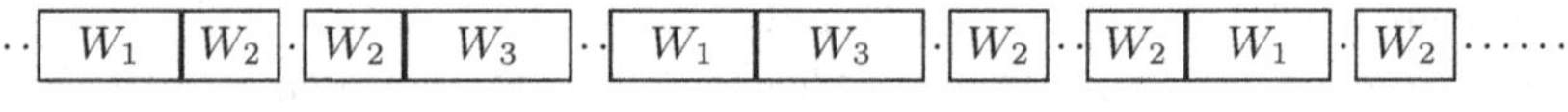

$$\cdot\cdot\boxed{W_1}\boxed{W_2}\cdot\boxed{W_2}\boxed{W_3}\cdot\cdot\boxed{W_1}\boxed{W_3}\cdot\boxed{W_2}\cdot\cdot\boxed{W_2}\boxed{W_1}\cdot\boxed{W_2}\cdots\cdots$$

Figure 3.2 The structure of the block B.

For every B we fix one such representation and we let k_B be the number of component blocks in it. For each i let p_i be the frequency of W_i in the selected concatenation representing B, i.e., the number of components equal to W_i divided by k_B. By (3.3.2) and Lemma 1.1.13, no matter what probability vector $\mathbf{p}_B = (p_i)$ is obtained, its entropy does not exceed $\mathbf{n}_{\mathbf{p}_B}(c + 2\varepsilon)$ (recall that $\mathbf{n}_{\mathbf{p}_B}$ represents the weighted average length with respect to the probability vector $\mathbf{p}_B$). Obviously, $\mathbf{n}_{\mathbf{p}_B} \leq m/k_B$, hence $H(\mathbf{p}_B) \leq c(k_B)$ where $c(k_B) = \frac{m}{k_B}(c + 2\varepsilon)$. Every B can be identified with a block of length k_B over the finite alphabet $\mathcal{W}$ with insertions of no more than $2m\delta$ symbols from $\mathcal{P}$. The cardinality $\mathbf{C}$ of cylinders B covering X' can be thus estimated as follows

$$\mathbf{C} \leq \sum_{k_B=1}^{m} \mathbf{C}[1, k_B, c(k_B)] \cdot \binom{m}{2\delta m} \cdot l^{2\delta m},$$

where $\mathbf{C}[1, k_B, c(k_B)]$ counts the blocks of length k_B over $\mathcal{W}$ with $H(\mathbf{p}_B) \leq c(k_B)$ (recall the notation (2.8.1)), the "choose" symbol bounds the number of ways the insertions can be distributed, and the last term counts the number of ways these insertions can be filled with symbols. This sum is highly exaggerated; k_B has a much smaller range.

By the elementary estimate (2.8.5) (and replacing the error term by 1), we have

$$\log \mathbf{C}[1, k_B, c(k_B)] \leq k_B(c(k_B) + 1) = m(c + 2\varepsilon + \tfrac{k_B}{m}).$$

We have $k_B/m \leq 1/n_0 < \varepsilon$. Combining the last two displayed inequalities we get

$$\log \mathbf{C} \leq \log m + m(c + 3\varepsilon) + mH(2\delta, 1 - 2\delta) + 2\delta m \log l \leq m(c + 4\varepsilon),$$

for δ sufficiently small and large m. Since the complement of X' is covered by no more than l^m cylinders of length m, and by (1.4.3), the total entropy $H(\mathcal{P}^m)$ does not exceed

$$(1 - \delta)m(c + 4\varepsilon) + \delta m \log l + H(\delta, 1 - \delta) < m(c + 5\varepsilon).$$

Because $H(\mathcal{P}^m) \geq mh(\mathcal{P})$ for every m, and ε is arbitrary, we have $c \geq h(\mathcal{P})$.

Now, if $\mathcal{P}$ is infinite countable, still with finite static entropy, we invoke the finite partitions $\mathcal{P}_{(m)}$. For each n the information function $I_{\mathcal{P}^n_{(m)}}(x)$ is dominated by $I_{\mathcal{P}^n}(x)$, thus $\liminf_n \frac{1}{n} I_{\mathcal{P}^n}(x)$ is larger than or equal to the term $\sup_m \liminf_n \frac{1}{n} I_{\mathcal{P}^n_{(m)}}(x)$, which, by the already proved part for finite partitions and by Fact 2.4.12, is at least $h(\mathcal{P})$ almost everywhere.

We will now proceed with proving that $C \leq h(\mathcal{P})$. At first we will show that

$$C \leq H(\mathcal{P}) \quad \mu\text{-a.e.} \tag{3.3.3}$$

The idea is to indicate a partition into cylinders of variable lengths whose length-entropy is close to C. Then we will change the measure, so that the same partition (with the same lengths) receives masses depending exponentially on the lengths times $H(\mathcal{P})$. Then we use the second part of Lemma 1.1.13.

Fix some $\delta > 0$ and $n_0 \in \mathbb{N}$. By the Ergodic Theorem, $\frac{1}{n}\sum_{j=0}^{n-1} I_{\mathcal{P}}(T^j x) \to H(\mathcal{P})$ μ-almost surely, i.e., for almost every x,

$$\frac{1}{n}\sum_{j=0}^{n-1} I_{\mathcal{P}}(T^j x) < H(\mathcal{P}) + \delta \tag{3.3.4}$$

holds for any sufficiently large n. On the other hand, by the definition of C, for almost every x,

$$\tfrac{1}{n}I_{\mathcal{P}^n}(x) > C - \delta, \tag{3.3.5}$$

for arbitrarily large n. Let n_x be the smallest choice of an integer for which both (3.3.4) and (3.3.5) are fulfilled at x and let W_x denote the cylinder of length n_x containing x. Since both (3.3.4) and (3.3.5) depend on the initial n coordinates of the $\mathcal{P}$-name of x, it is clear that n_x is constant on W_x. This implies that $\{W_x\}$ is in fact a partition of X into cylinders of variable lengths (clearly, such partition is at most countable). We denote this partition by $\{W_i\}$ and the corresponding lengths by n_i. The inequality (3.3.5) becomes

$$-\tfrac{1}{n_i}\log\mu(W_i) > C - \delta,$$

which implies that the length-entropy $H(\mathbf{p},\mathbf{n})$, where $\mathbf{p} = (\mu(W_i))$ and $\mathbf{n} = (n_i)$, is larger than $C - \delta$.

Now we apply a different measure ν on the symbolic space $\mathcal{P}^{\mathbb{N}_0}$. We let ν be the product measure in which each symbol $A \in \mathcal{P}$ maintains the measure value $\mu(A)$, but measures of longer blocks are computed by multiplication of the measures of the symbols in the block. The inequality (3.3.4) applied for n_i and any point in W_i says directly that

$$-\tfrac{1}{n_i}\log(\nu(W_i)) < H(\mathcal{P}) + \delta,$$

i.e., that $\nu(W_i) > 2^{-n_i(H(\mathcal{P})+\delta)}$. Because the cylinders W_i are disjoint, their measures ν are summable to a number not exceeding 1 (these cylinders cover a set of full measure μ, but perhaps not of full measure ν), all the more, the sum of the numbers $2^{-n_i(H(\mathcal{P})+\delta)}$ does not exceed 1. Now, the second assertion of Lemma 1.1.13 implies that $H(\mathcal{P}) + \delta + 1/n_0 > H(\mathbf{p},\mathbf{n}) > C - \delta$. Since δ and $1/n_0$ are arbitrarily small, we get $C \le H(\mathcal{P})$, as claimed.

In order to replace $H(\mathcal{P})$ in (3.3.3) by (a possibly smaller term) $h(\mathcal{P})$ recall that for sufficiently large n_0, $\frac{1}{n_0}H(\mathcal{P}^{n_0}) \le h(\mathcal{P}) + \varepsilon$, so it suffices to show that

$$n_0 C \le H(\mathcal{P}^{n_0}).$$

Consider the power process $(X, \mathcal{P}^{n_0}, \mu, T^{n_0}, \mathbb{S})$. For a point $x \in X$, the cylinder of length mn_0 containing x in the original process is (as a set, not as a block) the same as the cylinder of length m containing x in the power process. We can write this as

$$I_{\mathcal{P}^{mn_0}}(x) = I_{(\mathcal{P}^{n_0})^m}(x),$$

where $(\mathcal{P}^{n_0})^m$ denotes the partition obtained through m steps in the power process. For $n = mn_0 - r$ $(0 \le r < n_0)$ we have, by inclusion of the corresponding sets,

$$\tfrac{1}{n} I_{\mathcal{P}^n}(x) \le \tfrac{1}{n} I_{\mathcal{P}^{mn_0}}(x) = \tfrac{mn_0}{n} \tfrac{1}{mn_0} I_{\mathcal{P}^{mn_0}}(x),$$

so, at almost every point, the upper limit $C = \limsup_n \frac{1}{n} I_{\mathcal{P}^n}(x)$ is attained along a subsequence of mn_0, and then it equals

$$\frac{1}{n_0} \limsup_{m\to\infty} \frac{1}{m} I_{(\mathcal{P}^{n_0})^m}(x).$$

We have proved that the upper limit analogous to C, computed for the power process is constant almost everywhere and equals $n_0 C$.

Although the power process need not be ergodic, the measure μ has at most n_0 ergodic components $\mu^{(i)}$ supported by disjoint sets X_i (in fact of equal measures, but here it is inessential), and $\mu = \sum_i \mu(X_i)\mu^{(i)}$. Denote by C_i the constant analogous to C computed for $\mu^{(i)}$ in the (now ergodic) power process. The inequality (3.3.3) implies that

$$C_i \le H(\mu^{(i)}, \mathcal{P}^{n_0}).$$

As we know, the function $\mu \mapsto H(\mu, \mathcal{P}^{n_0})$ is concave (see (1.3.5)), while the information function is convex (see (1.3.4)), and this property passes via $\limsup$. So,

$$n_0 C \le \sum_i \mu(X_i) C_i \le \sum_i \mu(X_i) H(\mu^{(i)}, \mathcal{P}^{n_0}) \le H(\mu, \mathcal{P}^{n_0}),$$

and we are done. □

The Shannon–McMillan–Breiman Theorem also has a conditional version. Since it occurs in the literature usually in restricted generality, we have decided to present the full version with a complete proof. The formulation involves the notion of the conditional information function given a sigma-algebra which relies on conditional expectation. The Martingale Convergence Theorem is used to switch between this and a more elementary phrasing, in which we use only the conditional information function given a partition.

Let $\mathcal{P}$ and $\mathfrak{B}$ be a countable partition and a sigma-algebra, respectively. Recall (1.5.1) for the conditional information function $x \mapsto I_{\mathcal{P}|\mathfrak{B}}(x)$. If $H(\mathcal{P}|\mathfrak{B})$ (which is the integral of the conditional information function) is finite, then the Martingale Convergence Theorem allows one to replace this function by the almost everywhere limit over any sequence of partitions Q generating $\mathfrak{B}$:

$$I_{\mathcal{P}|\mathfrak{B}} = \lim_{Q\to\mathfrak{B}} I_{\mathcal{P}|Q}. \tag{3.3.6}$$

This is how we will understand the conditional information function in the theorem below. The proof is fairly long and rather technical, but follows the same scheme as the unconditional proof. We have chosen to skip it in the main course of the book. For interested readers we attach the proof in the Appendix.

Theorem 3.3.7 *Let $(X, \mathcal{P}, \mu, T, \mathbb{S})$ be an ergodic process on finitely or countably many states and let $\mathfrak{B}$ be a subinvariant (or invariant) sigma-algebra. Assume that $H(\mathcal{P}|\mathfrak{B}) < \infty$. Then for μ-almost every point x we have:*

$$\lim_n \tfrac{1}{n} I_{\mathcal{P}^n|\mathfrak{B}}(x) = h(\mu, T, \mathcal{P}|\mathfrak{B}).$$

Proof See Theorem B.0.1 in Appendix B. □

3.4 The Ornstein–Weiss Return Times Theorem

We continue to investigate the process $(X, \mathcal{P}, \mu, T, \mathbb{S})$ determined by a countable partition $\mathcal{P}$ with finite static entropy. Recall that A_x^n denotes the cylinder of $\mathcal{P}^n$ containing x. For each $x \in X$ we define its *first return time to the nth cylinder* as

$$\mathrm{R}_n(x) = \mathrm{R}_{A_x^n}(x) = \min\{k > 0 : T^k x \in A_x^n\}.$$

By the Poincaré Recurrence Theorem, this function is defined almost everywhere (even without ergodicity). In the language of blocks, $\mathrm{R}_n(x)$ is the first positive coordinate where a repetition of the block $x[0, n-1]$ begins in the $\mathcal{P}$-name of x. Notice that in the ergodic case, in order to acquire the measure of the cylinder A_x^n (hence the information function $I_{\mathcal{P}^n}(x)$) one needs to know all the forward return times, to compute their density along $\mathbb{N}$. Knowing only finitely many return times allows one to obtain an approximation, while the inverse of the first return time can be considered only a very crude estimate. The theorem below comes from [Ornstein and Weiss, 1993]. It asserts that this crude estimate is sufficiently good to have the same logarithmic asymptotic behavior as $\mu(A_x^n)$, allowing the entropy to be calculated. It allows us to compute the approximate value of the dynamical entropy by examining a long enough finite portion of a single $\mathcal{P}$-name. We do not even need to know the measures of the cylinders required in the Shannon–McMillan–Breiman Theorem.

Theorem 3.4.1 (Ornstein–Weiss) *If $(X, \mathcal{P}, \mu, T, \mathbb{S})$ is ergodic, then*

$$\lim_n \tfrac{1}{n} \log \mathrm{R}_n(x) = h(\mu, T, \mathcal{P}) \ \ \mu\text{-a.e.}$$

Proof Let B denote a block of length n (and its cylinder set). By ergodicity, the skyscraper over B (the union of trajectories of points from B until their first returns to B) covers a set of measure one. This can be written as

$$\int_B \mathrm{R}_n(x)\,d\mu = 1 \quad \text{or} \quad \int_B \mathrm{R}_n(x)\,d\mu_B = \frac{1}{\mu(B)}$$

(the above is also known as the Kac Theorem, see Theorem 4.3.4). Thus $\mathrm{R}_n(x) > \exp(n\varepsilon)/\mu(B)$ may hold on a subset of B of measure μ_B at most $\exp(-n\varepsilon)$. By the Law of Total Probability, the same estimate holds globally on X, which can be written as

$$\mu\{x : \tfrac{1}{n}\log \mathrm{R}_n(x) > I_{\mathcal{P}^n}(x) + \varepsilon\} < \exp(-n\varepsilon).$$

Applying the Borel–Cantelli Lemma [see e.g. Feller, 1968], the Shannon–McMillan–Breiman Theorem and because ε is arbitrary, we have proved that

$$\limsup_{n\to\infty} \tfrac{1}{n}\log \mathrm{R}_n(x) \le h(\mu, T, \mathcal{P})$$

μ-almost everywhere.

Because $\mathrm{R}_n(x) \ge \mathrm{R}_{n-1}(Tx)$, the function $\liminf_n \frac{1}{n}\log \mathrm{R}_n(x)$ is subinvariant, hence equal to a constant c. We need to show that $c \ge h(\mathcal{P})$. First of all, notice that if $\mathcal{P}$ is replaced by $\mathcal{P}_{(m)}$, the corresponding return times can only become shorter. Because of that and since $h(\mathcal{P}_{(m)}) \nearrow h(\mathcal{P})$, it suffices to prove the inequality $c \ge h(\mathcal{P})$ for finite partitions. In the remaining part of the proof l denotes the finite cardinality of $\mathcal{P}$.

Fix an $\varepsilon > 0$ and $\delta > 0$. For almost every x there is an $n_x > 1/\delta$ such that $\mathrm{R}_{n_x}(x) \le 2^{n_x(c+\varepsilon)}$. Thus, there is a set Z of measure smaller than $\delta/2$ such that $N_0 = \max\{n_x : x \notin Z\}$ is finite. By the Ergodic Theorem, for m sufficiently large, all points x from a set X' of measure $1-\delta$ visit Z no more than $m\delta/2$ times within the first m iterates. By taking m large enough we may assume that $2^{N_0(c+\varepsilon)} < m\delta/2$. Then X' can be covered by some number $\mathbf{D}$ of cylinders B of length m with the following structure: each B can be represented (perhaps in several ways) as a concatenation of some blocks W_i of lengths n_i ranging between $1/\delta$ and N_0, and no more than $m\delta$ single entries including $2^{N_0(c+\varepsilon)}$ entries at the end, in such a way that each of the blocks W_i is repeated in B to the right at a distance r_i not larger than $2^{n_i(c+\varepsilon)}$. Figure 3.3 shows the structure of B. We count the number $\mathbf{D}$ of blocks B with such a structure.

Because all blocks W_i are at least $1/\delta$ long, there are no more than $m\delta$ of them and hence their initial positions can be distributed in at most

$$2^{mH(\delta,1-\delta)} \tag{3.4.2}$$

different ways (approximately, see (2.8.4)).

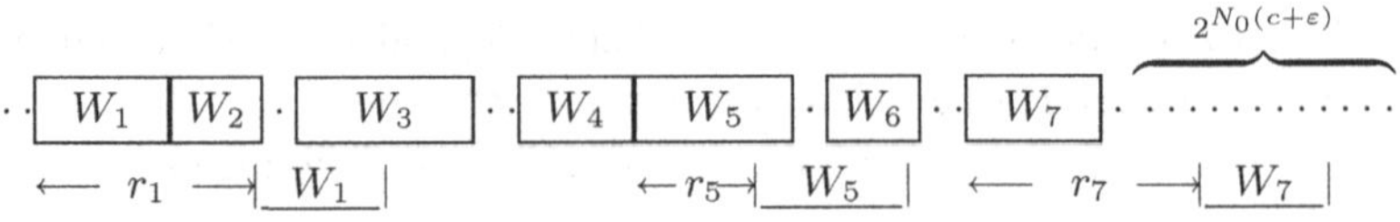

Figure 3.3 The structure of the block B. Only some repetitions are shown.

Similarly, there are no more than

$$2^{mH(\delta,1-\delta)}$$

ways of distributing the single entries over B and no more than

$$l^{m\delta}$$

ways of filling them with symbols. Now, imagine that the initial positions of the blocks W_i are set, as well as all the places outside these blocks, and that all positions outside these blocks, including $2^{N_0(c+\varepsilon)}$ positions at the end, are filled. Notice that now also the lengths n_i of the blocks W_i are determined. What is missing to determine B completely is the contents of the blocks W_i. In a moment we will explain that, instead, it suffices to know the distances to their first repetitions.

For each i the distance between W_i and its first repetition to the right assumes an integer value r_i not larger than $2^{n_i(c+\varepsilon)}$. Globally this makes

$$2^{\sum_i n_i(c+\varepsilon)} \leq 2^{m(c+\varepsilon)} \tag{3.4.3}$$

possible choices of the distances r_i (jointly for all blocks i).

Once these choices have been made, the block B is completely determined, because each entry of each block W_i (proceeding from right to left) can be copied from the entry r_i positions to the right, in the already determined part of B. Multiplying the displayed estimates (3.4.2) through (3.4.3) we obtain the upper bound:

$$\mathbf{D} \leq 2^{2mH(\delta,1-\delta)} \cdot l^{m\delta} \cdot 2^{m(c+\varepsilon)},$$

i.e.,

$$\log \mathbf{D} \leq 2mH(\delta, 1-\delta) + m\delta \log l + m(c+\varepsilon) \leq m(c+2\varepsilon),$$

for an appropriate *a priori* choice of δ. Because the complement of X' is covered by no more than l^m cylinders of length m, the entropy of $\mathcal{P}^m$ does not exceed

$$(1-\delta)m(c+2\varepsilon)+\delta m\log l+H(\delta,1-\delta)<m(c+3\varepsilon).$$

Since $H(\mathcal{P}^m)\geq mh(\mathcal{P})$ for every m, and ε is arbitrary, it must hold that $c\geq h(\mathcal{P})$. □

3.5 Horizontal data compression

A *data compression algorithm* in information theory is an algorithm (described in a finite number of instructions) allowing long blocks B to be replaced by (desirably) shorter blocks $\phi(B)$ in an injective way. It is convenient to assume that the output blocks are binary (i.e., they consist of zeros and ones). Then the length of the compressed block represents the effective information content in bits. The *compression rate* achieved on a block B is the ratio

$$\mathsf{CR}(B)=\frac{|\phi(B)|}{|B|\log\#\Lambda}$$

between the "size" of the block after and before the compression. Usually, such an algorithm divides B into a concatenation of relatively short subblocks A (of constant or variable lengths) and builds a one-to-one correspondence between the blocks A and some binary blocks $\Phi(A)$, trying to save on the length as much as possible. Eventually the image of B is obtained as the order preserving concatenation of the images $\Phi(A)$. The two most commonly known such algorithms are the Huffman algorithm and the Lempel–Ziv algorithm. The first one has been briefly described in the Introduction. It divides B into equal length subblocks A, checks their frequencies in B and assigns images of variable lengths, so that the most frequent A's receive the shortest images. The Lempel–Ziv algorithm uses component subblocks A of variable lengths and does not require examining their frequencies in B (thus is much faster). In order for the code to be reversible, not all blocks can be used in the role of $\Phi(A)$, because one has to be able to locate the "cutting places" in the concatenated image. For that, a special family of image blocks is selected, a so-called *prefix-free family*, with the property that no block in this family is a prefix (beginning) of another. Within such a family, the cardinality of all blocks of length n, for large n, nearly equals 2^n (in the sense that one-nth of log of that cardinality is close to 1), so the calculations of the compression rate are nearly the same as if all blocks were used.

The data compression code can also have the form of a program (finite set of instructions) which generates long output blocks when fed by some (shorter) input blocks. In fact, such a program is *decoding* (i.e., decompressing); the output blocks are our "originals," while the input blocks are their compressed counterparts. One needs to include both the length of the program and the length of the input block in the calculations of the compression rate of the output block. This leads to the notion of *Kolmogorov complexity* of a long block B, as the length of the shortest program (+ input) able to generate it. We refer the reader to the rich literature in information theory for precise description of many variants of lossless data compression codes, Kolmogorov complexity and related notions [see e.g. Cover and Thomas, 1991]. In this book we will concentrate on theoretical aspects of data compression and its connection with entropy. For an extended exposition see also [Shields, 1996].

Formally, a *data compression algorithm* is any injective function

$$\phi : \bigcup_{m=m_0}^{\infty} \Lambda^m \to \bigcup_{m=1}^{\infty} \{0,1\}^m.$$

Often one is interested in finding a code which highly compresses not all blocks, only a selected family of blocks, for example, all sufficiently long blocks appearing with positive probabilities in some ergodic process. From the counting argument in Theorem 2.8.9 it follows immediately that even with such a specialized algorithm, for a fixed and big enough n, the majority of blocks (appearing in the selected process) of sufficiently large length m cannot achieve the compression rate essentially better than their nth combinatorial entropy (which is close to the entropy of the process), simply because the cardinality of such blocks is too big to be injectively encoded by shorter blocks. One easily derives the following

Theorem 3.5.1 *Let ϕ be a compression algorithm that applies to all sufficiently long blocks appearing in some ergodic process $(X, \mathcal{P}, \mu, T, \mathbb{S})$ of entropy $h = h(\mu, T, \mathcal{P})$. Let n be so large that $H(\mu, \mathcal{P}^n) < n(h + \varepsilon)$. Then the joint measure of all blocks B of length m whose compression rate is smaller than $H_n(B)/\log \#\Lambda$ tends to zero with m.*

While the proof of this theorem is left to the reader (Exercise 3.9), we give a stronger almost everywhere result whose proof uses the Shannon–McMillan–Breiman Theorem. It says that the compression rate of the initial block $x[0, m-1]$ of a "typical" x cannot be better than the entropy of the process, if m is large enough.

Theorem 3.5.2 (The Data Compression Theorem) *Let ϕ be a data compression algorithm applied to blocks over a finite alphabet Λ. Let μ be a shift-invariant ergodic measure on $\Lambda^{\mathbb{S}}$ of entropy $h = h(\mu, \sigma, \mathcal{P}_\Lambda)$. For $x \in \Lambda^{\mathbb{S}}$*

let

$$\underline{\mathsf{CR}}(x) = \liminf_{m\to\infty} \mathsf{CR}(x[0, m-1]).$$

Then, for μ-almost every x, $\underline{\mathsf{CR}}(x) \geq h/\log\#\Lambda$.

Proof Fix some $\varepsilon > 0$. For $m_0 \in \mathbb{N}$ let X_{m_0} be the set where $I_m(x) \leq h - \varepsilon/2$ for all $m \geq m_0$. In other words, for $x \in X_{m_0}$, $\mu(x[0, m-1]) \leq 2^{-m(h-\frac{\varepsilon}{2})}$ (here $x[0, m-1]$ denotes the cylinder corresponding to the initial block of x). For any m, the number of all blocks of length m compressed to lengths smaller than $m(h-\varepsilon)$ (i.e., with compression rate smaller than $\frac{h-\varepsilon}{\log\#\Lambda}$) is clearly not larger than $2^{m(h-\varepsilon)}$. Within X_{m_0}, the cylinders corresponding to such blocks occupy jointly no more than a subset of measure

$$2^{-m(h-\frac{\varepsilon}{2})} \cdot 2^{m(h-\varepsilon)} \leq 2^{-m\frac{\varepsilon}{2}}.$$

By summability over m (and the Borel–Cantelli Lemma), the subset of X_{m_0} where $\underline{\mathsf{CR}}(x) \leq \frac{h-\varepsilon}{\log\#\Lambda}$ has measure zero. Because, by the Shannon–McMillan–Breiman Theorem, the sets X_{m_0} grow to a set of full measure, the entire set of points x satisfying $\underline{\mathsf{CR}}(x) \leq \frac{h-\varepsilon}{\log\#\Lambda}$ has measure zero. The assertion now follows by uniting such sets over a decreasing to zero sequence of epsilons. □

The Data Compression Theorem 3.5.2 fails for some very exceptional sequences, even ones which satisfy the Ergodic Theorem for μ. Below is an example.

Example 3.5.3 Consider the binary sequence obtained by concatenating consecutively all blocks of length 1 followed by all blocks of length 2, etc., each time ordered lexicographically, as shown below (the commas are added only to show the structure):

$$x = 0, 1, 00, 01, 10, 11, 000, 001, 010, 011, 100, 101, 110, 111, 0000, 0001\ldots$$

It is not hard to see that the frequency of any block A in this sequence is the same as its independent uniform measure μ (i.e., $\mu(A) = 2^{-n}$, where n is the length of A). In other words, x satisfies the Ergodic Theorem for every cylinder A and the measure μ whose entropy is 1. So, according to the assertion of the Data Compression Theorem 3.5.2, x should not allow for any compression. On the other hand, since it can be completely determined in a finite set of instructions (see Exercise 3.10), the Kolmogorov complexities of the blocks $x[0, m-1]$ tend to zero. This example shows that Kolmogorov complexity takes into account a much wider variety of structural regularities than just the ones based on frequencies of subblocks. Nevertheless, the Data Compression Theorem says that the collection of sequences with regularities undetected by the frequency-based codes has measure zero.

We only remark that there exist many "optimal" data compression algorithms which realize the inequality $\mathsf{CR}(B) \leq \frac{H_n(B)+\varepsilon}{\log\#\Lambda}$ for all sufficiently long blocks (this fact for the Lempel–Ziv algorithm was proved in [Ornstein and Weiss, 1993]).

A reflection: How is it AT ALL possible to create a data compression algorithm which replaces all blocks by blocks of at most the same length (and over the same alphabet), in an injective way, where at least one block is replaced by a strictly shorter one? This sounds like an impossible task! Such a replacement should lead inevitably to loss of information. Well... we apply the code not exactly to all blocks, only to blocks of lengths at least m_0, while we allow the coded images to be arbitrary. This creates a small space allowing us to move some of the blocks "down the scale." But is there much room indeed? Of course not! Data compression is an illusion. An illusion with spectacular practical results. Because we do not compress blocks shorter than m_0, we save $\sum_{m<m_0} 2^m = 2^{m_0}$ shortest blocks to be used as compressed images of some longer blocks. For a length m significantly larger than m_0 ANY compression with rate smaller than 1 may concern at most the relatively tiny amount 2^{m_0} out of the huge amount 2^m of blocks. A randomly chosen block of length m will usually turn out incompressible. Luckily, most computer files, due to their organized form, fall into the tiny collection of compressible blocks. This is why data compression works.

Exercises

3.1 Let $(X, \mathcal{P}, \mu, T, \mathbb{S})$ be an ergodic process with positive (and finite) entropy. Show that for large enough n the power process $(X, \mathcal{P}^n, \mu, T^n, \mathbb{S})$ is $n\varepsilon$-entropy independent. Is this independence trivial (see Remark 3.1.8)? Does this translate to δ-independence for some small δ?

3.2 Give an example that the sequence $H(\mathcal{P}^n) - H(\mathcal{P}^n|\mathcal{Q}^n)$ occurring in the definition of mutually ε-independent processes (formula (3.1.16)) need not have descending $n$ths. Notice that this strengthens Exercise 1.6.

3.3 Prove that the Pinsker sigma-algebra in an independent process is trivial.

3.4 Show that the bilaterally deterministic process constructed in Example 3.2.5 has positive entropy.

3.5 Fact 3.2.8 shows that the reversed inequality (2.4.18) need not hold even for the actions of $\mathbb{Z}$ (and invariant sigma-algebras). Provide a much simpler example for the actions of $\mathbb{N}_0$ (and subinvariant sigma-algebras).

3.6 Show that the Shannon–McMillan–Breiman Theorem for an independent process (with not necessarily uniform measure on the symbols) is equivalent to the Strong Law of Large Numbers [see e.g. Feller, 1968].

3.7 Define the kth return time to the nth cylinder as

$$\mathrm{R}_n^{(k)}(x) = \mathrm{R}_{A_x^n}^{(k)}(x) = \min\Big\{j > 0 : \sum_{i=1}^{j} \mathbb{I}_{A_x^n}(T^i x) = k\Big\}.$$

Prove that in an ergodic process $(X, \mathcal{P}, \mu, T, \mathbb{S})$ the variables $\mathrm{R}_n^{(k)}$ fulfill the assertion of the Ornstein–Weiss Theorem, i.e., that

$$\lim_n \tfrac{1}{n} \log \mathrm{R}_n^{(k)}(x) = h(\mathcal{P}) \;\; \mu\text{-a.e.}$$

3.8 Given an alphabet Λ of cardinality l, and $\varepsilon > 0$, create a prefix-free family of blocks over Λ in which the cardinality of blocks of length n exceeds $2^{n(\log l - \varepsilon)}$. Hint: Choose carefully a block W which cannot occur with "selfoverlapping" and compute the cardinality of all blocks of length n in which W occurs exactly one time – at the right end. By the way, notice that a prefix-free family of blocks is simply any collection of blocks with variable lengths, disjoint when treated as cylinders starting at the coordinate zero.

3.9 Prove Theorem 3.5.1.

3.10 Write a program that generates the sequence x of Example 3.5.3.

4
Kolmogorov–Sinai Entropy

4.1 Entropy of a dynamical system

Let $(X, \mathfrak{A}, \mu, T, \mathbb{S})$ be a dynamical system. The space X can be partitioned into finitely or countably many sets in many different ways producing many processes $(X, \mathcal{P}, \mu, T, \mathbb{S})$. We will consider only partitions with finite static entropy, since other partitions provide no information about the complexity of the dynamics.

Definition 4.1.1 The *dynamical* or *Kolmogorov–Sinai entropy* of the system $(X, \mathfrak{A}, \mu, T, \mathbb{S})$ is defined as follows:

$$h(\mu, T) = h(\mathfrak{A}) = h(\mu) = \sup\{h(\mathcal{P}) : H(\mathcal{P}) < \infty\}.$$

By Fact 2.4.12, it suffices to take the supremum over finite partitions $\mathcal{P}$. Moreover, using continuity of the dynamical entropy among finite partitions one can prove that

$$h(\mathfrak{A}) = \lim_k h(\mathcal{P}_k), \tag{4.1.2}$$

where $(\mathcal{P}_k)$ is a refining sequence of finite partitions which generates $\mathfrak{A}$ (the proof is left to the reader as Exercise 4.1). The notation used will depend on which of the parameters are fixed and selfunderstood in the context, and which ones are treated as variables.

Now consider two dynamical systems $(X, \mathfrak{A}, \mu, T, \mathbb{S})$ and $(Y, \mathfrak{B}, \nu, S, \mathbb{S})$ such that the latter is a factor of the former via a map π. This is to say, we identify $\mathfrak{B}$ with the (sub)invariant sigma-algebra $\pi^{-1}(\mathfrak{B}) \preccurlyeq \mathfrak{A}$. We have

Fact 4.1.3 *If $(Y, \mathfrak{B}, \nu, S, \mathbb{S})$ is a factor of $(X, \mathfrak{A}, \mu, T, \mathbb{S})$, then $h(\mu, T) \geq h(\nu, S)$. This can be written in short as*

$$\mathfrak{A} \succcurlyeq \mathfrak{B} \implies h(\mathfrak{A}) \geq h(\mathfrak{B}). \tag{4.1.4}$$

If two systems are isomorphic, then their entropies are equal.

Proof Every partition $\mathcal{Q}$ of Y with finite entropy $H(\nu, \mathcal{Q})$ lifts by π^{-1} to a partition $\mathcal{P} = \pi^{-1}(\mathcal{Q})$ of X, with the same static entropy $H(\mu, \mathcal{P}) = H(\nu, \mathcal{Q})$. Then, for each $n \in \mathbb{N}$, by equivariance, $\pi^{-1}(\mathcal{Q}^n) = \mathcal{P}^n$, and again, the partitions $\mathcal{Q}^n$ and $\mathcal{P}^n$ have the same static entropy in respective spaces. So, the dynamical entropy of $\mathcal{P}$ and $\mathcal{Q}$ are equal. Thus, the supremum defining $h(\mu, T)$ is not smaller than that of $h(\nu, S)$. The statement concerning isomorphic systems is now immediate, as each of the systems is a factor of the other. □

Definition 4.1.5 We define the *conditional entropy of a dynamical system given its factor* as follows

$$h(\mu, T|\nu, S) = h(\mathfrak{A}|\mathfrak{B}) = h(\mu|\nu) = \sup_{\mathcal{P}} h(\mathcal{P}|\mathfrak{B}).$$

The last notation will be used frequently in Part II of this book, where this kind of entropy is studied as a function of invariant measures. For now however, since we will work within a fixed measure space with a fixed transformation, and all factors are identified with (sub)invariant sigma-algebras, we will mainly use the second notation. We have

Fact 4.1.6

$$h(\mathfrak{A}|\mathfrak{B}) + h(\mathfrak{B}) = h(\mathfrak{A}) \quad (\textit{or} \quad h(\mu|\nu) + h(\nu) = h(\mu)).$$

Proof The equality is trivial when $h(\mathfrak{B}) = \infty$. Otherwise we need to prove the subtractive formula $h(\mathfrak{A}|\mathfrak{B}) = h(\mathfrak{A}) - h(\mathfrak{B})$. Let $\mathcal{P}$ and $\mathcal{Q}$ range over all finite $\mathfrak{A}$-measurable partitions and all finite $\mathfrak{B}$-measurable partitions, respectively. We have, using (2.4.4) and (2.4.17) in appropriate places,

$$\begin{aligned} h(\mathfrak{A}) - h(\mathfrak{B}) = \sup_{\mathcal{P}} h(\mathcal{P}) - \sup_{\mathcal{Q}} h(\mathcal{Q}) = \inf_{\mathcal{Q}} \sup_{\mathcal{P}} [h(\mathcal{P}) - h(\mathcal{Q})] = \\ \inf_{\mathcal{Q}} \sup_{\mathcal{P}} [h(\mathcal{P} \vee \mathcal{Q}) - h(\mathcal{Q})] \geq \sup_{\mathcal{P}} \inf_{\mathcal{Q}} [h(\mathcal{P} \vee \mathcal{Q}) - h(\mathcal{Q})] = \\ \sup_{\mathcal{P}} \inf_{\mathcal{Q}} [h(\mathcal{P}|\mathcal{Q})] = \sup_{\mathcal{P}} h(\mathcal{P}|\mathfrak{B}) = h(\mathfrak{A}|\mathfrak{B}). \end{aligned}$$

On the other hand,

$$h(\mathfrak{A}|\mathfrak{B}) = \sup_{\mathcal{P}} \inf_{\mathcal{Q}} [h(\mathcal{P} \vee \mathcal{Q}) - h(\mathcal{Q})] \geq \sup_{\mathcal{P}} \inf_{\mathcal{Q}} [h(\mathcal{P}) - h(\mathcal{Q})] = h(\mathfrak{A}) - h(\mathfrak{B}).$$

□

If $\mathfrak{B}$ is generated by a finite partition $\mathcal{Q}$, then the infimum over $\mathcal{Q}$ can be skipped. The alternative notation for $h(\mu, T|\nu, S)$ in this case is $h(\mu, T|\mathcal{Q})$ or simply $h(\mathfrak{A}|\mathcal{Q})$.

Now consider several factors of the master system $(X, \mathfrak{A}, \mu, T, \mathbb{S})$, represented by (sub)invariant sigma-algebras, for example $\mathfrak{B}$ and $\mathfrak{C}$. It is clear that $\mathfrak{B} \vee \mathfrak{C}$ is also a (sub)invariant sigma-algebra, hence represents a factor (it is a kind of a joining, see Section 4.4 for more). The familiar list of monotonicity and subadditivity properties holds (compare Fact 2.4.2):

Fact 4.1.7 *Let $\mathfrak{B}, \mathfrak{C}, \mathfrak{D}$ be subinvariant sigma-algebras. Then*

$$h(\mathfrak{B} \vee \mathfrak{C}|\mathfrak{D}) = h(\mathfrak{B}|\mathfrak{C} \vee \mathfrak{D}) + h(\mathfrak{C}|\mathfrak{D}), \tag{4.1.8}$$

$$\mathfrak{B} \succcurlyeq \mathfrak{C} \implies h(\mathfrak{B}|\mathfrak{D}) \geq h(\mathfrak{C}|\mathfrak{D}), \tag{4.1.9}$$

$$\mathfrak{C} \succcurlyeq \mathfrak{D} \implies h(\mathcal{B}|\mathfrak{C}) \leq h(\mathcal{B}|\mathfrak{D}), \tag{4.1.10}$$

$$h(\mathfrak{B} \vee \mathfrak{C}|\mathfrak{D}) \leq h(\mathfrak{B}|\mathfrak{D}) + h(\mathfrak{C}|\mathfrak{D}), \tag{4.1.11}$$

$$h(\mathfrak{B}|\mathfrak{D}) \leq h(\mathfrak{B}|\mathfrak{C}) + h(\mathfrak{C}|\mathfrak{D}). \tag{4.1.12}$$

Proof First of all, by conditioning all expressions in the proof of Fact 4.1.6 one obtains its conditional version

$$h(\mathfrak{A}|\mathfrak{B} \vee \mathfrak{D}) + h(\mathfrak{B}|\mathfrak{D}) = h(\mathfrak{A}|\mathfrak{D}). \tag{4.1.13}$$

Since for any $\mathfrak{C}$-measurable partition $\mathcal{R}$ we have $h(\mathcal{P} \vee \mathcal{R}|\mathfrak{C}) = h(\mathcal{P}|\mathfrak{C})$ (use (2.4.3) and (1.6.28)), we also have $h(\mathfrak{B} \vee \mathfrak{C}|\mathfrak{C} \vee \mathfrak{D}) = h(\mathfrak{B}|\mathfrak{C} \vee \mathfrak{D})$. Now (4.1.8) follows from (4.1.13) applied to $\mathfrak{C}$ as a factor of $\mathfrak{B} \vee \mathfrak{C}$:

$$h(\mathfrak{B} \vee \mathfrak{C}|\mathfrak{C} \vee \mathfrak{D}) + h(\mathfrak{C}|\mathfrak{D}) = h(\mathfrak{B} \vee \mathfrak{C}|\mathfrak{D}).$$

Monotonicity (4.1.9) follows from (4.1.8). The reversed monotonicity (4.1.10) is obvious by definition. Just like in all preceding similar lists of properties, the last two subadditivity statements are direct consequences of the first three (compare Exercise 1.3). □

The next property we prove is the *power rule*:

Fact 4.1.14 *For each $n \in \mathbb{S}$ we have $h(\mu, T^n) = |n| h(\mu, T)$.*

Proof By Fact 2.4.19, for each partition $\mathcal{P}$ we have

$$h(\mu, T^n, \mathcal{P}) \leq h(\mu, T^n, \mathcal{P}^{|n|}) = |n| h(\mu, T, \mathcal{P}).$$

Taking the supremum over all $\mathcal{P}$ we get

$$h(\mu, T^n) \leq |n| h(\mu, T) = \sup_{\mathcal{Q} = \mathcal{P}^{|n|}} h(\mu, T^n, \mathcal{Q}) \leq h(\mu, T^n).$$

□

It is clear that any measure on a countable space is atomic, hence any dynamics on such a space has entropy zero. This fact has also a conditional version: every countable-to-one extension has conditional entropy zero. The proof relies on disintegration and the conditional Shannon–McMillan–Breiman Theorem (Theorem 3.3.7). With this price paid the proof is easy. Without this machinery it is relatively easy to prove a weaker statement that the fibers are infinite (Exercise 4.5).

Theorem 4.1.15 *Let $(Y, \mathfrak{B}, \nu, S, \mathbb{S})$ be a factor of an ergodic system $(X, \mathfrak{A}, \mu, T, \mathbb{S})$ via a factor map $\pi : X \to Y$. Suppose that $h(\mathfrak{A}|\mathfrak{B}) > 0$. Then, for ν-almost every y, $\pi^{-1}(y)$ is uncountable.*

Proof We have

$$h(\mathfrak{A}|\mathfrak{B}) = \sup_{\mathcal{P}} h(\mathcal{P}|\mathfrak{B}) > 0,$$

where $\mathcal{P}$ ranges over all finite $\mathfrak{A}$-measurable partitions of X. Thus, there exists a finite partition $\mathcal{P}$ such that $h(\mathcal{P}|\mathfrak{B}) > 0$. We restrict our attention to the system generated jointly by $\mathcal{P}$ and the factor on Y, i.e., we set $\mathfrak{A} = \mathcal{P}^{\mathbb{S}} \vee \mathfrak{B}$. We will prove that almost every y has an uncountable preimage already in this system.

This follows immediately from the conditional Shannon–McMillan–Breiman Theorem 3.3.7. For almost every x in the extension the conditional information function $\frac{1}{n} I_{\mathcal{P}^n|\mathfrak{B}}(x)$ converges to the conditional entropy, where $I_{\mathcal{P}^n|\mathfrak{B}}(x) = -\log \mu_y(A_x^n)$, where y is the factor image of x, μ_y is the disintegration of μ and A_x^n is the cylinder over $\mathcal{P}$ of length n containing x. Whenever μ_y has an atom at x the measures of the cylinders A_x^n containing x do not decrease to zero, so their minus logarithms are bounded, hence the sequence $\frac{1}{n} I_{\mathcal{P}^n|\mathfrak{B}}(x)$ decreases to zero. Thus, if the conditional entropy is positive, such atoms may occur only for y with ν-probability zero. □

4.2 Generators

Throughout this section we will work with dynamical systems with finite Kolmogorov–Sinai entropy. Measure-theoretically, a *generator* is any partition $\mathcal{P}$ such that $\mathcal{P}^{\mathbb{S}}$ equals $\mathfrak{A}$ (up to measure). Nevertheless, in systems with finite Kolmogorov–Sinai entropy we will require that the static entropy of $\mathcal{P}$ is finite.

Definition 4.2.1 A countable partition $\mathcal{P}$ will be called a *generator* in a system $(X, \mathfrak{A}, \mu, T, \mathbb{S})$ if it has finite static entropy and the full history $\mathcal{P}^{\mathbb{S}}$ equals $\mathfrak{A}$ (after completing).

Existence of generators is one of the issues for which the kind of the acting semigroup may be of decisive importance. Thus we must distinguish between *unilateral* generators, i.e., generating under the action of $\mathbb{N}_0$: $\mathcal{P}^{\mathbb{N}_0} = \mathfrak{A}$ and *bilateral* generators, i.e., generating under the action of $\mathbb{Z}$: $\mathcal{P}^{\mathbb{Z}} = \mathfrak{A}$.

There are two key theorems concerning generators: the *Kolmogorov–Sinai Theorem* [Sinai, 1959] and the *Krieger Generator Theorem* (or simply *Krieger Theorem*) [Krieger, 1970].

Theorem 4.2.2 (Kolmogorov–Sinai) *If $\mathcal{P}$ is a generator (unilateral or bilateral) for a dynamical system $(X, \mathfrak{A}, \mu, T, \mathbb{S})$, then $h(\mu, T) = h(\mu, T, \mathcal{P})$.*

Proof For unilateral generators this is an immediate consequence of Fact 2.4.1, while for bilateral generators one needs the Exercise 2.6. □

The next theorem allows one to view every invertible system ($\mathbb{Z}$-action) of finite entropy as a process over a finite alphabet, i.e., as a symbolic system. It plays a crucial role in our interpretation of entropy, as the amount of information passing per unit of time. Although various proofs may be found in many textbooks, for sake of completeness of this book we provide a full proof below.

Theorem 4.2.3 (Krieger) *Let T be an ergodic automorphism of the standard probability space $(X, \mathfrak{A}, \mu)$ with $h(\mu, T) < \infty$. Then $(X, \mathfrak{A}, \mu, T, \mathbb{Z})$ has a finite bilateral generator of any cardinality $l > 2^{h(\mu,T)}$. Moreover, $l = 2^{h(\mu,T)}$ is possible if and only if $2^{h(\mu,T)}$ is an integer and the system is isomorphic to the Bernoulli shift on l symbols with equal measures.*

Remark 4.2.4 The Krieger Theorem solves the question about the vertical data compression, i.e., it allows the smallest alphabet which losslessly encodes the system in real time to be determined. Also, it provides an interpretation of Kolmogorov–Sinai entropy in terms of the vertical data compression: if l_n denotes the cardinality of the smallest alphabet sufficing to encode the action of T^n, then $h(\mu, T) = \lim_n \frac{1}{n} \log l_n$.

We will prove the Krieger Theorem in two steps. At first, we prove the assertion assuming that the system has a countable generator (with finite static entropy). The main proof will then reduce to finding a countable generator in a $\mathbb{Z}$-action with finite Kolmogorov–Sinai entropy. The proofs rely on cutting the $\mathcal{P}$-names into blocks at places that can be determined from the $\mathcal{P}$-name (i.e., using a $\mathcal{P}^{\mathbb{Z}}$-measurable procedure) without knowing the position of the coordinate zero (i.e., in a shift-equivariant procedure).

Lemma 4.2.5 *Let $(\Lambda^{\mathbb{Z}}, \mu, \sigma, \mathbb{Z})$ be an ergodic symbolic system over a countable alphabet such that $H(\mathcal{P}_\Lambda) < \infty$. Denote $h = h(\mu, \sigma, \mathcal{P}_\Lambda)$. Let l be such* *that* $\log l > h$. *Then the system* $(\Lambda^{\mathbb{Z}}, \mu, \sigma, \mathbb{Z})$ *is isomorphic to another symbolic system,* $(\Delta^{\mathbb{Z}}, \nu, \sigma, \mathbb{Z})$, *where* $\#\Delta = l$.

Proof We let Δ be an alphabet of cardinality l. We will replace each Λ-name by a Δ-name in an (almost surely) injective and shift-equivariant way. Let $\varepsilon = \frac{\log l - h}{2}$. As we know (see Exercise 3.8) we can select one block over Δ, say W, and find n_0 such that for every $n \geq n_0$ the cardinality of all blocks of length n over Δ in which W occurs exactly one time – at the right end (such blocks constitute a prefix-free family), is at least $2^{n(\log l - \varepsilon)}$ (which equals $2^{n(h+\varepsilon)}$).

Using the Shannon–McMillan–Breiman Theorem we can enlarge n_0 so that for every x in a set $C \subset \Lambda^{\mathbb{Z}}$ of measure larger than $1/2$ and any $n \geq n_0$ the cylinder $x[0, n-1] \in \Lambda^n$ has measure within the range $2^{-n(h \pm \varepsilon)}$. Notice that for every $n \geq n_0$ there are at most $2^{n(h+\varepsilon)}$ cylinders of length n which intersect C. Now we invoke the Kakutani–Rokhlin Lemma [see e.g. Petersen, 1983]: a set of measure at least $3/4$ in $\Lambda^{\mathbb{Z}}$ is occupied by a tower of height n_0, i.e., by a sequence of disjoint sets $A, T(A), \ldots, T^{n_0-1}(A)$. Notice that for each $i = 0, \ldots n_0 - 1$ the first return time to $T^i(A)$ is never smaller than n_0; each orbit must leave the tower through the top and re-enter through the base. Of course, the set $B = C \cap T^i(A)$ has positive measure for at least one i. In this way we have selected a set B (which we call a *marker*) with two properties: the return time R_B is at least n_0 and each value n of R_B is represented by at most $2^{n(h+\varepsilon)}$ cylinders of length n (intersecting B). Let $\mathcal{R}_n$ denote the family of blocks over Λ corresponding to these cylinders. For each n there is a 1-1 map Φ_n from $\mathcal{R}_n$ into the family of all blocks of length n over Δ ending with W and in which W does not occur otherwise (because, as we have noted, there are sufficiently many such blocks).

Now fix an $x \in \Lambda^{\mathbb{Z}}$ whose orbit visits B infinitely many times in both the past and the future (by the Ergodic Theorem almost every point has this property). At times of the visits of the orbit of x to B the sequence x is cut into a concatenation of blocks belonging to $\bigcup_{n \geq n_0} \mathcal{R}_n$. We define the map π as the code replacing each block R from this concatenation by its image $\Phi_n(R)$, where n is the length of R. It is immediate to see that π is defined almost everywhere, it is measurable, shift-equivariant and invertible (where defined): the cutting places are determined in the image sequences by the occurrences of W. □

In the above proof dealing with bilateral sequences is essential: in a unilateral sequence the block between the coordinate zero and the first visit in B need not belong to the family $\bigcup_{n \geq n_0} \mathcal{R}_n$, so we may not know how to encode it.

We can now prove the full version of the Krieger Theorem.

Proof of Theorem 4.2.3 Consider a system $(X, \mathfrak{A}, \mu, T, \mathbb{Z})$ with finite entropy $h = h(\mu, T)$. We need to find any (countable) generator. There exists a sequence $(\mathcal{P}_k)_{k\in\mathbb{N}}$ of finite partitions such that $\mathcal{P}_{k+1} \succcurlyeq \mathcal{P}_k$ for each k which together generate the sigma-algebra $\mathfrak{A}$ (without even using the action of T). Then $h_k = h(\mu, T, \mathcal{P}_k) \nearrow h$ (see (4.1.2)). We let Δ be an alphabet of cardinality l such that $\log l > h$. In the coding, we will use the countable alphabet: $\Lambda = \Delta \cup \{0, 1, 2, \ldots\}$. We fix a decreasing to zero sequence ε_k such that $\sum_k \varepsilon_k = \frac{\log l - h}{2}$. From now on we proceed inductively. The first step is almost identical as in the preceding proof. We choose a marker set B_1 so that each value n of the return time R_{B_1} is represented by at most $2^{n(h_1+\varepsilon_1)}$ cylinders in the $\mathcal{P}_1$-names. The map Φ_n sends these cylinders into blocks over Λ of length $n\frac{h_1+\varepsilon_1}{\log l} < n$ (there is enough of them). Next, we extend every image block to the right by attaching $n\frac{\log l - h_1 - \varepsilon_1}{\log l} - 1$ zeros and the terminal symbol "1," so that the image now has the same length n as the original. The code π_1 is constructed as follows: we cut each $\mathcal{P}_1$-name at the times of the visits to B_1 into blocks R and replace each R by $\Phi_n(R)$, where n is the length of R.

We now describe the second step, and we will skip a completely analogous description of the further steps. Find $n_2 > n_1$ such that on a set $C \subset B_1$ of measure at least half of $\mu(B_1)$ the following condition holds: for every $x \in C$ and $n \geq n_2$ the measure of the cylinder R_2 corresponding to the block $x[0, n-1]$ in the $\mathcal{P}_2$-name x is within the range $2^{-n(h_2 \pm \varepsilon_2)}$. Since the corresponding initial cylinder R_1 of length n in the $\mathcal{P}_1$-name of x has measure between $2^{-n(h_1 \pm \varepsilon_1)}$ (because $x \in B_1$), the intersection $R_1 \cap C$ splits into at most $2^{n(h_2 - h_1 + \varepsilon_1 + \varepsilon_2)}$ cylinders of the same length over $\mathcal{P}_2$. So, they can be injectively mapped into blocks over Λ of length $n\frac{h_2 - h_1 + \varepsilon_1 + \varepsilon_2}{\log l}$. We fix such a map (separately for each block R_1) and call it Φ_{R_1}.

As in the preceding proof (and step) we now select a set $B_2 \subset C$ of positive measure with the additional property that the return time R_{B_2} assumes only values larger than or equal to n_2. This is our new marker set. Now take any point $x \in X$. The times of visits to B_2 cut both its $\mathcal{P}_1$-name and $\mathcal{P}_2$-name into blocks of lengths $n \geq n_2$. Let R_1 and R_2 be a pair of blocks that appear between a fixed pair of markers, in these two names, respectively. Because $B_2 \subset B_1$, R_1 is a concatenation of (finitely many) blocks R used in the preceding step. The code π_1 replaces R_1 by a concatenation of blocks over Λ in such a way that there is a fraction

$$\frac{\log l - h_1 - \varepsilon_1}{\log l}$$

of zeros (we ignore the fraction occupied by the terminal symbols 1; it can be included in the error term ε_1). We use the smaller fraction $\frac{h_2 - h_1 + \varepsilon_1 + \varepsilon_2}{\log l}$

of these cells to encode R_2 *given* R_1. We simply "write" the block $\Phi_{R_1}(R_2)$ into the empty places proceeding from left to right, still leaving some number of the empty cells unused (see below), and we put the symbol "2" at the end of the image of R_2 so obtained, replacing the "1" put there by the code π_1. This concludes the description of the code π_2. Notice that there is still a fraction

$$\frac{\log l - h_2 - 2\varepsilon_1 - \varepsilon_2}{\log l}$$

of unused empty cells left. This is exactly what we need to repeat in the following steps: after step k there will be a fraction

$$\frac{\log l - h_k - 2\varepsilon_1 - 2\varepsilon_2 - \cdots - 2\varepsilon_{k-1} - \varepsilon_k}{\log l}$$

of zeros, of which we will use

$$\frac{h_{k+1} - h_k + \varepsilon_k + \varepsilon_{k+1}}{\log l}$$

in the following step.

Notice that throughout all countably many steps every coordinate is changed at most once, except the terminal integers which can only grow. Thus the limit code π is well defined (the symbol ∞ may occur). Clearly, it is measurable and shift-equivariant (as a limit of such). It is also invertible; the cutting places in step k can be found by locating, in the image, all integer symbols larger than or equal to k (including ∞). Then we know which symbols in this image were used by the code π_1, so we can reverse it and reconstruct the $\mathcal{P}_1$-name of the original. Next we know which symbols in the image were used by the code π_2 and knowing already the $\mathcal{P}_1$-name of the original, we can reconstruct its $\mathcal{P}_2$-name. And so on. By reversing all the codes π_k we reverse π.

The last thing to notice is that the new generating partition $\mathcal{P}_\Lambda$ so obtained has finite static entropy. The finitely many symbols from Δ have finite entropy. The other symbols $k \in \mathbb{N}_0$ have probabilities not larger than $1/n_k$, respectively (the symbol k occurs in the image with gaps at least n_k; in particular ∞ has probability zero). We can easily arrange that the expectation $\sum k \frac{1}{n_k}$ is finite. This implies finite static entropy (see Fact 1.1.4).

We have completed the proof of the main statement of the Krieger Theorem. It remains to check the case of equality. Clearly, if the system is isomorphic to a Bernoulli shift on finitely many symbols of equal measures, then the entropy of this system equals the logarithm of the cardinality of the generator transported from the Bernoulli shift.

Conversely, suppose a system with entropy $h = \log l$ has an l-element generator $\mathcal{P}$. Because $\mathcal{P}$ is a generator, we have $h(\mathcal{P}) = h$. The static entropy $H(\mathcal{P})$ is, on one hand, not larger than $\log l$, on the other, not smaller than the dynamical entropy $h(\mathcal{P})$. So, we have equality $h(\mathcal{P}) = H(\mathcal{P})$. This condition implies that the process is independent (see Definition 2.3.15) and since $H(\mathcal{P}) = \log(\#\mathcal{P})$, $\mathcal{P}$ is a partition into sets of equal measures. □

Note that the Krieger Theorem does not apply to actions of $\mathbb{N}_0$. For example, if T happens to be invertible, then, by Fact 2.3.12, the existence of a unilateral generator is possible only when the dynamical entropy is zero. Even when T is evidently not invertible, still, it may possess an invertible factor of positive entropy, which also makes the existence of a unilateral generator impossible.

We shall say a few words about the Pinsker factor of a dynamical system. This notion has been defined for processes (in Section 3.2), now we are one step away from extending it to general dynamical systems.

Definition 4.2.6 Let $(X, \mathfrak{A}, \mu, T, \mathbb{S})$ be a dynamical system. The *Pinsker sigma-algebra* is defined as $\Pi_\mu = \bigvee_{\mathcal{P}} \Pi_{\mathcal{P}}$ where $\mathcal{P}$ ranges over all countable partitions with finite static entropy of X.

Clearly, this is an invariant sigma-algebra. The associated factor is called the *Pinsker factor*.

Remark 4.2.7 The same Pinsker sigma-algebra will be obtained as $\bigvee_{k=1}^{\infty} \Pi_{\mathcal{P}_k}$ where $(\mathcal{P}_k)$ is a refining sequence of finite partitions that generate $\mathfrak{A}$. If $\mathcal{P}$ is a generator in $(X, \mathfrak{A}, \mu, T, \mathbb{S})$, then $\Pi_\mu = \Pi_{\mathcal{P}}$. The proof is left to the reader as Exercise 4.4.

Remark 4.2.8 In a system $(X, \mathfrak{A}, \mu, T, \mathbb{S})$ consider the "full remote future" $\mathfrak{A}^\infty = \bigcap_{n=1}^{\infty} T^{-n}(\mathfrak{A})$. It is trivially observed that for automorphisms $\mathfrak{A}^\infty$ equals $\mathfrak{A}$, while for systems having a unilateral generator $\mathcal{P}$ (with finite entropy) $\mathfrak{A}^\infty$ coincides with the Pinsker sigma-algebra. The sigma-algebra $\mathfrak{A}^\infty$ represents the largest invertible factor of the system. The conditional entropy $h(\mu|\mathfrak{A}^\infty)$ is interpreted as the entropy coming from the noninvertibile dynamics and is sometimes called the (measure-theoretic) *preimage entropy* [Cheng and Newhouse, 2005]. We refer to Section 6.10 where the connection between measure-theoretic and topological preimage entropy is discussed.

Theorem 4.2.9 *A factor of $(X, \mathfrak{A}, \mu, T, \mathbb{S})$ has entropy zero if and only if the sigma-algebra corresponding to this factor is contained in Π_μ.*

Proof Let $\mathfrak{B}$ be a sigma-algebra associated with some factor. Suppose the entropy of the factor is zero. Let $\mathcal{P}$ be a $\mathfrak{B}$-measurable partition of finite static entropy. The process $(X, \mathcal{P}, \mu, T, \mathbb{S})$ has entropy zero, thus, by Theorem 3.2.2, $\mathcal{P} \preccurlyeq \Pi_{\mathcal{P}} \preccurlyeq \Pi_\mu$. Thus $\mathfrak{B} \preccurlyeq \Pi_\mu$.

Conversely, suppose $\mathfrak{B} \preccurlyeq \Pi_\mu$ and let $\mathcal{Q}$ be a finite $\mathfrak{B}$-measurable partition. By Remark 4.2.7, for every ε the partition $\mathcal{Q}$ can be approximated up to ε in d_1 by a partition $\mathcal{Q}'$ of the same cardinality as $\mathcal{Q}$, measurable with respect to a finite join $\bigvee_{i=1}^k \Pi_{\mathcal{P}_i}$ for some partitions $\mathcal{P}_i$. It is easy to see that this last join equals $\Pi_{\mathcal{P}}$, where $\mathcal{P} = \bigvee_{i=1}^k \mathcal{P}_i$. By Theorem 3.2.2 again, $h(\mathcal{Q}') = 0$ and, by continuity of the dynamical entropy for partitions of bounded cardinality (Fact 2.4.13), $h(\mathcal{Q}) = 0$. This implies that the entropy of the factor associated with $\mathfrak{B}$ is zero (in the supremum defining the entropy of the factor system it suffices to use finite partitions). □

4.3 The natural extension

There is a technique which allows us to generalize many results from automorphisms to endomorphisms, especially those concerning entropy. This technique is called *the natural extension* and we will briefly describe it in this section.

Definition 4.3.1 Let $(X, \mathfrak{A}, \mu, T, \mathbb{N}_0)$ be a dynamical system. The *natural extension* is the system $(X', \mathfrak{A}', \mu', T', \mathbb{Z})$ (or $(X', \mathfrak{A}', \mu', T', \mathbb{N}_0)$, depending on the needs) defined as follows: $X' = X^{\mathbb{Z}}$, $\mathfrak{A}'$ is the product sigma-algebra, T' is the shift map and μ' is defined on measurable cylinders

$$C = C(A_{-n}, \dots, A_{n-1}) = \{(x(i))_{i\in\mathbb{Z}} \in X^{\mathbb{Z}} : \forall_{i\in[-n,n-1]}\, x(i) \in A_i\}$$

by

$$\mu'(C) = \mu\left(\bigcap_{i=-n}^{n-1} T^{-n-i}(A_i)\right).$$

It is elementary to verify that the natural extension is a well-defined automorphism, i.e., T' is invertible. The projection π_0 on the coordinate zero is a factor map from $(X', \mathfrak{A}', \mu', T', \mathbb{N}_0)$ to $(X, \mathfrak{A}, \mu, T, \mathbb{N}_0)$.

For unilateral symbolic systems the natural extension has a simpler (isomorphic) form. We just mention here the modification: X' is defined as the set of all bilateral sequences over the same alphabet as the original system. The measure is defined for cylinders over blocks over the alphabet by first shifting them so that all coordinates become positive, then applying the measure from the unilateral space.

It is obvious that for an invertible map T the dynamical entropy of a partition is the same in the $\mathbb{Z}$ and $\mathbb{N}_0$ action. The following fact is crucial:

Fact 4.3.2 *For a unilateral symbolic system* $(\Lambda^{\mathbb{N}_0}, \mu, \sigma, \mathbb{N}_0)$ *the dynamical entropy is the same as that of the natural extension* $(\Lambda^{\mathbb{Z}}, \mu', \sigma', \mathbb{N}_0)$. *More generally, the Kolmogorov–Sinai entropy of a dynamical system* $(X, \mathfrak{A}, \mu, T, \mathbb{N}_0)$ *is the same as that of the natural extension* $(X', \mathfrak{A}', \mu', T', \mathbb{N}_0)$. *The conditional entropy* $h(\mathfrak{A}'|\mathfrak{A})$ *is zero.*

Proof The first statement is obvious, as the dynamical entropy depends exclusively on the future of the process, i.e., on the partitions $\mathcal{P}^n$, $n \geq 0$, which generate the same probability vectors in both processes. For the latter statements we note that $(X, \mathfrak{A}, \mu, T, \mathbb{N}_0)$ is a factor of $(X', \mathfrak{A}', \mu', T', \mathbb{N}_0)$, so the entropy of the latter system is not smaller. Now, take any measurable partition $\mathcal{P}'$ of X' of finite static entropy. Then this partition can be approximated in d_1 by a finite partition $\mathcal{Q}'$ measurable with respect to the sigma-algebra $\mathfrak{A}^{[-n,n-1]}$, where $\mathfrak{A}$ is identified with the zero-coordinate sigma-algebra. We have

$$\tfrac{1}{m} H(\mathcal{Q}'^m|\mathfrak{A}^+) \leq \tfrac{1}{m} H(\mathcal{Q}'^n|\mathfrak{A}^+) + \tfrac{1}{m} H(\mathcal{Q}'^{[n,m-1]}|\mathfrak{A}^+).$$

The last term, by applying T'^{-n}, becomes $\frac{1}{m} H(\mathcal{Q}'^{m-n}|\mathfrak{A}^{[-n,\infty)})$, which is zero, because the partition is measurable with respect to the conditioning sigma-algebra. The first term on the right-hand side decreases to zero with m. We have obtained that $h(\mu', T', \mathcal{Q}'|\mathfrak{A}^+) = 0$. Since $\mathfrak{A}$ is subinvariant, $\mathfrak{A}^+$ equals $\mathfrak{A}$. By lower semicontinuity in d_1 of $h(\cdot|\mathfrak{A})$ on countable partitions with finite static entropy (see Fact 2.4.13), the conditional dynamical entropy of $\mathcal{P}'$ given $\mathfrak{A}$ is zero. Taking the supremum over $\mathcal{P}'$ we obtain $h(\mathfrak{A}'|\mathfrak{A}) = 0$, as claimed. As a consequence, $h(\mathfrak{A}') = h(\mathfrak{A})$. □

For a dynamical system $(X, \mathfrak{A}, \mu, T, \mathbb{S})$ and a set $B \in \mathfrak{A}$, recall Definition 3.1.10 of the system induced on B. We conclude this section with the Abramov Theorem, allowing us to compute the entropy of the induced system [Abramov, 1959].

Theorem 4.3.3 (Abramov) *Let* $(X, \mathfrak{A}, \mu, T, \mathbb{S})$ *be an ergodic system and let* $B \in \mathfrak{A}$ *satisfy* $\mu(B) > 0$. *Then*

$$h(\mu_B, T_B) = \frac{h(\mu, T)}{\mu(B)}.$$

The proof relies on Kac's theorem [Kac, 1947]. We give a proof which uses natural extensions.

Theorem 4.3.4 (Kac) *Let $(X, \mathfrak{A}, \mu, T, \mathbb{S})$ be an ergodic system and let $B \in \mathfrak{A}$ satisfy $\mu(B) > 0$. Then*

$$\int \mathrm{R}_B(x)\, d\mu_B = \frac{1}{\mu(B)}.$$

Proof We start by assuming T to be an automorphism. By ergodicity, $X = \bigcup_n \bigcup_{i=0}^{n-1} T^i(\{x \in B : \mathrm{R}_B(x) = n\})$, the union being disjoint and with the sets in the inner union having equal measures (we are using images of measurable sets, so invertibility is essential). Thus

$$1 = \sum_n n \cdot \mu(\{x \in B : \mathrm{R}_B(x) = n\}) = \int_B \mathrm{R}_B(x)\, d\mu.$$

The assertion is now obtained by dividing by $\mu(B)$.

For a noninvertible T we invoke the natural extension $(X', \mathfrak{A}', \mu', T', \mathbb{N}_0)$. Take a set $B \in \mathfrak{A}$ and then let $B' = \pi_0^{-1}(B) \in \mathfrak{A}'$. Notice that for every $x' \in B'$, $\mathrm{R}_{B'}(x') = \mathrm{R}_B(\pi_0(x'))$. Because $\pi\mu' = \mu$, we have

$$\int_B \mathrm{R}_B\, d\mu = \int_{B'} \mathrm{R}_{B'}\, d\mu' = 1.$$

Since $\mu'(B') = \mu(B)$, the integrals remain equal after appropriate normalizations. □

Proof of Theorem 4.3.3 A point $x \in B$ visits B at times $\ldots n_{-2} < n_{-1} < n_0 = 0 < n_1 < n_2 \ldots$, (or just $n_0 = 0 < n_1 < n_2 \ldots$ for $\mathbb{S} = \mathbb{N}_0$), where $n_{i+1} - n_i = \mathrm{R}_B(T_B^i x)$. Let $\mathcal{P}$ be a finite partition of X. The $\mathcal{P}$-name of x can be broken as the concatenation $\cdots x[n_{-1}, n_0)x[n_0, n_1)x[n_1, n_2)\cdots$ (or just $x[n_0, n_1)x[n_1, n_2)\cdots$, for $\mathbb{S} = \mathbb{N}_0$). If we treat the component blocks as symbols, we obtain a symbolic representation in which the shift corresponds to the induced map T_B (see Figure 4.1). The new symbols correspond to atoms of the countable partition $\overline{\mathcal{Q}}$ of B obtained in the following two steps: First, we partition B into the sets $Q_n = \{x \in B : \mathrm{R}_B(x) = n\}$ and we denote this partition by $\mathcal{Q}$, then we refine $\mathcal{Q}$ by applying to each Q_n the partition $\mathcal{P}^n$ (matching the index n). This refined partition $\overline{\mathcal{Q}}$ has finite static entropy. Indeed, by the Kac Theorem, R_B has finite expected value, i.e., $\sum_n n\mu(Q_n) < \infty$. By Fact 1.1.4, $\mathcal{Q}$ has finite entropy. Now

$$H(\overline{\mathcal{Q}}) = H(\overline{\mathcal{Q}}|\mathcal{Q}) + H(\mathcal{Q}) = \sum_n \mu(Q_n) H_{Q_n}(\mathcal{P}^n) + H(\mathcal{Q}).$$

The last term is finite and the sum is dominated by $\log \#\mathcal{P} \sum_n n\mu(Q_n)$, which is finite, as well.

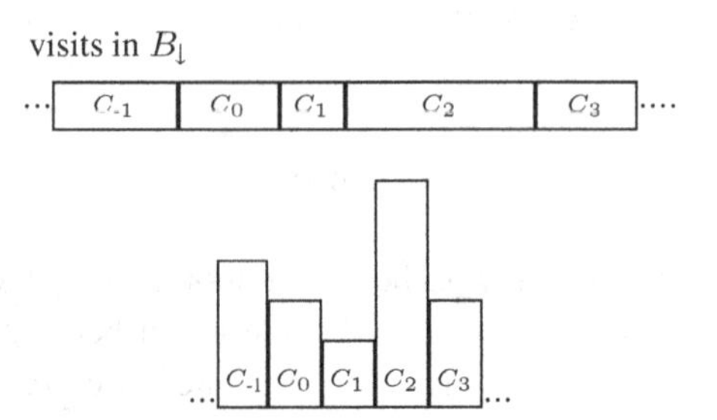

Figure 4.1 Symbolic representation of the process generated by $\mathcal{P}$ (above) and of the process induced on B generated by $\overline{\mathcal{Q}}$ (below).

By the Shannon–McMillan–Breiman Theorem for both the master process and the induced process, there exists in B a cylinder A over $\overline{\mathcal{Q}}$ such that

$$-\tfrac{1}{m}\log(\mu(A)) \approx h(\mu, T, \mathcal{P}) \quad \text{and} \quad -\tfrac{1}{l}\log(\mu_B(A)) \approx h(\mu_B, T_B, \overline{\mathcal{Q}}),$$

where l denotes the length of A (under T_B) and m is the length of A treated as a cylinder over $\mathcal{P}$ (under T). This implies

$$\frac{h(\mu, T, \mathcal{P})}{h(\mu_B, T_B, \overline{\mathcal{Q}})} \approx \frac{l}{m}\frac{\log(\mu(A))}{\log(\mu_B(A))}.$$

By the Ergodic Theorem, we also have

$$\frac{l}{m} \approx \mu(B),$$

while $\frac{\log(\mu(A))}{\log(\mu_B(A))} = 1 + \frac{\log(\mu(B))}{\log(\mu_B(A))}$, which is close to 1 when the conditional measure of A is small. Because the inaccuracies can be made arbitrarily small, we have proved that $h(\mu, T, \mathcal{P}) = \mu(B)h(\mu_B, T_B, \overline{\mathcal{Q}})$. Clearly, with $\mathcal{P}$ ranging over all finite partitions of X, the partitions $\overline{\mathcal{Q}}$ generate in the induced system, so taking the supremum over all $\mathcal{P}$ completes the proof. □

The following two examples teach us caution with handling induced maps, especially when computing the entropy of a system from an induced one.

Example 4.3.5 Consider the system constructed in the following way: We start with any system $(X, \mathfrak{A}, \mu, T, \mathbb{S})$ and we select a set $A \in X$ with $0 < \mu(A) < 1$. We partition A into countably many sets A_n so that $\sum_n n\mu(A_n) < \infty$. Above each A_n we imagine n copies of A_n (so-called "spacers") equipped with the measure copied from A_n. Let Y be the union of X and the spacers with the normalized measure denoted by ν and denote $c = \nu(X)$. The transformation S on Y is defined as follows: each point in A_n goes "vertically up" n times, until it reaches the top spacer, then it returns to Tx. Other points in X are mapped directly to Tx. The dynamics on X advances only "from time to time" with probability c, and at other times it "stays suspended."

Notice that the system induced on X is the same as the original system $(X, \mathfrak{A}, \mu, T, \mathbb{S})$, hence its entropy equals $h(\mu, T)$. By the Abramov Theorem, the entropy of the system $(Y, \mathfrak{B}, \nu, S, \mathbb{S})$ is

$$h(\nu, S) = c \cdot h(\mu, T).$$

Example 4.3.6 Now consider a very similar looking example. In addition to the system $(X, \mathfrak{A}, \mu, T, \mathbb{S})$ take any process on two symbols, i.e., $Z = \{A, B\}^{\mathbb{S}}$ with the shift map σ and some shift-invariant measure ξ. Construct $(Y, \mathfrak{B}, \nu, S, \mathbb{S})$ as the skew product, as follows: $Y = Z \times X$, $\nu = \xi \times \mu$, $S(z, x) = (\sigma(z), T^z(x))$, where $T^z = T$ when $z_0 = A$ and $T^z = \text{id}$ when $z_0 = B$. Let $c = \xi(A)$ (here A denotes the cylinder over the block of length 1). Again, the dynamics on X advances only "from time to time" with probability c, and at other times it "stays suspended." But in the previous example, the partition into the sets with the two types of behavior was determined within X (i.e., was measurable with respect to $\mathfrak{A}$), now it is not – it is determined within Z, hence is independent of $\mathfrak{A}$. The entropy of the system can be computed as follows: The transformation induced on $A \times X$ is the direct product of $(X, \mathfrak{A}, \mu, T, \mathbb{S})$ with the system induced from Z on A. The entropy of the latter induced system is $h(\xi, \sigma)/c$. So, the overall entropy is:

$$h(\nu, S) = c\Big(h(\mu, T) + \frac{h(\xi, \sigma)}{c}\Big) = c \cdot h(\mu, T) + h(\xi, \sigma).$$

We can use the Abramov Theorem to solve the maximization problem for static entropy announced in Section 1.1, after Fact 1.1.4.

Fact 4.3.7 *Fix some $p \in (0, 1]$. Among all countable probability vectors $\mathbf{p} = (p_i)_{i \in \mathbb{N}}$ with expected value $\sum_{i=1}^{\infty} i p_i = \frac{1}{p}$ the maximal entropy $\frac{1}{p} H(p, 1-p)$ is attained on the geometric distribution with parameter p: $p_i = p(1-p)^{i-1}$ (and only on this distribution).*

Proof Take a probability vector $\mathbf{p} = (p_i)_{i \in \mathbb{N}}$ with expected value $\frac{1}{p}$. Consider an independent (symbolic) process (say, bilateral) on countably many symbols a_i whose probabilities are p_i, respectively. Denote by B the corresponding symbolic space, by T_B the shift map, and by μ_B the shift-invariant measure of this process, and let $\mathcal{P}$ be the zero-coordinate partition. Note that $H(\mathbf{p}) = h(\mu_B, T_B, \mathcal{P})$. Now construct a skyscraper X over the base B having exactly $i - 1$ floors above each zero-coordinate cylinder set a_i. The total measure of the skyscraper equals the expected value of $\mathbf{p}$, i.e., $1/p$, so, after normalization, the base B becomes a set of measure p. The transformation T on X is standard; each point goes up until it reaches the top floor, then it returns to the base according to T_B. Ergodicity of T with respect to the normalized measure (denoted μ) follows easily from the ergodicity of μ_B with respect to T_B and the fact that X is the skyscraper over B. Notice that $\mathcal{P}$ coincides with the partition of B determined by the values of the return time variable. Consider the two-set partition $\mathcal{Q}$ of X into B and its complement. The generated process, after inducing on B is precisely the process generated by the partition into level sets of the return time, i.e., by the partition $\mathcal{P}$. By the Abramov formula, its dynamical entropy (which is $H(\mathbf{p})$) equals $\frac{1}{p}h(\mu, T, \mathcal{Q})$. Now, $h(\mu, T, \mathcal{Q}) \leq H(\mathcal{Q}) = H(p, 1-p)$, which proves the desired inequality.

Conversely, note that if we start the construction in the opposite order, with $(X, \mathcal{Q}, \mu, T, \mathbb{Z})$ being the independent process on two symbols $0, 1$, whose probabilites are $1 - p$ and p, respectively, and we define B as the cylinder 1, then the distribution $\mathbf{p}$ of the return time to B is geometric with the parameter p and has expected value $1/p$. The preceding argument now yields $H(\mathbf{p}) = \frac{1}{p}H(p, 1-p)$.

The uniqueness is obvious, because the equality implies $h(\mu, T, \mathcal{Q}) = H(\mathcal{Q})$, i.e., that the process on $\mathcal{Q}$ is independent and then the return times to B have the geometric distribution. □

4.4 Joinings

The notion of a joining is one of the most important in both ergodic theory and topological dynamics. We refer the reader to the book by Eli Glasner [Glasner, 2003] for an extensive study of joinings in ergodic theory. Here we concentrate on a few facts concerning the entropy. A joining of two systems $(X, \mathfrak{A}, \mu, T, \mathbb{S})$ and $(Y, \mathfrak{B}, \nu, S, \mathbb{S})$ is their common extension $(Z, \mathfrak{C}, \xi, R, \mathbb{S})$ with the additional property that the two factors together exhaust it, i.e., that $\mathfrak{C} = \mathfrak{A} \vee \mathfrak{B}$ (where now we mean the lifted sigma-algebras). In this sense, the notion of a joining becomes synonymous with the join of (sub)invariant sigma-algebras. The difference is in our approach: when talking about subinvariant sigma-algebras we usually work within a fixed "master" system, where we consider many possible factors. Now we fix two abstract systems and we treat their joinings (common extensions) as a varying object. The environment where all possible joinings can be found (up to isomorphism) is the product space. Here is the formal definition.

Definition 4.4.1 Let $(X, \mathfrak{A}, \mu, T, \mathbb{S})$ and $(Y, \mathfrak{B}, \nu, S, \mathbb{S})$ be two dynamical systems. By a *joining* of these systems (or equivalently of the measures μ and ν) we will mean the dynamical system $(X \times Y, \mathfrak{A} \otimes \mathfrak{B}, \xi, T \times S, \mathbb{S})$, where ξ is any probability measure on the product sigma-algebra, invariant under $T \times S$, such that its marginals (projections onto X and Y) equal μ and ν, respectively. Because the only variable parameter above is the measure ξ, the name "joining" often refers to this measure rather than the entire system.

We skip the purely measure-theoretic proof of the fact connecting the above two approaches to joinings:

Fact 4.4.2 *Let $(Z, \mathfrak{C}, \xi, R, \mathbb{S})$ be a dynamical system with two subinvariant sigma-algebras $\mathfrak{A}$ and $\mathfrak{B}$ contained in $\mathfrak{C}$. Then the factor corresponding to $\mathfrak{A} \vee \mathfrak{B}$ is a joining of the factors corresponding to $\mathfrak{A}$ and $\mathfrak{B}$.* □

An example of a joining is the *independent joining*, where ξ is the product measure $\mu \times \nu$. Systems admitting only one joining – the product measure, are called *disjoint* (in the sense of Furstenberg) [see Furstenberg, 1967].

We briefly describe joinings over a common factor. Suppose $(X, \mathfrak{A}, \mu, T, \mathbb{S})$ $(Y, \mathfrak{B}, \nu, S, \mathbb{S})$ admit a common factor $(Z', \mathfrak{C}', \xi', R', \mathbb{S})$. Let π_1 and π_2, be the respective factor maps. The product space $X \times Y$ maps onto $Z' \times Z'$ by the map $\pi_1 \times \pi_2$. The preimage of the diagonal is the set $Z = \{(x, y) : \pi_1(x) = \pi_2(y)\}$. Any joining of μ and ν supported by Z is called *a joining over the common factor* $(Z', \mathfrak{C}', \xi', R', \mathbb{S})$. Note that the map $\pi(z) = \pi_1(x)$ (where $z = (x, y)$) factors Z onto Z' and since the same map is obtained as $\pi_2(y)$, the sigma-algebra $\mathfrak{C}$ (lifted to Z) is contained in both (lifted) $\mathfrak{A}$ and (lifted) $\mathfrak{B}$. Joinings over a common factor are exactly such joinings inside which the two systems and their common factor correspond to three (sub)invariant sigma-algebras, $\mathfrak{A}, \mathfrak{B}$ and $\mathfrak{C}' \preccurlyeq \mathfrak{A} \cap \mathfrak{B}$.

Because joinings are simply joins of (sub)invariant sigma-algebras, all the properties listed in Fact 4.1.7 apply. We translate some of them into the language of joinings. The product rule follows directly from (1.6.16) and (1.4.3).

Fact 4.4.3 *Let ξ be a joining of μ and ν. Then*

$$h(\xi) \leq h(\mu) + h(\nu) \quad \textit{and} \quad h(\xi|\mu) \leq h(\nu).$$

For a joining ξ over a common factor ξ', we have

$$h(\xi|\xi') \leq h(\mu|\xi') + h(\nu|\xi'). \tag{4.4.4}$$

For independent joinings, we have the product rule

$$h(\mu \times \nu, T \times S) = h(\mu, T) + h(\nu, S). \tag{4.4.5}$$

□

(Such equality does not imply independence, for example it holds whenever one of the processes has entropy zero.)

More interesting is the following representation statement for $\mathbb{Z}$-actions:

Theorem 4.4.6 *Let $\mathfrak{B}$ be an invariant sigma-algebra in an ergodic system $(X, \mathfrak{A}, \mu, T, \mathbb{Z})$ of finite entropy. Then, for every $\varepsilon > 0$ there exists an invariant sigma-algebra $\mathfrak{C}$ such that $\mathfrak{A} = \mathfrak{B} \vee \mathfrak{C}$ and $h(\mathfrak{C}) \leq h(\mathfrak{A}|\mathfrak{B}) + \varepsilon$.*

By the Krieger Theorem the above fact can be rephrased as follows:

Theorem 4.4.7 *Consider an ergodic process $(X, \mathcal{P}, \mu, T, \mathbb{Z})$ (where $H(\mathcal{P}) < \infty$) and its factor generated by a partition $\mathcal{Q}$ with $H(\mathcal{Q}) < \infty$. Then there exists a partition $\mathcal{R}$ of X, such that $\mathcal{P}^{\mathbb{Z}} = (\mathcal{Q} \vee \mathcal{R})^{\mathbb{Z}}$ (i.e., the process generated by $\mathcal{P}$ is a joining of those generated by $\mathcal{Q}$ and $\mathcal{R}$), and the processes generated by $\mathcal{Q}$ and $\mathcal{R}$ are ε-entropy limit-independent of each other.*

Before the proof we state a corollary concerning a Bernoulli process in the role of the factor (here it is inessential whether it is a Bernoulli shift or a process isomorphic to one, because we are free to choose the generator). In this corollary we replace ε-entropy limit-independence by ε-entropy independence. This is possible, due to Fact 3.1.18. Recall that, by Fact 3.1.3, such independence translates to the genuine (stochastic) ε-independence of the processes. We remark that every invertible system of positive entropy h has Bernoulli factors with entropies ranging in $(0, h]$; this fact is the Sinai Theorem (Theorem 4.5.1).

Corollary 4.4.8 *Let $(X, \mathcal{P}, \mu, T, \mathbb{Z})$ be an ergodic process ($H(\mathcal{P}) < \infty$) and let $(X, \mathcal{Q}, \mu, T, \mathbb{Z})$ (where $\mathcal{Q}$ is $\mathcal{P}^{\mathbb{Z}}$-measurable and has finite entropy) be a Bernoulli factor. Then, for every $\varepsilon > 0$, there exists a partition $\mathcal{R}$ such that $\mathcal{P}^{\mathbb{Z}} = (\mathcal{Q} \vee \mathcal{R})^{\mathbb{Z}}$ and the processes generated by $\mathcal{Q}$ and $\mathcal{R}$ are ε-entropy independent. In other words, every bilateral process is an ε-independent joining of any of its Bernoulli factors (with something).* □

Proof of Theorem 4.4.6 We prove the statement in the version 4.4.7 (for processes). The assertion holds trivially if $h(\mathcal{Q}) = 0$ (then we can take $\mathcal{R} = \mathcal{P}$). We can thus assume that $h(\mathcal{Q}) > 0$. This eliminates periodic processes in the role of the factor. It suffices to isomorphically represent the process $(X, \mathcal{P}, \mu, T, \mathbb{S})$ in a two-row symbolic form, whose first row is the process generated by $\mathcal{Q}$, while the projection onto the second row is some process of entropy not exceeding $h(\mathcal{P}|\mathcal{Q}) + \varepsilon$. Using the Krieger Theorem we can replace $\mathcal{P}$ and $\mathcal{Q}$ by finite partitions which generate the same processes. By replacing, if necessary, $\mathcal{P}$ by $\mathcal{P} \vee \mathcal{Q}$, we can assume that $\mathcal{P} \succcurlyeq \mathcal{Q}$.

The construction relies on the Shannon–McMillan–Breiman Theorem. Find n such that for points x in a set X', whose complement has measure smaller than some preassigned δ (which we specify later), the following holds

$$\mu(B_x^n) \leq 2^{-n(h(\mathcal{Q})-\delta)} \quad \text{and} \quad \mu(A_x^n) \geq 2^{-n(h(\mathcal{P})+\delta)},$$

where A_x^n and B_x^n are the cylinders defined by the inclusions $x \in A_x^n \in \mathcal{P}^n$ and $x \in B_x^n \in \mathcal{Q}^n$. This implies that, within X', each cylinder from $\mathcal{Q}^n$ splits into at most $2^{n(h(\mathcal{P}|\mathcal{Q})+2\delta)}$ cylinders from $\mathcal{P}^n$. Let Λ be a set of cardinality

$\lceil 2^{n(h(\mathcal{P}|\mathcal{Q})+2\delta)} \rceil$, whose elements we will call *labels*. These labels must be different from the labels assigned (earlier) to the elements of $\mathcal{P}$. We will also need an extra symbol (say 0) appearing neither in Λ nor in the labeling of $\mathcal{P}$. Inside each cylinder B^n from $\mathcal{Q}^n$ intersected with X' we can injectively label all cylinders from $\mathcal{P}^n$ using the symbols from Λ. We fix such a labeling (separately in every B^n).

By the Rokhlin Lemma (and aperiodicity), we can find a set F measurable with respect to $\mathcal{Q}^{\mathbb{Z}}$ whose (at least) n/δ preimages by consecutive iterates of T are pairwise disjoint. By ergodicity, almost every point x visits F infinitely many times, with gaps at least n/δ. We will now select some blocks of length n in the $\mathcal{Q}$-name of x and call them *the marked blocks*. We will do that separately between every pair of consecutive visits of the orbit of x to F. Proceeding from a time (say m) of such a visit, we seek for the nearest time $n_1 \geq m$ for which $T^{n_1}x \in X'$. We mark the block $x[n_1, n_1 + n - 1]$. Then we seek for the smallest $n_2 \geq n_1 + n$ such that $T^{n_2}x \in X'$ and we mark the block $x[n_2, n_2 + n - 1]$. We continue in this manner until we reach the next visit to F at time, say, m' (we do not mark the last block if it covers the position m'). Notice that the positions not contained in the marked blocks have, in a typical x, the density at most equal to the measure of $X \setminus X'$ plus n times the measure of F (the sections right before the terminal m'), i.e., 2δ.

We are in a position to define the second row in the two-row representation of X, which, together with the first row (containing the $\mathcal{Q}$-name), completely determines the original $\mathcal{P}$-name of every x. To this end, under each marked block of length n, say $x[k, k + n - 1]$, in the $\mathcal{Q}$-name of x we put the block $[a, 0, 0, 0, \dots, 0]$ (of length n), where $a \in \Lambda$ is the label assigned to the $\mathcal{P}^n$-cylinder containing $T^k x$ (within the corresponding $\mathcal{Q}^n$-cylinder $x[k, k+n-1]$ intersected with X'; notice that since the block $x[k, k+n-1]$ is marked, $T^k x$ does belong to X'). Under the remaining (unmarked) positions in the $\mathcal{Q}$-name of x we put the original labels of the elements of $\mathcal{P}$ appearing at these places in the $\mathcal{P}$-name of x. It is clear that so defined two-row representation allows us to reconstruct the original $\mathcal{P}$-name of x, so we have constructed an isomorphic two-row representation of the process generated by $\mathcal{P}$. It remains to estimate the entropy of the second row. In a typical point we have here: the original labels of $\mathcal{P}$ appearing along a set of density at most 2δ, the symbols from Λ appearing with density at most $\frac{1}{n}$, and zeros. The entropy of such a symbolic system is at most $2\delta \log \#\mathcal{P} + \frac{1}{n} \log \#\Lambda + H(2\delta, \frac{1}{n}, 1 - (2\delta + \frac{1}{n}))$ (the last term comes from dividing the time into the three cases). The middle term does not exceed $h(\mathcal{P}|\mathcal{Q}) + 2\delta$. By choosing δ small (and n large) enough, the entropy of the second row can be made smaller than $h(\mathcal{P}|\mathcal{Q}) + \varepsilon$, as needed. □

Remark 4.4.9 Theorem 4.4.7 fails for actions of $\mathbb{N}_0$, even if both the system and its factor are processes, i.e., have unilateral generators (recall that an $\mathbb{N}_0$-action with finite entropy need not have a unilateral generator). Take the unilateral Bernoulli shift generated by $\mathcal{P}$ and its factor generated by $\mathcal{Q} = T^{-1}(\mathcal{P})$. If $\mathcal{R}$ is such that $(\mathcal{R} \vee \mathcal{Q})^{\mathbb{N}_0}$ (which equals $\mathcal{R}^{\mathbb{N}_0} \vee \mathcal{P}^{\mathbb{N}}$) equals $\mathcal{P}^{\mathbb{N}_0} = \mathcal{P} \vee \mathcal{P}^{\mathbb{N}}$, then, since $\mathcal{P}$ is independent of $\mathcal{P}^{\mathbb{N}}$, it must be that $\mathcal{R}^{\mathbb{N}_0} \succcurlyeq \mathcal{P}$. But since $\mathcal{R}^{\mathbb{N}_0}$ is subinvariant, we get $\mathcal{R}^{\mathbb{N}_0} = (\mathcal{R}^{\mathbb{N}_0})^{\mathbb{N}_0} \succcurlyeq \mathcal{P}^{\mathbb{N}_0}$, i.e., the process generated by $\mathcal{R}$ is the whole process, so it cannot have small entropy.

Question 4.4.10 Our theorem does not cover the case of such $\mathbb{Z}$-actions that both the extension and the factor have infinite entropies yet the conditional entropy is finite. We leave this case open.

4.5 Ornstein Theory*

Classical Ornstein Theory is concerned with bilateral Bernoulli processes (or invertible Bernoulli systems) and their entropy. By a *Bernoulli process* we will understand any process $(X, \mathcal{P}, \mu, T, \mathbb{Z})$ measure-theoretically isomorphic to an independent process (i.e., to a Bernoulli shift). The term *Bernoulli system* is used for a dynamical system $(X, \mathfrak{A}, \mu, T, \mathbb{Z})$ isomorphic to a Bernoulli process. The differences between Bernoulli systems, processes, and shifts are in the choice of a generator: when speaking about a Bernoulli system we do not fix any generator, in a Bernoulli process we fix some (arbitrary) generator, and in a Bernoulli shift we choose a generator for which the process is independent. To make this terminology applicable in the infinite entropy case, we admit Bernoulli shifts generated by partitions with infinite static entropy (which in other situations are usually ruled out).

The central role in the theory is played by three types of statement:

- characterization of Bernoulli processes (systems);
- existence of Bernoulli factors in arbitrary positive entropy systems;
- establishing the Kolmogorov–Sinai entropy as a *complete* invariant in the class of Bernoulli systems.

The first type identifies a number of properties necessary and sufficient for a process $(X, \mathcal{P}, \mu, T, \mathbb{Z})$ to be isomorphic to a Bernoulli shift. As this subject is outside the scope of the book, we only mention the terminology skipping the definitions which require introducing more background. And so, we have *finitely determined* processes, *weakly Bernoulli* processes and *very weakly*

Bernoulli processes. These characterizations allow one to establish some natural examples of dynamical systems to be Bernoulli, for instance ergodic automorphisms of compact groups, geodesic flows on manifolds of negative curvature, mixing Markov shifts, and many more. The reader is referred to the book of Paul Shields [Shields, 1996] for more information on various characterizations of Bernoulli processes. The second type can be summarized in the two theorems quoted below. The first of them belongs to Yakov Sinai [Sinai, 1962] and precedes the Ornstein theory, nevertheless, from today's perspective, it is considered a part of this theory. The second theorem identifies factors of Bernoulli processes [Ornstein, 1970b].

Theorem 4.5.1 (Sinai) *Let $(\Delta^{\mathbb{Z}}, \nu, \sigma, \mathbb{Z})$ denote a bilateral Bernoulli shift, i.e., ν is the product measure $\mathbf{p}^{\mathbb{Z}}$, where $\mathbf{p}$ is a probability distribution on Δ (we admit infinite entropy $H(\mathbf{p})$). Let $(X, \mathcal{P}, \mu, T, \mathbb{Z})$ be any bilateral process with dynamical entropy larger than or equal to $h(\nu)$. Then $(\Delta^{\mathbb{Z}}, \nu, \sigma, \mathbb{Z})$ is a measure-theoretic factor of $(X, \mathcal{P}, \mu, T, \mathbb{Z})$.* □

Theorem 4.5.2 *Any nontrivial measure-theoretic factor of a Bernoulli system is a Bernoulli system.* □

A consequence of the Sinai Theorem is the so-called *weak isomorphism theorem*: two Bernoulli processes of equal entropies are weakly isomorphic, i.e., each is a factor of the other (this does not imply isomorphism). This statement was essentially strengthened by Donald Ornstein in 1970 in what is now known as the *Ornstein Theorem* (representing the third type of statement on our list).

Theorem 4.5.3 (Ornstein) *Two Bernoulli systems of equal (finite or infinite) entropies are isomorphic.* □

Historically, the first proof of an isomorphism between Bernoulli shifts of equal entropies goes back to Lev Mešalkin [Mešalkin, 1959], who considered the independent processes with measures $\{\frac{1}{4}, \frac{1}{4}, \frac{1}{4}, \frac{1}{4}\}^{\mathbb{Z}}$ and $\{\frac{1}{2}, \frac{1}{8}, \frac{1}{8}, \frac{1}{8}, \frac{1}{8}\}^{\mathbb{Z}}$ (note that both have entropy $2\log 2$). Then Ornstein gave a proof for all Bernoulli shifts on finite alphabets [Ornstein, 1970a], which was later generalized by Smorodinsky [Smorodinsky, 1972] for countable alphabets with finite entropy. This was complemented by Ornstein again [Ornstein, 1970c], who established that any two Bernoulli shifts of infinite entropy were isomorphic.

The original (Ornstein's) proof of Theorem 4.5.3 is very complicated and many alternative proofs have occurred thereafter. The most popular one belongs to Mike Keane and Meir Smorodinsky [Keane and Smorodinsky, 1979]. It provides an effective description of an isomorphism between two finite state Bernoulli processes of equal entropies in terms of a *finitary code*, i.e., a map allowing one to determine (almost surely) the zero-coordinate entry in one process by viewing only a finite block around zero in the other process.

Another class of proofs relies on investigating the space of all joinings between a given process and a Bernoulli process and on a category argument for the existence of factor maps (or isomorphisms). A short and elegant exposition can be found in a paper by Robert Burton, Mike Keane and Jacek Serafin [Burton *et al.*, 2000]. They use a unified approach via joinings, which allows stronger versions of both the Krieger and Sinai Theorems to be obtained, while the Ornstein Theorem becomes an immediate consequence of their version of the Sinai Theorem.

We remark that part of the Ornstein Theory for endomorphisms also exists, although it is much less known. The Sinai Theorem in its original version implicitly applies to endomorphisms giving the existence of a unilateral Bernoulli factor (however, we recommend the proof given in [Ornstein and Weiss, 1975]). Andres del Junco proved that two unilateral Bernoulli shifts with equal entropies, each on at least three states, were finitarily weakly isomorphic [del Junco, 1981]. Ornstein Theorem fails for unilateral shifts, for example, the unilateral Bernoulli shifts $\{\frac{1}{4},\frac{1}{4},\frac{1}{4},\frac{1}{4}\}^{\mathbb{N}_0}$ and $\{\frac{1}{2},\frac{1}{8},\frac{1}{8},\frac{1}{8},\frac{1}{8}\}^{\mathbb{N}_0}$, although they have equal entropies, cannot be isomorphic, because almost every point in the first process has 4 preimages, and 5 in the other.

[1]We now reproduce from [Downarowicz and Serafin, in print] a new, fairly short and elementary, proof of the "Residual Sinai Theorem" (and of the resulting "Residual Ornstein Theorem"). In principle, it follows the lines of [Burton *et al.*, 2000], but it avoids any substantial quotations, in particular, invoking explicitly any characterizations of Bernoulli systems; it relies only on standard facts in ergodic theory. The proof below extends that of [Downarowicz and Serafin, in print] beyond the systems with finite entropy.

Consider two ergodic dynamical systems $(X,\mathfrak{A},\mu,T,\mathbb{Z})$ and $(Y,\mathfrak{B},\nu,S,\mathbb{Z})$ realized as topological dynamical systems, i.e., so that X and Y are compact metric spaces, while T and S are homeomorphisms (the existence of such realizations with X and Y zero-dimensional, without requiring minimality or unique ergodicity, is a standard and very easy fact). We let $\mathcal{J}(\mu,\nu)$ denote the set of all joinings of μ and ν. It is nonempty (contains the product measure), compact in the weak-star topology (see Appendix A.2.2) and convex, and its extreme points are precisely the ergodic joinings, whose collection we denote by $\mathcal{J}_{\mathsf{erg}}(\mu,\nu)$. So, $\mathcal{J}_{\mathsf{erg}}(\mu,\nu)$ is of type G_δ relatively in $\mathcal{J}(\mu,\nu)$ (see Appendix A.2.4), hence it is a Polish space, and the Baire Category Theorem

[1] The rest of this section was added at the stage of author's proofs. The author thanks the editors for allowing such an extensive "correction."

applies to its subsets. We remark that, if $\phi : X \to Y$ is a (measure-theoretic) factor map then there is a canonical map from X to $X \times Y$ defined by $x \mapsto (x, \phi x)$. The image of μ by this map is a joining (supported by the graph of ϕ) of μ and ν. We will treat such a joining and the factor map as one object.

In the formulation below, $(X, \mathfrak{A}, \mu, T, \mathbb{Z})$ is realized as a zero-dimensional topological dynamical system with a selected invariant measure, while the Bernoulli shift is equipped with the standard product topology, where the alphabet is either finite or it is the one-point compactification of the positive integers, where the symbol ∞ has measure zero.

Theorem 4.5.4 (Residual Sinai) *Let $(X, \mathfrak{A}, \mu, T, \mathbb{Z})$ be an ergodic system of positive (possibly infinite) entropy $h(\mu)$ and let $(\Delta^{\mathbb{Z}}, \nu, \sigma, \mathbb{Z})$ be a Bernoulli shift on a finite or countable alphabet, of entropy $h(\nu) \le h(\mu)$. Then the set of factors from μ to ν is residual among all ergodic joinings of μ and ν.*

Before we proceed with the proof, let us deduce a strengthening of the Ornstein Theorem:

Theorem 4.5.5 (Residual Ornstein) *Let $(\Lambda^{\mathbb{Z}}, \mu, \sigma, \mathbb{Z})$ and $(\Delta^{\mathbb{Z}}, \nu, \sigma, \mathbb{Z})$ be two Bernoulli shifts on finite or countable alphabets, with equal entropies (finite or infinite). The set of isomorphisms between μ and ν is residual among all ergodic joinings of μ and ν.*

Proof Isomorphisms between these processes are characterized as joinings of μ and ν which are factors in both directions. By the Residual Sinai Theorem, they are members of an intersection of two residual sets in a Polish space, so they also constitute a residual set. □

Proof of Theorem 4.5.4 It is well known (and easy to see) that a joining ξ of μ and ν is a factor from μ to ν if and only if for every $b \in \Delta$ (treated as a cylinder set of length 1) there exists a set $A_b \in \mathfrak{A}$ such that $X \times b = A_b \times \Delta^{\mathbb{Z}}$ mod ξ. We will also use an approximate version of this condition:

Definition 4.5.6 A joining ξ of μ and ν will be called an *ε-factor* (from μ to ν) if for every $b \in \Delta$ there exists a set $A_b \in \mathfrak{A}$ with $\xi((X \times b) \ominus (A_b \times \Delta^{\mathbb{Z}})) < \varepsilon$ ($\ominus$ is the symmetric difference).

Note that the set $\mathcal{F}^{\varepsilon}$ of all ε-factors from μ to ν is open in the weak-star topology: if ξ is an ε-factor, then for each b there exists an $\varepsilon_b < \varepsilon$ such that $\xi((X \times b) \ominus (A_b \times \Delta^{\mathbb{Z}})) < \varepsilon_b$. The characteristic function of A_b can be approximated in $L^1(\mu)$ up to $\varepsilon'_b = (\varepsilon - \varepsilon_b)/2$ by a continuous function f_b, and then $\int |f_b(x) - 1\!\!1_b(y)|\, d\xi < \varepsilon_b + \varepsilon'_b = \varepsilon - \varepsilon'_b$. The latter condition is open in the weak-star topology (as $1\!\!1_b$ is continuous as well) and any joining of μ and ν which satisfies it for these finitely many symbols $b \in \Delta$ for which $\nu(b) \geq \varepsilon$ is easily seen to be an ε-factor.

It is clear that the set $\mathcal{F}$ of factors from μ to ν is a countable intersection of sets of the form $\mathcal{F}^\varepsilon$ and that all factor joinings are ergodic. Passing to the relative topology on $\mathcal{J}_{\mathsf{erg}}(\mu, \nu)$, we get that $\mathcal{F}$ is a countable intersection of open sets $\mathcal{F}^\varepsilon_{\mathsf{erg}}$ of ergodic ε-joinings. We will show that each $\mathcal{F}^\varepsilon_{\mathsf{erg}}$ is also dense in $\mathcal{J}_{\mathsf{erg}}(\mu, \nu)$. The assertion will then follow by the Baire Theorem.

From now on we assume that Δ is finite and that $h(\mu) = h(\nu) = h$ ($h < \infty$); the reduction to this case will be provided at the end. By the Krieger Theorem 4.2.3, $(X, \mathfrak{A}, \mu, T, \mathbb{Z})$ can be represented as a symbolic system $(\Lambda^{\mathbb{Z}}, \mu, \sigma, \mathbb{Z})$, where Λ is finite. The product space now consists of bilateral two-row sequences; the rows are over Λ and Δ, respectively. Given an arbitrary $\xi \in \mathcal{J}_{\mathsf{erg}}(\mu, \nu)$, we need to find an ergodic ε-factor ξ'' nearby. We will do that in three steps. In step 1 we construct a factor joining ξ' from μ to a measure ν' on $\Delta^{\mathbb{Z}}$ with $h(\nu')$ almost as large as h, and such that ξ' is close to ξ in the weak-star topology of measures on $\Lambda^{\mathbb{Z}} \times \Delta^{\mathbb{Z}}$. We do not hope to get $\nu' = \nu$ yet (usually ν' is not even Bernoulli), so by considering ξ' we are driven outside $\mathcal{J}_{\mathsf{erg}}(\mu, \nu)$ into the larger space of shift-invariant measures on $\Lambda^{\mathbb{Z}} \times \Delta^{\mathbb{Z}}$. In step 2 we approximate ξ' (again, in this larger space) by a (not necessarily ergodic) ε-factor $\bar{\zeta}$ from μ to ν, and in step 3 we replace $\bar{\zeta}$ by an ergodic ε-factor ξ'', i.e., by a member of $\mathcal{F}^\varepsilon_{\mathsf{erg}} \subset \mathcal{J}_{\mathsf{erg}}(\mu, \nu)$.

Step 1. Given $\varepsilon > 0$, we let $\varepsilon_0 < \varepsilon/3$ and n_0 be such that shift-invariant measures on $\Lambda^{\mathbb{Z}} \times \Delta^{\mathbb{Z}}$, which agree up to $5\varepsilon_0 + \delta_0$ on all two-row blocks of length n_0, are less than ε apart in the metric d^* compatible with the weak-star topology (consult (7.3.1), the version for symbolic systems). The term δ_0 appearing above is defined as either $4\varepsilon_0/(h(\xi) - h)$ or, if $h(\xi) = h$, as $6\varepsilon_0/h$ (in both cases δ_0 tends to zero with ε_0). Next, by a straightforward application of the Ergodic Theorem (to ξ) and of the Shannon–McMillan–Breiman Theorem (to ξ, μ and ν), we can find an $N_0 > n_0/\varepsilon_0$ and a set $G \subset \Lambda^{\mathbb{Z}} \times \Delta^{\mathbb{Z}}$ being a union of two-row cylinders D of length $N = N_0/(1 - \delta_0)$ (slightly larger than N_0), satisfying $\xi(G) > 1 - \varepsilon_0$ and the properties 1 and 2 listed below, where the following notation is used: D_Λ and D_Δ are the single rows of D while $\bar{D}$, $\bar{D}_\Lambda$ and $\bar{D}_\Delta$ are the prefixes of length N_0 of D, D_Λ and D_Δ, respectively. Recall also that $\mathsf{fr}_A(B)$ is the frequency of a block B in a longer block A.

1 For all two-row blocks B of length n_0 and any $D \subset G$ we have

$$|\mathsf{fr}_{\bar{D}}(B) - \xi(B)| < \varepsilon_0. \tag{4.5.7}$$

2 For any $D \subset G$ we have (where exp is the exponential function to base 2)

$$\exp(-N(h(\xi)+\varepsilon_0)) \leq \xi(D) \leq \exp(-N(h(\xi)-\varepsilon_0)), \quad (4.5.8)$$
$$\exp(-N(h+\varepsilon_0)) \leq \mu(D_\Lambda), \quad (4.5.9)$$
$$\nu(D_\Delta) \leq \exp(-N(h-\varepsilon_0)), \quad (4.5.10)$$

and analogous inequalities hold for $\bar{D}$, $\bar{D}_\Lambda$, $\bar{D}_\Delta$ and N_0.

We now apply to $(\Lambda^{\mathbb{Z}}, \mu, \sigma, \mathbb{Z})$ a variant of the Kakutani–Rokhlin Lemma (we skip the easy proof; see [Downarowicz and Serafin, in print, Lemma 2.5] for a hint): we find a positive measure set M (a "marker"), contained in the projection G_Λ of G, such that the sets $M, \sigma(M), \ldots, \sigma^{N-1}(M)$ are pairwise disjoint and the complement of their union has measure not exceeding $2\varepsilon_0$. Let $\mathcal{D}_0$ be the family of all two-row blocks of the form $(x,y)[0, N-1]$ with $x \in M$. The Ergodic Theorem implies that ξ-almost every (x,y) breaks as a concatenation of the blocks $D \in \mathcal{D}_0$ separated here and there by some insertions of joint density not exceeding $2\varepsilon_0$.

Let $\mathcal{D} \subset \mathcal{D}_0$ denote the subfamily consisting of blocks D contained in G. Dividing the left-hand side of (4.5.9) by the right-hand side of (4.5.8) we obtain that every first row D_Λ ($D \in \mathcal{D}$) splits (as a cylinder set) into at least $\exp(N(h(\xi)-h-2\varepsilon_0))$ blocks $D \in \mathcal{D}$. In other words, every first row D_Λ appears in $\mathcal{D}$ "paired" with at least $\exp(N(h(\xi)-h-2\varepsilon_0))$ different second rows D_Δ. Clearly, the prefix $\bar{D}_\Lambda$ is "paired" with even more blocks D_Δ. Next, dividing the right-hand side of (4.5.10) (the version for $\bar{D}_\Delta$) by the left-hand side of (4.5.8) (the version for $\bar{D}$) we obtain that every $\bar{D}_\Delta$ appears in $\mathcal{D}$ "paired" with at most $\exp(N_0(h(\xi)-h+2\varepsilon_0))$ blocks $\bar{D}_\Lambda$. Clearly, every D_Δ (as a set smaller than $\bar{D}_\Delta$) is "paired" with even less blocks $\bar{D}_\Lambda$. The reader will verify that, since $N_0 = N(1-\delta_0)$ and by the choice of δ_0, in case $h(\xi)-h>0$ we have $N(h(\xi)-h-2\varepsilon_0) \geq N_0(h(\xi)-h+2\varepsilon_0)$. This enables us to apply the Marriage Lemma A.3.5 to the set of prefixes $\bar{D}_\Lambda$, the set of full second rows D_Δ and the relation of being "paired" within $\mathcal{D}$, providing an injection Φ assigning to every $\bar{D}_\Lambda$ a D_Δ "paired" with it.

If $h(\xi) = h$, we must count slightly differently. Dividing the left-hand side of (4.5.9) (the version for $\bar{D}_\Lambda$) by the right-hand side of (4.5.8) we get that every $\bar{D}_\Lambda$ extends to at least $\exp(N(\delta_0 h - 2\varepsilon_0))$ full two-row blocks D. Dividing the right-hand side of (4.5.10) by the left-hand side of (4.5.8) we get that at most $\exp(N(2\varepsilon_0))$ of the D's share a common second row D_Δ. Thus every prefix $\bar{D}_\Lambda$ is "paired" with at least $\exp(N(\delta_0 h - 4\varepsilon_0))$ different second rows D_Δ. The other part of the counting is the same as before and now yields that every second row D_Δ is "paired" with at most $\exp(N_0(2\varepsilon_0)) < \exp(N(2\varepsilon_0))$ prefixes $\bar{D}_\Lambda$. The choice of δ_0 in this case gives that $\delta_0 h - 4\varepsilon_0 \geq 2\varepsilon_0$, and the

Marriage Lemma applies. In either case, it is very important that since $M \subset G_\Lambda$, the set $\{\bar{D}_\Lambda : D \in \mathcal{D}\}$ (the domain of Φ) is the same as $\{\bar{D}_\Lambda : D \in \mathcal{D}_0\}$.

We can now build the factor joining ξ'. In (μ-almost) every $x \in \Lambda^{\mathbb{Z}}$ we first locate the markers (the times of visits to M), then we observe the blocks of length N following these markers. Every such block equals D_Λ for some $D \in \mathcal{D}_0$. Creation of the joining ξ' consists in placing in the second row, below every such D_Λ, the image by Φ of the prefix $\bar{D}_\Lambda$ of D_Λ. At all remaining positions of the second row we place one fixed symbol from the alphabet Δ. Since the second row is uniquely determined by the first one, the joining is a factor map ϕ from μ to some measure ν' supported by $\Delta^{\mathbb{Z}}$.

By the Ergodic Theorem (applied to ξ'), the measure of a two-row block B can be evaluated as the density of its occurrence in the two-row sequence $(x, \phi x)$ belonging to a set of full measure ξ'. We will verify that the distance between ξ and ξ' is smaller than ε. Let B be a two-row block of length n_0. Since the two-row sequence $(x, \phi x)$ is almost entirely covered by the blocks $\bar{D}$ ($D \in \mathcal{D}$), the density of occurrence of B in $(x, \phi x)$ is nearly a weighted average of its frequencies in the blocks $\bar{D}$. By (4.5.7), any such average differs from $\xi(B)$ by less than ε_0. Further, $\xi'(B)$ differs from such an average by at most $\delta_0 + 2\varepsilon_0 + 2n_0/N_0$, where $\delta_0 + 2\varepsilon_0$ estimates the density of the portion of $(x, \phi x)$ not covered by the blocks $\bar{D}$, and $2n_0/N_0 \leq 2\varepsilon_0$ estimates the occurrences of B overlapping with, but not covered by, these blocks. So, $|\xi(B) - \xi'(B)| < 5\varepsilon_0 + \delta_0$, implying $d^*(\xi, \xi') < \varepsilon$.

We now estimate $h(\nu')$ from below. We know that $h(\mu) - h(\nu') = h(\mu|\nu')$ is not larger than the entropy of any process generated by a partition $\mathcal{Q}$ of $\Lambda^{\mathbb{Z}}$ such that $\mathcal{Q} \vee \Delta$ (here Δ stands for $\phi^{-1}(\Delta)$) generates the full process $(\Lambda^{\mathbb{Z}}, \mu, \sigma, \mathbb{Z})$ (see Lemma 4.4.3). Let $\mathcal{Q}$ consist of $F = M \cup \sigma(M) \cup \cdots \cup \sigma^{N_0-1}(M)$ and the sets $a \cap F^c$ ($a \in \Lambda$). To see that $\mathcal{Q}$ joined with Δ generates the full process, note that the symbols "F" occur in the $\mathcal{Q}$-name of (μ-almost) every x in groups of length N_0, the starting places of these groups allow us to locate the markers in x. The blocks of length N_0 following the markers in the Λ-name of x can be recovered from $y = \phi x$ by injectivity of the code Φ. The remaining symbols in the Λ-name of x occur at times when x visits F^c, and these are provided directly by the $\mathcal{Q}$-name of x. Now we estimate

$$h(\mathcal{Q}) \leq H(\mathcal{Q}) \leq H(\mu(F), \mu(F^c)) + \mu(F^c) \log \#\Lambda. \tag{4.5.11}$$

Note that F^c consists of the "upper floors" $\sigma^{N_0}(M), \ldots, \sigma^{N-1}(M)$ of joint measure not exceeding $(N - N_0)/N = \delta_0$ united with the complement of the tower, of measure at most $2\varepsilon_0$. The right-hand side of (4.5.11) tends to zero as $\mu(F^c) \to 0$, so, by an appropriate choice of ε_0 (and the resulting δ_0) we can arrange that $h(\mathcal{Q}) < \varepsilon$ implying $h(\nu') > h - \varepsilon$.

Step 2. We will now approximate the joining ξ' (of μ and ν') by an ε-factor $\bar{\zeta}$ from μ to the Bernoulli measure ν. While ε retains its fixed value, we assume that the first step has been performed for a much smaller parameter ε_1 (in place of ε). And so, ξ' is very close to ξ (less than ε_1 apart). By projecting onto $\Delta^{\mathbb{Z}}$ we get that ν' is very close to ν. This implies, in particular, that the probability vector $\mathbf{q}$ assigned by ν' to the symbols in Δ is very close to the analogous vector $\mathbf{p}$ of ν. By continuity of the static entropy on probability vectors of a fixed finite dimension, we have $H(\nu', \Delta) = H(\mathbf{q}) \approx H(\mathbf{p}) = h(\nu) = h < h(\nu') + \varepsilon_1$, where the last inequality has been arranged in step 1. We have derived that the process associated with the measure ν' is $(\varepsilon_1 + \delta_1)$-entropy independent (recall Definition 3.1.7) for some small δ_1 decreasing to zero with ε_1. This we can write as $H(\nu', \Delta|\Delta^-) \geq H(\nu', \Delta) - \varepsilon_1 - \delta_1$ or, using the formula (1.5.4) (this is an "asterisk section", so we feel free to use disintegration) , as

$$\int H(\nu'_{y^-}, \Delta)\, d\nu'(y^-) \geq H(\nu', \Delta) - \varepsilon_1 - \delta_1,$$

where ν'_{y^-} is the disintegration measure of ν' on the atom y^- of the past Δ^- (such an atom can be identified with a unilateral sequence $y(-\infty, -1])$. This can be further rewritten as

$$\int H(\mathbf{q}_{y^-})\, d\nu'(y^-) \geq H(\mathbf{q}) - \varepsilon_1 - \delta_1,$$

where $\mathbf{q}_{y^-}$ is the probability vector assigned by ν_{y^-} to the symbols in Δ. By the uniformly strict concavity of static entropy (Fact 1.1.11) combined with the rectangle rule (Fact A.3.3), we deduce that (for small enough ε_1), on a set of y's of large measure ν', the vectors $\mathbf{q}_{y^-}$ are close to $\mathbf{q}$, and hence also to $\mathbf{p}$. Close distributions on Δ admit a *maximal coupling* which is almost supported by the diagonal (see Lemma A.3.6). Altogether, given $\varepsilon_2 > 0$, by a good choice of ε_1 (and the associated δ_1) we can assure that, for y's from a set of measure ν' at least $1 - \varepsilon_2$, there is a coupling $\boldsymbol{\xi}_{y^-}$ of $\mathbf{q}_{y^-}$ with $\mathbf{p}$ giving the diagonal in $\Delta \times \Delta$ a mass at least $1 - \varepsilon_2$. Since the assignment $y \mapsto \mathbf{q}_{y^-}$ is ν'-measurable, it is easy to make the assignment $y \mapsto \boldsymbol{\xi}_{y^-}$ measurable as well (for instance, by using each time the particular coupling as in the proof of Lemma A.3.6).

Postponing the specification of ε_2 we will now create the ε-factor $\bar{\zeta}$. We begin by defining a *coupling* ζ of ν' and ν (i.e., a not necessarily shift-invariant measure on $\Delta^{\mathbb{Z}} \times \Delta^{\mathbb{Z}}$ with marginals ν' and ν). We take the projections ν'^- and ν^- of ν' and ν onto the "past" Δ^-, respectively, and on the "joint past" $\Delta^- \times \Delta^-$ we define ζ as $\nu'^- \times \nu^-$. Then, on each atom (y^-, z^-) of the "joint past", we apply on the coordinate zero the maximal coupling $\boldsymbol{\xi}_{y^-}$ in the role of $\zeta_{(y^-, z^-)}$ restricted to $\Delta \times \Delta$. In this manner, we have extended the definition of ζ to $\Delta^{(-\infty,0]} \times \Delta^{(-\infty,0]}$, the sigma-algebra determined by the coordinates from $-\infty$ to 0 in both rows. Inductively, once we have defined the coupling on the sigma-algebra determined by the coordinates from $-\infty$ to $n-1$, for each pair of "semi-trajectories" $(y(-\infty, n-1], z(-\infty, n-1])$, we apply on the coordinate n the coupling $\boldsymbol{\xi}_{(\sigma^n y)^-}$ (note that $(\sigma^n y)^-$ is $y(-\infty, n-1]$ with indexation reset to $(\infty, -1])$. Eventually we have determined a measure ζ on the entire product sigma-algebra, and it is obvious from the construction that its marginals are, respectively, ν' and the product measure $\mathbf{p}^{\mathbb{Z}} = \nu$.

By elementary integration we get, for every $j \geq 0$,

$$\zeta(\{(y,z) : y(j) = z(j)\}) \geq (1-\varepsilon_2)^2 \geq 1 - 2\varepsilon_2.$$

Furthermore, estimating the measure of an intersection by 1 minus the sum of the measures of the complements, we obtain

$$\zeta(\{(y,z) : y[j, j+n-1] = z[j, j+n-1]\}) \geq 1 - 2n\varepsilon_2. \tag{4.5.12}$$

We now specify $\varepsilon_2 = \varepsilon_0/2n_0$, and then, for any $n \leq n_0$, the right-hand side above is at least $1 - \varepsilon_0$.

The coupling ζ can be written as $\int \zeta_y \, d\nu'(y)$, where ζ_y is the disintegration measure of ζ given $y \in \Delta^{\mathbb{Z}}$ on the first coordinate. We can now *lift ζ against* the factor map ϕ as follows: $\zeta' = \int \zeta_{\phi x} \, d\mu(x)$. It is elementary to verify that ζ' is a coupling of μ and ν and that it projects to ζ by the map $\phi \times \mathrm{Id}$ from $\Lambda^{\mathbb{Z}} \times \Delta^{\mathbb{Z}}$ to $\Delta^{\mathbb{Z}} \times \Delta^{\mathbb{Z}}$ (Id denotes the identity map). The inequality (4.5.12) implies that

$$\zeta'(\{(x,z) : \phi x[j, j+n-1] = z[j, j+n-1]\}) \geq 1 - \varepsilon_0,$$

for any $n \leq n_0$, which can be written as

$$\sigma^j(\zeta')(\{(x,z) : \phi x[0, n-1] = z[0, n-1]\}) \geq 1 - \varepsilon_0. \tag{4.5.13}$$

That is to say, after discarding a set of measure $\sigma^j(\zeta')$ at most ε_0, all two-row blocks $(x,z)[0,n-1]$ are the same as $(x, \phi x)[0, n-1]$. For a given x, the latter is the unique block admitted at this place by the joining ξ'. This easily implies that for any two-row cylinder B of length n_0, we have

$$|\sigma^j(\zeta')(B) - \xi'(B)| \leq \varepsilon_0. \tag{4.5.14}$$

Also, applying (4.5.13) for $n = 1$ we get that for every $b \in \Delta$,

$$\sigma^j(\zeta')\big((\Lambda^{\mathbb{Z}} \times b) \ominus (\phi^{-1}(b) \times \Delta^{\mathbb{Z}})\big) \leq \varepsilon_0. \tag{4.5.15}$$

In order to replace the coupling ζ' of μ and ν by a joining, we apply the standard averaging procedure:

$$\bar{\zeta} = \lim_k \frac{1}{n_k} \sum_{j=0}^{n_k-1} \sigma^j(\zeta'),$$

where the limit is in the weak-star topology and the existence of a convergent subsequence follows from compactness. Clearly, the inequalities (4.5.14) and (4.5.15) are maintained by convex combinations. Since (4.5.14) concerns two-row cylinders (and weak inequality), it also passes via the weak-star limit and we obtain $|\bar{\zeta}(B) - \xi'(B)| \leq \varepsilon_0$. We have proved that $d^*(\bar{\zeta}, \xi') < \varepsilon$, hence $d^*(\bar{\zeta}, \xi) < 2\varepsilon$. In order to pass with (4.5.15) over the limit, note that the set $\phi^{-1}(b) \times \Delta^{\mathbb{Z}}$, although not necessarily a finite union of cylinders, depends only on the first row, where all measures in the sequence project to the same measure μ. Now it suffices to approximate this set (say, up to ε_0) by a finite union of cylinders $A \times \Delta^{\mathbb{Z}}$ to get (4.5.15) up to $3\varepsilon_0$ for the limit measure $\bar{\zeta}$. Since $\varepsilon_0 < \varepsilon/3$, $\bar{\zeta}$ is an ε-factor from μ to ν.

Step 3. But $\bar{\zeta}$ need not be ergodic. We will now pick ξ'' from the support of the ergodic decomposition of $\bar{\zeta}$. For that we must, as we already did once, maintaining the value of ε, assume that all above procedures have been performed for a much smaller ε_3. We now invoke continuity of the barycenter map on compact convex sets (see Appendix A.2.3) and upper semicontinuity of the fiber partition of a continuous map (see Appendix A.1.3). The ergodic joining ξ, being extreme, is the barycenter of a unique probability distribution supported by $\mathcal{J}(\mu, \nu)$, namely of $\boldsymbol{\delta}_\xi$ (the point mass at ξ). So, any joining sufficiently close to ξ has its fiber via the barycenter map (in particular the ergodic decomposition) contained in a small neighborhood of $\boldsymbol{\delta}_\xi$. Since the mass assigned by a distribution to an open set is a lower semicontinuous function of the distribution, any distribution sufficiently close to $\boldsymbol{\delta}_\xi$ gives a selected neighborhood of ξ a mass close to 1. Summarizing, we can choose ε_3 so small that the ergodic decomposition distribution of any joining $2\varepsilon_3$-close to ξ (in particular that of $\bar{\zeta}$) is supported mainly (say, with the contribution of "more than half" of its mass) by ergodic joinings situated within the ε-neighborhood of ξ. Further, the fact that $\bar{\zeta}$ is an ε_3-factor means ε_3-smallness of one set (a symmetric difference) per symbol in Δ. By the rectangle rule (Fact A.1.1), each of these sets may be larger than $2\#\Delta \cdot \varepsilon_3$ (which we now declare smaller than ε) only for ergodic components contributing to $\bar{\zeta}$ at most $1/2\#\Delta$ of its mass. Jointly, there is still "at least half" of the components for which ε-smallness of the corresponding set holds for all symbols $b \in \Delta$, i.e., which are ε-factors. Intersecting this "at least half" with the preceding "more than half," we find an ergodic component ξ'' within ε from ξ, and which is an ε-factor. This ends the proof for finite Δ and equal entropies.

We continue with the remaining cases. If $h(\nu) < h(\mu) < \infty$, then we take the product of ν with a Bernoulli shift ν_0 on a finite alphabet satisfying $h(\nu_0) = h(\mu) - h(\nu)$. Now $\nu \times \nu_0$ is a Bernoulli shift of entropy equal to $h(\mu)$. The set $\mathcal{J}_{\mathsf{erg}}(\mu, \nu \times \nu_0)$ maps (by projection) onto $\mathcal{J}_{\mathsf{erg}}(\mu, \nu)$, the projection preserves the property of being a factor and is continuous. Thus the set of factors from μ to ν (which we have shown to be always of type G_δ) is also dense, as the image of a dense set (of factors from μ to $\nu \times \nu_0$) via a continuous surjection.

Let us now assume that Δ is essentially infinite while $(X, \mathfrak{A}, \mu, T, \mathbb{Z})$ is arbitrary (realized on a zero-dimensional space) with $h(\nu) \leq h(\mu) \leq \infty$. Let $\Delta_{(m)}$ be obtained by uniting all but the largest $m - 1$ symbols, where each of the united symbols has measure smaller than ε. Clearly, $\Delta_{(m)}$ generates a Bernoulli shift with entropy strictly smaller than $h(\mu)$. There exists a finite partition $\mathcal{P}$ of X into closed-and-open sets, and an $m \in \mathbb{N}$, such that the ε-closeness in the weak-star distance on $\mathcal{J}(\mu, \nu)$ can be determined by examining only the values assumed by these measures on cylinders over the product partition $\mathcal{P} \otimes \Delta_{(m)}$. We let μ' and ν' denote the projections of μ and ν to the processes generated by $\mathcal{P}$ and $\Delta_{(m)}$. By refining $\mathcal{P}$ (if necessary), we can arrange that $h(\mu') \geq h(\nu')$. Now we argue as follows: a given joining ξ of μ and ν projects to a joining ξ' of μ' and ν'. By the version already proved for finite alphabets, we can find a factor ζ' from μ' to ν' which is ε-close to ξ'. We can lift ζ' to an ergodic joining ζ between μ and ν, and since ζ maintains the values on cylinders over $\mathcal{P} \otimes \Delta_{(m)}$, it is ε-close to ξ. All large symbols in Δ are, modulo ζ, measurable with respect to $\mathfrak{A}$, while the small symbols are smaller than ε, so ζ is an ε-factor. We have proved the density of $\mathcal{F}^{\varepsilon}_{\mathsf{erg}}(\mu, \nu)$.

In the last remaining case Δ is finite while $h(\mu) = \infty$. We argue as above, with Δ in place of $\Delta_{(m)}$. Since we have strict inequality $h(\nu) < h(\mu)$ we will be able to refine the partition $\mathcal{P}$ so that $h(\mu') \geq h(\nu)$. □

Exercises

4.1 Prove the formula (4.1.2) and its conditional analog,

$$h(\mathfrak{A}|\mathfrak{B}) = \lim_k \uparrow h(\mathcal{P}_k|\mathfrak{B}),$$

where $(\mathcal{P}_k)$ is a refining sequence of finite partitions that generate $\mathfrak{A}$.

4.2 Prove that the Kolmogorov–Sinai entropy and conditional entropy given a fixed subinvariant sigma-algebra are affine functions of the measure.

4.3 Prove a conditional version of the power rule Fact 4.1.14.

4.4 Prove Remark 4.2.7 (without using Theorem 4.2.9).

4.5 Let $(Y, \mathfrak{B}, \nu, S, \mathbb{S})$ be a factor of an ergodic system $(X, \mathfrak{A}, \mu, T, \mathbb{S})$ via a factor map $\pi : X \to Y$. Suppose that $h(\mathfrak{A}|\mathfrak{B}) > 0$. Using elementary methods prove that $\pi^{-1}(y)$ is infinite for ν-almost every y.

4.6 In a dynamical system $(X, \mathfrak{A}, \mu, T, \mathbb{S})$ given are two countable partitions $\mathcal{P}$ and $\mathcal{Q}$. Prove the following inequality (compare (1.4.3)):

$$h(\mu, T, \mathcal{P}|\mathcal{Q}) \leq \sum_{B \in \mathcal{Q}} \mu(B) h(\mu_B, T_B, \mathcal{P}).$$

Give an example in which the inequality is strict.

4.7 Illustrate the following phenomenon: We start with an endomorphism $(X, \mathfrak{A}, \mu, T, \mathbb{N}_0)$ of finite entropy. Its natural extension is an automorphism, so we can consider it as a $\mathbb{Z}$-action $(X', \mathfrak{A}', \mu', T', \mathbb{Z})$. It has the same entropy, so, by the Krieger Theorem it has a finite generator, thus is isomorphic to a process $(X', \mathcal{P}', \mu', T', \mathbb{Z})$. The factor of this process determined by the subinvariant sigma-algebra $\mathcal{P}'^{\mathbb{N}_0}$ is an endomorphism (a unilateral process) which has the same same natural extension. Nevertheless, it need not be isomorphic to the the initial system.

4.8 Two sigma-algebras $\mathfrak{A}$ and $\mathfrak{B}$ are *relatively independent* over a third sigma-algebra $\mathfrak{C}$ if for any finite $\mathfrak{A}$-measurable partition $\mathcal{P}$ and any finite $\mathfrak{B}$-measurable partition $\mathcal{Q}$ we have $H(\mathcal{P} \vee \mathcal{Q}|\mathfrak{C}) = H(\mathcal{P}|\mathfrak{C}) + H(\mathcal{Q}|\mathfrak{C})$. A joining of two systems over a common factor is *relatively independent* (over that factor) when the sigma-algebras $\mathfrak{A}$ and $\mathfrak{B}$ corresponding to the systems within the joining system are relatively independent over the sigma-algebra $\mathfrak{C}'$ corresponding to the common factor. Prove that then we have equality in (4.4.4).

4.9 Prove that every automorphism $(X, \mathfrak{A}, \mu, T, \mathbb{Z})$ of finite entropy admits, for every $\varepsilon > 0$, a generator $\mathcal{P}$ which is ε-entropy independent.

4.10 Use the Sinai Theorem to show that every system $(X, \mathfrak{A}, \mu, T, \mathbb{S})$ admits a countable partition $\mathcal{P}$ achieving the full dynamical entropy. Moreover, if the entropy is finite, $\mathcal{P}$ can be chosen finite.

5
The Ergodic Law of Series*

In this chapter we describe a relatively new result – a consequence of positive dynamical entropy of a process. It concerns the behavior of the return time random variables $\mathrm{R}_n(x)$ for large n, the same as treated by the Ornstein–Weiss Return Times Theorem, but in a complementary manner. The theorem has a very interesting interpretation, easy to articulate in a language accessible also to nonspecialists. Yet, as usual on such occasions, one has to be very cautious and not get enticed into pushing the conclusions too far. We begin this chapter with a short historical note concerning the debate on the Law of Series in the colloquial meaning. We explain how the *Ergodic Law of Series* contributes to this debate. Then we pass to the mathematical proof preceded by introducing a number of ergodic-theoretic tools.

5.1 History of the Law of Series

In the colloquial language, a "series" happens when a random event, usually extremely rare, is observed surprisingly often throughout a period of time. Even two repetitions, one shortly after another, are often interpreted as a "series." The Law of Series is the belief that such series happen more often than they should by "pure chance" (whatever that means). This belief is usually associated with another; that there exists some unexplained force or rule behind this "law." A number of idioms, such as "run of good luck" or "run of misfortune," or proverbs like "misfortune never comes alone," exist in nearly all languages, which confirms that people have been noticing this kind of mystery for a long time. The most commonly known examples of "series" are runs of good luck in gambling with the famous case of Charles Wells taking the lead (see e.g. *Charles Wells (gambler)* on Wikipedia).

Serial occurrences of certain types of events is perfectly understandable as a result of physical dependence. For example, volcanic eruptions appear in series during periods of increased tectonic activity. Another good example is when series of people fall ill due to a contagious disease, or very simply, the return of certain motifs in fashion design. The dispute around the Law of Series clearly concerns only such events for which there are no obvious clustering mechanisms, and which are expected to appear completely independently from each other, and yet, they do appear in series. With this restriction the Law of Series belongs to the category of unexplained mysteries, such as synchronicity, telepathy or even Murphy's Law, and is often considered a manifestation of paranormal forces that exist in our world and escape scientific explanation. This might be the reason why, after the first burst of interest, serious scientists and journals refused to get involved in the investigations of this and related topics. Below we review the list of selected scientists involved in the debate.

Kammerer. An Austrian biologist *Paul Kammerer* (1880–1926) was the first scientist to study the Law of Series (law of seriality, in some translations). His book *Das Gesetz der Serie* [Kammerer, 1919] contains many examples from his own life and the lives of his relatives and friends. Richard von Mises in his book [von Mises, 1981] describes that Kammerer conducted many (rather naive) experiments, spending hours in parks noting occurrences of pedestrians with certain features (glasses, umbrellas, etc.), or in shops, noting precise times of arrivals of clients, and the like. Kammerer "discovered" that the number of time intervals (of a fixed length) in which the number of objects under observation agrees with the average is much smaller than the number of intervals, where that number is either zero or larger than the average. This, he argued, provided evidence for clustering. From today's perspective, Kammerer merely noted the perfectly normal spontaneous clustering of signals in the Poisson process. Nevertheless, Kammerer's book attracted some attention from the public, and even from some serious scientists, toward the phenomenon of clustering. Kammerer himself lost authority due to accusations of manipulating his biological experiments (unrelated to our topic), which eventually drove him to suicide.

Pauli and Jung. Examples of series are, in the popular culture, mixed with examples of other kinds of "unbelievable" coincidences. Pioneer theories about coincidences (including series) were postulated not only by Kammerer but also by a noted Swiss psychologist Carl Gustav Jung (1875–1961) and a Nobel prize winner in physics, Austrian, Wolfgang Pauli (1900–1958). They believed that there exist undiscovered physical "attracting" forces driving objects that are alike, or have common features, closer together in time and space (so-called synchronicity) [see e.g. Jung and Pauli, 1955; Jung, 1977].

Moisset. The Law of Series and synchronicity interests the investigators of spirituality, magic and parapsychology. It fascinates with its potential to generate "meaningful coincidences." Frenchman Jean Moisset (born 1924), a self-educated specialist in parapsychology, wrote a number of books on synchronicity, Law of Series, and similar phenomena. He connects the Law of Series with psychokinesis and claims that it is even possible to use it for a purpose [Moisset, 2000].

Skeptics: Weaver, Kruskall, Diaconis and others. In opposition to the theory of synchronicity is the belief, represented by many statisticians, among others by Warren Weaver (closely collaborating with Claude Shannon), that any series, coincidences and the like, appear exclusively by pure chance and that there is no mysterious or unexplained force behind them. People's perception has the tendency to ignore all those sequences of events which do not possess the attribute of being unusual, so that we largely underestimate the size of the sample space, where the "unusual events" are observed. Human memory registers coincidences as more frequent simply because they are more distinctive. This is the "mysterious force" behind synchronicity.

With regard to series of repetitions of identical or similar events, the skeptics' argumentation refers to the effect of spontaneous clustering. For an event, to repeat in time by "pure chance" means to follow a trajectory of a Poisson process. In a typical realization of a Poisson process the distribution of signals along the time axis is far from being uniform; the gaps between signals are sometimes bigger, sometimes smaller. Places where several smaller gaps accumulate (which obviously happens here and there along the time axis) can be interpreted as "spontaneous clusters" of signals. It is nothing but these natural clusters that are being observed and over-interpreted as the mysterious "series." Richard von Mises clearly indicates that it is this kind of "seriality" that has been seen by Kammerer in most of his experiments.

Yet another "cool-minded" explanation of synchronicity (including the Law of Series) asserts that very often events that seem unrelated (hence should appear independently of each other) are in fact strongly related. Many "accidental" coincidences or series of similar events, after taking a closer look at the mechanisms behind them, can be logically explained as "not quite accidental." Ordinary people simply do not bother to seek the logical connection. After all, it is much more exciting to "encounter the paranormal." This point of view is neatly described by Robert Matthews in some of his essays. Criticism of the ubiquitous assumption of independence in various experiments can be found in works of William Kruskal [e.g. Kruskal, 1988]. Percy Diaconis is famous for proving that coin tosses in reality do not represent an i.i.d. process [e.g. Diaconis *et al.*, 2007].

Summarizing, the debate concentrates around the major question:

- *Does there indeed exist a Law of Series or is it just an illusion, a matter of our selective perception or memory?*

So far, this debate has avoided strict scientific language; even its subject is not precisely defined, and it is difficult to imagine appropriate repetitive experiments in a controlled environment. Thus, in this approach, the dispute is probably fated to remain an exchange of speculations.

Law of series in ergodic theory. Below we describe a rigorous approach embedded in ergodic theory. Surprisingly, the study of stochastic processes supports the Law of Series against the skeptic point of view, of course, subject to correct interpretation.

We begin with definitions of *attracting* and *repelling*, the tools allowing us to formalize the subject of study. Using entropy theory we prove that in nondeterministic processes, for events of certain type (long cylinder sets), attracting prevails, while repelling (almost) does not exist – this is exactly how we understand the Ergodic Law of Series.

One has to be very wary about the applicability of this theory in reality. It concerns only events of a specific form (long cylinders) and it gives no quantitative lower bound on the time perspective at which the phenomenon becomes observable. Perhaps it might be applied in genetics, computer science, or in data transmission, where one deals with really long blocks of symbols, but again, with extreme caution. The theory does not explain "runs of good luck," or why "misfortune never comes alone," because such "series" are not repetitions of one and the same long cylinder set. Nonetheless it contributes to the general debate at the philosophic level: Properly understood Law of Series is neither an illusion nor a paranormal phenomenon, but a rigorous mathematical law.

5.2 Attracting and repelling in signal processes

By a *signal process* we will understand a continuous time (also discrete time, when the increment of time is very small) stochastic process $(\mathrm{X}_t)_{t\geq 0}$ defined on a probability space $(\Omega, \mathfrak{A}, \mu)$ and assuming integer values, such that $\mathrm{X}_0 = 0$ a.s., and with nondecreasing and right-continuous trajectories $t \mapsto \mathrm{X}_t(\omega)$. We say that (for given $\omega \in \Omega$) a *signal* (or several simultaneous signals) occurs at time t if the trajectory $\mathrm{X}_t(\omega)$ jumps by a unit (or several units) at t.

Definition 5.2.1 A signal process is *homogeneous* if, for every $t_0 \geq 0$ and every finite collection $0 \leq t_1 < t_2 < \cdots < t_n$, the joint distribution of the increments

$$X_{t_2} - X_{t_1},\ X_{t_3} - X_{t_2},\ \ldots,\ X_{t_n} - X_{t_{n-1}} \tag{5.2.2}$$

is the same as that of

$$X_{t_2+t_0} - X_{t_1+t_0},\ X_{t_3+t_0} - X_{t_2+t_0},\ \ldots,\ X_{t_n+t_0} - X_{t_{n-1}+t_0}.$$

Assume that X_1 has an expected value $\mathsf{E}(X_1) = \lambda \in (0, \infty)$, which we call the *intensity* of the signals. Using homogeneity and a standard divisibility and monotonicity argument, one shows that then $\mathsf{E}(X_t) = t\lambda$ for every $t \in \mathbb{R}$.

With a homogeneous signal process we associate a random variable defined on Ω and called the *waiting time*:

$$W(\omega) = \min\{t : X_t(\omega) \geq 1\}.$$

The most basic example of a homogeneous signal process is the *Poisson process* [see e.g. Feller, 1968]). It is characterized by two properties: 1. the increments as described in (5.2.2) are independent, and 2. jumps by more than one unit have probability zero. These properties imply that the distribution of X_t is the Poisson distribution with the parameter λt, i.e., $P\{X_t = k\} = e^{-\lambda t}\frac{(\lambda t)^k}{k!}$, $k = 0, 1, \ldots$, where $\lambda > 0$ coincides with the intensity. The waiting time in a Poisson process has the exponential distribution with the distribution function

$$\mathsf{F}(t) = 1 - e^{-\lambda t}.$$

The independence between the increments means that the signals arriving before some fixed time do not influence the future signals, i.e., the signals arrive "independently from one another." This pattern of signal arrivals is exactly what is intuitively described as "by pure chance." The Poisson process is the reference point while defining any deviation from the "by pure chance" scheme.

We will consider two such deviations: *attracting* and *repelling*. Intuitively, the signals *attract* each other if they have the tendency to occur in groups (also called *clusters* or *series*), separated by periods of absence. Likewise, the signals *repel* each other if they have the tendency to occur more evenly distributed along the time. We will put this intuition into a rigorous form. It turnes out that these properties depend solely on the distribution of the waiting time.

Definition 5.2.3 We say that the signals *attract* each other from a distance $t > 0$, if

$$\mathsf{F}_W(t) < 1 - e^{-\lambda t}.$$

where F_W is the distribution function of the waiting time W and $\boldsymbol{\lambda}$ is the intensity. Analogously, the signals *repel* each other from a distance t, if

$$\mathsf{F}_\mathrm{W}(t) > 1 - e^{-\boldsymbol{\lambda} t}.$$

The difference $|1 - e^{-\boldsymbol{\lambda} t} - \mathsf{F}_\mathrm{W}(t)|$ is called the *force* of attracting (or repelling) at t.

Why is attracting and repelling defined in this way? Consider the random variable X_t (the number of signals in the time period $(0, t]$). As we know, $\mathsf{E}(\mathrm{X}_t) = \boldsymbol{\lambda} t$. On the other hand, $P\{\mathrm{X}_t > 0\} = P\{\mathrm{W} \le t\} = \mathsf{F}_\mathrm{W}(t)$. Thus

$$\frac{\boldsymbol{\lambda} t}{\mathsf{F}_\mathrm{W}(t)} = \mathsf{E}(\mathrm{X}_t | \mathrm{X}_t > 0)$$

represents the conditional expected number of signals in the interval $(0, t]$ for these ω for which at least one signal occurs there. Attracting from the distance t, as defined above, means that $\mathsf{F}_\mathrm{W}(t)$ is smaller than the analogous distribution function (at t) evaluated for the reference Poisson process. This implies that the above conditional expected number is larger in our process than in the Poisson process (the numerators $\boldsymbol{\lambda} t$ are the same for both processes). This fact can be further expressed as follows: If we observe the signal process for time t and we happen to observe at least one signal, then the expected number of all observed signals is larger than as if they arrived "by pure chance." The first signal "attracts" further signals (within time length t). By homogeneity, the same happens in any interval $(s, s + t]$ of length t, contributing to an increased clustering effect. Repelling is the converse: the first signal lowers the expected number of signals in the observation period, contributing to a decreased clustering, and a more uniform distribution of signals in time, see Figure 5.1.

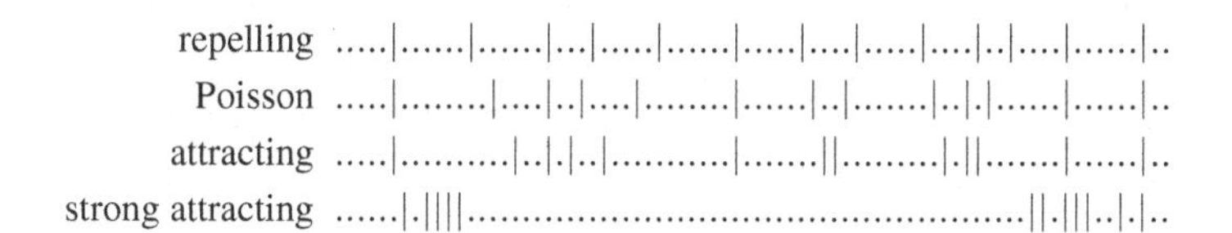

Figure 5.1 The distribution of signals along the time in processes with the same intensity.

The force of attracting can be arbitrarily close to 1, which happens when the distribution function F_W remains near zero until large values of t (this implies attracting from all distances, except very small and very large ones, where marginal repelling can occur). Such $\mathsf{F_W}$ indicates that for most ω the waiting time is very long. In particular, $X_1(\omega) = 0$. Because the intensity $\mathsf{E}(X_1)$ is a fixed number λ, there must be a small part of the space Ω, where many signals arrive within a unit of time. In other words, we observe two types of behavior: long lasting silence observed with very high probability and rarely a swarm of signals. This kind of behavior will be called *strong attracting* (we neglect to put sharp formal bounds on $\mathsf{F_W}$ for this new term).

On the other hand, it is not hard to see that the distribution function $\mathsf{F_W}$ cannot exceed the function $\min\{\lambda t, 1\}$ $(t \geq 0)$, which is attained for the process in which the signals arrive periodically in time (with gaps equal to $1/\lambda$). This is the maximally repelling process, and the maximal force of repelling occurs at $t = 1/\lambda$ and equals e^{-1} (see Figure 5.2 below).

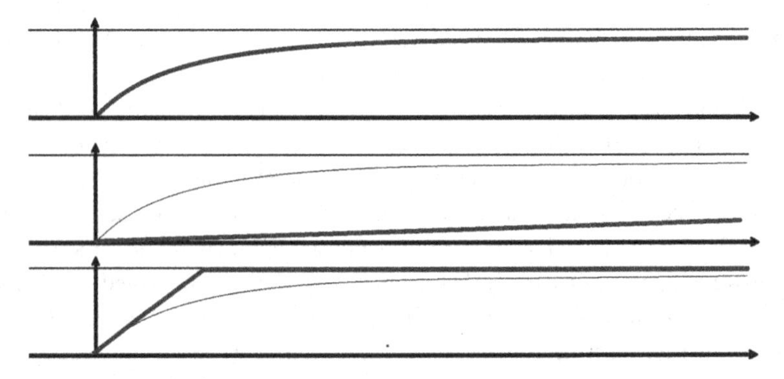

Figure 5.2 The distribution function $\mathsf{F_W}$ in the Poisson, strongly attracting and strongly repelling processes.

If a given process reveals attracting from some distance and repelling from another, the tendency to clustering is not clear and depends on the applied time perspective. However, if there is only attracting (without repelling), then at any time scale we shall see the increased clustering. This type of behavior is our subject of interest:

Definition 5.2.4 A homogeneous signal process *obeys the Law of Series* if

$$\mathsf{F_W}(t) \leq 1 - e^{-\lambda t},$$

for all t, and the two functions are not equal.

In other words, the Law of Series is the conjunction of the following two postulates:

1. There is no repelling from any distance, and
2. there is attracting from at least one distance.

In practice, we agree to accept the presence of some "marginal" repelling with a force much smaller than the force of the existing attracting as shown in the Figure 5.3. Let us explain at this point that the distribution function of the waiting time is always concave (this will become clear e.g. from the integral formula (5.3.2)), hence it cannot be drawn as just any distribution function.

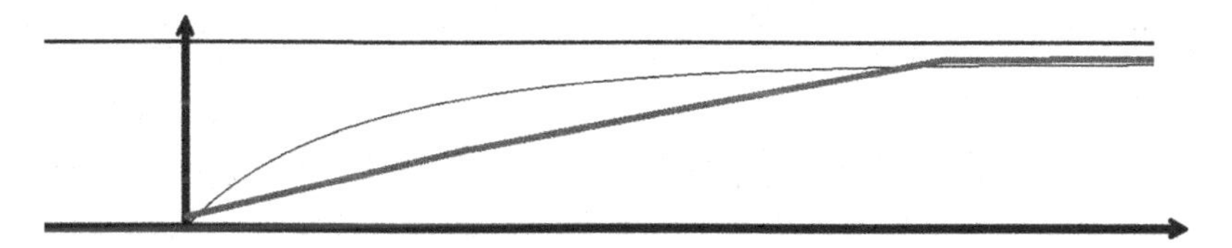

Figure 5.3 The distribution F_W in a process that "nearly" obeys the Law of Series.

5.3 Decay of repelling in positive entropy

In an ergodic nonperiodic process $(X, \mathcal{P}, \mu, T, \mathbb{S})$ (with $\mathcal{P}$ finite) fix a measurable set B and consider the signal process defined on the probability space (X, μ), where signals are occurrences of the event B, i.e.,

$$\mathrm{X}_t(x) = \#\{n \in (0, t] : T^n x \in B\}.$$

This is a *discrete time homogeneous process*; the homogeneity (see Definition 5.2.1) holds for integer t_0. By the Ergodic Theorem, the intensity $\boldsymbol{\lambda}$ equals $\mu(B)$, and $\mathrm{E}(\mathrm{X}_t) = \boldsymbol{\lambda} t$ holds for integer t. Since every nonatomic standard probability space is isomorphic to the unit interval (and the measure in an ergodic nonperiodic process is nonatomic), we can draw B (equipped with the meaure μ_B) as the interval $[0, 1]$ and we can arrange that the return time $\mathrm{R}_B(x)$ (recall (3.1.11) for definition) increases from left to right. Then the graph of the return time R_B coincides with the roof of the skyscraper over B representing the entire space X. Now, the same graph reflected about the diagonal represents the distribution function G_B of R_B.

Notice that there is a relation between G_B and the distribution function F_B of the waiting time W_B in this process; by an elementary consideration of the skyscraper (which we leave to the reader) one easily verifies that, for any integer t,

$$\mathsf{F}_B(t) = \mu(B)\sum_{i\le t}(1 - \mathsf{G}_B(i)) \tag{5.3.1}$$

(thus $\mathsf{G}_B(t) = 1 - \frac{\mathsf{F}_B(t)-\mathsf{F}_B(t-1)}{\mu(B)}$). Both functions are determined by their values at integer arguments. Thus it is completely equivalent whether we study the distribution of the return time variable (defined on B), or of the waiting time variable (defined on X).

The Law of Series in occurrences of the event B can be nicely expressed in terms of the shape of the skyscraper above B; the formula (5.3.1) translates the inequality $\mathsf{F}_B \le 1 - e^{-\boldsymbol{\lambda} t}$ into the following property of the shape of the skyscraper:

- At any point $t \in B$ the area above the graph of $-\frac{\log(1-s)}{\boldsymbol{\lambda}}$ and below the roof function to the left of t (i.e., for $s \le t$) must not exceed the area below the graph of $-\frac{\log(1-s)}{\boldsymbol{\lambda}}$ and above the roof function to the left of t.

This property is explained graphically in Figure 5.4. In particular, the graph

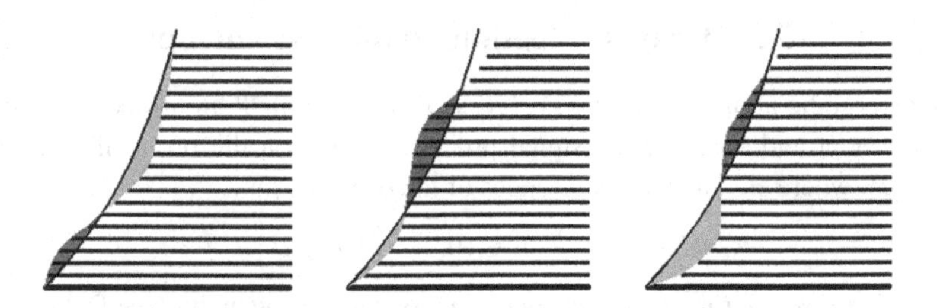

Figure 5.4 The first two skyscrapers are not admitted by the Law of Series, the last one is. The dark-grey area must be smaller than or equal to the light-grey area to the left.

of the roof function must start at zero tangentally to or below the line $s \mapsto s/\boldsymbol{\lambda}$. For instance, the return time cannot be bounded below by a positive value.

Although the Ornstein–Weiss Theorem (Theorem 3.4.1) provides some information about the return time R_B, where B is a "typical" long cylinder, its precise distribution on B, i.e., the shape of the skyscraper over B is by no means captured. Small deviations of the value $\frac{1}{n}\log \mathrm{R}_B(x)$ as x ranges over B (allowed in the Ornstein–Weiss Theorem) mean, for large n, huge deviations of $\log \mathrm{R}_B(x)$ i.e., huge freedom in the proportions between $\mathrm{R}_B(x)$ (hence also of W_B) at different points. In order to be able to compare the distribution function of W_B with the exponential distribution function $1 - e^{-\lambda t}$ we will need completely different tools.

First of all, it will be convenient to change the time unit to $1/\lambda$, i.e., to replace R_B by $\overline{\mathsf{R}}_B = \mu(B)\mathsf{R}_B$ (and W_B by $\overline{\mathsf{W}}_B = \mu(B)\mathsf{W}_B$). We call this step *normalization* because the *normalized return time* $\overline{\mathsf{R}}_B$ has expected value 1 (although the *normalized waiting time* $\overline{\mathsf{W}}_B$ may even have infinite expected value). This trick has many advantages: (1) the signal process in this new time scale has intensity 1, hence the parameter λ disappears from the calculations, (2) the time of the signal process becomes nearly continuous (the increment of time is now $\lambda = \mu(B)$, which is very small), (3) the formula (5.3.1) takes on, for the distribution functions $\overline{\mathsf{F}}_B$ of $\overline{\mathsf{W}}_B$ and $\overline{\mathsf{G}}_B$ of $\overline{\mathsf{R}}_B$, the integral form

$$\overline{\mathsf{F}}_B(t) \approx \int_0^t 1 - \overline{\mathsf{G}}_B(s)\,ds \tag{5.3.2}$$

(up to accuracy $\mu(B)$) and (4) we can compare the behaviors of signal processes obtained for sets B of different measures. In particular, we can see what happens in the limit when B represents longer and longer cylinders.

A rich literature is devoted to the subject of the limit distributions of the normalized return (and waiting) time variables as the lengths of the cylinders grow, in specific types of processes [see Coelho, 2000; Abadi, 2001; Abadi and Galves, 2001; Durand and Maass, 2001; Hirata *et al.*, 1999; Haydn *et al.*, 2005, and the references therein]. Here we will be mainly interested in consequences of the sole assumption of positive entropy. For each x define

$$\mathsf{Rep}_n(x) = \sup_{t\geq 0}(\overline{\mathsf{F}}_{A_x^n}(t) - 1 + e^{-t}),$$

the maximal force of repelling of the cylinder $A_x^n \in \mathcal{P}^n$ containing x. The main theorem of this chapter is this [Downarowicz and Lacroix, 2011]:

Theorem 5.3.3 (The Ergodic Law of Series) *Let $(X, \mathcal{P}, \mu, T, \mathbb{S})$ be an ergodic process with positive entropy, where $\mathcal{P}$ is finite. Then*

$$\mathsf{Rep}_n \underset{n\to\infty}{\longrightarrow} 0 \quad \text{in } L^1(\mu).$$

Because for functions bounded by a common bound the L^1-convergence is the same as the convergence in measure, the above can be equivalently expressed as follows: for every $\varepsilon > 0$ the measure of the union of all blocks of length n, $B \in \mathcal{P}^n$ which repel with force ε, converges to zero as n grows to infinity.

The above theorem asserts that the majority of sufficiently long cylinders reveals almost no repelling, in which they satisfy the first postulate of the Law of Series (phrased next to Definition 5.2.4). Examples show that arbitrarily strong attracting is admitted by such cylinders, (and it is proved that in the majority of processes it indeed occurs; see the last section of this chapter), hence they satisfy also the second postulate.

Question 5.3.4 It is unknown whether Theorem 5.3.3 holds also in the almost everywhere convergence.

5.3.1 The idea of the proof and the basic lemma

Before we turn to the formal proof of Theorem 5.3.3 we would like to fill in some of the details of the idea behind it. First of all, by applying the natural extension, we will assume that the process is invertible, i.e., its symbolic representation is bilateral. We intend to estimate (from above, by $1 - e^{-t} + \varepsilon$) the function $\overline{\mathsf{F}}_B$ for a long cylinder $B \in \mathcal{P}^n$. Instead of B, we can consider a concatenation $BA \in \mathcal{P}^{[-n,r)}$ (i.e., the cylinder set $B \cap A$ with $B \in \mathcal{P}^{-n}, A \in \mathcal{P}^r$), where the "positive" part A has a fixed length r, while we allow the "negative" part B to be (sufficiently) long.

There are two key ingredients leading to the estimation. The first one, contained in Lemma 5.3.11, is the observation that for a fixed typical $B \in \mathcal{P}^{-n}$ the process induced on B (with the conditional measure μ_B) generated by the partition $\mathcal{P}^r$ is not only a β-independent[1] process but also it is "nearly" β-independent of the process on $(B, \mathcal{Q}, \mu_B, T_B, \mathbb{Z})$ generated by the partition $\mathcal{Q}$ depending on the return time (see Figure 5.5). The precise meaning of "nearly" will be explained later.

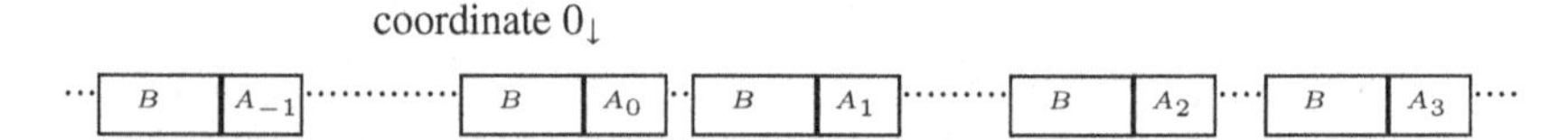

Figure 5.5 The process $\dots A_{-1}A_0A_1A_2\dots$ of blocks of length r following the copies of B is a β-independent process with additional β-independence properties of the positioning of the copies to B.

In addition to the random variables of the absolute and normalized return times R_B and $\overline{\mathrm{R}}_B$ let us introduce the notation for the *kth return time*

$$\mathrm{R}_B^{(k)}(x) = \min\{i : \#\{0 < j \le i : T^j x \in B\} = k\},$$

and of the *normalized kth return time* $\overline{\mathrm{R}}_B^{(k)} = \mu(B)\mathrm{R}_B^{(k)}$ (both defined on B), with $\overline{\mathsf{G}}_B^{(k)}$ always denoting the distribution function of the latter. Because

[1] β-independent means the same as ε-independent; we have changed the letter because throughout this chapter ε is used for the force of repelling.

$\mathsf{R}_B^{(k)}(x) = \mathsf{R}_B(x) + \mathsf{R}_B(T_B(x)) + \cdots + \mathsf{R}_B(T_B^{k-1}(x))$, and T_B preserves μ_B, by the Kac Theorem, the expected value of $\mathsf{R}_B^{(k)}$ is $k/\mu(B)$, and that of $\overline{\mathsf{R}}_B^{(k)}$ equals k.

The above-mentioned β-independences allow us to decompose (with high accuracy) the distribution function $\overline{\mathsf{G}}_{BA}$ of the normalized return time to BA as follows:

$$\overline{\mathsf{G}}_{BA}(t) = \mu_{BA}\{\overline{\mathsf{R}}_{BA} \le t\} = \mu_{BA}\{\mathsf{R}_{BA} \le \tfrac{t}{\mu(BA)}\} =$$
$$\sum_{k\ge1} \mu_{BA}\{\mathsf{R}_A^{(B)} = k, \mathsf{R}_B^{(k)} \le \tfrac{t}{p\mu(B)}\} \approx \sum_{k\ge1} \mu_{BA}\{\mathsf{R}_A^{(B)} = k\} \cdot \mu_B\{\overline{\mathsf{R}}_B^{(k)} \le \tfrac{t}{p}\} \approx$$
$$\sum_{k\ge1} p(1-p)^{k-1} \cdot \overline{\mathsf{G}}_B^{(k)}(\tfrac{t}{p}), \quad (5.3.5)$$

where $\mathsf{R}_A^{(B)}$ denotes the return time of A in the process generated by $\mathcal{P}^r$ induced on B, and $p = \mu_B(A)$. Because this last process is β-independent, the distribution of the kth return time is nearly geometric with parameter p – this explains the occurrence of the term $p(1-p)^{k-1}$ above.

The second key observation is contained in the elementary Lemma 5.3.6 below. We assume, for simplicity, exact equalities in (5.3.5) and (5.3.2). The idea behind this lemma is as follows: The strongest repelling for BA occurs when the repelling of B is the strongest, i.e., when B occurs periodically. But if B does appear periodically, the return time of BA has nearly the geometric distribution, because it is a return time in a β-independent process (only the increment of time is now equal to the constant gap between the occurrences of B). If p is small, this geometric distribution, after normalization, is nearly the exponential law $1-e^{-t}$. Later, in Lemma 5.3.9, we will regulate the smallness of p by the choice of the parameter r.

Lemma 5.3.6 *Fix some $p \in (0,1)$. Let $\overline{\mathsf{G}}^{(k)}$ ($k \ge 1$) be a sequence of distribution functions on $[0,\infty)$ such that the expected value of the distribution associated with $\overline{\mathsf{G}}^{(k)}$ equals k. Define*

$$\overline{\mathsf{G}}(t) = \sum_{k\ge1} p(1-p)^{k-1}\overline{\mathsf{G}}^{(k)}(\tfrac{t}{p}), \quad \textit{and} \quad \overline{\mathsf{F}}(t) = \int_0^t 1 - \overline{\mathsf{G}}(s)\, ds.$$

Then

$$\overline{\mathsf{F}}(t) \le \frac{1}{\log e_p}(1 - e_p^{-t}), \quad \textit{where} \quad e_p = (1-p)^{-\frac{1}{p}}.$$

Proof We have

$$\overline{\mathsf{F}}(t) = \sum_{k\geq 1} p(1-p)^{k-1} \int_0^t 1 - \overline{\mathsf{G}}^{(k)}(\tfrac{s}{p})ds.$$

We know that $\overline{\mathsf{G}}^{(k)}(t) \in [0,1]$ and that $\int_0^\infty 1 - \overline{\mathsf{G}}^{(k)}(s)\,ds = k$ (the expected value). With such constraints, it is the indicator function $1_{[k,\infty)}$ that maximizes the integrals from 0 to t simultaneously for every t (because the "mass" k above the graph is, for such choice of the function $\overline{\mathsf{G}}^{(k)}$, swept maximally to the left). The rest follows by direct calculations:

$$\overline{\mathsf{F}}(t) \leq \sum_{k\geq 1} p(1-p)^{k-1} \int_0^t 1_{[0,k)}(\tfrac{s}{p})\,ds = \int_0^t \sum_{k=\lceil \frac{s}{p}\rceil}^{\infty} p(1-p)^{k-1}ds =$$

$$\int_0^t (1-p)^{\lceil \frac{s}{p}\rceil}ds \leq \frac{(1-p)^{\frac{t}{p}} - 1}{\log(1-p)^{\frac{1}{p}}} = \frac{1-e_p^{-t}}{\log e_p}.$$

□

Notice that the maximizing distribution functions $\overline{\mathsf{G}}_B^{(k)} = 1_{[k,\infty)}$ occur, for the normalized return time of a set B, precisely when B is visited periodically. This is exactly what was said at the beginning of the description of the idea of the proof.

We can now pass to the complete rigorous proof of Theorem 5.3.3.

5.3.2 The proof of Theorem 5.3.3

In course of the proof, we will make frequent use of a certain lengthy condition, abbreviated in the following definition.

Definition 5.3.7 Given a finite partition $\mathcal{P}$ of a space with a probability measure μ and $\delta > 0$, we will say that a property $\Phi(B)$ *holds for* $B \in \mathcal{P}$ *with* μ*-tolerance* δ if

$$\mu\left(\bigcup\{B \in \mathcal{P} : \Phi(B)\}\right) \geq 1 - \delta.$$

We recall an elementary estimate, which has been assigned as Exercise 1.4: For each cell A of a finite partition $\mathcal{P}$ we have

$$H(\mathcal{P}) \leq (1 - \mu(A)) \log \#\mathcal{P} + 1. \tag{5.3.8}$$

We will frequently use the "rectangle rule" (see Fact A.1.2), and we give up recalling it each time; it is accompanied by the appearance of square roots.

Throughout the sequel we assume ergodicity and that the entropy h of the process is positive. We begin our computations with an auxiliary lemma allowing us to assume (by replacing $\mathcal{P}$ by some $\mathcal{P}^r$) that the elements of the "present" partition are small, relatively in most of $B \in \mathcal{P}^{-n}$ and for every n. Note that the Shannon–McMillan–Breiman Theorem is insufficient: for the conditional measure the error term in that theorem depends increasingly on n, which we do not fix.

Lemma 5.3.9 *For each δ there exists an $r \in \mathbb{N}$ such that for every $n \in \mathbb{N}$ the following holds for $B \in \mathcal{P}^{-n}$ with μ-tolerance δ:*

$$\textit{for every } A \in \mathcal{P}^r, \ \ \mu_B(A) \leq \delta.$$

Proof Let α be so small that

$$\sqrt{\alpha} \leq \delta \ \text{ and } \ \tfrac{h-3\sqrt{\alpha}}{h+\alpha} \geq 1 - \tfrac{\delta}{2}$$

and set $\gamma = \alpha / \log \#\mathcal{P}$. Let r be so big that

$$\tfrac{1}{r} \leq \alpha, \ \ \tfrac{1}{r(h+\alpha)} \leq \tfrac{\delta}{2},$$

and that there exists a collection $\overline{\mathcal{P}^r}$ of no more than $2^{r(h+\alpha)} - 1$ elements of $\mathcal{P}^r$ whose joint measure μ exceeds $1-\gamma$ (by the Shannon–McMillan–Breiman Theorem).

Let $\widetilde{\mathcal{P}^r}$ denote the partition into the elements of $\overline{\mathcal{P}^r}$ and the complement of their union, and let $\mathcal{R}$ be the partition into the remaining elements of $\mathcal{P}^r$ and the complement of their union, so that $\mathcal{P}^r = \widetilde{\mathcal{P}^r} \vee \mathcal{R}$. By the power rule (2.4.19) (in the form of Exercise 2.2) we can write $rh = H(\mathcal{P}^r|\mathcal{P}^-)$. Further, for any n we have

$$rh = H(\mathcal{P}^r|\mathcal{P}^-) \leq H(\mathcal{P}^r|\mathcal{P}^{-n}) = H(\widetilde{\mathcal{P}^r} \vee \mathcal{R}|\mathcal{P}^{-n}) =$$
$$H(\widetilde{\mathcal{P}^r}|\mathcal{R} \vee \mathcal{P}^{-n}) + H(\mathcal{R}|\mathcal{P}^{-n}) \leq H(\widetilde{\mathcal{P}^r}|\mathcal{P}^{-n}) + H(\mathcal{R}) \leq$$
$$\sum_{B \in \mathcal{P}^{-n}} \mu(B) H_B(\widetilde{\mathcal{P}^r}) + \gamma r \log \#\mathcal{P} + 1$$

(we have used (5.3.8) for the last passage). After dividing by r we obtain

$$\sum_{B \in \mathcal{P}^{-n}} \mu(B) \tfrac{1}{r} H_B(\widetilde{\mathcal{P}^r}) \geq h - \gamma \log \#\mathcal{P} - \tfrac{1}{r} \geq h - 2\alpha.$$

Because each term $\frac{1}{r} H_B(\widetilde{\mathcal{P}^r})$ is not larger than $\frac{1}{r} \log \#\widetilde{\mathcal{P}^r}$, which was set to be at most $h + \alpha$, we deduce that

$$\tfrac{1}{r} H_B(\widetilde{\mathcal{P}^r}) \geq h - 3\sqrt{\alpha}$$

for $B \in \mathcal{P}^{-n}$ with μ-tolerance $\sqrt{\alpha}$, hence also with μ-tolerance δ. On the other hand, by (5.3.8) again, for any B and $A \in \widetilde{\mathcal{P}^r}$ it holds that

$$H_B(\widetilde{\mathcal{P}^r}) \leq (1 - \mu_B(A)) \log \#\widetilde{\mathcal{P}^r} + 1 \leq (1 - \mu_B(A))r(h + \alpha) + 1.$$

Combining the last two displayed inequalities we establish that, with μ-tolerance δ for $B \in \mathcal{P}^{-n}$ and then for every $A \in \widetilde{\mathcal{P}^r}$,

$$1 - \mu_B(A) \geq \tfrac{h-3\sqrt{\alpha}}{h+\alpha} - \tfrac{1}{r(h+\alpha)} \geq 1 - \delta.$$

So, $\mu_B(A) \leq \delta$. Because $\mathcal{P}^r$ refines $\widetilde{\mathcal{P}^r}$, the elements of $\mathcal{P}^r$ are also not larger than δ. □

We continue the proof with a lemma which could be also deduced from [Rudolph, 1978, Lemma 3], nevertheless we choose to provide a direct proof. For $\alpha > 0$ and $M \in \mathbb{N}$ we define a special periodic subset of $\mathbb{Z}$

$$\mathbb{D}(M, \alpha) = \bigcup_{m \in \mathbb{Z}} [mM + \alpha M, (m+1)M - \alpha M) \cap \mathbb{Z}.$$

Lemma 5.3.10 *For fixed α and r there exists $M_0 \in \mathbb{N}$ such that, for every $M \geq M_0$,*

$$H(\mathcal{P}^r | \mathcal{P}^- \vee \mathcal{P}^{\mathbb{D}(M,\alpha)}) \geq rh - \alpha$$

(see Figure 5.6).

************oo..************.........************.........************.........

Figure 5.6 The circles indicate the coordinates 0 through $r - 1$, the conditioning sigma-algebra is over the coordinates marked by stars, which includes the entire past and part of the future with gaps of size $2\alpha M$ repeated periodically with period M (the first gap is half the size and is partly covered by the circles).

Proof First assume that $r = 1$. Denote also

$$\mathbb{D}'(M, \alpha) = \bigcup_{m \in \mathbb{Z}} [mM + \alpha M, (m+1)M) \cap \mathbb{Z}.$$

Let M be so large that $H(\mathcal{P}^{(1-\alpha)M}) < (1-\alpha)M(h+\gamma)$, where $\gamma = \frac{\alpha^2}{2(1-\alpha)}$. Then, for any $m \geq 1$,

$$H(\mathcal{P}^{\mathbb{D}'(M,\alpha)\cap[0,mM)} | \mathcal{P}^-) \leq H(\mathcal{P}^{\mathbb{D}'(M,\alpha)\cap[0,mM)}) < (1-\alpha)mM(h+\gamma).$$

Because $H(\mathcal{P}^{[0,mM)}|\mathcal{P}^-) = mMh$, the complementary part of entropy must exceed $mMh - (1-\alpha)mM(h+\gamma)$ (which equals $\alpha mM(h-\alpha/2)$), i.e., we have

$$H(\mathcal{P}^{[0,mM)\setminus\mathbb{D}'(M,\alpha)}|\mathcal{P}^- \vee \mathcal{P}^{\mathbb{D}'(M,\alpha)\cap[0,mM)}) > \alpha mM(h - \tfrac{\alpha}{2}).$$

Expressing the last entropy term as a sum over $j \in [0,mM) \setminus \mathbb{D}'(M,\alpha)$ of the conditional entropies of $T^{-j}(\mathcal{P})$ given the sigma-algebra over all coordinates left of j and all coordinates from $\mathbb{D}'(M,\alpha) \cap [0,mM)$ right of j, and because every such term is at most h, we deduce that more than half of these terms reach or exceed $h-\alpha$. So, a term not smaller than $h-\alpha$ occurs for a j within one of the gaps in the left half of $[0,mM)$. Shifting by j we obtain

$$H\big(\mathcal{P}|\mathcal{P}^- \vee T^i(\mathcal{P}^{\mathbb{D}'(M,\alpha)\cap[0,\frac{mM}{2})})\big) \geq h - \alpha,$$

where $i \in [0,\alpha M)$ denotes the relative position of j in the gap. As we increase m, one value i repeats in this role infinitely many times, say, along a subsequence m'. The partitions $\mathcal{P}^- \vee T^i(\mathcal{P}^{\mathbb{D}'(M,\alpha)\cap[0,\frac{m'M}{2})})$ increase with m' to the sigma-algebra $\mathcal{P}^- \vee T^i(\mathcal{P}^{\mathbb{D}'(M,\alpha)})$ and conditional static entropy passes via increasing limits of the conditioning sigma-algebras (see (1.7.13)), hence $H(\mathcal{P}|\mathcal{P}^- \vee T^i(P^{\mathbb{D}'(M,\alpha)})) \geq h-\alpha$. The assertion now follows because $\mathbb{D}(M,\alpha)$ is contained in $\mathbb{D}'(M,\alpha)$ shifted to the left by any $i \in [0,\alpha M)$.

Finally, if $r > 1$, we can simply argue for $\mathcal{P}^r$ replacing $\mathcal{P}$. This will impose that M_0 and M are divisible by r, but it is not hard to see that for large M the argument works without divisibility at a cost of a slight adjustment of α. □

For a block $B \in \mathcal{P}^{-n}$ consider the process $(B, \mathcal{P}^r, \mu_B, T_B, \mathbb{Z})$ generated by $\mathcal{P}^r$ under the induced transformation T_B (and with the measure μ_B). Adapting Theorem 3.1.13 by reversing the time, replacing $\mathcal{P}$ by $\mathcal{P}^r$, and $\mathcal{P}^{[1,n]}$ by $\mathcal{P}^{-n}$, we can see that for a fixed $\beta > 0$ and n large enough, the above is a β-independent process for $B \in \mathcal{P}^{-n}$ with μ-tolerance β. The following lemma shows that it is also "nearly" β-independent of the induced process generated by the return times, more precisely, it is β-independent of the entire past and a finite number of future return times. This fact is crucial and the most difficult item in the proof of Theorem 5.3.3.

Lemma 5.3.11 *For every $\beta > 0$, $r \in \mathbb{N}$ and $K \in \mathbb{N}$ there exists n_0 such that for every $n \geq n_0$, with μ-tolerance β for $B \in \mathcal{P}^{-n}$, with respect to μ_B, $\mathcal{P}^r$ is β-independent of jointly the past $\mathcal{P}^-$ and the first K return times, $\mathrm{R}_B^{(k)}$ ($k \in [1,K]$), to the set B.*

Proof We choose $\gamma > 0$ so that

$$H(\mathcal{P}^r|\mathcal{Q}) \geq H(\mathcal{P}^r) - \gamma \implies \mathcal{P}^r \perp^{\beta} \mathcal{Q}$$

for any partition $\mathcal{Q}$ (see Fact 3.1.3). Let α satisfy

$$0 < \tfrac{2\alpha}{h-\alpha} < 1, \;\; 18K\sqrt{\alpha} < 1, \;\; \sqrt{2\alpha} < \gamma, \;\; K\sqrt[4]{\alpha} < \tfrac{\beta}{2}.$$

Applying the power rule (in the form of Exercise 2.2) and the Ornstein–Weiss Theorem 3.4.1 (in its version for the kth return time, Exercise 3.7), we can find n_0 so large that for every $n \geq n_0$ both $H(\mathcal{P}^r|\mathcal{P}^{-n}) < rh + \alpha$ and that for every $k \in [1, K]$ with μ-tolerance α for $B \in \mathcal{P}^{-n}$ it holds that

$$\mu_B\{2^{n(h-\alpha)} \leq \mathrm{R}_B^{(k)} \leq 2^{n(h+\alpha)}\} > 1 - \alpha.$$

Let $M_0 \geq 2^{n_0(h-\alpha)}$ be so large that the assertion of Lemma 5.3.10 holds for α, r and M_0, and that for every $M \geq M_0$,

$$(M+1)^{1+\frac{2\alpha}{h-\alpha}} < \alpha M^2 \;\text{ and }\; \tfrac{\log(M+1)}{M(h-\alpha)} < \alpha.$$

We can now redefine (enlarge) n_0 and M_0 so that $M_0 = \lfloor 2^{n_0(h-\alpha)} \rfloor$. Similarly, for each $n \geq n_0$ we set $M_n = \lfloor 2^{n(h-\alpha)} \rfloor$. Observe that the interval, where the first K returns of most blocks B of length n may occur (up to probability α), is contained in $[M_n, \alpha M_n^2]$ (because $2^{n(h+\alpha)} \leq (M_n+1)^{1+\frac{2\alpha}{h-\alpha}} < \alpha M_n^2$).

At this point we fix some $n \geq n_0$. The idea is to carefully select an M between M_n and $2M_n$ (hence not smaller than M_0), such that the initial K returns of nearly every block of length n happen most likely inside (with all its n symbols) the set $\mathbb{D}(M, \alpha)$, so that they are "controlled" by the sigma-algebra $\mathcal{P}^{\mathbb{D}(M,\alpha)}$. Let $\alpha' = \alpha + n/M_n$, so that every block of length n overlapping with $\mathbb{D}(M, \alpha')$ is completely covered by $\mathbb{D}(M, \alpha)$. By the definition of M_n, we have $n \leq \frac{\log(M_n+1)}{h-\alpha}$, hence $\frac{n}{M_n} \leq \frac{\log(M_n+1)}{M_n(h-\alpha)}$, which is smaller than α. Thus $\alpha' < 2\alpha$. To define M we will invoke the "triple Fubini Theorem" (in the completely trivial version for discrete measures, i.e., for sums). Fix $k \in [1, K]$ and consider the probability space

$$\mathcal{P}^{-n} \times [M_n, 2M_n] \times \mathbb{N}$$

equipped with the (discrete) measure Prob whose marginal on $\mathcal{P}^{-n} \times [M_n, 2M_n]$ is the product of μ (more precisely, of its projection onto $\mathcal{P}^{-n}$) with the uniform probability distribution on the integers in $[M_n, 2M_n]$, while, for fixed B and M, the conditional measure on the corresponding $\mathbb{N}$-section is the distribution of the random variable $\mathrm{R}_B^{(k)}$. In this space let $\mathbb{D}$ be the set whose $\mathbb{N}$-section for a fixed M (and any fixed B) is the set $\mathbb{D}(M, \alpha')$. We claim that for every $l \in [M_n, \alpha M_n^2] \cap \mathbb{N}$ (and any fixed B) the $[M_n, 2M_n]$-section of $\mathbb{D}$ has measure exceeding $1 - 16\alpha$. This is quite obvious (even for every $l \in [M_n, \infty)$ and with $1 - 15\alpha$) if $[M_n, 2M_n]$ is equipped with the normalized Lebesgue measure. The details of this estimate are provided in the description of Figure 5.7.

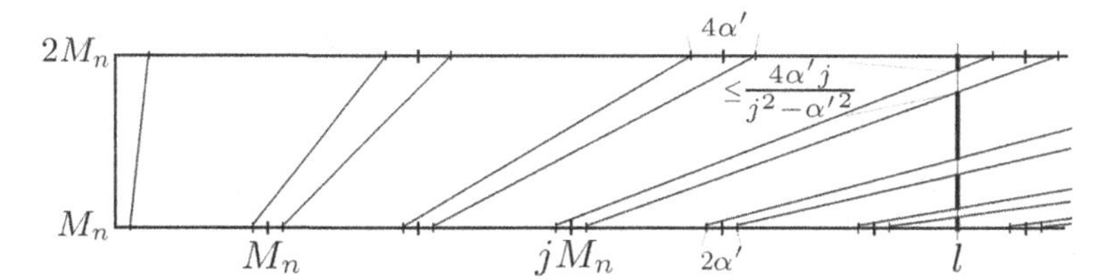

Figure 5.7 The complement of $\mathbb{D}$ splits into thin skew strips shown in the picture. The normalized Lebesgue measure of any vertical section of the jth strip (starting at jM_n with $j \geq 1$) is at most $\frac{4\alpha' j}{j^2-\alpha'^2} \leq \frac{5\alpha'}{j} \leq \frac{10\alpha}{j}$. Each vertical line at $l \geq M_n$ intersects strips with indices $j, j+1, j+2$ up to at most $2j$ (for some j), so the joint measure of the complement of the section of $\mathbb{D}$ does not exceed $10\alpha(\frac{1}{j} + \frac{1}{j+1} + \cdots + \frac{1}{2j}) \leq 15\alpha$.

In the discrete case, however, it might happen that the integers along some $[M_n, 2M_n]$-section often "miss" the section of $\mathbb{D}$ leading to a decreased measure value. (For example, it is easy to see that for $l = (2M_n)!$ the measure of the section of $\mathbb{D}$ is zero.) But because we restrict to $l \leq \alpha M_n^2$, the discretization does not affect the measure of the section of $\mathbb{D}$ by more than α, and the estimate with $1 - 16\alpha$ holds (see Figure 5.8 and its description for details).

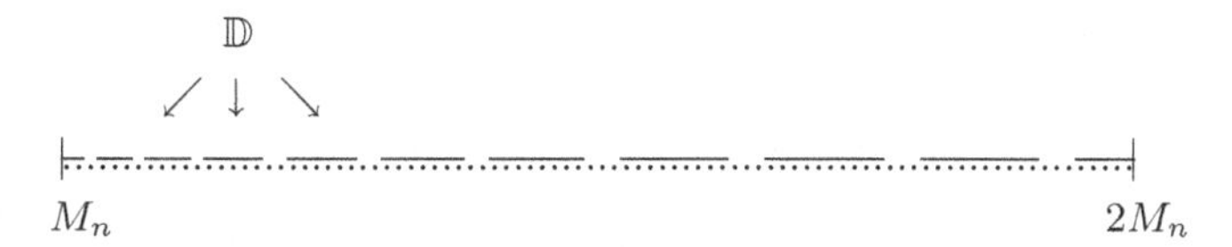

Figure 5.8 The discretization replaces the Lebesgue measure by the uniform measure on M_n integers, thus the measure of any interval can deviate from its Lebesgue measure by at most $1/M_n$. For $l \leq \alpha M_n^2$ the corresponding section of $\mathbb{D}$ (in this picture drawn horizontally) consists of at most αM_n intervals, so its measure can deviate by no more than α.

Taking into account all other inaccuracies (the smaller than α part of $\mathbb{D}$ outside $[M_n, \alpha M_n^2]$ and the smaller than α part of $\mathbb{D}$ projecting onto blocks B which do not obey the Ornstein–Weiss return time estimate) we have proved that

$$\mathsf{Prob}(\mathbb{D}) > 1 - 18\alpha.$$

This implies that for every M from a set of measure at least $1 - 18\sqrt{\alpha}$ the measure of the $(\mathcal{P}^{-n} \times \mathbb{N})$-section of $\mathbb{D}$ is larger than or equal to $1 - \sqrt{\alpha}$. For every such M, with μ-tolerance $\sqrt[4]{\alpha}$ for $B \in \mathcal{P}^{-n}$, the probability μ_B that the

kth repetition of B falls in $\mathbb{D}(M, \alpha')$ (hence with all its n terms inside the set $\mathbb{D}(M, \alpha)$) is at least $1 - \sqrt[4]{\alpha}$.

Because $18K\sqrt{\alpha} < 1$, there exists at least one M for which the above holds for every $k \in [1, K]$. This is our final choice of M which from now on remains fixed. For this M, and for cylinders B chosen with μ-tolerance $K\sqrt[4]{\alpha}$, each of the considered K returns of B with probability (meaning μ_B) $1 - \sqrt[4]{\alpha}$ falls (with all its coordinates) inside $\mathbb{D}(M, \alpha)$. Thus, for such a B, with probability $1 - K\sqrt[4]{\alpha}$ the same holds simultaneously for all K return times. In other words, there is a set $B' \subset B$ of measure μ_B not exceeding $K\sqrt[4]{\alpha}$ outside of which (in B) $\mathrm{R}_B^{(k)} = \tilde{\mathrm{R}}_B^{(k)}$, where $\tilde{\mathrm{R}}_B^{(k)}$ is defined as the time of the kth fully visible inside $\mathbb{D}(M, \alpha)$ return of B. Notice that $\tilde{\mathrm{R}}_B^{(k)}$ is $\mathcal{P}^{\mathbb{D}(M,\alpha)}$-measurable.

Let us go back to our entropy estimates. We have, by Lemma 5.3.10,

$$\sum_{B \in \mathcal{P}^{-n}} \mu(B) H_B(\mathcal{P}^r | \mathcal{P}^- \vee \mathcal{P}^{\mathbb{D}(M,\alpha)}) = H(\mathcal{P}^r | \mathcal{P}^{-n} \vee \mathcal{P}^- \vee \mathcal{P}^{\mathbb{D}(M,\alpha)}) =$$

$$H(\mathcal{P}^r | \mathcal{P}^- \vee \mathcal{P}^{\mathbb{D}(M,\alpha)}) \geq rh - \alpha \geq H(\mathcal{P}^r | \mathcal{P}^{-n}) - 2\alpha =$$

$$\sum_{B \in \mathcal{P}^{-n}} \mu(B) H_B(\mathcal{P}^r) - 2\alpha.$$

Because $H_B(\mathcal{P}^r | \mathcal{P}^- \vee \mathcal{P}^{\mathbb{D}(M,\alpha)}) \leq H_B(\mathcal{P}^r)$ for every B, we deduce that with μ-tolerance $\sqrt{2\alpha}$ for $B \in \mathcal{P}^{-n}$ it must hold that

$$H_B(\mathcal{P}^r | \mathcal{P}^- \vee \mathcal{P}^{\mathbb{D}(M,\alpha)}) \geq H_B(\mathcal{P}^r) - \sqrt{2\alpha} \geq H_B(\mathcal{P}^r) - \gamma.$$

Combining this with the preceding arguments, with μ-tolerance $K\sqrt[4]{\alpha} + \sqrt{2\alpha} < \beta$ for $B \in \mathcal{P}^{-n}$ both the above entropy inequality holds and we have the estimate $\mu_B(B') \leq K\sqrt[4]{\alpha}$. By the choice of γ, we obtain that with respect to μ_B, $\mathcal{P}^r$ is jointly $\frac{\beta}{2}$-independent of the past and the modified return times $\tilde{\mathrm{R}}_B^{(k)}$ ($k \in [1, K]$). Because $\mu_B(B') \leq K\sqrt[4]{\alpha} < \frac{\beta}{2}$, this clearly implies β-independence if each $\tilde{\mathrm{R}}_B^{(k)}$ is replaced by $\mathrm{R}_B^{(k)}$. □

To complete the proof of Theorem 5.3.3 it now remains to put the items together.

Proof of Theorem 5.3.3 Fix an $\varepsilon > 0$. On $[0, \infty)$, the functions

$$g_p(t) = \min\{1, \tfrac{1}{\log e_p}(1 - e_p^{-t}) + pt\},$$

where $e_p = (1-p)^{-\frac{1}{p}}$, decrease uniformly to $1 - e^{-t}$ as $p \to 0^+$. So, let δ be such that $g_\delta(t) \leq 1 - e^{-t} + \varepsilon$ for every t. We also assume that

$$(1 - 2\delta)(1 - \delta) \geq 1 - \varepsilon.$$

Let r be specified by Lemma 5.3.9, so that $\mu_B(A) \le \delta$ for every $n \ge 1$, every $A \in \mathcal{P}^r$ and for $B \in \mathcal{P}^{-n}$ with μ-tolerance δ. On the other hand, once r is fixed, the partition $\mathcal{P}^r$ has at most $(\#\mathcal{P})^r$ elements, so with μ_B-tolerance δ for $A \in \mathcal{P}^r$, $\mu_B(A) \ge \delta(\#\mathcal{P})^{-r}$. Let $\mathcal{A}_B$ be the subfamily of $\mathcal{P}^r$ (depending on B) where this inequality holds. Let K be so large that for any $p \ge \delta(\#\mathcal{P})^{-r}$,

$$\sum_{k=K+1}^{\infty} p(1-p)^k < \tfrac{\delta}{2},$$

and choose $\beta < \delta$ so small that

$$(K^2 + K + 1)\beta < \tfrac{\delta}{2}.$$

The application of Lemma 5.3.11 now provides an n_0 such that for any $n \ge n_0$, with μ-tolerance β for $B \in \mathcal{P}^{-n}$, the process induced on B generated by $\mathcal{P}^r$ has the desired β-independence properties involving the initial K return times of B. So, with tolerance $\delta + \beta < 2\delta$ we have both, the above β-independence and the estimate $\mu_B(A) < \delta$ for every $A \in \mathcal{P}^r$. Let $\mathcal{B}_n$ be the subfamily of $\mathcal{P}^{-n}$ where these two conditions hold. Fix some $n \ge n_0$.

Let us consider a cylinder set $BA \in \mathcal{P}^{[-n,r)}$ where $B \in \mathcal{B}_n$, $A \in \mathcal{A}_B$. The length of BA is $n + r$, which represents an arbitrary integer larger than $n_0 + r$. Notice that the family of such sets BA covers more than $(1-2\delta)(1-\delta) \ge 1-\varepsilon$ of the space.

We will examine the distribution of the normalized first return time for BA. Recall that $\mathrm{R}_A^{(B)}$ denotes the return time to A in the induced process on B, i.e., a variable defined on BA, counting the number of visits to B until the first return to BA. Let $p = \mu_B(A)$ (recall, this is not smaller than $\delta(\#\mathcal{P})^{-r}$). We have

$$\overline{\mathrm{G}}_{BA}(t) = \mu_{BA}\{\overline{\mathrm{R}}_{BA} \le t\} = \mu_{BA}\{\mathrm{R}_{BA} \le \tfrac{t}{\mu(BA)}\} =$$
$$\sum_{k\ge 1} \mu_{BA}\{\mathrm{R}_A^{(B)} = k, \mathrm{R}_B^{(k)} \le \tfrac{t}{p\mu(B)}\}.$$

The kth term of this sum equals

$$\tfrac{1}{p}\mu_B(\{A_k = A\} \cap \{A_{k-1} \ne A\} \cap \cdots \cap \{A_1 \ne A\} \cap \{A_0 = A\} \cap \{\mathrm{R}_B^{(k)} \le \tfrac{t}{p\mu(B)}\}),$$

where A_i is the block of length r following the ith copy of B (the counting starts from 0 at the copy of B positioned at $[-n,-1]$).

By Lemma 5.3.11, for $k \le K$, in this intersection of sets each term is β-independent of the intersection to its right. So, proceeding from the left, we can replace the probabilities of the intersections by products of probabilities, allowing an error of β (multiplied by some number not exceeding 1). Note that the last term equals $\mu_B\{\overline{\mathsf{R}}_B^{(k)} \leq \frac{t}{p}\} = \overline{\mathsf{G}}_B^{(k)}(\frac{t}{p})$. Jointly, the inaccuracy will not exceed $(K+1)\beta$:

$$\left|\mu_{BA}\{\mathsf{R}_A^{(B)} = k, \mathsf{R}_B^{(k)} \leq \tfrac{t}{p\mu(B)}\} - p(1-p)^{k-1}\overline{\mathsf{G}}_B^{(k)}(\tfrac{t}{p})\right| \leq (K+1)\beta.$$

Similarly, we also have $\left|\mu_{BA}\{\mathsf{R}_A^{(B)} = k\} - p(1-p)^{k-1}\right| \leq K\beta$, hence the tail of the series $\mu_{BA}\{\mathsf{R}_A^{(B)} = k\}$ summed over $k \geq K+1$ is smaller than $K^2\beta$ plus the tail of the geometric series $p(1-p)^{k-1}$, which, by the fact that $p \geq \delta(\#\mathcal{P})^{-r}$, is smaller than $\delta/2$. Therefore

$$\overline{\mathsf{G}}_{BA}(t) \approx \sum_{k\geq 1} p(1-p)^{k-1}\overline{\mathsf{G}}_B^{(k)}(\tfrac{t}{p}),$$

up to $(K^2+K+1)\beta + \delta/2 \leq \delta$, uniformly for every t. By the application of the elementary Lemma 5.3.6, $\overline{\mathsf{F}}_{BA}$ satisfies

$$\overline{\mathsf{F}}_{BA}(t) \leq \min\{1, \tfrac{1}{\log e_p}(1-e_p^{-t}) + \delta t\} \leq g_\delta(t) \leq 1 - e^t + \varepsilon$$

(because $p \leq \delta$). We have proved that for our choice of ε and an arbitrary length $m \geq n_0 + r$, with μ-tolerance ε for the cylinders $BA \in \mathcal{P}^m$, the force of repelling (from any distance t) of the visits to *BA* is at most ε. This concludes the proof of Theorem 5.3.3. □

5.4 Typicality of attracting for long cylinders

We have included this section in order to complete the picture of the Ergodic Law of Series. We skip the proofs of the cited theorems, as they do not use entropy. In passing we prove a statement about typicality of positive entropy.

The preceding section provides evidence that in positive entropy processes the occurrences of a selected long cylinder, in principle, do not repel. This corresponds to the first postulate in the interpretation of Definition 5.2.4 of the Law of Series. As to the second postulate (presence of attracting), of course, it cannot be satisfied by long cylinders in all positive entropy processes. For example, in the independent process all long cylinders occur with neither attracting nor repelling. The same holds in sufficiently fast mixing processes (see [Abadi, 2001] or [Hirata *et al.*, 1999]). But such processes are in fact exceptional; in a "typical" process many blocks reveal strong attracting. We know that a fixed dynamical system $(X, \mathfrak{A}, \mu, T, \mathbb{S})$ gives rise to many processes $(X, \mathcal{P}, \mu, T, \mathbb{S})$, each generated by some partition $\mathcal{P}$. We can thus parametrize the processes by the partitions and use the complete metric structure that exists on the space of partitions to determine the meaning of "typicality":

Definition 5.4.1 We say that a property Υ of a process is *typical* (*Rokhlin-typical* or *typical among finitely generated processes*) in a certain class of measure-preserving transformations, if for every $(X, \mathfrak{A}, \mu, T, \mathbb{S})$ in this class, the set of partitions $\mathcal{P}$ such that the generated process $(X, \mathcal{P}, \mu, T, \mathbb{S})$ has the property Υ, is *residual* (i.e., contains a dense G_δ set) in the space $\mathfrak{P}_{\aleph_0}$ of all countable partitions endowed with the metric d_1 (respectively, in the space $\mathfrak{P}_{\mathrm{R}}$ of all countable partitions with finite static entropy endowed with the Rokhlin metric, or, for every natural $m \geq 2$, in the space $\mathfrak{P}_m$ of all partitions into at most m elements endowed with either metric).

Notice that the spaces $\mathfrak{P}_m$ are nowhere dense in both $\mathfrak{P}_{\mathrm{R}}$ and $\mathfrak{P}_{\aleph_0}$, while $\mathfrak{P}_{\mathrm{R}}$ is a first category subset of $\mathfrak{P}_{\aleph_0}$, thus there is no implication between the notions of typicality in the above three senses.

The theorem below captures the typicality of strong attracting:

Theorem 5.4.2 *The following property of a process is typical in all three senses in the class of all ergodic measure-preserving transformations: There exists a set of lengths $N \subset \mathbb{N}$ with upper density 1, such that for every ε and sufficiently large $n \in N$, every block of length n reveals strong attracting (with force $1 - \varepsilon$) of its occurrences.* □

We skip the proof, which can be found in [Downarowicz *et al.*, 2010]. Recall that strong attracting automatically eliminates repelling other than marginal. So, this theorem alone, implies that all blocks of selected lengths obey the Law of Series. Nevertheless, blocks of other lengths may strongly repel (but only if the entropy is zero). Examples of such systems have been built by Paulina Grzegorek and Michal Kupsa [Grzegorek and Kupsa, 2009]. In such systems, in the overall picture, where all long cylinders are taken into account, we can still see a mixed behavior without decisive domination of attracting over repelling.

Remark 5.4.3 The typicality of strong attracting has been recently extended in [Downarowicz *et al.*, 2010] also to events which are not single cylinders but unions of cylinders over blocks differing at a small percentage of coordinates (δ-balls in Hamming distance). An open question remains, whether an analog of Theorem 5.3.3 holds for such events in positive entropy processes.

Now we prove the following fact concerning entropy:

Theorem 5.4.4 *Positive entropy is Rokhlin-typical and typical among finitely generated processes in the class of measure-preserving transformations with positive Kolmogorov–Sinai entropy.*

Proof The set of partitions with finite static and positive dynamical entropy is open (in both $\mathfrak{P}_{\mathrm{R}}$ and $\mathfrak{P}_m$), which follows immediately from continuity of dynamical entropy with respect to the Rokhlin metric (Fact 2.4.13). We need to prove that it is dense. Because the dynamical system has positive entropy, there exists a partition $\mathcal{P}$ which generates positive dynamical entropy. For large enough n, all cells of $\mathcal{P}^n$ are smaller than δ in measure. At least one of these cells, say A, is not measurable with respect to the Pinsker sigma-algebra $\Pi_{\mathcal{P}}$, otherwise the process would have entropy zero. The two-element partition $\mathcal{Q} = \{A, A^c\}$ generates positive dynamical entropy (otherwise, as a zero-entropy factor of the process generated by $\mathcal{P}$ it would have to be measurable with respect to $\Pi_{\mathcal{P}}$). The static entropy $H(\mathcal{Q})$ is at most $H(\delta, 1-\delta)$, which is arbitrarily small, say, smaller than ε. Now, every other partition can be perturbed by at most ε in the Rokhlin metric, to a partition with positive dynamical entropy by simply joining it with the partition $\mathcal{Q}$ (this increases the cardinality; we leave fixing this problem to the reader). □

Combining the above two facts (recall that the intersection of two residual sets is residual) with Theorem 5.3.3 of the preceding section we obtain that in the class of ergodic measure-preserving transformations with positive entropy, in a typical finitely generated process, long cylinders reveal almost no repelling, while many of them reveal strong attracting. This time we do have decisive domination of attracting over repelling. This is the full strength of the Ergodic Law of Series.

The following example shows how the Ergodic Law of Series can manifest itself in reality. Of course, it should be treated with due reserve.

Example 5.4.5 Consider the experiment of randomly generating independent ASCII characters (the monkey typing[2]). In theory this is an independent process hence every possible long block should appear with positive probability and it should reveal neither repelling nor attracting. In reality, however, the independence of the consecutive outcomes is imperfect (there is no perfect physical independence between any events in reality). We can thus consider the process as being generated by a slightly perturbed partition corresponding to the alphabet. Then there are high chances that the process falls in the class of typical processes (of positive entropy) described in the above

[2] This kind of experiment has fascinated people since a long time. The reader can look it up under "The infinite monkey theorem". The idea goes back to Aristotle and, in a more contemporary setting, to Émile Borel and his 1913 essay "Mécanique Statistique et Irréversibilité". The long block in question is usually either Shakespeare's *Hamlet* or the entire book collection of the British Museum put into one long string of letters. Of course, it was merely the possibility of randomly generating such a block that fascinated, not the Law of Series. Hard to believe, but there have been not only attempts to simulate this on a computer, but also experiments with real macaques!

theorems. If so, then majority of blocks will obey the Law of Series and if we focus on one particular long block (say the tex file of this book) it is quite likely that once it occurs it will occur again very "soon" (compared with the expected waiting time, which is unimaginably large).

Part II

Entropy in topological dynamics

6
Topological entropy

6.1 Three definitions of topological entropy

By a *topological dynamical system* we understand the triple $(X, T, \mathbb{S})$, where X is a compact metric space, $T : X \to X$ is continuous and $\mathbb{S} \in \{\mathbb{N}_0, \mathbb{Z}\}$ is the semigroup acting on X via the iterates of T. Of course, $\mathbb{Z}$ is available only when T is a homeomorphism.

Just like in the measure-theoretic case, we are interested in a notion of entropy that captures the complexity of the dynamics, interpreted as the amount of information transmitted by the system per unit of time. Again, the initial state carries complete information about the evolution (forward, or both forward and backward in time, depending on the acting semigroup $\mathbb{S}$), but the observer cannot "read" all this information immediately. Since we do not fix any particular measure, we want to use the metric (or, more generally, the topology) to describe the "amount of information" about the initial state, acquired by the observer in one step (one measurement). A reasonable interpretation relies on the notion of *topological resolution.* Intuitively, resolution is a parameter measuring the ability of the observer to distinguish between points. A resolution is topological, when this ability agrees with the topological structure of the space. The simplest such resolution is based on the metric and a positive number ε: two points are "indistinguishable" if they are less than ε apart. Another way to define a topological resolution (applicable in all topological spaces) refers to an open cover of X. Points cannot be distinguished when they belong to a common cell of the cover.

By compactness, the observer is able to "see" only some finite number N of "classes of indistinguishability" and classify the current state of the system to one of them. The logarithm to base 2 of N roughly corresponds to the number of binary questions, answering which is equivalent to what the observer has learned, i.e., to the amount of acquired information. The static entropy, instead of an expectation (which requires a measure), will now be replaced by the supremum over the space of this information. The rest is done just like in the measure-theoretic case; we define the topological dynamical entropy with respect to a resolution as the average (along the time) information acquired per step. Finally we pass to the supremum as the resolution refines. Multiple ways of understanding topological resolution lead to multiple ways of defining topological entropy.

Notice that "indistinguishability" is not an equivalence relation; the "classes" often overlap without being equal. This makes the interpretation of a topological resolution a bit fuzzy and its usage in rigorous computations – rather complicated.[1] This difficulty has not occurred in measurable dynamics, where "classes of indistinguishability" were simply the cells of a partition. Only in zero-dimensional topological spaces do we have the comfort that arbitrarily fine topological resolutions can be defined as partitions. Zero-dimensional spaces provide a bridge between measure-theoretic and topological dynamics, and we will learn later how this bridge is created.

6.1.1 The metric definition via separated orbits

This and the next section describe topological entropy in the sense of Dinaburg and Bowen, using the metric [comp. Dinaburg, 1970; Bowen, 1971]. Let X be endowed with a metric d. For $n \in \mathbb{N}$, by d^n we will mean the metric

$$d^n(x, y) = \max\{d(T^i x, T^i y) : i = 0, \dots, n-1\}.$$

Of course $d^1 = d$, $d^{n+1} \geq d^n$ for each natural n, and, by compactness of X, all these metrics are pairwise uniformly equivalent.

Following the concept of distinguishability in the resolution determined by a distance $\varepsilon > 0$, a set $F \subset X$ is said to be (n, ε)-*separated* if the distances between distinct points of F in the metric d^n are at least ε:

$$\forall_{x,y \in F}\ d^n(x, y) \geq \varepsilon.$$

By compactness, the cardinalities of all (n, ε)-separated sets in X are finite and bounded. By $s(n, \varepsilon)$ we will denote the maximal cardinality of an (n, ε)-separated set:

$$s(n, \varepsilon) = \max\{\#F : F \text{ is } (n, \varepsilon)\text{-separated}\}.$$

[1] The phenomenon that "indistinguishability" is not an equivalence relation leads to many misunderstandings (and abuses) in our everyday life. A typical example is that parents do not notice how their children grow day after day. They feel surprised when some relatives exclaim how much they have grown! Another example: Some firms try to increase profits by lowering the quality of their products. But they do it so gradually that the regular clients do not notice. Casual clients do.

It is clear that $s(n, \varepsilon)$ (hence also the first two terms defined below) depends decreasingly on ε. So, we can apply the general scheme:

$$\mathbf{H}_1(n,\varepsilon) = \log s(n,\varepsilon),$$
$$\mathbf{h}_1(T,\varepsilon) = \limsup_{n\to\infty} \tfrac{1}{n}\mathbf{H}_1(n,\varepsilon) \quad \text{(alternatively } \liminf\text{)},$$
$$\mathbf{h}_1(T) = \lim_{\varepsilon\to 0} \uparrow \mathbf{h}_1(T,\varepsilon).$$

We will soon explain why the choice between lim sup and lim inf is inessential.

For the interpretation, suppose we observe the system through a device whose resolution is determined by the distance ε. Then we can distinguish between two n-orbits $(x, Tx, \ldots, T^{n-1}x)$ and $(y, Ty, \ldots, T^{n-1}y)$ if and only if for at least one $i \in \{0, \ldots, n-1\}$ the points $T^i x, T^i y$ can be distinguished (which means their distance is at least ε), i.e., when the points x, y are (n, ε)-separated. Thus, $s(n, \varepsilon)$ is the *maximal* number of pairwise distinguishable n-orbits that exist in the system. The term $\mathbf{h}_1(T, \varepsilon)$ is hence the *rate of the exponential growth of the number of ε-distinguishable n-orbits.*

6.1.2 The metric definition via spanning orbits

Let $B^n(x, \varepsilon)$ denote the ε-ball around x in the metric d^n. We will call it the *(n, ε)-ball* (around x). For $n = 1$ we will simply write $B(x, \varepsilon)$. Notice that

$$B^n(x,\varepsilon) = \bigcap_{i=0}^{n-1} T^{-i}(B(T^i x, \varepsilon)).$$

A set F is called *(n, ε)-spanning* if it intersects every (n, ε)-ball in X. Since X is totally bounded, there exists a finite (n, ε)-spanning set in X. The number $r(n, \varepsilon)$ is defined as the smallest cardinality of an (n, ε)-spanning set:

$$r(n,\varepsilon) = \min\{\#F : F \text{ is } (n,\varepsilon)\text{-spanning}\}.$$

The number $r(n, \varepsilon)$ can be interpreted as the minimal number of n-orbits representing up to indistinguishability all possible n-orbits (easy examples show that this is not the same as $s(n, \varepsilon)$). Again, the dependence on ε is decreasing. Then we follow the scheme:

$$\mathbf{H}_2(n,\varepsilon) = \log r(n,\varepsilon),$$
$$\mathbf{h}_2(T,\varepsilon) = \limsup_{n\to\infty} \tfrac{1}{n}\mathbf{H}_2(n,\varepsilon) \quad \text{(or } \liminf\text{)},$$
$$\mathbf{h}_2(T) = \lim_{\varepsilon\to 0} \uparrow \mathbf{h}_2(T,\varepsilon).$$

6.1.3 The topological definition via covers

By a *cover* $\mathcal{U}$ we will understand an arbitrary family of open sets whose union is X. A cover $\mathcal{V}$ is a *refinement* of another cover $\mathcal{U}$, which we write as $\mathcal{V} \succcurlyeq \mathcal{U}$, if every element of $\mathcal{V}$ is contained in an element of $\mathcal{U}$. Unlike for partitions, it no longer holds that each element of $\mathcal{U}$ is then a union of some elements of $\mathcal{V}$. A *join* of two covers $\mathcal{U} \vee \mathcal{V}$ is defined the same way as it was done for partitions (compare (1.2.1)):

$$\mathcal{U} \vee \mathcal{V} = \{U \cap V : U \in \mathcal{U}, V \in \mathcal{V}\}.$$

Clearly, such a join refines both $\mathcal{U}$ and $\mathcal{V}$. A *subcover* of a cover $\mathcal{U}$ is any subfamily $\mathcal{V} \subset \mathcal{U}$ which is also a cover. Note that a subcover of $\mathcal{U}$ is its refinement, which might be a bit counterintuitive, because we are accustomed to thinking of a refinement (of a partition) as having larger cardinality. By compactness, every cover has a finite subcover. For a cover $\mathcal{U}$ we let $N(\mathcal{U})$ denote the minimal cardinality of a subcover. A subcover of this cardinality will be referred to as *optimal*.

Let $T : X \to X$ be a continuous transformation. By continuity of T, if $\mathcal{U}$ is a cover, then $T^{-1}(\mathcal{U}) = \{T^{-1}(U) : U \in \mathcal{U}\}$ is also a cover. The map T^{-1} acting on covers preserves the relation $\succcurlyeq$ and that of being a subcover. In particular, $N(T^{-1}(\mathcal{U})) \leq N(\mathcal{U})$.

Like for partitions, but only for a finite set $\mathbb{D} \in \mathbb{S}$, we will denote

$$\mathcal{U}^{\mathbb{D}} = \bigvee_{i \in \mathbb{D}} T^{-i}(\mathcal{U}),$$

and we will abbreviate $\mathcal{U}^{[0,n)}$ as $\mathcal{U}^n$. It is easily verified that

$$\mathcal{U} \succcurlyeq \mathcal{V} \implies \mathcal{U}^n \succcurlyeq \mathcal{V}^n. \tag{6.1.1}$$

In this subsection we introduce the topological entropy in the sense of Adler, Konheim and McAndrew [comp. Adler *et al.*, 1965]. It relies on treating a cover $\mathcal{U}$ as a topological resolution. Open covers form a directed family with respect to the partial order $\succcurlyeq$, hence can be used to index nets (see Appendix A.1.3).

We follow the scheme:

$$\begin{aligned}
\mathbf{H}(\mathcal{U}) &= \log N(\mathcal{U}),\\
\mathbf{h}(T,\mathcal{U}) &= \lim_n \tfrac{1}{n}\mathbf{H}(\mathcal{U}^n),\\
\mathbf{h}(T) &= \lim_{\mathcal{U}} \uparrow \mathbf{h}(T,\mathcal{U}).
\end{aligned}$$

The interpretation of this definition is similar to the preceding one: $N(\mathcal{U}^n)$ is the minimal number of n-orbits which represent up to distinguishability all n-orbits in the system. The monotonicity of the last limit (over the net of all covers) follows from (6.1.1) and the first observation below, while the second limit (in n) exists by subadditivity stated in the last observation:

Fact 6.1.2 *For open covers $\mathcal{U}$ and $\mathcal{V}$, we have*

$$\mathcal{U} \succcurlyeq \mathcal{V} \implies N(\mathcal{U}) \geq N(\mathcal{V}), \tag{6.1.3}$$

$$N(\mathcal{U} \vee \mathcal{U}) = N(\mathcal{U}), \tag{6.1.4}$$

$$N(\mathcal{U} \vee \mathcal{V}) \leq N(\mathcal{U})N(\mathcal{V}), \tag{6.1.5}$$

$$\text{the sequence } \mathbf{H}(\mathcal{U}^n) \text{ is subadditive.} \tag{6.1.6}$$

Proof If $\mathcal{U}_o$ is an optimal subcover of $\mathcal{U}$, then for each $U \in \mathcal{U}_o$ there exists $V_U \in \mathcal{V}$ such that $U \subset V_U$. Let $\mathcal{V}_o = \{V_U : U \in \mathcal{U}_o\}$. This is a subcover of $\mathcal{V}$ and its cardinality is not larger than that of $\mathcal{U}_o$. This proves (6.1.3). Although the join $\mathcal{U} \vee \mathcal{U}$ usually does not equal $\mathcal{U}$, it is elementary to see that it both refines and is refined by $\mathcal{U}$. Now (6.1.4) follows from (6.1.3). For (6.1.5) let $\mathcal{U}_o$ be an optimal subcover of $\mathcal{U}$, likewise, let $\mathcal{V}_o$ be an optimal subcover of $\mathcal{V}$. Then $\mathcal{U}_o \vee \mathcal{V}_o$ is a subcover of $\mathcal{U} \vee \mathcal{V}$ (perhaps not even optimal) and its cardinality is at most the product of the cardinalities of $\mathcal{U}_o$ and $\mathcal{V}_o$. Subadditivity is an immediate consequence of (6.1.5) and the inequality $N(T^{-1}(\mathcal{U})) \leq N(\mathcal{U})$, noted earlier. □

Remark 6.1.7 In metric spaces we can always find a sequence $(\mathcal{U}_k)$ of open covers which eventually refine every cover. Replacing $\mathcal{U}_k$ by $\bigvee_{i=1}^{k} \mathcal{U}_i$ we can have a sequence with the additional property $\mathcal{U}_{k+1} \succcurlyeq \mathcal{U}_k$ for all k. A sequence of covers $(\mathcal{U}_k)$ with both the above properties will be called *refining* or we will say that the sequence $(\mathcal{U}_k)$ *refines in* X. The limit over the net of all covers in the last definition can be replaced by the limit over a refining sequence of covers. Still, in some situations, it will be better to use nets anyway.

6.1.4 Relations between the above notions

The fact below was observed already in [Bowen, 1971]:

Theorem 6.1.8 *In metric spaces* $\mathbf{h}_1(T) = \mathbf{h}_2(T) = \mathbf{h}(T)$. *In particular, the Bowen–Dinaburg definition does not depend on the metric.*

This allows us to define our main notion:

Definition 6.1.9 *Topological entropy* $\mathbf{h}(T)$ of T is defined as the common value $\mathbf{h}_1(T) = \mathbf{h}_2(T) = \mathbf{h}(T)$.

Before we proceed with the proof, we introduce two parameters related to an open cover: $\mathsf{diam}(\mathfrak{U})$ and $\mathsf{Leb}(\mathfrak{U})$. The first one denotes the maximal diameter of an element of $\mathfrak{U}$ and we will call it simply the *diameter* of $\mathfrak{U}$. This, of course, is bounded by the (finite) diameter of X. The second one is called the *Lebesgue number* of $\mathfrak{U}$ and it is defined as the maximal number ε such that every open ball of radius ε is contained in an element of $\mathfrak{U}$. It is a standard fact in metric topology that this number is positive.

Proof of Theorem 6.1.8 A set is (n, ε)-spanning if and only if the family of the (n, ε)-balls around its members is a subcover of the cover $\mathfrak{U}_{(n,\varepsilon)}$ by all (n, ε)-balls which, in turn, is a subcover of $\mathfrak{U}^n_{(1,\varepsilon)}$. By the definition of the Lebesgue number, $\mathfrak{U}_{(1,\varepsilon)} \succcurlyeq \mathfrak{U}$ whenever $\varepsilon \leq \mathsf{Leb}(\mathfrak{U})$. By (6.1.1) and (6.1.3), we get

$$r(n, \varepsilon) = N(\mathfrak{U}_{(n,\varepsilon)}) \geq N(\mathfrak{U}^n_{(1,\varepsilon)}) \geq N(\mathfrak{U}^n). \tag{6.1.10}$$

Note that any (n, ε)-separated set F of maximal cardinality must be (n, ε)-spanning, otherwise there would exist a point whose distances to all members of F in the metric d^n were larger than or equal to ε. The set F enhanced by such a point would remain (n, ε)-separated, contradicting the maximality of F. This implies that

$$s(n, \varepsilon) \geq r(n, \varepsilon). \tag{6.1.11}$$

Now take an open cover $\mathcal{V}$ with $\mathsf{diam}(\mathcal{V}) < \varepsilon$. Let F be an (n, ε)-separated set. Then every cell of $\mathcal{V}^n$ contains at most one element of F. On the other hand, any subcover of $\mathcal{V}^n$ covers all elements of F. Thus

$$N(\mathcal{V}^n) \geq s(n, \varepsilon).$$

Combining the above displayed formulae, we conclude

$$\mathbf{h}(T, \mathcal{V}) \geq \mathbf{h}_1(T, \varepsilon) \geq \mathbf{h}_2(T, \varepsilon) \geq \mathbf{h}(T, \mathfrak{U}). \tag{6.1.12}$$

The proof is completed by passing to the limit over a refining sequence of covers $\mathfrak{U}_k$, letting $\varepsilon_k = \mathsf{Leb}(\mathfrak{U}_k)$ and choosing a refining sequence of covers $\mathcal{V}_k$ with $\mathsf{diam}(\mathcal{V}_k) \leq \varepsilon_k$. □

At this point we notice that the above argument works regardless of whether we use lim sup or lim inf in either definition involving ε, i.e., that of $\mathbf{h}_1(T, \varepsilon)$ and that of $\mathbf{h}_2(T, \varepsilon)$. In each case, depending on this choice, we may obtain two slightly different values for $\mathbf{h}_1(T, \varepsilon)$ and two different values for $\mathbf{h}_2(T, \varepsilon)$, but the inequalities (6.1.12) hold in any case and the differences disappear in the last limit passage (in which $\varepsilon \to 0$), always leading to the same value of topological entropy.

6.2 Properties of topological entropy

The facts stated in this section show that the behavior of topological entropy is, in many aspects, the same as that of measure-theoretic dynamical entropy. This concerns the behavior with respect to the factor-extension relation, product systems and power systems. In the proofs we will use the version of the definition which is most convenient.

A topological dynamical system $(Y, S, \mathbb{S})$ is a *subsystem* of $(X, T, \mathbb{S})$ when Y is a closed T-invariant subset of X (meaning $T(Y) \subset Y$ or $T(Y) = Y$ depending on whether $\mathbb{S} = \mathbb{N}_0$ or $\mathbb{Z}$, respectively) and $S = T|_Y$.

Fact 6.2.1 *If $(Y, S, \mathbb{S})$ is a subsystem of $(X, T, \mathbb{S})$, then $\mathbf{h}(S) \leq \mathbf{h}(T)$.*

Proof Every maximal (n, ε)-separated set in Y is (n, ε)-separated in X (and here perhaps not maximal). □

A topological dynamical system $(Y, S, \mathbb{S})$ is a *topological factor* of $(X, T, \mathbb{S})$ if there exists a continuous and equivariant surjection $\pi : X \to Y$ (recall that equivariant means $\pi \circ T = S \circ \pi$). We also call $(X, T, \mathbb{S})$ a *topological extension* of $(Y, S, \mathbb{S})$. Two systems are *topologically conjugate* (we just say *conjugate*) if the above π is a homeomorphism.

Fact 6.2.2 *If $(Y, S, \mathbb{S})$ is a topological factor of $(X, T, \mathbb{S})$, then $\mathbf{h}(S) \leq \mathbf{h}(T)$. Conjugate systems have the same topological entropy (we say that topological entropy is an* invariant of topological conjugacy*).*

Proof This is obvious, since every open cover $\mathcal{V}$ of Y lifts against the factor map π to an open cover $\mathcal{U} = \pi^{-1}(\mathcal{V})$ of X, and the numbers $N(\mathcal{V}^n)$ in the system $(Y, S, \mathbb{S})$ and $N(\mathcal{U}^n)$ in $(X, T, \mathbb{S})$ coincide (by surjectivity), so that $\mathbf{h}(S, \mathcal{V}) = \mathbf{h}(T, \mathcal{U})$. Now, the limit in the definition of $\mathbf{h}(T)$ via covers equals the supremum over all covers of X, including those lifted from Y, hence the desired inequality follows. The last statement is now trivial. □

By a *power system* we will understand the system (X, T^n), where $n \in \mathbb{S}$ is fixed. We have the following *power rule* for topological entropy:

Fact 6.2.3

$$\mathbf{h}(T^n) = |n|\mathbf{h}(T).$$

Proof For $n = 0$ the map T^0 is the identity and the statement holds trivially. Now take $n \in \mathbb{N}$. For an open cover $\mathcal{U}$ let $\mathcal{V} = \mathcal{U}^n$ (under the action of T). Then $\mathcal{U}^{nm}$ (under the action of T) equals $\mathcal{V}^m$ under the action of T^n. Thus,

$$\mathbf{h}(T^n, \mathcal{V}) = \lim_m \tfrac{1}{m}\mathbf{H}(\mathcal{U}^{nm}) = n\mathbf{h}(T, \mathcal{U}).$$

Since $\mathcal{V} \succcurlyeq \mathcal{U}$ it suffices to take the supremum over all covers $\mathcal{U}$ (i.e., over all covers $\mathcal{V}$ of the form $\mathcal{U}^n$) to obtain the topological entropy of the power system.

It remains to prove that for homeomorphisms $\mathbf{h}(T) = \mathbf{h}(T^{-1})$. We have, for any open cover $\mathcal{U}$, that $\mathcal{U}^n$ (under the action of T^{-1}) equals $T^n(\mathcal{U}^n)$ (here $\mathcal{U}^n$ is under the action of T). Because T is a homeomorphism, both covers have the same smallest cardinality of a subcover. This implies that $\mathbf{h}(T, \mathcal{U}) = \mathbf{h}(T^{-1}, \mathcal{U})$. Taking the supremum over all $\mathcal{U}$ on both sides, we complete the proof. □

On the other hand, the following easy fact holds (compare Fact 2.4.1). The proof is left to the reader as Exercise 6.2.

Fact 6.2.4 *For any $n \in \mathbb{N}$ we have $\mathbf{h}(T, \mathcal{U}^n) = \mathbf{h}(T, \mathcal{U})$.* □

Definition 6.2.5 A cover $\mathcal{U}$ is called a (unilateral) *topological generator* of $(X, T, \mathbb{S})$ if $\mathcal{U}^n$ is a refining sequence of covers (in the sense of Remark 6.1.7). For homeomorphisms we can also define *bilateral generators*; we require the sequence $\mathcal{U}^{[-n,n]}$ to refine.

It follows from Fact 6.2.4 that if a cover $\mathcal{U}$ is a generator of $(X, T, \mathbb{S})$, then $\mathbf{h}(T, \mathcal{U}) = \mathbf{h}(T)$. The same easily generalizes to bilateral generators.

Definition 6.2.6 A system $(X, T, \mathbb{S})$ is *expansive* when for any $x \neq y \in X$ there exists an $n \in \mathbb{S}$ with $d(T^n x, T^n y) \geq M$, where $M > 0$ is constant.

Expansiveness strongly depends upon the acting semigroup, for example the shift map on $\{0, 1\}^{\mathbb{Z}}$ is $\mathbb{Z}$-expansive but not $\mathbb{N}_0$-expansive. Contrary to how it is defined, expansiveness does not depend on the metric and is an invariant of topological conjugacy. Every expansive system has a topological generator (unilateral or bilateral, respectively to the meaning of expansiveness). The converse implication fails, for example, the irrational rotation of the circle has a generator without being expansive.

The last thing we examine in this section is how topological entropy behaves as the transformation varies on a fixed space.

Fact 6.2.7 *Let $C(X, X)$ denote the set of all continuous transformations of X endowed with the supremum metric. Fix an open cover $\mathcal{U}$ of X. Then the map $T \mapsto \mathbf{h}(T, \mathcal{U})$ is upper semicontinuous on $C(X, X)$. The map $T \mapsto \mathbf{h}(T)$ is of Young class* LU *(see Definition A.1.23 in the Appendix A.1).*

Proof Fix an $n \geq 1$ and let $\mathcal{U}_o$ be an optimal (of the smallest possible cardinality) subcover of $\mathcal{U}^n$ (here n refers to the action of T). Let ε be the Lebesgue number of $\mathfrak{U}_o$ in the metric d^n (again, n refers to the action of T). There is a $\delta > 0$ such that if $T' \in C(X, X)$ is within δ from T, then T'^i is within ε from T^i for $i = 1, \ldots, n-1$. Fix one such map T' and let $\mathfrak{U}'_o$ be the family of sets obtained by the same intersections as $\mathfrak{U}_o$, but with the action of T replaced by the action of T'. We claim that $\mathfrak{U}'_o$ covers X. Indeed, take an $x \in X$. Then x belongs to some member of $\mathfrak{U}_o$, say $\bigcap_{i=0}^{n-1} T^{-i}(U_i)$ together with the (n, ε)-ball around x. In other words, for each $i = 0, \ldots, n-1$ not only $T^i x \in U_i$, but also $y \in U_i$ whenever $d(y, T^i x) < \varepsilon$, in particular, y may be taken $T'^i x$. This implies that $x \in \bigcap_{i=0}^{n-1} T'^{-i}(U_i)$, and this set, by definition, is a member of $\mathfrak{U}'_o$. So $\mathfrak{U}'_o$ covers X. Of course, $\mathfrak{U}'_o$ is a subcover (not necessarily optimal) of $\mathfrak{U}^n$ where this time n refers to the action of T'. We have proved that the (implicit) dependence of $\mathbf{H}(\mathfrak{U}^n)$ on the map T is upper semicontinuous. Since $\mathbf{h}(T, \mathfrak{U})$ is the infimum over n of $\frac{1}{n}\mathbf{H}(\mathfrak{U}^n)$, this function of the map T is also upper semicontinuous. The last assertion is now obvious, since the function $T \mapsto \mathbf{h}(T)$ is an increasing limit of $T \mapsto \mathbf{h}(T, \mathfrak{U}_k)$, where $\mathfrak{U}_k$ is a refining sequence of covers. □

6.3 Topological conditional and tail entropies

Copying the measure-theoretic notions, we now introduce the *topological conditional entropy given a cover* and *given a topological factor*. Although the last notion will become useful much later, we include it here in order to gather similar ideas in one place. We begin with the appropriate static notions.

Definition 6.3.1 Let $\mathfrak{U}$ and $\mathcal{V}$ be open covers of X and let $F \subset X$. We denote

$$\mathbf{H}(\mathfrak{U}|F) = \log(\min\{\#\mathfrak{U}_F : \mathfrak{U}_F \subset \mathfrak{U},\ F \subset \bigcup \mathfrak{U}_F\}),$$

the logarithm of the smallest cardinality of a subfamily of $\mathfrak{U}$ covering F. Next we let

$$\mathbf{H}(\mathfrak{U}|F, \mathcal{V}) = \max\{\mathbf{H}(\mathfrak{U}|F \cap V) :\ V \in \mathcal{V}\},$$
$$\mathbf{H}(\mathfrak{U}|\mathcal{V}) = \mathbf{H}(\mathfrak{U}|X, \mathcal{V}).$$

The following properties are almost the same as those of the Shannon entropy (compare (1.6.5), (1.6.6), (1.6.7), (1.6.3), (1.6.9), (1.6.10), (1.6.11) and (1.6.12)). The set F will play a role much later. In the applications to this section the set F will simply be X.

Fact 6.3.2 *For any covers $\mathcal{U}, \mathcal{U}', \mathcal{V}, \mathcal{V}', \mathcal{W}$ and a set F, the following hold*

$$\mathcal{U} \succcurlyeq \mathcal{V} \iff \mathbf{H}(\mathcal{V}|F, \mathcal{U}) = 0, \tag{6.3.3}$$

$$\mathcal{U} \succcurlyeq \mathcal{V} \implies \mathbf{H}(\mathcal{U}|F, \mathcal{W}) \geq \mathbf{H}(\mathcal{V}|F, \mathcal{W}),\ \mathbf{H}(\mathcal{U}) \geq \mathbf{H}(\mathcal{V}), \tag{6.3.4}$$

$$\mathcal{V} \succcurlyeq \mathcal{W} \implies \mathbf{H}(\mathcal{U}|F, \mathcal{V}) \leq \mathbf{H}(\mathcal{U}|F, \mathcal{W}), \tag{6.3.5}$$

$$\mathbf{H}(\mathcal{U} \vee \mathcal{V}|F, \mathcal{W}) \leq \mathbf{H}(\mathcal{U}|F, \mathcal{V} \vee \mathcal{W}) + \mathbf{H}(\mathcal{V}|F, \mathcal{W}), \tag{6.3.6}$$

$$\mathbf{H}(\mathcal{U} \vee \mathcal{V}|F, \mathcal{W}) \leq \mathbf{H}(\mathcal{U}|F, \mathcal{W}) + \mathbf{H}(\mathcal{V}|F, \mathcal{W}), \tag{6.3.7}$$

$$\mathbf{H}(\mathcal{U} \vee \mathcal{V}) \leq \mathbf{H}(\mathcal{U}) + \mathbf{H}(\mathcal{V}), \tag{6.3.8}$$

$$\mathbf{H}(\mathcal{U} \vee \mathcal{U}'|F, \mathcal{V} \vee \mathcal{V}') \leq \mathbf{H}(\mathcal{U}|F, \mathcal{V}) + \mathbf{H}(\mathcal{U}'|F, \mathcal{V}'), \tag{6.3.9}$$

$$\mathbf{H}(\mathcal{U}|F, \mathcal{W}) \leq \mathbf{H}(\mathcal{U}|F, \mathcal{V}) + \mathbf{H}(\mathcal{V}|F, \mathcal{W}), \tag{6.3.10}$$

$$\mathbf{H}(\mathcal{U} \vee \mathcal{V}|F, \mathcal{V}) = \mathbf{H}(\mathcal{U}|F, \mathcal{V}). \tag{6.3.11}$$

Proof The first three implications are easily seen directly from the definition. We pass to (6.3.6). Let W be the element of $\mathcal{W}$ such that the optimized subfamily of $\mathcal{U} \vee \mathcal{V}$ needed to cover $F \cap W$ is the largest. Let $\mathcal{V}_{F \cap W}$ be an optimal subfamily of $\mathcal{V}$ covering $F \cap W$. The cardinality of $\mathcal{V}_{F \cap W}$ does not exceed $2^{\mathbf{H}(\mathcal{V}|F,\mathcal{W})}$. For each V from this family, $F \cap V \cap W$ can be covered by a family of at most $2^{\mathbf{H}(\mathcal{U}|F,\mathcal{V}\vee\mathcal{W})}$ elements of $\mathcal{U}$. Replacing each U in this family by $U \cap V$ we obtain (without increasing the cardinality) a subfamily of $\mathcal{U} \vee \mathcal{V}$ covering $F \cap V \cap W$. The union of these latter families over V has cardinality at most $2^{\mathbf{H}(\mathcal{U}|F,\mathcal{V}\vee\mathcal{W})+\mathbf{H}(\mathcal{V}|F,\mathcal{W})}$ and covers (by elements of $\mathcal{U} \vee \mathcal{V}$) the "most demanding set" $F \cap W$, so the assertion follows.

The remaining statements are direct consequences of (6.3.6) and the monotonicities (6.3.4) and (6.3.5). □

Remark 6.3.12 Note that, unlike in case of Shannon entropy, the equality $\mathbf{H}(\mathcal{U} \vee \mathcal{V}) = \mathbf{H}(\mathcal{U}|\mathcal{V}) + \mathbf{H}(\mathcal{V})$ need not hold. By (6.3.6) for trivial $\mathcal{W}$, one inequality does hold, but it is not enough to prove any of the inequalities with nontrivial $\mathcal{W}$ (compare Exercise 1.3). This is the reason why the proof of (6.3.6) is different and does not use the version with trivial $\mathcal{W}$.

We continue the preparations to define the topological conditional entropy.

Fact 6.3.13 *The sequence $\mathbf{H}(\mathcal{U}^n|\mathcal{V}^n)$ is subadditive.*

Proof This follows immediately from the equality $\mathcal{U}^{m+n} = \mathcal{U}^m \vee T^{-m}(\mathcal{U}^n)$, the analogous equality for $\mathcal{V}$, (6.3.9) and the easy observation that for any covers $\mathcal{U}, \mathcal{V}$ we have $\mathbf{H}(T^{-1}(\mathcal{U})|T^{-1}(\mathcal{V})) \leq \mathbf{H}(\mathcal{U}|\mathcal{V})$. □

The above obtained subadditivity implies that the limit in the definition below exists and equals the infimum:

Definition 6.3.14 The *topological conditional entropy* of $\mathcal{U}$ *given* $\mathcal{V}$ is defined as

$$\mathbf{h}(T,\mathcal{U}|\mathcal{V}) = \lim_n \tfrac{1}{n}\mathbf{H}(\mathcal{U}^n|\mathcal{V}^n).$$

The *topological conditional entropy* of the system $(X, T, \mathbb{S})$ *given* $\mathcal{V}$ is defined as

$$\mathbf{h}(T|\mathcal{V}) = \sup_{\mathcal{U}} \mathbf{h}(T,\mathcal{U}|\mathcal{V}). \tag{6.3.15}$$

The *topological tail entropy* of the system $(X, T, \mathbb{S})$ is defined as

$$\mathbf{h}^*(T) = \inf_{\mathcal{V}} \mathbf{h}(T|\mathcal{V}). \tag{6.3.16}$$

Both the supremum in (6.3.15) and the infimum in (6.3.16) can be replaced by monotone limits along the net of all covers or a refining sequence of covers. It is immediate to see that

$$\mathbf{h}^*(T) \leq \mathbf{h}(T),$$

and that $\mathbf{h}(T) = \infty \implies \mathbf{h}^*(T) = \infty$.

Remark 6.3.17 The notion $\mathbf{h}^*(T)$ is a manifestation of an essential difference between measure-theoretic dynamical conditional entropy and topological conditional entropy. Due to the formula $h(\mathcal{P}|\mathcal{Q}) = h(\mathcal{P} \vee \mathcal{Q}) - h(\mathcal{Q})$ (for finite partitions) an analog of $\mathbf{h}^*(T)$ for the dynamical entropy is either zero (for finite entropy systems) or infinity (otherwise). It is the failure of this formula for topological conditional entropy, which makes $\mathbf{h}^*(T)$ a meaningful invariant.

Remark 6.3.18 In the older literature the topological tail entropy is called, as the inventor M. Misiurewicz first called it, the "topological conditional entropy" [Misiurewicz, 1976]. It is clear from the definition that this is not a very fortunate choice of a name, as the tail entropy does not depend on any conditioning parameter any more. Moreover, we need the term "topological conditional entropy" in the meaning used in the book (i.e., *given* a cover or *given* a factor). In such a meaning it stands in perfect correspondence with the measure-theoretic conditional entropy and it is hard to imagine using a different terminology here.

Unlike a partition in the measure-theoretic case, an open cover does not necessarily lead to a topological factor. Still, we can use topological factors as the conditioning object, as it is defined below. But first we will establish a simplification in our notation. Suppose that $\pi : X \to Y$ is a continuous map, and $A \subset Y$. Then, when the choice of π is unambiguous, we will write $\mathbf{H}(\mathcal{U}|A)$

replacing what should be formally written as $\mathbf{H}(\mathcal{U}|\pi^{-1}(A))$. Similarly, for a cover $\mathcal{V}$ of Y, we will write $\mathbf{H}(\mathcal{U}|\mathcal{V})$ and $\mathbf{h}(\mathcal{U}|\mathcal{V})$ instead of $\mathbf{H}(\mathcal{U}|\pi^{-1}(\mathcal{V})))$ and $\mathbf{h}(\mathcal{U}|\pi^{-1}(\mathcal{V}))$.

Definition 6.3.19 Let $\pi_1 : X \to Y$ be a topological factor map between the systems $(X, T, \mathbb{S})$ and $(Y, S, \mathbb{S})$. The *topological conditional entropy* of $\mathcal{U}$ *given* $(Y, S, \mathbb{S})$ is defined as

$$\mathbf{h}(T, \mathcal{U}|S) = \inf_{\mathcal{V}} \mathbf{h}(T, \mathcal{U}|\mathcal{V}),$$

where $\mathcal{V}$ ranges over all covers of Y. Then the *topological conditional entropy of T given the factor* $(Y, S, \mathbb{S})$ is

$$\mathbf{h}(T|S) = \sup_{\mathcal{U}} \mathbf{h}(T, \mathcal{U}|S),$$

where $\mathcal{U}$ ranges over all covers of X. Similarly, if $\pi_2 : X \to Z$ is a factor map between $(X, T, \mathbb{S})$ and another factor $(Z, R, \mathbb{S})$, then we define the *topological conditional entropy of the factor* $(Z, R, \mathbb{S})$ *given the factor* $(Y, S, \mathbb{S})$:

$$\mathbf{h}(R|S) = \sup_{\mathcal{W}} \mathbf{h}(T, \mathcal{W}|S),$$

where $\mathcal{W}$ ranges over all covers of Z.

In all cases, the above infima and suprema can be replaced by monotone limits along the appropriate nets or refining sequences of covers.

Notice that $\mathbf{h}(T|T) = 0$ (where the factor map is the identity); supremum and infimum are switched compared to the definition of $\mathbf{h}^*(T)$.

Among the above notions most important are two: the tail entropy and the conditional entropy given a factor. Like the (unconditional) topological entropy, they are both subject of variational principles, which will be proved in the subsequent sections. Variational principles shed a lot of light on the properties of these notions. Meanwhile, we discuss some more elementary issues, which do not invoke invariant measures.

When either of the two above parameters (the topological tail entropy or the topological conditional entropy given a factor) equals zero, we are dealing with rather distinguished situations, implying a number of further special properties (which will be provided later). Now just the definitions:

Definition 6.3.20 The system $(X, T, \mathbb{S})$ is called *asymptotically h-expansive* if $\mathbf{h}^*(T) = 0$.

A trivial example of an asymptotically h-expansive system is an expansive system (in particular a subshift, see the next chapter); any cover of diameter smaller than the expansive constant M is a topological generator, hence the conditional entropy given this (and any finer) cover is zero. There is also a notion of *h-expansive* systems, i.e., such that $\mathbf{h}(T|\mathcal{V}) = 0$ for some finite open cover $\mathcal{V}$. This property is weaker than expansiveness and stronger than asymptotic h-expansiveness. It will play no essential role in this book.

Definition 6.3.21 Let $\pi : X \to Y$ be a factor map between the systems $(X, T, \mathbb{S})$ and $(Y, S, \mathbb{S})$. We say that $(Y, S, \mathbb{S})$ is a *principal factor* of $(X, T, \mathbb{S})$, or that $(X, T, \mathbb{S})$ is a *principal extension* of $(Y, S, \mathbb{S})$, if $\mathbf{h}(T|S) = 0$.

6.4 Properties of topological conditional entropy

This section describes the most elementary relations between topological conditional entropy, topological entropy and topological tail entropy. We begin with the easiest ones. The familiar list is obtained directly from Fact 6.3.2 via the limit passage.

Fact 6.4.1 *For any covers $\mathcal{U}, \mathcal{V}, \mathcal{W}$ we have*

$$\mathcal{U} \succcurlyeq \mathcal{V} \implies \mathbf{h}(T, \mathcal{U}|\mathcal{W}) \geq \mathbf{h}(T, \mathcal{V}|\mathcal{W}), \tag{6.4.2}$$

$$\mathcal{V} \succcurlyeq \mathcal{W} \implies \mathbf{h}(T, \mathcal{U}|\mathcal{V}) \leq \mathbf{h}(T, \mathcal{U}|\mathcal{W}), \tag{6.4.3}$$

$$\mathbf{h}(T, \mathcal{U} \vee \mathcal{V}|\mathcal{W}) \leq \mathbf{h}(T, \mathcal{U}|\mathcal{V} \vee \mathcal{W}) + \mathbf{h}(T, \mathcal{V}|\mathcal{W}), \tag{6.4.4}$$

$$\mathbf{h}(T, \mathcal{U} \vee \mathcal{V}|\mathcal{W}) \leq \mathbf{h}(T, \mathcal{U}|\mathcal{W}) + \mathbf{h}(T, \mathcal{V}|\mathcal{W}), \tag{6.4.5}$$

$$\mathbf{h}(T, \mathcal{U}|\mathcal{W}) \leq \mathbf{h}(T, \mathcal{U}|\mathcal{V}) + \mathbf{h}(T, \mathcal{V}|\mathcal{W}), \tag{6.4.6}$$

$$\mathbf{h}(T, \mathcal{U} \vee \mathcal{V}|\mathcal{V}) = \mathbf{h}(T, \mathcal{U}|\mathcal{V}). \tag{6.4.7}$$

□

We pass to analogous properties involving factors.

Fact 6.4.8 *Let $(X, T, \mathbb{S})$, $(Y, S, \mathbb{S})$, $(Z, R, \mathbb{S})$ be factors of some common extension (we do not need to denote it). Then*

$$T \mapsto S \implies \mathbf{h}(T|R) \geq \mathbf{h}(S|R), \tag{6.4.9}$$

$$S \mapsto R \implies \mathbf{h}(T|S) \leq \mathbf{h}(T|R), \tag{6.4.10}$$

$$\mathbf{h}(T|R) \leq \mathbf{h}(T|S) + \mathbf{h}(S|R), \tag{6.4.11}$$

$$\mathbf{h}(T) \leq \mathbf{h}(T|S) + \mathbf{h}(S), \tag{6.4.12}$$

(an arrow between factors indicates that one of them is a factor of another).

Proof Let $\mathcal{U}$, $\mathcal{V}$ and $\mathcal{W}$ be some covers of X, Y and Z, respectively. Then, we apply Fact (6.4.2), (6.4.3), (6.4.6), and refine the covers in the following order: first $\mathcal{W}$, then $\mathcal{V}$, and in the end $\mathcal{U}$. The last inequality is (6.4.11) for the trivial (one-point) factor $(Z, R, \mathbb{S})$. □

Fact 6.4.13 *Let $\pi : X \to Y$ be a factor map from $(X,T,\mathbb{S})$ to $(Y,S,\mathbb{S})$. Then*

$$\mathbf{h}^*(S) \leq \mathbf{h}(T|S) + \mathbf{h}^*(T) \quad \textit{and} \quad \mathbf{h}^*(T) \leq \mathbf{h}(T|S) + \mathbf{h}^*(S), \qquad (6.4.14)$$

(or simply $|\mathbf{h}^*(T) - \mathbf{h}^*(S)| \leq \mathbf{h}(T|S)$ if either $\mathbf{h}^*(S)$ or $\mathbf{h}^*(T)$ is finite).

Proof Consider four covers of X: two arbitrary covers $\mathcal{U}' \succcurlyeq \mathcal{U}$, and two (lifted) covers $\mathcal{V}' \succcurlyeq \mathcal{V}$ of Y. We assume that also $\mathcal{U}' \succcurlyeq \mathcal{V}'$. We have

$$\mathbf{h}(T, \mathcal{V}'|\mathcal{V}) \leq \mathbf{h}(T, \mathcal{U}'|\mathcal{V}) \leq \mathbf{h}(T, \mathcal{U}'|\mathcal{U}) + \mathbf{h}(T, \mathcal{U}|\mathcal{V}).$$

We disregard the middle term. Now we let these covers refine in the following order: first $\mathcal{U}'$, next $\mathcal{V}'$, then $\mathcal{V}$, and finally $\mathcal{U}$. This yields the first inequality in (6.4.14).

Next, instead of $\mathcal{U}' \succcurlyeq \mathcal{V}'$, we assume $\mathcal{U} \succcurlyeq \mathcal{V}$ and we write

$$\mathbf{h}(T, \mathcal{U}'|\mathcal{U}) \leq \mathbf{h}(T, \mathcal{U}'|\mathcal{V}) \leq \mathbf{h}(T, \mathcal{U}'|\mathcal{V}') + \mathbf{h}(T, \mathcal{V}'|\mathcal{V}).$$

We can ignore the middle term, and let these covers refine, this time in another order: first $\mathcal{V}'$, next $\mathcal{U}'$, then $\mathcal{U}$, and finally $\mathcal{V}$. This yields the last inequality in (6.4.14). □

Corollary 6.4.15 (comp. [Ledrappier, 1979]) *Let $\pi : X \to Y$ be a principal factor map (i.e., such that $\mathbf{h}(T|S) = 0$). Then $\mathbf{h}(S) = \mathbf{h}(T)$ and $\mathbf{h}^*(S) = \mathbf{h}^*(T)$. In other words, principal factors (or extensions) preserve topological entropy and topological tail entropy, in particular, they preserve asymptotic h-expansiveness. Also, the composition of principal factor maps is principal.*

Remark 6.4.16 In general, the fact that $(Y, S, \mathbb{S})$ is a factor of $(X, T, \mathbb{S})$ does not imply any inequality between $\mathbf{h}^*(S)$ and $\mathbf{h}^*(T)$. In particular, a factor of an asymptotically h-expansive system need not be asymptotically h-expansive. For instance, a non-asymptotically h-expansive system may have a symbolic (hence asymptotically h-expansive) extension. This statement will become clear in the chapter devoted to symbolic extensions.

6.5 Topological joinings

Definition 6.5.1 Given two topological dynamical systems, $(X, T, \mathbb{S})$ and $(Y, S, \mathbb{S})$, their *direct product* is the dynamical system $(X \times Y, T \times S, \mathbb{S})$, where $T \times S$ is defined by

$$(T \times S)(x, y) = (Tx, Sy).$$

By a *topological joining* of $(X, T, \mathbb{S})$ and $(Y, S, \mathbb{S})$ we understand any subsystem (a closed invariant subset) of the product system $(X \times Y, T \times S)$, whose projections on the first and second axis are surjections onto X and Y, respectively.

Notice that a joining of two systems is an extension of both of them (via the projection maps). If two systems $(X, T, \mathbb{S})$ and $(Y, S, \mathbb{S})$ are already factors of some system $(Z, R, \mathbb{S})$, say, by factor maps π_1 and π_2, then, just like in the measure-theoretic case (see Fact 4.4.2), one of their joinings is naturally realized (as we shall see) within the common extension:

Definition 6.5.2 With the denotation of the paragraph above, the joining of the factors $(X, T, \mathbb{S})$ and $(Y, S, \mathbb{S})$ obtained as the set of pairs

$$\{(\pi_1(z), \pi_2(z)) : z \in Z\} \subset X \times Y$$

with the action of the restriction of $T \times S$, will be called *the joining within* $(Z, R, \mathbb{S})$ and denoted by $T \vee S$.

This is a rather imperfect notation as it does not indicate the common extension. In fact the mappings π_1, π_2 should be indicated because a pair of systems may "sit" in one extension in many different ways creating nonconjugate joinings. We will keep this notation, but we will use it exclusively when a common extension and the pair of factor maps is clear from the context.

Fact 6.5.3 *The above joining $T \vee S$ is a factor of $(Z, R, \mathbb{S})$, and the covers $\mathcal{U} \vee \mathcal{V}$, where $\mathcal{U}$, $\mathcal{V}$ range over the covers lifted from X and Y, respectively, refine in this joining.*

Proof Indeed, the assignment $z \mapsto (\pi_1(z), \pi_2(z))$ sends Z onto the joining. A pair of covers $\mathcal{U}$ of X and $\mathcal{V}$ of Y lifts in the product space to the covers $\{U \times Y : U \in \mathcal{U}\}$ and $\{X \times V : V \in \mathcal{V}\}$. The join of these lifts coincides with the product cover $\mathcal{U} \otimes \mathcal{V} = \{U \times V : U \in \mathcal{U}, V \in \mathcal{V}\}$. It is clear that such covers refine in the product space, hence also relatively in the joining. □

Now we can generalize the remaining statements of Fact 6.4.1 involving joinings. The proof is obvious by refining the covers.

Fact 6.5.4 *Let $(X, T, \mathbb{S})$, $(Y, S, \mathbb{S})$ and $(Z, R, \mathbb{S})$ be factors of a common extension (which we do not denote). The joinings below refer to the joinings* *within this common extension. We have*

$$\mathbf{h}(T \vee S|R) \leq \mathbf{h}(T|S \vee R) + \mathbf{h}(S|R), \tag{6.5.5}$$

$$\mathbf{h}(T \vee S|R) \leq \mathbf{h}(T|R) + \mathbf{h}(S|R), \tag{6.5.6}$$

$$\mathbf{h}(T \vee S) \leq \mathbf{h}(T) + \mathbf{h}(S), \tag{6.5.7}$$

$$\mathbf{h}(T \vee S|S) = \mathbf{h}(T|S) \leq \mathbf{h}(T). \tag{6.5.8}$$

□

It is seen from (6.5.8) that if $(X, T, \mathbb{S})$ has entropy zero, then its joining with any other system is a principal extension of that system.

Topological entropy also satisfies the *product rule*:

Fact 6.5.9 *Given two systems $(X, T, \mathbb{S})$ and $(Y, S, \mathbb{S})$, consider their product $(X \times Y, T \times S, \mathbb{S})$. Then*

$$\mathbf{h}(T \times S) = \mathbf{h}(T) + \mathbf{h}(S). \tag{6.5.10}$$

Proof The inequality "$\leq$" is (6.5.7) because the product is a joining. The other inequality is best proved using (n, ε)-separated sets.[2] In $X \times Y$ we will use the maximum metric $d((x, y), (x', y')) = \max\{d_X(x, x'), d_Y(y, y')\}$. Then any (n, ε)-separated set of maximal cardinality in X producted with any (n, ε)-separated set of maximal cardinality in Y is an (n, ε)-separated set in $X \times Y$ (perhaps not even maximal), thus the term $\log s(n, \varepsilon)$ for the product action is at least the sum of the same terms evaluated for $(X, T, \mathbb{S})$ and $(Y, S, \mathbb{S})$. Recall that in the definition of $\mathbf{h}_1(T, \varepsilon)$ we can choose between lim sup and lim inf (which produces possibly two different values of $\mathbf{h}_1(T, \varepsilon)$, but the difference disappears in the limit over ε defining the topological entropy). We choose lim inf because it is *superadditive*: lim inf of a sum of two sequences is larger than or equal to the sum of the corresponding lim inf's. The rest of the proof is obvious. □

We remark that all statements of Facts 6.4.8 and 6.5.4 and the product rule can be quickly proved using the Conditional Variational Principle (Theorem 6.8.8 below) and the properties of the measure-theoretic conditional dynamical entropy, but the topological proofs are more direct.

We devote a few lines to countable joinings. Let $(X_k, T_k, \mathbb{S})$ be a sequence of dynamical systems. Their *countable joining* is any subsystem of the countable product system, of which all coordinate projections are surjective. A special type of a countable joining is an inverse limit.

[2] The seemingly obvious equality $N(\mathcal{U} \otimes \mathcal{V}) = N(\mathcal{U})N(\mathcal{V})$ is false (see Exercise 6.11). This mistake was made (and published) by R. Bowen in his first attempt to prove the product rule.

Definition 6.5.11 Let $(X_k, T_k, \mathbb{S})$ be a sequence of dynamical systems such that for each k $(X_k, T_k, \mathbb{S})$ is a topological factor of $(X_{k+1}, T_{k+1}, \mathbb{S})$ via a map π_k. By the *inverse limit* of this sequence we shall mean their countable joining $(X, T, \mathbb{S})$, where

$$X = \{(x_k)_{k\in\mathbb{N}} : \forall_k\ x_k = \pi_k(x_{k+1})\}.$$

The inverse limit (as a dynamical system) is denoted by

$$(X, T, \mathbb{S}) = \overleftarrow{\lim_k}(X_k, T_k, \mathbb{S}, \pi_k)$$

(with the tendency to skip the maps π_k in the denotation).

Every countable joining $(X, T, \mathbb{S})$ of a sequence of systems $(X_k, T_k, \mathbb{S})$ can be represented as (is conjugate to) the inverse limit $\overleftarrow{\lim_k}(Y_k, S_k, \mathbb{S}, \pi_k)$, where $(Y_k, S_k, \mathbb{S})$ is the finite joining of $(X_i, T_i, \mathbb{S})$ with $i = 1, \ldots, k$ within $(X, T, \mathbb{S})$, while π_k is the projection of $(Y_{k+1}, S_{k+1}, \mathbb{S})$ onto the first k coordinates.

Fact 6.5.12 *Let $(X, T, \mathbb{S})$ denote an inverse limit $\overleftarrow{\lim_k}(X_k, T_k, \mathbb{S})$. Then the topological entropy of $(X, T, \mathbb{S})$ equals the limit of the topological entropies: $\mathbf{h}(T) = \lim_k \uparrow \mathbf{h}(T_k)$.*

Proof For each k let $\mathcal{U}_{k,n}$ be a refining sequence of open covers in X_k. By lifting, these covers become a double sequence of open covers $\mathcal{U}_{k,n}$ in X. Clearly, this double sequence refines in X. Thus

$$\mathbf{h}(T) = \sup_{k,n} \mathbf{h}(T, \mathcal{U}_{k,n}) = \lim_k \uparrow \lim_n \uparrow \mathbf{h}(T_k, \mathcal{U}_{k,n}) = \lim_k \uparrow \mathbf{h}(T_k).$$

□

6.6 The simplex of invariant measures

This section supplies necessary information about the simplex of invariant measures and the behavior of the Kolmogorov–Sinai entropy as a function on this simplex. Elementary facts not related to entropy will be stated without proofs. We will frequently refer to the material gathered in Appendix A.2. We recommend that the reader becomes acquainted with that appendix before proceeding.

If X is a compact metric space, then by $\mathcal{M}(X)$ we denote the set of all Borel probability measures on X endowed with the weak-star topology. Any measurable, bounded from at least one side, function f on X can be lifted as a function (admitting infinite values) on $\mathcal{M}(X)$ by assigning $\mu \mapsto \int f\,d\mu$. Such a lift maintains continuity or semicontinuity of f and is always harmonic (see Appendices A.2.2 and A.2.3).

Now, let $(X, T, \mathbb{S})$ be a topological dynamical system. Then T induces a continuous map on $\mathcal{M}(X)$ by the formula $T\mu(B) = \mu(T^{-1}(B))$ (B is a Borel subset of X). Also notice that if B belongs to the completed (with respect to μ) Borel sigma-algebra, then $T^{-1}(B)$ belongs to the Borel sigma-algebra completed with respect to $T\mu$. Indeed, B lies between two Borel sets, say $A \subset B \subset C$, where $\mu(A) = \mu(C)$. Then $T^{-1}(B)$ is between the Borel sets $T^{-1}(A)$ and $T^{-1}(C)$, which also have equal measures. From now on, whenever we say "a Borel measure," we always mean the measure prolonged onto the completed Borel sigma-algebra. Although the completion depends on the measure, the transformation T is measurable with respect to the corresponding completions for μ and $T\mu$. A measure μ is called T-invariant if $T\mu = \mu$. It is well known that the set $\mathcal{M}_T(X)$ of all T-invariant probability measures on X is nonempty, compact, convex, and that ergodic measures are exactly the extreme points i.e. the elements of $\mathrm{ex}\mathcal{M}_T(X)$ [Kryloff and Bogoliouboff, 1937; Oxtoby, 1952].

For $n \in \mathbb{N}$ we define a continuous map $\mathrm{M}_n : \mathcal{M}(X) \to \mathcal{M}(X)$ by

$$\mathrm{M}_n(\mu) = \frac{1}{n}\sum_{i=0}^{n-1} T^i\mu.$$

Then, as can be easily verified, for any subsequence (n_k) such that, for each k, n_{k+1} is a multiple of n_k, the sets $\mathrm{M}_{n_k}(\mathcal{M}(X))$ decrease and their (nonempty) intersection is contained in $\mathcal{M}_T(X)$. By an easy compactness argument, this implies:

Fact 6.6.1 *If $U \supset \mathcal{M}_T(X)$ is an open set in $\mathcal{M}(X)$, then $\mathrm{M}_n(\mathcal{M}(X)) \subset U$ for all sufficiently large n.* □

The following fact is classical in topological dynamics [e.g. Walters, 1982, §6.2] (consult also Appendix A.2.4 for a background on Choquet simplices):

Theorem 6.6.2 *Let $(X, T, \mathbb{S})$ be a topological dynamical system. The set $\mathcal{M}_T(X)$ of all T-invariant Borel probability measures on X, endowed with the weak-star topology, is a Choquet simplex.* □

For $\mu \in \mathcal{M}_T(X)$ the formula $\mu = \int \nu\,d\xi^\mu$, where ξ^μ is the unique probability distribution supported on $\mathrm{ex}\mathcal{M}_T(X)$ and with barycenter at μ (see Appendix A.2.4), coincides with the ergodic decomposition of μ.

Definition 6.6.3 Let $(X, T, \mathbb{S})$ be a topological dynamical system. By the *entropy function* we mean the function $h : \mathcal{M}_T(X) \to [0, \infty]$, where $h(\mu)$

denotes the Kolmogorov–Sinai entropy $h(\mu, T)$ in the measure-preserving system $(X, \mathfrak{A}_\mu, \mu, T, \mathbb{S})$ and $\mathfrak{A}_\mu$ stands for the (completed with respect to μ) Borel sigma-algebra in X.

We have the following fundamental properties of the entropy function:

Fact 6.6.4 *The entropy function on $\mathcal{M}_T(X)$ is Borel-measurable and harmonic.*

Proof Measurability is obvious by the definition of the dynamical entropy, which involves several limits applied to sums of functions of the form $\eta(\mu(A))$, where the sets A are Borel-measurable in X (although we deal with the completed Borel sigma-algebra, which depends on μ, in Definition 4.1.1 it suffices to take the supremum over all genuine Borel-measurable partitions $\mathcal{P}$, and these do not depend on μ). That this function is harmonic, follows directly from Theorem 2.6.4 and Fact A.2.15. Alternatively, it suffices to show that h is an increasing limit of upper semicontinuous affine (hence harmonic, see Fact A.2.10) functions. This will be proved later in Section 8.4 (Fact 8.4.5), where, for affinity of the approximating functions, we will use Theorem 2.5.1 (which requires much simpler tools than Theorem 2.6.4). □

Often we will be interested in continuity or semicontinuity of the entropy function (under additional assumptions on the space or on a partition). The following lemma is the key:

Lemma 6.6.5 *Fix a Borel set $A \subset X$. Then the function $\mu \mapsto \mu(A)$ defined on $\mathcal{M}(X)$ is continuous at μ if and only if $\mu(\partial A) = 0$ (the boundary of A has measure zero).*

Proof The functions $\mu \mapsto \mu(\overline{A})$ and $\mu \mapsto \mu(\mathrm{int}A)$ (measures of the closure and of the interior of A) are upper and lower semicontinuous, respectively (see Fact A.2.7), the latter being dominated by the former. Thus

$$\mu(\overline{A}) \geq \limsup_{\nu \to \mu} \nu(\overline{A}) \geq \limsup_{\nu \to \mu} \nu(A) \geq \liminf_{\nu \to \mu} \nu(A) \geq \liminf_{\nu \to \mu} \nu(\mathrm{int}A) \geq \mu(\mathrm{int}A).$$

If $\mu(\partial A) = 0$, then $\mu(\overline{A}) = \mu(\mathrm{int}A) = \mu(A)$, which implies equalities in the line above. We skip the (easy) proof of the other implication, as we shall never use it. □

Fact 6.6.6 *For one or even countably many measures μ_i ($i \in \mathbb{N}$) there always exist arbitrarily fine partitions $\mathcal{P}$ such that the boundaries of the cells have measure zero for all measures μ_i.*

Proof We sketch the standard construction: Fix a point x. The boundary of the ε-ball around x is contained in the ε-sphere, and such spheres are disjoint for different radii. Thus, only countably many of these balls may have boundaries of positive measure for some μ_i. By compactness, we can choose a cover of X by finitely many balls $B_1, B_2, \dots, B_k$ of radii smaller than some fixed ε and whose boundaries have measure zero for each μ_i. The partition $B_1, B_2 \setminus B_1, B_3 \setminus (B_1 \cup B_2), \dots, B_k \setminus (B_1 \cup B_2 \cup \dots \cup B_{k-1})$ has boundaries as desired and is fine in the sense that the cells have diameters at most 2ε. □

Lemma 6.6.7 *If $\mathcal{P}$ is a finite Borel-measurable partition of X and μ_0 satisfies $\mu_0(\partial A) = 0$ for all $A \in \mathcal{P}$, then*

(a) μ_0 is a continuity point of the function $\mu \mapsto H(\mu, \mathcal{P})$ defined on $\mathcal{M}(X)$.
(b) If $\pi : X \to Y$ is continuous and $\mathfrak{A}_Y$ is the Borel sigma-algebra of Y lifted to X, then μ_0 is a continuity point of the function $\mu \mapsto H(\mu, \mathcal{P}|\mathfrak{A}_Y)$ defined on $\mathcal{M}(X)$.
(c) If μ_0 is T-invariant, then the function $\mu \mapsto h(\mu, T, \mathcal{P})$ defined on $\mathcal{M}_T(X)$ is upper semicontinuous at μ_0.
(d) If $(Y, S, \mathbb{S})$ is a topological factor of $(X, T, \mathbb{S})$ via a factor map π, and $\mathfrak{A}_Y$ is as above, then the function $\mu \mapsto h(\mu, T, \mathcal{P}|\mathfrak{A}_Y)$ defined on $\mathcal{M}_T(X)$ is upper semicontinuous at μ_0.

Proof Item (a) is an immediate consequence of Lemma 6.6.5 and continuity of the function $\eta(t) = -t \log t$ on $[0, 1]$. For the dynamical entropy in item (c) we must also use the fact that if μ is invariant and $\mu(\partial A) = 0$ for all $A \in \mathcal{P}$, then the same holds for all $A \in \mathcal{P}^n$, for any $n \in \mathbb{N}$. We also need to remember that $\frac{1}{n}H(\mu, \mathcal{P}^n)$ converge decreasingly to $h(\mu, T, \mathcal{P})$ (and use Fact A.1.11).

We pass to proving items (b) and (d). Let ν_0 denote the image of μ_0 by π. By Fact 6.6.6, there exists a refining sequence of partitions $\mathcal{Q}_k$ of Y, all having the boundaries of measure zero for ν_0. Then the lifted partitions have boundaries of measure zero for μ_0. Thus, each of the functions $\mu \mapsto H(\mu, \mathcal{P}|\mathcal{Q}_k)$ (and also $\mu \mapsto \frac{1}{n}H(\mu, \mathcal{P}^n|\mathcal{Q}_k)$ in the case of item (d)) are continuous at μ_0. With increasing k these functions decrease to $H(\mu, \mathcal{P}|\mathfrak{A}_Y)$, in this manner shown to be upper semicontinuous as claimed in (b). Further, the upper semicontinuous functions $\frac{1}{n}H(\mu, \mathcal{P}^n|\mathfrak{A}_Y)$ decrease with n to $h(\mu, T, \mathcal{P}|\mathfrak{A}_Y)$, hence (d) holds. □

We remark that if $\mathcal{P}$ is a partition into sets with *small boundary*, i.e., with $\mu(\partial A) = 0$ for all invariant measures, then the above lemma holds for all $\mu \in \mathcal{M}_T(X)$. It is so, for example, if $\mathcal{P}$ is a partition into *clopen sets* (i.e., sets with empty boundary). The existence of arbitrary fine partitions with small boundaries is called the *small boundary property*. Not every dynamical system admits such partitions, for instance, the identity map on the interval does not. The reader will find sufficient conditions in the work of Elon Lindenstrauss [Lindenstrauss, 1999].

At the end we recall the important fact that, if $(Y, S, \mathbb{S})$ is a topological factor of $(X, T, \mathbb{S})$ via a factor map π, then the dual map π on invariant measures is a surjection from $\mathcal{M}_T(X)$ onto $\mathcal{M}_S(Y)$.

6.7 Topological fiber entropy

Topological fiber entropy is a notion intermediate between measure-theoretic fiber entropy and topological conditional entropy given a factor. It will be useful as a tool in proving the variational principles in the next sections. It has special meaning also in the discussion of the zero-dimensional case and in the theory of symbolic extensions.

Let $\pi : X \to Y$ be a topological factor map from $(X, T, \mathbb{S})$ onto $(Y, S, \mathbb{S})$.

Definition 6.7.1 For a cover $\mathcal{U}$ of X and a point $y \in Y$ we define successively

1. $\mathbf{H}(\mathcal{U}|y) = \mathbf{H}(\mathcal{U}|\{y\})$ (by convention $\mathbf{H}(\mathcal{U}|\pi^{-1}(y))$, see Definition 6.3.1),
2. $\mathbf{h}(T, \mathcal{U}|y) = \limsup_{n\to\infty} \frac{1}{n}\mathbf{H}(\mathcal{U}^n|y)$,
3. $\mathbf{h}(T|y) = \sup_{\mathcal{U}} \mathbf{h}(T, \mathcal{U}|y)$.

For a (not necessarily S-invariant) measure $\nu \in \mathcal{M}(Y)$ on Y, we let

4. $\mathbf{H}(\mathcal{U}|\nu) = \int \mathbf{H}(\mathcal{U}|y)\, d\nu(y)$,
5. $\mathbf{h}(T, \mathcal{U}|\nu) = \inf_n \frac{1}{n}\mathbf{H}(\mathcal{U}^n|\nu)$,
6. $\mathbf{h}(T|\nu) = \sup_{\mathcal{U}} \mathbf{h}(T, \mathcal{U}|\nu)$.

The terms defined in items 2 and 5 above are called *the topological fiber entropy* of $\mathcal{U}$ given the point y and *given the measure* ν, respectively. The terms defined in items 3 and 6 are called *the topological fiber entropy (of T) given y* and *given ν*.

Since the partition of X into the fibers is upper semicontinuous (see Fact A.1.4), it is not hard to see that the function $y \mapsto \mathbf{H}(\mathcal{U}|y)$ is upper semicontinuous on Y. This implies that also the functions $\nu \mapsto \mathbf{H}(\mathcal{U}|\nu)$ and $\nu \mapsto \mathbf{h}(T, \mathcal{U}|\nu)$ are upper semicontinuous on the set $\mathcal{M}(Y)$. Clearly, all four notions involving the cover $\mathcal{U}$ defined above increase when the cover refines (while y or ν are fixed). Since $S^{-1}(Sy) \ni y$, we easily verify that

$$\mathbf{H}(T^{-1}(\mathcal{U})|y) \leq \mathbf{H}(\mathcal{U}|Sy). \tag{6.7.2}$$

We will prove the following:

Fact 6.7.3 *For any cover $\mathcal{U}$ the sequence of functions $(\mathbf{H}(\mathcal{U}^n|\cdot))_n$ defined on Y is a subadditive cocycle (see Definition 2.1.3).*

Proof We note that $\mathcal{U}^{m+n} = \mathcal{U}^m \vee T^{-m}(\mathcal{U}^n)$, then we apply (6.3.7) (for $A = \{y\}$ and trivial $\mathcal{W}$), and finally the inequality (6.7.2). □

Note that $0 \le \mathbf{H}(\mathcal{U}|y) \le a$, where $a = \mathbf{H}(\mathcal{U})$. Thus, we can apply the Subadditive Ergodic Theorem 2.1.4 (together with the preceding remark), which leads to the following:

Corollary 6.7.4

(a) *If ν is invariant under S, then the sequence $\mathbf{H}(\mathcal{U}^n|\nu)$ is subadditive, so the* $\inf_n$ *in Definition 6.7.1 item 5 is in fact the limit.*

(b) *For an ergodic ν (i.e., $\nu \in \mathrm{ex}\mathcal{M}_S(Y)$), the* $\limsup_n$ *in Definition 6.7.1 item 2 is ν-almost surely the limit and it equals $\mathbf{h}(T, \mathcal{U}|\nu)$.*

(c) *Now, applying the ergodic decomposition to any $\nu \in \mathcal{M}_S(Y)$, we obtain that for ν-almost every y, in Definition 6.7.1 item 2 convergence holds. Further, by the dominated convergence theorem, we get*

$$\mathbf{h}(T, \mathcal{U}|\nu) = \int \mathbf{h}(T, \mathcal{U}|y)d\nu \quad (\textit{only for } \nu \in \mathcal{M}_S(Y)). \tag{6.7.5}$$

In particular, $\mathbf{h}(T, \mathcal{U}|\cdot)$ is a harmonic function on $\mathcal{M}_S(Y)$ (see Fact A.2.15).

(d) *The suprema in Definition 6.7.1 items 3 and 6 can be realized as monotone limits over a fixed refining sequence of covers, thus the functions $\mathbf{h}(T|\cdot)$ are Borel-measurable both on Y and on $\mathcal{M}(Y)$, and*

$$\mathbf{h}(T|\nu) = \int \mathbf{h}(T|y)d\nu,$$

for any $\nu \in \mathcal{M}_S(Y)$ (thus the function $\nu \mapsto \mathbf{h}(T|\nu)$ is harmonic on $\mathcal{M}_S(Y)$), and

$$\mathbf{h}(T|y) = \mathbf{h}(T|\nu) \tag{6.7.6}$$

ν-almost surely, for an ergodic measure $\nu \in \mathcal{M}_S(Y)$. □

We remark that although $\mathbf{H}(\mathcal{U}|y) = \mathbf{H}(\mathcal{U}|\boldsymbol{\delta}_y)$ for any $y \in Y$ (here $\boldsymbol{\delta}_y$ means the measure concentrated at y), we must not confuse $\mathbf{h}(T, \mathcal{U}|y)$ with $\mathbf{h}(T, \mathcal{U}|\boldsymbol{\delta}_y)$ where the latter may be smaller.

It follows from Corollary 6.7.4 item (d) and Exercise 6.8 that if $\mathcal{U}$ is a topological generator, then it realizes the suprema in Definition 6.7.1 items 3 and 6. In particular, the function $\nu \mapsto \mathbf{h}(T|\nu)$ is then upper semicontinuous on $\mathcal{M}_S(Y)$.

We prove the fact below:

Fact 6.7.7 *If* $\nu \in \mathcal{M}_S(Y)$*, then* $\mathbf{h}(T, T^{-1}(\mathcal{U})|\nu) = \mathbf{h}(T, \mathcal{U}|\nu)$.

Proof Indeed, we have, by (6.7.2) and (6.3.7),

$$\mathbf{H}(T^{-1}(\mathcal{U}^n)|y) \le \mathbf{H}(\mathcal{U}^n|Sy) = \mathbf{H}(\mathcal{U} \vee T^{-1}(\mathcal{U}^{n-1})|Sy) \le \mathbf{H}(\mathcal{U}|Sy) + \mathbf{H}(T^{-1}(\mathcal{U}^{n-1})|Sy).$$

Integrating with respect to the invariant measure ν on Y, dividing by n and letting $n \to \infty$, we verify the assertion. □

6.8 The major Variational Principles

Variational principles equate topological notions of entropy with maxima of the corresponding entropy functions defined (usually) on the simplex of invariant measures. There are nearly as many variational principles as there are entropy notions, including those not discussed in this book, such as pressure. This section covers four major variational principles: the "usual" Variational Principle and the Inner, Outer and Conditional Variational Principles. Only the Outer and Inner Variational Principles need proofs, the other two are their immediate consequences. For an elegant direct proof of the "usual" Variational Principle, we refer the reader to Petersen's book [Petersen, 1983] (the proof is due to Misiurewicz). In Chapter 7 we will give another, much simpler and probably not existing in the literature, proof of the Variational Principle for zero-dimensional systems. Let us remark that the Inner Variational Principle has been first proved (in a more general version involving pressure) by François Ledrappier and Peter Walters [Ledrappier and Walters, 1977]. In [Downarowicz and Serafin, 2002] the reader will find a version valid just for entropy but also in nonmetrizable spaces, as well as the Outer Variational Principle.

In Chapter 8 we shall also prove the Tail Entropy Variational Principle, but prior to that we need to introduce several notions related to the entropy structure. In Chapter 9, we will also prove a variational principle for the topological symbolic extension entropy. Finally in Part III the reader will find one direction of the variational principle for Markov operators.

Theorem 6.8.1 (The Variational Principle) *Let* $(X, T, \mathbb{S})$ *be a topological dynamical system. Then*

$$\mathbf{h}(T) = \sup\{h(\mu, T) : \mu \in \mathcal{M}_T(X)\} = \sup\{h(\mu, T) : \mu \in \mathsf{ex}\mathcal{M}_T(X)\}.$$

Proof This is a particular case of the Inner Variational Principle (Theorem 6.8.4 below) for $(Y, S, \mathbb{S})$ being the trivial (one-point) factor. □

Before we pass to the Outer Variational Principle we introduce an alternative formula for the topological conditional entropy, involving the fibers.

Lemma 6.8.2 *Let $\pi : X \to Y$ be factor map from $(X,T,\mathbb{S})$ onto $(Y,S,\mathbb{S})$. Let $\mathcal{U}$ be an open cover of X. Then there exists an open cover $\mathcal{V}$ of Y such that for each $V \in \mathcal{V}$ there is some $y \in V$ with*

$$\mathbf{H}(\mathcal{U}|V) = \mathbf{H}(\mathcal{U}|y).$$

As a consequence,

$$\mathbf{h}(T|S) = \sup_{\mathcal{U}} \inf_n \sup_{y\in Y} \tfrac{1}{n}\mathbf{H}(\mathcal{U}^n|y).$$

Proof It is clear that regardless of the choice of the cover $\mathcal{V}$, one has $\mathbf{H}(\mathcal{U}|V) \geq \mathbf{H}(\mathcal{U}|y)$, whenever $y \in V \in \mathcal{V}$. Now, fix some $y \in Y$ and pick an optimal (i.e., of smallest cardinality) family $\mathcal{U}_y \subset \mathcal{U}$ covering $\pi^{-1}(y)$. The set $\bigcup \mathcal{U}_y$ is open, hence, by Fact A.1.4, $\mathcal{U}_y$ covers the preimage by π of an open set, say V_y, containing y. The cover $\mathcal{V}$ of Y by the sets V_y $(y \in Y)$ so obtained satisfies the first assertion.

We pass to proving the second assertion. For a fixed cover $\mathcal{U}$ and a natural n, the first assertion implies that

$$\sup_{y\in Y} \tfrac{1}{n}\mathbf{H}(\mathcal{U}^n|y) = \inf_{\mathcal{V}} \tfrac{1}{n}\mathbf{H}(\mathcal{U}^n|\mathcal{V})$$

(we have proved the inequality "$\geq$", while "$\leq$" is obvious for covers $\mathcal{V}$ of Y). Since $\mathcal{V}^n \succcurlyeq \mathcal{V}$, the right-hand side does not increase if $\mathcal{V}$ is replaced by $\mathcal{V}^n$. On the other hand, such a substitution corresponds to taking the infimum over a smaller set of covers (only the "nth powers" of covers), thus it cannot decrease its value. Thus, we can write

$$\inf_n \sup_{y\in Y} \tfrac{1}{n}\mathbf{H}(\mathcal{U}^n|y) = \inf_n \inf_{\mathcal{V}} \tfrac{1}{n}\mathbf{H}(\mathcal{U}^n|\mathcal{V}^n).$$

Exchanging the infima on the right leads to

$$\inf_n \sup_{y\in Y} \tfrac{1}{n}\mathbf{H}(\mathcal{U}^n|y) = \inf_{\mathcal{V}} \mathbf{h}(T,\mathcal{U}|\mathcal{V}) = \mathbf{h}(T,\mathcal{U}|S).$$

We can now apply the supremum over $\mathcal{U}$ on both sides, and we are done. □

Theorem 6.8.3 (The Outer Variational Principle) *Let $\pi : X \to Y$ be a factor map from $(X,T,\mathbb{S})$ onto $(Y,S,\mathbb{S})$. Then*

$$\mathbf{h}(T|S) = \sup_{y\in Y} \mathbf{h}(T|y) = \sup_{\nu\in\mathcal{M}_S(Y)} \mathbf{h}(T|\nu) = \sup_{\nu\in\mathsf{ex}\mathcal{M}_S(Y)} \mathbf{h}(T|\nu).$$

Proof This proof relies largely on exchanging suprema and infima. For the first inequality "$\geq$," by Lemma 6.8.2 and Definition 6.7.1, we need to show

$$\sup_{\mathcal{U}} \inf_{n} \sup_{y} \tfrac{1}{n}\mathbf{H}(\mathcal{U}^n|y) \geq \sup_{y} \sup_{\mathcal{U}} \limsup_{n\to\infty} \tfrac{1}{n}\mathbf{H}(\mathcal{U}^n|y).$$

By Fact 6.7.3, the sequence $\sup_y \mathbf{H}(\mathcal{U}^n|y)$ is subadditive thus the infimum on the left-hand side is in fact a limit, hence also $\limsup$. Thus the inequality follows trivially by moving the supremum over y to the left (see Appendix A.1.5).

The next inequality "$\geq$" is also elementary: by (6.7.5), the inequality

$$\sup_{y} \mathbf{h}(T, \mathcal{U}|y) \geq \sup_{\nu} \mathbf{h}(T, \mathcal{U}|\nu)$$

holds for every cover $\mathcal{U}$, so we can apply $\sup_{\mathcal{U}}$ on both sides, and then exchange the order of suprema.

Next, we show that $\sup_\nu \mathbf{h}(T|\nu) \geq \mathbf{h}(T|S)$, i.e. that

$$\sup_{\nu} \sup_{\mathcal{U}} \mathbf{h}(T, \mathcal{U}|\nu) \geq \sup_{\mathcal{U}} \inf_{n} \sup_{y} \tfrac{1}{n}\mathbf{H}(\mathcal{U}^n|y),$$

with the first supremum taken over all invariant measures on Y. Since we can exchange the suprema on the left-hand side, it suffices to show that, for a fixed cover $\mathcal{U}$,

$$\sup_{\nu} \mathbf{h}(T, \mathcal{U}|\nu) \geq \inf_{n} \sup_{y} \tfrac{1}{n}\mathbf{H}(\mathcal{U}^n|y).$$

Denote the right-hand side of the last inequality by C and suppose that the left-hand side is smaller than C minus some ε. Recall (see item (a) in Corollary 6.7.4) that on the compact set $\mathcal{M}_S(Y)$ the function $\nu \mapsto \mathbf{h}(T, \mathcal{U}|\nu)$ is realized as $\lim_{n} \frac{1}{n}\mathbf{H}(\mathcal{U}^n|\nu)$, where the sequence $\mathbf{H}(\mathcal{U}^n|\nu)$ is subadditive. Thus, this limit is decreasing along a subsequence indexed by (n_k) with each n_{k+1} being a multiple of n_k. Since each $\mathbf{H}(\mathcal{U}^n|\cdot)$ is an upper semicontinuous function, we conclude that

$$\tfrac{1}{n}\mathbf{H}(\mathcal{U}^n|\nu) < C - \varepsilon,$$

for some positive integer n and all invariant measures ν (see Fact A.1.14). Further, by upper semicontinuity of $\mathbf{H}(\mathcal{U}^n|\cdot)$ on the set $\mathcal{M}(Y)$ of all probability measures, the same holds on some neighborhood U of $\mathcal{M}_S(Y)$. Thus, for sufficiently large numbers m, the above inequality is valid for all measures of the form

$$\nu = \mathbf{M}_m(\boldsymbol{\delta}_y) = \frac{1}{m}\sum_{i=0}^{m-1} \boldsymbol{\delta}_{S^i y},$$

for all $y \in Y$ (see Fact 6.6.1). We can choose m so large that $2na/m \leq \varepsilon/2$, where $a = \mathbf{H}(\mathcal{U})$ bounds the first function $y \mapsto \mathbf{H}(\mathcal{U}|y)$ in the subadditive cocycle $(\mathbf{H}(\mathcal{U}^n|y))_n$ (see Fact 6.7.3). The inequality (2.1.7) reads

$$\mathbf{H}(\mathcal{U}^{m+n}|y) \leq \frac{1}{n}\sum_{i=0}^{m-1} \mathbf{H}(\mathcal{U}^n|S^i y) + 2na.$$

The sum on the right equals $m\mathbf{H}(\mathcal{U}^n|\mathrm{M}_m(\boldsymbol{\delta}_y))$, so, dividing the above inequality by m, we get

$$\tfrac{1}{m+n}\mathbf{H}(\mathcal{U}^{m+n}|y) \leq \tfrac{1}{m}\mathbf{H}(\mathcal{U}^{m+n}|y) \leq \tfrac{1}{n}\mathbf{H}(\mathcal{U}^n|\mathrm{M}_m(\boldsymbol{\delta}_y)) + \tfrac{\varepsilon}{2} \leq C - \tfrac{\varepsilon}{2},$$

for any y, that is, we can apply $\sup_y$ on the left-hand side. But by definition, $C \leq \sup_y \frac{1}{m+n}\mathbf{H}(\mathcal{U}^{m+n}|y)$, a contradiction.

That the supremum over all ergodic measures is not too small (it is obviously not too large) follows immediately from the harmonic property of the function $\nu \mapsto \mathbf{h}(T|\nu)$ on $\mathcal{M}_S(Y)$ (see (6.7.5)). This proves the last equality in the assertion of the theorem. □

Theorem 6.8.4 (The Inner Variational Principle) *Let $\pi : X \to Y$ be a topological factor map between topological dynamical systems $(X, T, \mathbb{S})$ and $(Y, S, \mathbb{S})$. For every $\nu \in \mathcal{M}_S(Y)$ we have*

$$\mathbf{h}(T|\nu) = \sup\{h(\mu, T|\nu, S) : \mu \in \mathcal{M}_T(X), \pi(\mu) = \nu\}.$$

Proof Since the functions $\mu \mapsto \mathbf{h}(T|\pi\mu)$ and $\mu \mapsto h(\mu|\pi\mu)$ are harmonic on $\mathcal{M}_T(X)$, it suffices to prove the inequality "$\geq$" for ergodic μ and $\nu = \pi\mu$. We will show that $\mathbf{h}(T|\nu) \geq h(\mu|\nu) - \varepsilon$, i.e., (by Definition 6.7.1 items 5 and 6, Corollary 6.7.4 item (a), and Definitions 4.1.5, 2.3.3 and 1.4.5), that

$$\sup_{\mathcal{U}} \lim_n \frac{1}{n}\int \mathbf{H}(\mathcal{U}^n|y)d\nu \geq \sup_{\mathcal{P}} \lim_n \frac{1}{n}\inf_{\mathcal{Q}} H(\mu, \mathcal{P}^n|\mathcal{Q}) - \varepsilon,$$

where $\mathcal{Q}$ and $\mathcal{P}$ range over all Borel measurable partitions of Y (lifted to X) and over all Borel measurable partitions of X, respectively. We do so by constructing, for each partition $\mathcal{P}$ of X, a cover $\mathcal{U}$ such that for every sufficiently large n there exists a partition $\mathcal{Q}$ of Y satisfying

$$\int \mathbf{H}(\mathcal{U}^n|y)\,d\nu \geq H(\mu, \mathcal{P}^n|\mathcal{Q}) - n\varepsilon. \tag{6.8.5}$$

Let $\mathcal{P} = \{A_1, \ldots, A_l\}$. By regularity of μ, we can enlarge each set A_j to an open set U_j so that $\mu(G) < \delta$, where G denotes $\bigcup_{j=1}^{l}(U_j \setminus A_j)$ and $\delta > 0$

will be specified later. We let $\mathcal{U}$ be the cover by the sets U_j. By the Ergodic Theorem, the sets

$$X_n = \{x : \tfrac{1}{n}\#\{i \in [0, n-1] : T^i x \in G\} < \delta\}$$

have measures tending to 1, so for n large enough, $\mu(X_n) > 1 - \delta$. We fix an n with this property. The partition $\mathcal{Q}$ of Y is obtained as follows: for each y we fix an optimal family $\mathcal{U}_y \subset \mathcal{U}^n$ covering $\pi^{-1}(y)$. Since $\mathcal{U}$ is finite, there are finitely many choices of these families. In this manner we classify the points y into finitely many disjoint sets Q (this defines the partition $\mathcal{Q}$), such that $\mathbf{H}(\mathcal{U}|y) = \mathbf{H}(\mathcal{U}|Q_y)$, where Q_y is determined by the inclusions $y \in Q_y \in \mathcal{Q}$. It remains to verify (6.8.5).

For an open cover $\mathcal{U}$ and a point x a $\mathcal{U}$-*name* of length n of x is any sequence $U_0, \dots, U_{n-1}$ of elements of $\mathcal{U}$ such that $x \in \bigcap_{i=0}^{n-1} T^{-i}(U_i)$. Unlike for partitions, each point may admit multiple $\mathcal{U}$-names. The value $\mathbf{H}(\mathcal{U}^n|V)$ ($V \subset X$) is the logarithm of the minimal number of $\mathcal{U}$-names of length n sufficing to "call" all elements of V. We are going to compare $\mathbf{H}(\mathcal{U}^n|\pi^{-1}(Q_y) \cap X_n)$ with an analogous value, denoted by $\mathbf{H}(\mathcal{P}^n|\pi^{-1}(Q_y) \cap X_n)$, the logarithm of the number of the $\mathcal{P}$-names of length n appearing in the same set. A $\mathcal{U}$-name of a point x translates to its $\mathcal{P}$-name (by replacing the symbols U_j by A_j) except at the coordinates n for which $T^n x \in G$. Because each point of X_n has an established frequency (not exceeding δ) of the visits in G in time $[0, n-1]$, each $\mathcal{U}$-name of length n appearing in X_n splits into at most $\binom{n}{\delta n} \cdot l^{\delta n}$ different $\mathcal{P}$-names of length n appearing in X_n (the former factor represents the number of possible distributions of the visits in G during the time, the latter is the number of possible configurations of $\mathcal{P}$-symbols at the corresponding positions). We can thus write:

$$\mathbf{H}(\mathcal{P}^n|\pi^{-1}(Q_y) \cap X_n) \leq \mathbf{H}(\mathcal{U}^n|\pi^{-1}(Q_y) \cap X_n) + nH(\delta, 1-\delta) + n\delta \log l.$$

Further, by the elementary estimate of the static entropy of a partition by the logarithm of the cardinality of this partition, we have

$$\mathbf{H}(\mathcal{P}^n|\pi^{-1}(Q_y) \cap X_n) \geq H_{\pi^{-1}(Q_y) \cap X_n}(\mathcal{P}^n),$$

where, by convention, the right-hand side is the entropy with respect to the conditional measure induced by μ on the set appearing in the subscript. This, combined with other obvious passages, yields the following calculation:

$$\int \mathbf{H}(\mathcal{U}^n|y)\,d\nu =$$
$$\int \mathbf{H}(\mathcal{U}^n|\pi^{-1}(Q_y))\,d\nu \geq \int \mathbf{H}(\mathcal{U}^n|\pi^{-1}(Q_y)\cap X_n)\,d\nu \geq$$
$$\int H_{\pi^{-1}(Q_y)\cap X_n}(\mathcal{P}^n)\,d\nu - n(H(\delta,1-\delta)+\delta\log l).$$

For an appropriate δ the error term does not exceed $n\varepsilon/2$. We continue to estimate the last integral:

$$\int H_{\pi^{-1}(Q_y)\cap X_n}(\mathcal{P}^n)\,d\nu = \sum_{Q\in\mathcal{Q}} \mu(\pi^{-1}(Q))H_{\pi^{-1}(Q)\cap X_n}(\mathcal{P}^n) \geq$$
$$\sum_{Q\in\mathcal{Q}} \mu(\pi^{-1}(Q)\cap X_n)H_{\pi^{-1}(Q)\cap X_n}(\mathcal{P}^n) + \mu(X_n^c)H_{X_n^c}(\mathcal{P}^n) - \delta n\log l \geq$$
$$H(\mu,\mathcal{P}^n|\mathcal{Q}\vee\mathcal{R}) - n\delta\log l \geq H(\mu,\mathcal{P}^n|\mathcal{Q}) - H(\mu,\mathcal{R}) - n\delta\log l,$$

where the partition $\mathcal{R} = \{X_n, X_n^c\}$ has entropy at most $H(\delta,1-\delta)$, which, together with the last error term, is smaller than $n\varepsilon/2$. The proof of the "easy" part of the Inner Variational Principle is completed.

The proof of the other inequality follows the standard line: we will construct an invariant measure μ on X lifting the given measure ν on Y, such that $h(\mu|\nu) \geq \mathbf{h}(T|\nu) - \varepsilon$. This measure will be obtained as a weak-star limit of certain atomic measures concentrated on (n,δ)-separated sets contained in the fibers. At some point we will need subadditivity of the cocycle $(\mu,n) \mapsto H(\mu,\mathcal{P}^n|\mathfrak{B})$ on the space $\mathcal{M}(X)$ of all probability measures on X, where $\mathfrak{B}$ is any subinvariant sigma-algebra. We prove it now. We have

$$H(\mu,\mathcal{P}^{m+n}|\mathfrak{B}) \leq H(\mu,\mathcal{P}^m|\mathfrak{B}) + H(\mu,T^{-m}(\mathcal{P}^n)|\mathfrak{B}) \leq$$
$$H(\mu,\mathcal{P}^m|\mathfrak{B}) + H(\mu,T^{-m}(\mathcal{P}^n)|T^{-m}(\mathfrak{B})) = H(\mu,\mathcal{P}^m|\mathfrak{B}) + H(T^m\mu,\mathcal{P}^n|\mathfrak{B}),$$

which is exactly the subadditivity condition for this cocycle, as defined in (2.1.3).

Notice that the function $\nu \mapsto f(\nu) = \sup\{h(\mu|\nu) : \pi\mu = \nu\}$ is supharmonic, i.e., its value at a measure ν cannot be smaller than the corresponding average with respect to the ergodic decomposition. Indeed, if $\int \nu_\omega\,d\xi^\nu(\omega)$ is the ergodic decomposition of ν (ν_ω are the ergodic measures on Y parametrized by $\omega\in\Omega$) and, for each parameter ω, μ_ω is a lift of ν_ω with $h(\mu_\omega|\nu_\omega) \geq f(\nu_\omega) - \varepsilon$, then $\mu = \int \mu_\omega\,d\xi^\nu(\omega)$ is a lift of ν, and its conditional entropy is at least $\int f(\nu_\omega)\,d\xi^\nu(\omega) - \varepsilon$. So, $f(\nu)$ cannot be smaller than this average.

Having verified this, recall that $\nu \mapsto \mathbf{h}(T|\nu)$ is harmonic (see Corollary 6.7.4 item (d)), hence it suffices to prove the inequality $f(\nu) \geq \mathbf{h}(T|\nu)$ for ergodic ν only. We fix some ergodic ν now and we fix an $\varepsilon > 0$. We choose an open cover $\mathcal{U}$ of X so that $\mathbf{h}(T, \mathcal{U}|\nu) > \mathbf{h}(T|\nu) - \varepsilon$. We let $\delta = \mathsf{Leb}(\mathcal{U})$.

It is known that for ν-almost every $y \in Y$ the measures

$$\nu_{m,y} = \frac{1}{m} \sum_{i=0}^{m-1} \boldsymbol{\delta}_{S^i y}$$

converge weakly-star to ν. We also know (see Corollary 6.7.4 item (b)) that for ν-almost every y, $\frac{1}{m}\mathbf{H}(\mathcal{U}^m|y)$ converges to $\mathbf{h}(T, \mathcal{U}|\nu)$. We fix some y with both of the above properties. For each m let E_y^m be the maximal (m, δ)-separated set in $\pi^{-1}(y)$, and let μ_y^m be the atomic probability measure equally distributed over E_y^m. We now define the measure μ as a weak-star accumulation point of the sequence

$$\mu_{m,y} = \frac{1}{m} \sum_{i=0}^{m-1} T^i(\mu_y^m).$$

Clearly, by being a limit of longer and longer averages along orbits of measures, μ is T-invariant. Since every μ_y^m projects by π to $\boldsymbol{\delta}_y$, $\mu_{m,y}$ projects to $\nu_{m,y}$ and, by continuity of π on measures, μ projects to ν. It remains to compare the conditional entropy of μ given the sigma-algebra lifted from Y with $\mathbf{h}(T|\nu)$.

Because $\delta = \mathsf{Leb}(\mathcal{U})$, we can apply (6.1.10) and (6.1.11) restricted to the fiber $\pi^{-1}(y)$, which yield

$$\tfrac{1}{m} \log \#E_y^m \geq \tfrac{1}{m}\mathbf{H}(\mathcal{U}^m|y).$$

The right-hand side converges to $\mathbf{h}(T, \mathcal{U}|\nu)$, so, for large m, it is larger than $\mathbf{h}(T|\nu) - \varepsilon$. We choose a finite partition $\mathcal{P}$ of X satisfying two conditions: $\mathsf{diam}(\mathcal{P}) < \delta$ and $\mu(\partial A) = 0$ for each element $A \in \mathcal{P}$. Such partitions exist, by elementary facts in measure theory (see Fact 6.6.6). Notice that, since the diameter of $\mathcal{P}$ is smaller than δ, every cell of $\mathcal{P}^{m+i}$ (for any $i \geq 0$) contains at most one element of any (m, δ)-separated set, in particular, of E_y^m. Thus, with regard to μ_y^m, the partition $\mathcal{P}^{m+i}$ has $\#E_y^m$ nonzero cells of equal measures, so that

$$\tfrac{1}{m} H(\mu_y^m, \mathcal{P}^{m+i}) = \tfrac{1}{m} \log \#E_y^m > \mathbf{h}(T|\nu) - \varepsilon.$$

Hence, for every $i \geq 0$,

$$\tfrac{1}{m} H(\mu_y^m, \mathcal{P}^{[i,m+i-1]}) \geq \tfrac{1}{m} H(\mu_y^m, \mathcal{P}^{m+i}) - \tfrac{1}{m} H(\mu_y^m, \mathcal{P}^i) > \mathbf{h}(T|\nu) - \varepsilon - \tfrac{i}{m} \log \#\mathcal{P}.$$

The term on the left equals $\frac{1}{m}H(T^i\mu_y^m, \mathcal{P}^m)$ and, since the measure is supported by one fiber (precisely, by $\pi^{-1}(S^i y)$), it equals $\frac{1}{m}H(T^i\mu_y^m, \mathcal{P}^m|\mathfrak{A}_Y)$, where $\mathfrak{A}_Y$ stands for the Borel sigma-algebra of Y lifted to X. We have proved that

$$\tfrac{1}{m}H(T^i\mu_y^m, \mathcal{P}^m|\mathfrak{A}_Y) \geq \mathbf{h}(T|\nu) - \varepsilon - \tfrac{i}{m}\log\#\mathcal{P}. \tag{6.8.6}$$

Averaging (6.8.6) over $i = 0, \ldots, n-1$ we get

$$\frac{1}{mn}\sum_{i=0}^{n-1} H(T^i\mu_y^m, \mathcal{P}^m|\mathfrak{A}_Y) \geq \mathbf{h}(T|\nu) - \varepsilon - \tfrac{n}{m}\log\#\mathcal{P}. \tag{6.8.7}$$

We now invoke the subadditivity of the cocycle $H(\cdot, \mathcal{P}^m|\mathfrak{A}_Y)$ on the space $\mathcal{M}(X)$ of all probability measures on X. The formula (2.1.6) yields that the left-hand side in the last formula is dominated by

$$\frac{1}{mn}\sum_{i=0}^{m-1} H(T^i\mu_y^m, \mathcal{P}^n|\mathfrak{A}_Y) + \tfrac{n}{m}\log\#\mathcal{P}.$$

By concavity of the conditional static entropy (see (1.4.8)), the above average is dominated by $\frac{1}{n}H(\mu_{m,y}, \mathcal{P}^n|\mathfrak{A}_Y)$. Plugging this into (6.8.7), we obtain

$$\tfrac{1}{n}H(\mu_{m,y}, \mathcal{P}^n|\mathfrak{A}_Y) \geq \mathbf{h}(T|\nu) - \varepsilon - \tfrac{2n}{m}\log\#\mathcal{P}.$$

Now we let m grow along the subsequence for which the measures $\mu_{m,y}$ converge to μ, while n remains fixed. Because μ is T-invariant and $\mathcal{P}$ has boundaries of measure zero, so does $\mathcal{P}^n$. This implies that the function $H(\cdot, \mathcal{P}^n|\mathfrak{A}_Y)$ is continuous at μ (see Lemma 6.6.7) and yields

$$\tfrac{1}{n}H(\mu, \mathcal{P}^n|\mathfrak{A}_Y) \geq \mathbf{h}(T|\nu) - \varepsilon.$$

Finally we can pass with n to infinity:

$$h(\mu, T, \mathcal{P}|\mathfrak{A}_Y) \geq \mathbf{h}(T|\nu) - \varepsilon.$$

The left-hand side is not smaller than $h(\mu|\mathfrak{A}_Y)$, which is another notation for $h(\mu|\nu)$. The proof is now complete. □

Combining the Inner and Outer Variational Principles, we immediately deduce:

Theorem 6.8.8 (The Conditional Variational Principle) *Let $\pi : X \to Y$ be a topological factor map between topological dynamical systems $(X, T, \mathbb{S})$ and $(Y, S, \mathbb{S})$. Then*

$$\mathbf{h}(T|S) = \sup\{h(\mu|\nu) : \mu \in \mathcal{M}_T(X), \nu = \pi\mu\} \ \left(= \sup_{\mu\in\mathcal{M}_T(X)} h(\mu, T|\mathfrak{A}_Y)\right).$$

□

Note that Fact 6.4.8, proved earlier without using measures, now appears as a consequence of the mere fact that the supremum of the sum of two functions does not exceed the sum of their suprema. We can draw numerous similar easy consequences of the variational principles, see Exercises 6.9 through 6.12.

An important consequence concerns the principal factors (extensions):

Corollary 6.8.9 *A factor map $\pi : X \to Y$ is principal if and only if*

$$h(\mu|\pi\mu) = 0, \ \textit{for every } \mu \in \mathcal{M}_T(X).$$

In case $(Y, S, \mathbb{S})$ has finite topological entropy, this is equivalent to

$$h(\mu) = h(\pi\mu), \ \textit{for every } \mu \in \mathcal{M}_T(X). \qquad \square$$

Finally, we can easily deduce preservation of topological entropy by the topological natural extension. Natural extensions have been introduced in Part I in the measure-theoretic context. The same construction works also in the topological context, however there is a subtlety concerning surjectivity:

Definition 6.8.10 Let $(X, T, \mathbb{N}_0)$ be a topological dynamical system with T surjective. By the *topological natural extension* we will mean the $\mathbb{N}_0$-action of the shift map T' on the space $X' \subset X^{\mathbb{Z}}$ (equipped with the product topology) defined by the rule $(x_n)_{n\in\mathbb{Z}} \in X' \iff \forall_n \, x_{n+1} = Tx_n$. Notice that T' is a homeomorphism and $(X', T', \mathbb{N}_0)$ factors onto $(X, T, \mathbb{N}_0)$ via the projection onto the coordinate zero.

In case T is not surjective, such a "perfect" natural extension need not exist. We must choose between either admitting T' to be not necessarily surjective or X' to factor not precisely onto X. We prefer the first option. And so, we first enlarge the space X (to some X_1) and prolong[3] T to the enlarged space in such a manner that it becomes surjective. For example, X_1 can be the one-point compactification of $X \times \mathbb{N}_0$ with T_1 defined by the rule

$$T_1(x, n) = \begin{cases} (x, n-1); & n \geq 1 \\ (Tx, 1); & n = 0 \end{cases},$$

and with the point at infinity being fixed. Such T_1 is clearly surjective, and the subsystem on $X \times \{0\}$ is conjugate to $(X, T, \mathbb{S})$. Next we apply the natural extension to this enlarged surjective system X_1: we obtain a space X_1', a homeomorphism T_1' and a factor map from X_1' onto X_1. Finally X' is defined as the preimage of X in this factor map. The transformation T' is the restriction of T_1' to X' (which is forward invariant).

[3] We prefer not to use the word "extend," as it is associated with "extensions."

Definition 6.8.11 We call $(X', T', \mathbb{N}_0)$ the *natural extension* of $(X, T, \mathbb{N}_0)$. Notice that T' is always injective, but not necessarily surjective.

We skip the proof of the fact that in either case (of T surjective and not surjective), each invariant measure $\mu \in \mathcal{M}_T(X)$ lifts to a unique measure $\mu' \in \mathcal{M}_{T'}(X')$ and that the measure-theoretic system $(X', \mathfrak{A}_{\mu'}, \mu', T', \mathbb{N}_0)$ is the natural extension (in the sense defined in Definition 4.3.1) of $(X, \mathfrak{A}_\mu, \mu, T, \mathbb{N}_0)$. By Fact 4.3.2, we can see that the natural extension preserves the entropy of each invariant measure, moreover, the relevant conditional entropy is zero. This, combined with Corollary 6.8.9, gives the following conclusion:

Fact 6.8.12 *The topological natural extension is principal.* □

6.9 Determinism in topological systems

In ergodic theory deterministic systems (with entropy zero) have many equivalent characterizations which can be easily deduced from the definition of Kolmogorov–Sinai entropy (Definition 4.1.1), the Krieger Generator Theorem 4.2.3, Fact 2.3.12 and Theorem 4.2.9:

1. The Kolmogorov–Sinai entropy of $(X, \mathfrak{A}, \mu, T, \mathbb{S})$ is zero.
2. The sigma-algebra $\mathfrak{A}$ equals the Pinsker sigma-algebra Π_μ (recall Definition 4.2.6 and Section 3.2: $\Pi_\mu = \bigvee_{\mathcal{P}} \Pi_{\mathcal{P}}$, $\Pi_{\mathcal{P}} = \bigcap_{n \geq 0} \mathcal{P}^{[n,\infty)}$).
3. The system occurs as factor of another system (extension), so that the (lifted) sigma-algebra $\mathfrak{A}$ is contained in the Pinsker sigma-algebra of the extension.
4. The system occurs as factor of a process over a finite partition $\mathcal{P}$, and the (lifted) sigma-algebra $\mathfrak{A}$ is contained in the Pinsker sigma-algebra $\Pi_{\mathcal{P}}$ of the extending process.
5. Every subinvariant sub-sigma-algebra of $\mathfrak{A}$ is invariant (i.e., the transformation in every factor of the forward action $(X, \mathfrak{A}, \mu, T, \mathbb{N}_0)$ is invertible).

There have been several attempts to create topological analogs of determinism and of the notion of the Pinsker factor. Depending on which one is treated as the starting point one obtains five classes of topological systems, analogs of determinism. We will prove that four out of five "determinism" conditions are mutually equivalent, providing three new characterizations of systems with topological entropy zero. Similarly, we will define four topological analogs of the Pinsker factor, this time however all four will turn out essentially different.

6.9.1 Topological analogs of determinism

The most obvious analog of a measure-theoretic deterministic system is a topological dynamical system with topological entropy zero. We denote the corresponding class by TEZ. It is obviously closed under taking factors.

Following Kaminski, Siemaszko and Szymanski [Kamiński *et al.*, 2003], we call a system $(X, T, \mathbb{S})$ *topologically deterministic* when in all factors $(Y, S, \mathbb{N}_0)$ of the system $(X, T, \mathbb{N}_0)$, S is a homeomorphism. The same authors showed that topologically deterministic systems had entropy zero. We will denote this class by TD. By definition, it is closed under taking factors. It can be considered the topological analog of the measure-theoretic class of deterministic systems defined via the last characterization (item 5 in the introduction to this section). Not all systems with topological entropy zero are topologically deterministic; for instance, there are noninvertible zero-entropy systems. In other words, we have the proper inclusion

$$\mathsf{TD} \subset \mathsf{TEZ}.$$

Before we introduce the next class of systems (and factors) we recall an elementary definition from topological dynamics:

Definition 6.9.1 A pair of distinct points x, y in a topological dynamical system $(X, T, \mathbb{S})$ is *(forward) asymptotic* if

$$\limsup_{n \to \infty} d(T^n x, T^n y) = 0.$$

Suppose we want to mimic the notion of the Pinsker sigma-algebra from ergodic theory. For a process generated by a finite partition $\mathcal{P}$, this sigma-algebra equals $\Pi_{\mathcal{P}} = \bigcap_{n \geq 1} \mathcal{P}^{[n,\infty)}$. If a function f on X is $\Pi_{\mathcal{P}}$-measurable, then its value $f(x)$ at $x = (x(n))_{n \in \mathbb{S}}$ is (almost surely) determined by the unilateral sequence $x[n, \infty)$ starting at any positive n.

In topological dynamics an analog of a process over a finite partition is a subshift over a finite alphabet Λ (see Definition 7.1.1 in the next chapter for details). The following definition attempts to copy the measure-theoretic concept of measurability (of a factor of a process generated by a finite partition $\mathcal{P}$) with respect to the Pinsker sigma-algebra $\Pi_{\mathcal{P}}$: the image of each point x via the topological factor map should be determined by the unilateral sequence $x[n, \infty)$ starting at any positive n.

Definition 6.9.2 Let $(X, T, \mathbb{S})$ be a subshift. A topological factor $(Y, S, \mathbb{S})$ of $(X, T, \mathbb{S})$ (with a factoring map $\pi : X \to Y$) is *Pinsker-like* if

$$\forall_{n \in \mathbb{N}} \, \forall_{x, x' \in X} \;\; x[n, \infty) = x'[n, \infty) \implies \pi x = \pi x'.$$

In other words, π collapses asymptotic pairs.

The last phrasing of this condition can be applied not only to subshifts: factors that collapse asymptotic pairs can be considered in any topological dynamical systems. They will also be called *Pinsker-like factors*. There seems to be an essential difference between Pinsker-like factor maps applied to subshifts and to arbitrary systems. A pair x, x' in a subshift is asymptotic whenever it is "ε-asymptotic," i.e, when $\limsup_n d(T^n x, T^n x') < \varepsilon$ for a sufficiently small epsilon. In general, asymptoticity cannot be weakened this way. The requirement that a factor map collapses all asymptotic pairs is stronger for subshifts than for general systems, because it means that all "ε-asymptotic" pairs are already collapsed. So, we will distinguish between two seemingly different classes of topological systems, as defined below, by analogy to the characterizations 4 and 3 of measure-theoretic determinism listed in the introduction to this section:

Definition 6.9.3 We will call a topological dynamical system $(X, T, \mathbb{S})$ *(strongly) Pinsker-like* if there exists a subshift, such that $(X, T, \mathbb{S})$ is its Pinsker-like factor. A system $(X, T, \mathbb{S})$ is *weakly Pinsker-like* if it occurs as a Pinsker-like factor of another topological dynamical system (not necessarily a subshift).

The classes PL of Pinsker-like and WPL of weakly Pinsker-like systems are both closed under taking factors, which follows from the completely trivial observation below:

Lemma 6.9.4 *Let π be a topological factor map (between two topological dynamical systems) which is a composition of several factor maps, at least one of which is Pinsker-like. Then π is Pinsker-like.*

Proof Just observe that any factor map sends an asymptotic pair either to an asymptotic pair or collapses it. □

We will now introduce yet another class of systems, defined by analogy to the second characterization of measure-theoretic determinism (item 2 in the introduction to this section). A measure-theoretic system is deterministic if it is its own Pinsker factor (via the identity map). In our analogy, this would mean that a topological system should be its own Pinsker-like factor via identity, i.e., that identity collapses asymptotic pairs. This is possible only in systems which simply do not have (distinct) asymptotic pairs, leading to the following class:

Definition 6.9.5 A topological dynamical system $(X, T, \mathbb{S})$ is called NAP (no asymptotic pairs) if it has no asymptotic pairs.

The class of NAP systems is not closed under taking factors. There is a quite complicated example in [Blanchard *et al.*, 2002]. Also, from the results given below, it follows that any nonperiodic subshift of entropy zero has a NAP extension, while nonperiodic subshifts are never NAP (this last elementary fact goes back to [Bryant and Walters, 1969]). Below we give a very simple explicit example:

Example 6.9.6 There exists a NAP-system $(X, T, \mathbb{Z})$ admitting a nontrivial factor $(Y, S, \mathbb{Z})$ with all distinct pairs in Y asymptotic. We begin by describing the factor system. We let $(Y, S, \mathbb{Z})$ be the one-point compactification of the integers with the map $n \mapsto n+1$ (and $\infty \mapsto \infty$). It is obvious that all distinct pairs in this system are asymptotic. The extension $(X, T, \mathbb{Z})$ is a subsystem of the product space $Y \times \mathbb{T}$, where $\mathbb{T}$ is the circle treated as the additive group $[0, 1)$ with addition modulo 1. On this space we introduce the following action: we fix an irrational number $\varrho \in (0, 1)$ and we define T by the formula

$$T(n, t) = (n + 1, t + \varrho + \tfrac{1}{n}) \quad \text{(for } n = 0 \text{ we simply skip } \tfrac{1}{n}\text{)},$$

and on the invariant circle $\{\infty\} \times \mathbb{T}$ we apply the irrational rotation by ϱ. We restrict the system to this invariant circle and the two-sided orbit of the point $x_0 = (0, 0)$. It is easy to see that we obtain a closed invariant set X on which T is a homeomorphism, extending $(Y, S, \mathbb{Z})$. It remains to show that there are no asymptotic pairs in X.

If a pair x, x' consists of two points from the invariant circle, then the distance between $T^n x$ and $T^n x'$ does not depend on n, and such a pair is not asymptotic. If x belongs to the circle and x' is on the single orbit outside the circle, then the projection of $T^n x'$ onto the circle rotates by the varying angle $\varrho + 1/n$ (while x rotates by the constant angle ϱ). The differences $1/n$ decrease to zero, but form a divergent series, so it is easy to see that this pair of points is not asymptotic either. Finally consider a pair x, x' where both points are outside the invariant circle. Then $x = T^m x_0$, $x' = T^{m+k} x_0$, for some $m \in \mathbb{Z}$ and a positive integer k. The projections of the points $T^n x = T^{m+n} x_0$ and $T^n x' = T^{m+n+k} x_0$ onto the circle differ by

$$k\varrho + \tfrac{1}{m+n} + \tfrac{1}{m+n+1} + \cdots + \tfrac{1}{m+n+k-1}.$$

The finite sum of the harmonic series visible in the above formula decreases to zero as n grows, hence the distance between such pair converges to $k\varrho$ (mod 1). Because ϱ is irrational, for any k this limit is positive. So such a pair is not asymptotic either.

Since the class of NAP systems is not closed under taking factors (which makes it a poor analog of the measure-theoretic class of deterministic systems), it is reasonable to enlarge the class by admitting all factors of NAP systems. So enlarged class, denoted FNAP, is going to be our last topological analog of determinism, corresponding to property 2 in the introduction.

The inclusion FNAP $\subset$ WPL is obvious: a factor of a NAP system is its factor via a map that collapses all asymptotic pairs (because there are none). The inclusion PL $\subset$ WPL is trivial.

The inclusion WPL $\subset$ TEZ has been proved in [Blanchard *et al.*, 2002]:

Theorem 6.9.7 *Weakly Pinsker-like systems have topological entropy zero.* □

The argument in [Blanchard *et al.*, 2002] uses invariant measures and measure-theoretic entropy. Its clue is an interesting observation, which we have decided to quote (without a proof). We remark that although $\Pi_\mu = \bigvee_k \bigcap_n \mathcal{P}_k^{[n,\infty)}$ (see Remark 4.2.7; $\mathcal{P}_k$ are as in the formulation below), in general one cannot reverse the order of the big operators; this makes the lemma nontrivial.

Lemma 6.9.8 *Let $(X,T,\mathbb{S})$ be a topological dynamical system and let μ be an ergodic measure on X. Fix a refining sequence of partitions $\mathcal{P}_k$ with diameters of the cells decreasing to zero. Then there exists a sequence n_k of natural numbers such that*

$$\bigcap_{n=1}^{\infty} \bigvee_{k=1}^{\infty} \mathcal{P}_k^{[n_k+n,\infty)} = \Pi_\mu. \qquad \square$$

Once the lemma is proved, it suffices to notice that for every n (after discarding a null set) any pair of points in the same atom of the sigma-algebra $\bigvee_{k=1}^{\infty} \mathcal{P}_k^{[n_k+n,\infty)}$ is asymptotic. Thus any measurable map collapsing asymptotic pairs must be constant on such atoms, and hence Π_μ-measurable. This implies Theorem 6.9.7 (via Theorem 4.2.9 and the Variational Principle).

We repeat after the authors that a purely topological proof would be desirable.

The inclusions TEZ $\subset$ PL and TEZ $\subset$ FNAP also hold. Since the proofs need a tool developed in the next chapter, we postpone them until the end of that chapter. Now just the formulations:

Theorem 6.9.9 *Every topological dynamical system $(X,T,\mathbb{S})$ with topological entropy zero is a Pinsker-like factor of a subshift of topological entropy zero.*

Theorem 6.9.10 *Every topological dynamical system $(X,T,\mathbb{S})$ with topological entropy zero is a factor of a zero-dimensional* NAP *system.*

Combining the inclusions provided above we obtain the main theorem of this section [Downarowicz and Lacroix, in print]

Theorem 6.9.11

$$\mathsf{FNAP} = \mathsf{WPL} = \mathsf{PL} = \mathsf{TEZ}. \qquad \square$$

6.9.2 Hierarchy of maximal factors

The *topological Pinsker factor* (denotation TPF) of a topological dynamical system is defined as the largest factor that has topological entropy zero. Every zero-entropy factor $(Y, S, \mathbb{S})$ of $(X, T, \mathbb{S})$ factors *through* the topological Pinsker factor of $(X, T, \mathbb{S})$. There is a nice characterization of the topological Pinsker factor by means of the so-called entropy pairs, defined by Francois Blanchard [Blanchard, 1993]:

Definition 6.9.12 A pair x, x' of points of X is an *entropy pair* if every open cover $\mathfrak{U}$ by two sets $\mathfrak{U} = \{U, V\}$, such that $x \in \mathrm{int}(U^c)$ and $x' \in \mathrm{int}(V^c)$, has positive topological entropy $\mathbf{h}(T, \mathfrak{U})$.

Just like in ergodic theory measure-theoretic factors correspond to subinvariant sub-sigma-algebras, in topological dynamics topological factors correspond to subinvariant closed equivalence relations (in $X \times X$). The factor map sends every point to its equivalence class. The analogy is rather distant, for instance, in ergodic theory a larger sigma-algebra produces a larger factor, while in topological dynamics the larger the relation the smaller the factor.

Theorem 6.9.13 *The topological Pinsker factor corresponds to the smallest subinvariant closed equivalence relation that contains all entropy pairs.* $\square$

We refer to the original paper [Blanchard and Lacroix, 1993] for the proof.

Corresponding to the notion of topologically deterministic systems one can define the *maximal topologically deterministic factor* (denoted MTDF). It arises as the smallest subinvariant (in fact invariant) equivalence relation such that all subinvariant closed equivalence relations containing it are invariant. The existence of such an equivalence relation is obvious: it is the intersection of a nonempty family of relations with properties preserved by intersections.

We also note that every system possesses the *maximal Pinsker-like factor* (denotation MPLF); it corresponds to the smallest subinvariant closed equivalence relation which contains all asymptotic pairs.

Given a topological dynamical system $(X, T, \mathbb{S})$, we can determine its *maximal* NAP *factor* (denotation MNAPF); the maximal factor which is NAP. At first glance it is not even clear that such an object is well defined. We only sketch the argument, which requires the Kuratowski–Zorn Lemma. First of all, the collection of NAP factors is nonempty; it contains at least the trivial one-point factor. Now, given an increasing chain (i.e., a linearly ordered family) (Y_κ) of NAP factors of $(X, T, \mathbb{S})$, their inverse limit Y (the construction of an inverse limit applies to chains as well) is a factor of $(X, T, \mathbb{S})$ and an extension of all the factors in the chain. This inverse limit is also NAP; any asymptotic pair projects in Y_κ to either an asymptotic pair (which is impossible) or it is collapsed. Thus such a pair is collapsed in every Y_κ, which means that the pair is in fact identical in the inverse limit space. We have shown that Y is NAP. By the Kuratowski–Zorn Lemma [see e.g. Ciesielski, 1997], the maximal NAP factor exists.

Unlike in the case of the corresponding classes of systems, none of the above four types of factors coincide, so this is where the analogy to ergodic theory ends.

Theorem 6.9.14 *We have the following factorization*

$$\mathsf{TPF} \mapsto \mathsf{MPLF} \mapsto \mathsf{MNAPF} \mapsto \mathsf{MTDF}.$$

These four types of factors are essentially different.

Proof Let us first explain the factorizations: The maximal Pinsker-like factor has entropy zero (Theorem 6.9.7), so it factors through the topological Pinsker factor. The maximal NAP factor is NAP, so the factor map leading to it must collapse all asymptotic pairs (the image of a not collapsed asymptotic pair would remain an asymptotic pair). So it is Pinsker-like, thus it factors through the MPLF. The MTD factor is deterministic, so it is NAP, and hence it factors through the maximal NAP factor.

The first arrow is not realized by the identity map in any zero-entropy system that possesses asymptotic pairs (for example in a nonperiodic subshift of entropy zero). The second arrow is not by identity in the Example 6.9.15 below. The third arrow is not by identity in any NAP system which is not deterministic (like the one in Example 6.9.6). □

Example 6.9.15 There exists a bilateral subshift $(X, \sigma, \mathbb{Z})$ such that the maximal Pinsker-like factor $(Y, S, \mathbb{Z})$ is not NAP.

Indeed, let $(X, T, \mathbb{S})$ be the orbit-closure (in the $\mathbb{Z}$-action) of the following (bilateral) sequence over two symbols:

$$x = \dots 000000011111000001110001011100011111000001111111\dots$$

In addition to the countable orbit of this sequence, the system contains also the points

$$a = \dots 000000111111\dots, \quad b = \dots 111111000000\dots$$

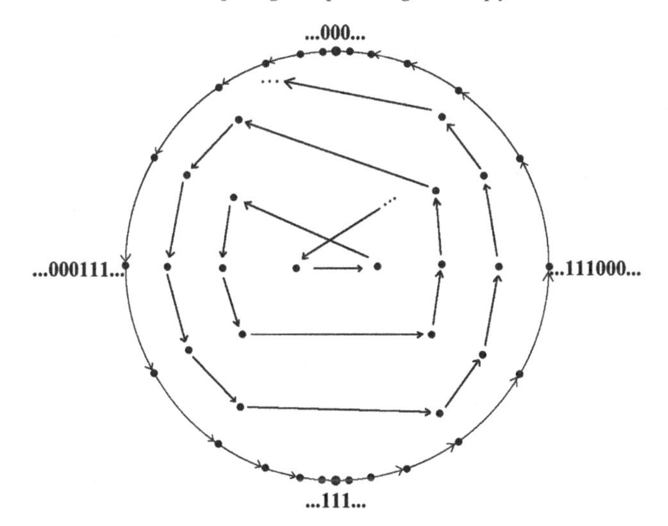

Figure 6.1 The dynamics in the example. The backward orbit of the central point is not shown. It is more or less symmetric to the forward orbit.

and their countable orbits, and the fixpoints

$$c = \ldots 000000\ldots\,, \quad d = \ldots 111111\ldots$$

The dynamics of this system is shown in Figure 6.1. It is elementary to see that all points in the orbit of a are asymptotic to the fixpoint d, and all points in the orbit of b are asymptotic to c. The maximal factor collapsing asymptotic pairs must also collapse the pair c, d, because the corresponding relation must be closed. So, all points a, b, c, d and their orbits are collapsed to one point. That is all. No other collapsing is necessary (we leave it to the reader). The factor so obtained looks exactly the same as the factor $(Y, S, \mathbb{Z})$ in Example 6.9.6: it is a one-point compactification (by a fixpoint) of a single discrete bilateral orbit. As before, all pairs in this factor are asymptotic, so this maximal factor is not NAP.

6.10 Topological preimage entropy*

A notion which attempts to capture the "purely noninvertible" complexity of the dynamics in a topological dynamical system is *topological preimage entropy*, as introduced by Mike Hurley [Hurley, 1995], then studied by Zbigniew Nitecki with coauthors [Nitecki and Przytycki, 1999; Fiebig *et al.*, 2003]. The idea is to count (n, ε)-separated sets in the nth preimage of a point. In fact, there are two versions, depending on the order of applying certain suprema. The third version has been defined and studied by Wen-Chao Cheng and Sheldon Newhouse [Cheng and Newhouse, 2005] where also the measure-theoretic notion is introduced.

Definition 6.10.1 Let $T : X \to X$ be a continuous map of a compact space. Let $\mathbf{H}_1(n, \varepsilon|F)$ denote the logarithm of the maximal cardinality of an (n, ε)-separated set contained in F. We define three versions of *topological preimage entropy* as

$$\mathbf{h}_{\mathsf{p}}(T) = \lim_{\varepsilon} \uparrow \sup_{x \in X} \limsup_{n \to \infty} \tfrac{1}{n} \mathbf{H}_1(n, \varepsilon|T^{-n}\{x\}), \tag{6.10.2}$$

$$\mathbf{h}_{\mathsf{m}}(T) = \lim_{\varepsilon} \uparrow \limsup_{n \to \infty} \tfrac{1}{n} \sup_{x \in X} \mathbf{H}_1(n, \varepsilon|T^{-n}\{x\}), \tag{6.10.3}$$

$$\mathbf{h}_{\mathsf{pre}}(T) = \lim_{\varepsilon} \uparrow \limsup_{n \to \infty} \tfrac{1}{n} \sup_{x \in X} \sup_{k \geq n} \mathbf{H}_1(n, \varepsilon|T^{-k}\{x\}). \tag{6.10.4}$$

Remark 6.10.5 Instead of counting (n, ε)-separated sets within a set we can as well count minimal subfamilies of $\mathcal{U}^n$ covering that set. Then $\mathbf{H}_1(n, \varepsilon|F)$ should be replaced by $\mathbf{H}(\mathcal{U}^n|F)$. It is easy to see (using the inequalities (6.1.10) and (6.1.11) relative to the set F) that such a change has no effect on the final notions in the above definition, as long as the limit along ε is replaced by the limit along the net of all covers.

It is elementary to see that

$$\mathbf{h}_{\mathsf{p}}(T) \leq \mathbf{h}_{\mathsf{m}}(T) \leq \mathbf{h}_{\mathsf{pre}}(T) \leq \mathbf{h}(T). \tag{6.10.6}$$

It has been proved in [Fiebig *et al.*, 2003] that in forward expansive systems, in particular in unilateral subshifts, all the above notions (including topological entropy) agree. Clearly, for homeomorphisms, all three notions of topological preimage entropy are equal to zero, hence can be strictly smaller than topological entropy. There are also examples for which $\mathbf{h}_{\mathsf{p}}(T) < \mathbf{h}_{\mathsf{m}}(T)$. We do not know about the middle inequality.

Cheng and Newhouse have also introduced a measure-theoretic notion of preimage entropy. For convenience, we repeat the definition, already cited in Remark 4.2.8. We let $\mathfrak{A}^\infty = \bigcap_{n \geq 1} T^{-n}(\mathfrak{A})$ and we define the measure-theoretic preimage entropy in the system $(X, \mathfrak{A}, \mu, T, \mathbb{S})$ as the usual conditional entropy $h(\mu, T|\mathfrak{A}^\infty)$. The paper [Cheng and Newhouse, 2005] provides a number of properties of this last notion, such as product rule, power rule, affinity as a function of the measure and a version of the Shannon–McMillan–Breiman Theorem. However, all the above are immediate consequences of the same properties known for the conditional entropy given a subinvariant sigma-algebra (including the conditional Shannon–McMillan–Breiman Theorem 3.3.7).

That paper, however, contains one very interesting result, a kind of variational principle:

Theorem 6.10.7 (Preimage Entropy Variational Principle) *Let* $(X, T, \mathbb{N}_0)$ *be a topological dynamical system and let* $\mathfrak{A}$ *denote the Borel sigma-algebra in* X. *Then*

$$\mathbf{h}_{\mathsf{pre}}(T) = \sup_{\mu \in \mathcal{M}_T(X)} h(\mu, T|\mathfrak{A}^\infty). \qquad \square$$

The right-hand side resembles that in the Conditional Variational Principle (Theorem 6.8.8) except that the sigma-algebra $\mathfrak{A}^\infty$ does not represent the Borel sigma-algebra lifted from any topological factor (at least it is not obtained in this manner). This provokes the following question:

Question 6.10.8 Can the Preimage Entropy Variational Principle be reduced to a variant of the Conditional Variational Principle (for instance valid for some specific Borel-measurable factors)?

Exercises

6.1 Prove (by example) that the sequence $\mathbf{H}(\mathcal{U}^n)$ need not have decreasing $n$ths.

6.2 Prove Fact 6.2.4.

6.3 Show that if T is Lipschitz, i.e., $d(Tx, Ty) \le cd(x, y)$ for some constant c, then $\mathbf{h}(T) \le \max\{0, \log c\}$.

6.4 Show that if $T : [0, 1] \to [0, 1]$ is piecewise monotone, with N branches of monotonicity, then $\mathbf{h}(T) \le \log N$.

6.5 Let $(X, T, \mathbb{S})$ and $(Y, S, \mathbb{S})$ be factors of some common extension. Assume they both have finite entropies. Prove that then

$$|\mathbf{h}^*(S) - \mathbf{h}^*(T)| \le \mathbf{h}(T|S) + \mathbf{h}(S|T).$$

6.6 Prove the power rule for tail entropy: $\mathbf{h}^*(T^n) = |n|\mathbf{h}^*(T) \quad (n \in \mathbb{S})$.

6.7 Prove the power rules for different kinds of fiber entropy: For $n \in \mathbb{S}$ we have $\mathbf{h}(T^n, \mathcal{U}^{|n|}|y) = |n|\mathbf{h}(T, \mathcal{U}|y)$, $\mathbf{h}(T^n|y) = |n|\mathbf{h}(T|y)$, $\mathbf{h}(T^n, \mathcal{U}^{|n|}|\nu) = |n|\mathbf{h}(T, \mathcal{U}|\nu)$ and $\mathbf{h}(T^n|\nu) = |n|\mathbf{h}(T|\nu)$.

6.8 For $n \in \mathbb{N}$ prove $\mathbf{h}(T, \mathcal{U}^n|y) = \mathbf{h}(T, \mathcal{U}|y)$ and similarly $\mathbf{h}(T, \mathcal{U}^n|\nu) = \mathbf{h}(T, \mathcal{U}|\nu)$.

6.9 Derive the inequality $\mathbf{h}(T|\nu) + h(\nu) \le \mathbf{h}(T)$.

6.10 Consider the composition of factor maps: $\phi : X \to Y$ and $\psi : Y \to Z$. Fix some invariant measure ξ on Z. Derive $\mathbf{h}(T|\xi) \leq \mathbf{h}(T|S) + \mathbf{h}(S|\xi)$.

6.11 Give an example of covers $\mathcal{U}$ and $\mathcal{V}$ (of two spaces) such that $N(\mathcal{U} \otimes \mathcal{V}) < N(\mathcal{U})N(\mathcal{V})$.

6.12 Let $(X, T, \mathbb{S})$ be a topological joining of $(Y, S, \mathbb{S})$ and $(Z, R, \mathbb{S})$. Show that then

1. $\mathbf{h}(T|z) \leq \mathbf{h}(S), \ (z \in Z)$;
2. $\mathbf{h}(T|\xi) \leq \mathbf{h}(S), \ (\xi \in \mathcal{M}_R(Z))$.

7

Dynamics in dimension zero

7.1 Zero-dimensional dynamical systems

Due to the existence of arbitrarily fine covers which are at the same time partitions (by disjoint open sets – we call them "clopen covers"), zero-dimensional dynamical systems allow us to switch between the topological and measure-theoretic dynamical notions easier than any other systems.

In order to understand zero-dimensional dynamical systems it suffices to understand subshifts and their countable joinings.

Definition 7.1.1 By a *subshift* we mean a topological dynamical system $(X, T, \mathbb{S})$, where X is a closed, shift-invariant subset of $\Lambda^{\mathbb{S}'}$, Λ is a finite set (called alphabet), and T denotes the shift map σ restricted to X. Both $\mathbb{S}'$ and $\mathbb{S}$ stand for either $\mathbb{N}_0$ or $\mathbb{Z}$ but we require that $\mathbb{S} \subset \mathbb{S}'$. Shift-invariance of X is understood as $\sigma(X) \subset X$ or $\sigma(X) = X$, depending on whether $\mathbb{S} = \mathbb{N}_0$ or $\mathbb{Z}$, respectively.

Note that $\mathbb{S}' = \mathbb{Z}$ implies that T is injective.

Definition 7.1.2 By a *symbolic array system* we will mean a topological dynamical system $(X, T, \mathbb{S})$, where X consists of symbolic arrays of the form $x = (x_{k,n})_{k \in \mathbb{N}, n \in \mathbb{S}'}$ ($\mathbb{S} \subset \mathbb{S}'$ like for subshifts), where for each k, n, $x_{k,n}$ belongs to a finite set Λ_k not depending on n or $x \in X$, called *the alphabet in row* k. The transformation T is the restriction to X of the left shift map on arrays

$$(Tx)_{k,n} = (x_{k,n+1}).$$

The above definition implicitly requires that X be closed and invariant under T (recall that the meaning of invariance depends on $\mathbb{S}$). The fact below is completely elementary.

Fact 7.1.3 *The following terms are synonyms (up to topological conjugacy): zero-dimensional dynamical system, countable joining of subshifts, inverse limit of subshifts, symbolic array system.* □

7.2 Topological entropy in dimension zero

We begin by providing a more direct formula for the topological entropy of a subshift and then of a general zero-dimensional topological dynamical system. It is based on the topological version of the definition of the topological entropy.

Fact 7.2.1 *If $(X, T, \mathbb{S})$ is a subshift over a finite alphabet Λ, then*

$$\mathbf{h}(T) = \lim_n \tfrac{1}{n} \log \#\{B \in \Lambda^n : B \cap X \neq \emptyset\}, \tag{7.2.2}$$

where the blocks B are identified with the corresponding cylinder sets over the coordinates $0, \ldots, n-1$.

The above cardinality is interpreted as the number of blocks of length n *appearing* in the subshift, by which we mean appearing as a subblock in some $x \in X$.

Proof In a symbolic system, the zero-coordinate partition $\mathcal{P}_\Lambda$ is clopen, so it is also a cover. The same is true for $\mathcal{P}_\Lambda^n$. Because this cover consists of disjoint sets, subcovers can be obtained only by discarding the empty sets. So, $\mathbf{H}(\mathcal{P}_\Lambda^n) = \log \#\mathcal{P}_\Lambda^n$ (where we ignore the empty sets). This is exactly the number of blocks of length n that appear in the system. We have proved that the right-hand side of (7.2.2) equals $\mathbf{h}(\mathcal{P}_\Lambda, T)$. The assertion is now a consequence of the remark following Definition 6.2.5 because in the symbolic space the cover $\mathcal{P}_\Lambda$ is a topological generator. □

We can now formulate the "topological vertical data compression" statement for subshifts:

Theorem 7.2.3 *A subshift $(X, T, \mathbb{S})$ over an alphabet Λ with topological entropy $\mathbf{h}$ has a principal extension $(Y, S, \mathbb{S})$ which is a bilateral subshift over an alphabet Λ_Y of cardinality $l_Y = \lfloor 2^{\mathbf{h}} \rfloor + 1$.*

Proof First observe that the natural extension of $(X, T, \mathbb{S})$ (recall Definitions 6.8.10 and 6.8.11; the action need not be surjective but it is always injective) can be realized as a bilateral shift (see Exercise 7.2). This extension is principal and has the same topological entropy $\mathbf{h}$. It remains to prove the statement for bilateral subshifts $(X, T, \mathbb{S})$.

Notice that $\log l_Y > \mathbf{h}$. Let n be such that

$$\log \#\{B \in \Lambda^n : B \cap X \neq \emptyset\} < n \log l_Y.$$

We replace $(X, T, \mathbb{S})$ by its direct product with the n-periodic orbit of the sequence obtained by repeating $000\ldots01$ (with $n-1$ zeros). We imagine this new subshift $(X', T', \mathbb{S})$ as having two rows. The collection $\mathcal{B}'$ of all blocks in X', of length n and starting with 1 in the second row, has the same cardinality as that of all blocks of length n in X. Thus there exists an injective map Φ from $\mathcal{B}'$ into Λ_Y^n and, if n is large enough, we can arrange that there exists a specific block W over Λ_Y such that every block in the range of Φ ends with W and W does not occur in it anywhere else (see Exercise 3.8; the same trick has already been used in the proof of Lemma 4.2.5). Now we can define a conjugacy ϕ between $(X', T', \mathbb{S})$ and a subshift $(Y, S, \mathbb{S})$ over Λ_Y: each element x' of X' is a concatenation of blocks $B' \in \mathcal{B}'$ and we let $\phi(x')$ be the corresponding concatenation of the blocks $\Phi(B')$. Invertibility of this map is easy to see; the block W allows us to determine the cutting places in every image, and then we can reverse the map Φ on each component block. □

We can now deduce semicontinuity properties of the entropy function on the invariant measures in dimension zero.

Fact 7.2.4 *If $(X, T, \mathbb{S})$ is a subshift, then the entropy function $\mu \mapsto h(\mu, T)$ on $\mathcal{M}_T(X)$ is upper semicontinuous. In particular, there exists an invariant measure of maximal entropy. If $(X, T, \mathbb{S})$ is zero-dimensional, then this function is of Young class* LU.

Proof We begin with a subshift $(X, T, \mathbb{S})$. Because the characteristic function of a clopen set is continuous, the functions $\mu \mapsto \frac{1}{n} H(\mathcal{P}_\Lambda^n)$ are continuous on $\mathcal{M}(X)$. On $\mathcal{M}_T(X)$, these functions decrease to the entropy function, implying the assertion.

For zero-dimensional systems the assertion now follows from the inverse limit representation, $(X, T, \mathbb{S}) = \overleftarrow{\lim_k} (X_k, T_k, \mathbb{S})$, where each $(X_k, T_k, \mathbb{S})$ is a subshift. Fact 6.5.12 yields that $\mu \mapsto h(\mu, T)$ is an increasing limit of upper semicontinuous functions. □

7.3 The invariant measures in dimension zero

This short section is not about entropy. It provides some elementary facts about invariant measures in zero-dimensional systems, especially in subshifts, which allows us to better understand the connection between measures and blocks (or rectangles) in the symbolic space.

We know that the weak-star topology on measures can be endowed with a metric, involving a linearly dense sequence of continuous functions on X (see (A.2.6)). In the zero-dimensional case, the role of this sequence is played by the characteristic functions of the cylinder sets corresponding to rectangular blocks in the symbolic array representation. More specifically, let (k_i, n_i) be a sequence of integer-valued vectors such that both coordinates increase to infinity as i grows, and let $\mathcal{R}_i$ denote the collection of all rectangles with k_i rows and n_i columns (recall that each row number k has its own finite alphabet Λ_k). Then we define

$$d^*(\mu, \nu) = \sum_i 2^{-i} \sum_{R \in \mathcal{R}_i} |\mu(R) - \nu(R)|. \tag{7.3.1}$$

In symbolic spaces (with one row only), we replace the rectangles by blocks of length n_i.

Now we will focus on subshifts. As it was explained in Section 2.8, with each block B we associate the invariant measure $\mu_{(B)}$ supported by the periodic orbit of the sequence $\ldots BBB \ldots$ (or $BBB\ldots$ for unilateral shifts). This measure assigns to a block A much shorter than B the value roughly equal to the frequency with which A occurs in B. If we want to acquire an approximative distance between $\mu_{(B)}$, where B is very long, and some other measure μ, we only need to compare the values for not too long blocks A, and for such blocks use the approximation of $\mu_{(B)}(A)$ by the frequencies. Below we state a number of facts concerning such measures and their distances. The proofs are standard applications of the above observation (on how to acquire an approximation of the distance), and will be skipped.

Fact 7.3.2 *We fix an alphabet Λ. For every $\varepsilon > 0$ and $r \in \mathbb{N}$ the following hold for n sufficiently large:*

1. $d^*(\mu_{(C)}, \frac{1}{q}\sum_{i=1}^{q} \mu_{B^{(i)}}) < \varepsilon$, *where $C = B^{(1)}B^{(2)}\ldots B^{(q)}$ is a concatenation of an arbitrary number q of blocks of length n;*
2. $d^*(\mu_{(B)}, \mu_{(C)}) < \varepsilon$, *where C is a subblock of length $n - r$ of a block B of length n;*
3. $d^*(\mu_{(B)}, \frac{1}{n}\sum_{i=0}^{n-1} \delta_{T^i x}) < \varepsilon$, *where $x \in \Lambda^{\mathbb{S}'}$, T is the shift map, and $B = x[0, n-1]$.* □

7.4 The Variational Principle in dimension zero

Although we have already proved the Variational Principle (as a special case of the Inner Variational Principle), we give its direct proof in dimension zero just because of its striking simplicity. We begin with subshifts.

Proof of the Variational Principle for subshifts Let $X \subset \Lambda^{\mathbb{S}'}$ and let μ be a shift-invariant measure on X. Clearly, for any n, $H(\mu, \mathcal{P}_\Lambda^n) \leq \log \#\mathcal{P}_\Lambda^n = \mathbf{H}(\mathcal{P}_\Lambda^n)$. Thus $h(\mu, T)$, which equals $h(\mu, T, \mathcal{P}_\Lambda)$, is not larger than $\mathbf{h}(T, \mathcal{P}_\Lambda)$, which equals $\mathbf{h}(T)$.

The converse inequality is just a bit harder. We will construct an invariant measure with the dynamical entropy equal to the topological entropy. Let $n \in \mathbb{N}$ and let $\mathcal{B}_n = \{B \in \Lambda^n : B \cap X \neq \emptyset\}$ be the collection of all blocks of length n appearing in the system. Let X_n be the collection of all (unilateral or bilateral, depending on the meaning of $\mathbb{S}'$) sequences concatenated of the blocks from $\mathcal{B}_n$, one such block starting at the coordinate zero. Clearly, X_n is closed and T^n-invariant, and it supports the T^n-invariant Bernoulli measure $\mu_{n,0}$ assigning equal masses to all blocks in $\mathcal{B}_n$. The entropy $h(\mu_{n,0}, T^n)$ equals $\log \#\mathcal{B}_n$. For $i \in \{1, 2, \ldots, n-1\}$, consider the set $T^i(X_n)$. This is again a T^n-invariant set and the map T^{n-i} factors the system $(T^i(X_n), T^n, \mathbb{S})$ back onto $(X_n, T^n, \mathbb{S})$. The measure $\mu_{n,0}$ lifts to some T^n-invariant measure $\mu_{n,i}$ supported by $T^i(X_n)$. Clearly, $h(\mu_{n,i}, T^n) \geq h(\mu_{n,0}, T^n) = \log \#\mathcal{B}_n$. Now, the measure $\mu_n = \frac{1}{n}\sum_{i=0}^{n-1} \mu_{n,i}$ is T-invariant and

$$h(\mu_n, T) = \frac{1}{n} h(\mu_n, T^n) = \frac{1}{n^2} \sum_{i=0}^{n-1} h(\mu_{n,i}, T^n) \geq \frac{1}{n} \log \#\mathcal{B}_n.$$

Let μ be an accumulation point of the sequence (μ_n) (say, along n_k) in the simplex of all shift-invariant measures of the full shift on $\Lambda^{\mathbb{S}'}$. By upper semicontinuity of the entropy function in symbolic systems,

$$h(\mu, T) \geq \limsup_{k \to \infty} h(\mu_{n_k}, T) \geq \lim_k \tfrac{1}{n_k} \log \#\mathcal{B}_{n_k} = \mathbf{h}(T).$$

The last thing to do is to verify that μ is supported by X. Consider an open set disjoint of X. This set is a union of countably many cylinders C not appearing in X. Let C be such a cylinder and let n_0 be its length. We need to show that $\mu(C) = 0$. The block C does not occur in any block $B \in \mathcal{B}_n$ for any length $n \geq n_0$. This implies that the cylinder over C is disjoint of X_n and of $T^i(X_n)$ for $0 \leq i \leq n - n_0$ (the block C can occur in concatenations of the blocks from $\mathcal{B}_n$ only at the contact places of the concatenated blocks). Thus $\mu_n(C) \leq n_0/n$. Since C is clopen, $\mu \mapsto \mu(C)$ is a continuous function on measures (see Fact 6.6.5), hence $\mu(C) = \lim_n \mu_n(C) = 0$, as needed. □

Proof of the Variational Principle in dimension zero Having proved the Variational Principle for subshifts, in order to generalize it to all zero-dimensional dynamical systems, we first apply Fact 6.5.12 to get that thc topological entropy equals the limit of $\mathbf{h}(T_k)$, the topological entropies of the subshifts of which the system is an inverse limit. Then we combine two other facts about the measure-theoretic entropy: (1) the entropy of every invariant measure is the limit of its projections to the involved subshifts, hence it does not exceed the topological entropy, and (2), the measure of maximal entropy on the subshift X_k lifts to some measure on the entire system and this lift has entropy at least $\mathbf{h}(T_k)$. Thus there are measures with entropies arbitrarily close to the topological entropy. (This time, however, the measure of maximal entropy need not exist.) □

7.5 Tail entropy and asymptotic h-expansiveness in dimension zero

The notion of tail entropy simplifies for zero-dimensional systems and asymptotic h-expansiveness admits a nice characterization for zero-dimensional $\mathbb{Z}$-actions. This is due, roughly speaking, to the fact that in dimension zero the system admits enough topological factors, which can replace the refining conditioning covers in the definition of the tail entropy.

Fact 7.5.1 *Let* $= (X, T, \mathbb{S}) = \overleftarrow{\lim_k}(X_k, T_k, \mathbb{S})$ *be a zero-dimensional dynamical system represented as an inverse limit of subshifts. Then*

$$\mathbf{h}^*(T) = \lim_k \downarrow \mathbf{h}(T|T_k) = \lim_k \downarrow \lim_j \uparrow \mathbf{h}(T_j|T_k). \tag{7.5.2}$$

Proof The first equality follows easily from the combination of the following facts:

1. The zero-coordinate partitions $\mathcal{P}_{\Lambda_k}$ are at the same time clopen covers and they have the property that for any ε there exist k and n such that the cover $\mathcal{P}^n_{\Lambda_k}$ has diameter smaller than ε. Hence every open cover $\mathcal{U}$ is refined by one of these covers, which implies that the topological conditional entropy given $\mathcal{U}$ is not larger than that given $\mathcal{P}^n_{\Lambda_k}$.
2. The topological conditional entropy of (X, T) given $\mathcal{P}^n_{\Lambda_k}$ is the same as that given $\mathcal{P}_{\Lambda_k}$.
3. For a cover consisting of disjoint clopen sets, the topological conditional entropy given this cover is the same as the topological conditional entropy given the factor generated by this cover.

The second equality in (7.5.2) follows from the definition of $\mathbf{h}(T|T_k)$ and the above argument 1: any open cover $\mathcal{U}$ is refined by some $\mathcal{P}^n_{\Lambda_j}$. □

Why is it so important to replace the topological conditional entropy given a cover by the topological conditional entropy given a factor? This replacement is specifically significant for bilateral shifts, for which it allows one to represent the system as a joining of the factor with an "almost independent" system, i.e., one of entropy just a bit larger than the conditional entropy (see Theorem 7.5.3 below). We recall that in the measure-theoretic setup an analogous representation-by-joining theorem holds for all systems with finite entropy (see Theorem 4.4.6). In topological dynamics such representation is possible (in general) for bilateral subshifts, and, like in the measure-theoretic case, it fails for the actions of $\mathbb{N}_0$. To see this take Example 4.4.9; we just need to replace the Bernoulli shift by a full shift. Eventually, Theorem 7.5.3 will be used to characterize zero-dimensional asymptotically h-expansive $\mathbb{Z}$-actions (Theorem 7.5.9).

Theorem 7.5.3 *Let $(Y, S, \mathbb{Z})$ be a subshift and a topological factor of another subshift $(X, T, \mathbb{Z})$. Then, for every $\varepsilon > 0$, the system $(X, T, \mathbb{Z})$ is conjugate to a topological joining of the factor subshift $(Y, S, \mathbb{Z})$ with another factor subshift $(Y', S', \mathbb{Z})$ of topological entropy not exceeding $\mathbf{h}(T|S) + \varepsilon$.*

The proof relies on a lemma, which replaces the Rokhlin Lemma in zero-dimensional topological dynamics. This is probably one of the most important observations concerning zero-dimensional dynamics. The lemma appears in print for the first time in a paper of M. Boyle [Boyle, 1983] (in a version for bilateral shifts), where it is attributed to W. Krieger. Some generalizations can be found in [Downarowicz, 2006, 2008].

Lemma 7.5.4 *Let $(X, T, \mathbb{Z})$ be a zero-dimensional topological dynamical system. For every $n \geq 1$ and $\varepsilon > 0$ there exists a clopen set F such that:*

(i) $T^{-i}(F)$ are pairwise disjoint for $i = 0, 1, \dots, n-1$, and
(ii) $T^{-i}(F)$ for $i = -n+1, \dots, 0, \dots, n-1$ cover the set $X \setminus P_n^\varepsilon$,

where P_n^ε denotes the ε-neighborhood of the set P_n of all periodic points with periods not exceeding n.

Proof The set P_n is closed, so replacing if necessary P_n^ε by a smaller neighborhood we can assume that P_n^ε is clopen. For a given n every point $x \in X \setminus P_n^\varepsilon$ belongs to a clopen set $U_x \subset X \setminus P_n^\varepsilon$ such that its n consecutive preimages $T^{-i}(U_x)$ $(0 \leq i \leq n-1)$ are pairwise disjoint. Choose a finite cover

$\mathfrak{U} = \{U_j\}_{1 \le j \le m}$ of $X \setminus P_n^\varepsilon$ consisting of some of the sets U_x. Define inductively

$$F_1 = U_1$$
$$F_j = F_{j-1} \cup \Big(U_j \setminus \bigcup_{-n+1 \le i \le n-1} T^i(F_{j-1})\Big) \quad (j = 2, \dots, m)$$

and define $F = F_m$. Because this construction involves finite set operations applied to clopen sets, all the sets F_j $(1 \le j \le m)$ are clopen. It is also clear that the sets F_j ascend. Every $x \in X \setminus P_n^\varepsilon$ belongs to some U_j $(1 \le j \le m)$. If $x \in F_j$, then $x \in T^{-i}(F)$ for $i = 0$. The only way x may not belong to F_j is that $j > 1$ and x belongs to $\bigcup_{-n+1 \le i \le n-1} T^i(F_{j-1})$. In such case, however, $x \in T^{-i}(F_{j-1}) \subset T^{-i}(F)$ for some $i \in [-n+1, n-1]$. In either case x belongs to $T^{-i}(F)$ with some $i \in [-n+1, n-1]$, as required in (ii).

For (i), suppose that $T^{-i}(F)$ and $T^{-i'}(F)$ are not disjoint for some $0 \le i < i' \le n-1$. Equivalently, F and $T^{i'-i}(F)$ are not disjoint. From now on i replaces $i' - i$ and ranges between 1 and $n-1$. Let $x \in F \cap T^{-i}(F)$. Let j be the smallest index for which $x \in F_j$ and let j' be the smallest index for which $T^i x \in F_{j'}$. Assume for a while that $j' \le j$. Then x belongs to U_j (because F_j is a union of F_{j-1} to which x does not belong, and a part of U_j). Since $F_{j'} \subset F_j$, we also have $T^i x \in F_j$. Here, $T^i x$ falls either in F_{j-1} or in a part of U_j, however, by the choice of the sets U_j, it cannot happen that both x and $T^i x$ belong to U_j. So, $T^i x \in F_{j-1}$. But then x falls into the set subtracted from U_j in the construction of F_j, a contradiction. If $j' \ge j$, we replace i by $-i$ (now ranging between $-n+1$ and 1), and switch the roles of x and $T^i x$, which brings us to the preceding case. □

Remark 7.5.5 The set F is called an *n-marker*. Because F is clopen, the factor generated by the partition into F and its complement is a topological factor in the form of a subshift over two symbols ("marker" and "no marker"). For $x \in X$ the markers occur exactly at the coordinates n such that $T^n x \in F$.

Corollary 7.5.6 *If $(X, T, \mathbb{Z})$ is a subshift, then, using any metric in the symbolic space, the statement of the above lemma implies that given natural n and k there exists a clopen set F such that for every $x \in X$ either $T^i x \in F$ (x has a marker at i) for some $i \in [-n+1, n-1]$ or the block $x[-n-k, n+k]$ is periodic with a period $p \le n$ (i.e., $x(i) = x(i+p)$ for $-n-k \le i \le n+k-p$). If there is a larger gap between the markers, say from a to $b > a + 2n + 2k$, then all subblocks of length $2n + 2k + 1$ of $x[a, b]$ are periodic, each with a period not exceeding n. It is an easy exercise to check that then the entire block $x[a, b]$ must be periodic with one such period.*

Proof of Theorem 7.5.3 Initially we represent $(X, T, \mathbb{Z})$ as a joining inside $(X, T, \mathbb{Z})$ of $(Y, S, \mathbb{Z})$ with some other subshift (for instance with $(X, T, \mathbb{Z})$), this joining realized as a two-row subshift with Y in the first row. Let l be the cardinality of the alphabet used in the second row. We can assume that this alphabet does not contain natural numbers. By the definition of the topological conditional entropy, there exists a length n_0 such that any block C of any length $n \geq n_0$ appearing in Y admits $m(C) \leq 2^{n(\mathbf{h}(T|S)+\frac{\varepsilon}{3})}$ different two-row completions appearing in X. For each C we order the corresponding blocks available in the second row and we denote them as $B_{C,j}$ $(1 \leq j \leq m(C))$. We would like to replace the completing blocks of the second row by blocks belonging to a smaller family. We apply the markers of Corollary 7.5.6 with respect to the system $(X, T, \mathbb{Z})$, n_0 and k satisfying

$$\tfrac{1}{k}(\log n_0 + n_0 \log l) \leq \tfrac{\varepsilon}{3}. \tag{7.5.7}$$

The gaps between the n_0-markers are never smaller than n_0 and, if a gap is longer than $2n_0$, the entire block (both in the first and second row) extending at least k positions to the left and right around the gap, is periodic with some period not exceeding n_0.

Now, we modify the second row of each x: if two markers appear at a distance smaller than or equal to $2n_0$, and C and $B_{C,j}$ are the blocks between these markers (say including a position of the marker on the left and excluding that on the right) in the first and second row, respectively, then we replace $B_{C,j}$ by the block $j00\ldots0$ (of length the same as that of C). At the remaining places we maintain the original symbols. The alphabet used in the new second row is enhanced by the symbol 0 and a finite number of integers j. It is obvious that the above procedure is a continuous invertible code (when applied to the two-row elements of X), hence it produces a conjugate two-row representation of $(X, T, \mathbb{Z})$. Notice that the markers coincide with the integers $j > 0$ in the new second row, so the two-symbol subshift of the markers is a topological factor of the subshift appearing in the second row. We will estimate the topological conditional entropy of the new second row given the process of markers. We count the number of possible blocks B of length k in the new second row assuming some fixed positions of the markers in the first row. Recall that if there exists a section of length larger than $2n_0$ without markers, then the entire block B is periodic with period not exceeding n_0 (a periodic section extends at least k positions in both directions from any place in the gap, hence it covers B). It is possible that inside B there are still pairs of markers less than $2n_0$ apart. In such places in the new second row we have introduced the blocks $j00\ldots0$, but everywhere else the new second row fits to one periodic pattern. Clearly, there exist at most $n_0 l^{n_0}$ periodic patterns of periods not exceeding n_0. Wherever we see a block $j00\dots0$, the integer j does not exceed $2^{n(\mathbf{h}(T|S)+\frac{\varepsilon}{3})}$, where n is the length of this block. Jointly, there are no more than $2^{k(\mathbf{h}(T|S)+\frac{\varepsilon}{3})}$ possibilities at all such places. Together, the number of all possible blocks of length k in the new second row, with a fixed structure of the markers does not exceed

$$n_0 l^{n_0} \cdot 2^{k(\mathbf{h}(T|S)+\frac{\varepsilon}{3})}.$$

Taking the logarithm and dividing by k and applying (7.5.7) we get that the topological conditional entropy of the second row given the factor of markers is at most

$$\mathbf{h}(T|S) + \tfrac{\varepsilon}{3} + \tfrac{1}{k}(\log n_0 + n_0 \log l) \leq \mathbf{h}(T|S) + \tfrac{2\varepsilon}{3}$$

In order to pass from the topological conditional to unconditional entropy of the subshift appearing in the new second row, we apply the formula (6.4.12) with regard to this subshift and the subshift of markers, which is its factor. The markers occur with gaps at least n_0, hence the topological entropy of the subshift of markers is at most $H(\frac{1}{n_0}, 1 - \frac{1}{n_0})$. If we start the construction by choosing n_0 large enough, we can assure that this number is smaller than $\varepsilon/3$. Then the entropy of the new second row does not exceed $\mathbf{h}(T|S) + \varepsilon$, as claimed. □

Question 7.5.8 We leave the validity of Theorem 7.5.3 for zero-dimensional $\mathbb{Z}$-actions other than subshifts open. It can be asked for systems with finite entropy or when the topological conditional entropy is finite.

Later we resolve the above question in two special cases: when either $(X, T, \mathbb{Z})$ is asymptotically h-expansive or when it is a principal extension of $(Y, S, \mathbb{Z})$ (see Theorem 7.5.10).

We will now provide the characterization of zero-dimensional asymptotically h-expansive $\mathbb{Z}$-actions. It originates from [Downarowicz, 2001] but is in fact a joint work with M. Boyle.

Theorem 7.5.9 *A zero-dimensional $\mathbb{Z}$-action $(X, T, \mathbb{Z})$ is asymptotically h-expansive if and only if it is conjugate to a countable joining of subshifts $(X'_k, T'_k, \mathbb{Z})$ ($k \in \mathbb{N}$) whose topological entropies form a summable sequence.*

Proof Suppose the latter condition holds. Letting $T_k = \bigvee_{j=1}^{k} T'_j$ we represent $(X, T, \mathbb{Z})$ as the inverse limit of subshifts $(X_k, T_k, \mathbb{Z})$. Using (6.4.11) and Fact 7.5.1 (second equality), we deduce asymptotic h-expansiveness of $(X, T, \mathbb{Z})$. (This works in fact also for actions of $\mathbb{N}_0$.)

We begin the proof of the other implication with fixing any inverse limit representation $(X,T,\mathbb{Z}) = \overleftarrow{\lim_k}(X_k,T_k,\mathbb{Z})$. By Fact 7.5.1, asymptotic h-expansiveness translates to

$$\lim_k \downarrow \mathbf{h}(T|T_k) = 0.$$

In the inverse limit representation we can replace the sequence of subshifts by any subsequence, and we can do it so that the entropies $\mathbf{h}(T|T_k)$ (along that subsequence) are summable. From now on, k indexes such a subsequence. The sequence $\mathbf{h}(T_{k+1}|T_k)$ is also summable. We can now define the subshifts $(X'_k,T'_k,\mathbb{Z})$ inductively. We let $(X'_1,T'_1,\mathbb{Z}) = (X_1,T_1,\mathbb{Z})$. Suppose that $(X_k,T_k,\mathbb{Z})$ is represented as a joining of subshifts $\bigvee_{j=1}^k T'_j$, satisfying, for every $2 \le j \le k$, the inequality $\mathbf{h}(T'_j) \le \mathbf{h}(T_j|T_{j-1}) + 2^{-j}$. Recall that $(X_k,T_k,\mathbb{Z})$ is a topological factor of $(X_{k+1},T_{k+1},\mathbb{Z})$, hence, by Theorem 7.5.3, the latter can be represented as a joining of $(X_k,T_k,\mathbb{Z})$ with a subshift $(X'_{k+1},T'_{k+1},\mathbb{Z})$ of entropy not exceeding $\mathbf{h}(T_{k+1}|T_k) + 2^{-k-1}$. Now $(X_{k+1},T_{k+1},\mathbb{Z})$ satisfies the inductive assumption for $k+1$. The whole inverse limit is hence a countable joining with summable entropies, as in the assertion. □

Necessity fails for actions of $\mathbb{N}_0$: the full bilateral shift (as an action of $\mathbb{N}_0$) is asymptotically h-expansive, and it can be represented as the inverse limit of unilateral shifts, where each of the bonding maps is the shift map. As indicated earlier, the corresponding extension cannot be replaced by a joining with a system of small entropy. Perhaps it can be represented as a countable joining of other unilateral subshifts, but never with decreasing entropies. We skip the full proof.

Now we return to Theorem 7.5.3 and prove it for zero-dimensional $\mathbb{Z}$-actions assuming that the extension is principal or that the larger system $(X,T,\mathbb{S})$ is asymptotically h-expansive.

Theorem 7.5.10 *Let $(X,T,\mathbb{Z})$ and $(Y,S,\mathbb{Z})$ be zero-dimensional systems such that $(Y,S,\mathbb{Z})$ is a topological factor of $(X,T,\mathbb{Z})$. Suppose that either $(X,T,\mathbb{Z})$ is a principal extension of $(Y,S,\mathbb{Z})$ or that $(X,T,\mathbb{Z})$ is asymptotically h-expansive. Then, for every $\varepsilon > 0$, the system $(X,T,\mathbb{Z})$ is conjugate to a topological joining of $(Y,S,\mathbb{Z})$ with another zero-dimensional system $(Y',S',\mathbb{Z})$ of topological entropy not exceeding $\mathbf{h}(T|S)+\varepsilon$.*

Proof Let $(X,T,\mathbb{Z}) = \overleftarrow{\lim_k}(X_k,T_k,\mathbb{Z})$ and $(Y,S,\mathbb{Z}) = \overleftarrow{\lim_k}(Y_k,S_k,\mathbb{Z})$ be some inverse limit representations. We have

$$\mathbf{h}(T|S) = \lim_k \uparrow \lim_j \downarrow \mathbf{h}(T_k|S_j). \tag{7.5.11}$$

If the extension is principal, then this iterated limit is zero, hence for every k the second limit is zero. Let ε_k be a sequence of positive numbers with sum $\varepsilon/2$. For each k let $j(k)$ be such that $\mathbf{h}(T_k|S_{j(k)}) < \varepsilon_k$. In the inverse limit representation of $(Y, S, \mathbb{Z})$ we can replace the sequence S_j by the subsequence $S_{j(k)}$ and call it S_k. Next, each $(X_k, T_k, \mathbb{Z})$ can be replaced by its joining with $(Y_k, S_k, \mathbb{Z})$ (this does not change the conditional entropy, see (6.5.8)), so we can assume that $(Y_k, S_k, \mathbb{Z})$ is a factor of $(X_k, T_k, \mathbb{Z})$ satisfying $\mathbf{h}(T_k|S_k) < \varepsilon_k$. By Theorem 7.5.3, the subshift $(X_k, T_k, \mathbb{Z})$ is a joining of $(Y_k, S_k, \mathbb{Z})$ with a subshift $(Y'_k, S'_k, \mathbb{Z})$ of entropy at most $2\varepsilon_k$. Thus, $(X, T, \mathbb{Z})$ is a joining of $(Y, S, \mathbb{Z})$ with the countable joining $(Y', S', \mathbb{Z})$ of the systems $(Y'_k, S'_k, \mathbb{Z})$. The entropy of $(Y', S', \mathbb{Z})$ clearly does not exceed $\sum_k 2\varepsilon_k = \varepsilon$.

Now assume that $(X, T, \mathbb{Z})$ is asymptotically h-expansive. By Theorem 7.5.9, we can represent $(X, T, \mathbb{Z})$ as a countable joining $\bigvee_{k=1}^{\infty} T'_k$ of subshifts with summable topological entropies. We can build an inverse limit representation of $(X, T, \mathbb{Z})$ by taking for $(X_k, T_k, \mathbb{Z})$ the finite joining $\bigvee_{j=1}^{k} T'_j$. We can pick k_0 so large that

$$\sum_{j=k_0+1}^{\infty} \mathbf{h}(T'_j) < \frac{\varepsilon}{4},$$

and, by (7.5.11), also that

$$\lim_j \downarrow \mathbf{h}(T_{k_0}|S_j) < \mathbf{h}(T|S) + \tfrac{\varepsilon}{4}.$$

We find j_0 such that $\mathbf{h}(T_{k_0}|S_{j_0}) < \mathbf{h}(T|S) + \varepsilon/2$, which, by (6.5.8), we can write as $\mathbf{h}(T_{k_0} \vee S_{j_0}|S_{j_0}) < \mathbf{h}(T|S) + \varepsilon/2$. Applying Theorem 7.5.3, we can represent $T_{k_0} \vee S_{j_0}$ as a joining of $(Y_{j_0}, S_{j_0}, \mathbb{Z})$ with another subshift (factor of $(X, T, \mathbb{Z})$), say $(Z, R, \mathbb{Z})$, of entropy not exceeding $\mathbf{h}(T|S) + 3\varepsilon/4$. Let $(Y', S', \mathbb{Z})$ be the joining inside $(X, T, \mathbb{Z})$ of $(Z, R, \mathbb{Z})$ with the countable joining $\bigvee_{j=k_0+1}^{\infty} T'_j$. This zero-dimensional system has entropy equal to at most the sum of the entropies of the joined systems, which does not exceed $\mathbf{h}(T|S) + \varepsilon$. The joining $S \vee S'$ extends $S_{j_0} \vee S'$ which equals

$$S_{j_0} \vee R \vee \bigvee_{j=k_0+1} T'_j \quad = \quad T_{j_0} \vee \bigvee_{j=k_0+1} T'_j \quad = \quad T.$$

So, the joining of $(Y, S, \mathbb{Z})$ with $(Y', S', \mathbb{Z})$ equals $(X, T, \mathbb{Z})$. □

7.6 Principal zero-dimensional extensions

In this section we prove, using elementary methods, that every topological dynamical system has a principal zero-dimensional extension. This fact has been long known for homeomorphism of finite entropy as a consequence of the theory of *mean dimension* created by Elon Lindenstrauss and Benjy Weiss [Lindenstrauss and Weiss, 2000; Lindenstrauss, 1999]. Briefly, any homeomorphism of finite entropy which admits a nonperiodic minimal factor (we refrain from discussing minimality here) has the small boundary property (this notion has been mentioned at the end of Section 6.6), and then it is very easy to construct a zero-dimensional extension which is not only principal, but even isomorphic for every invariant measure. The existence of a minimal factor can be easily achieved replacing the system by its product with any minimal system of entropy zero (such a product is a principal extension).

Here we will replicate from [Downarowicz and Huczek, in print] a new proof which produces just the principal extension (no small boundary property or isomorphic extension), but without assuming anything about the system.

Theorem 7.6.1 *Let $(X, T, \mathbb{S})$ be a topological dynamical system. Then there exists a zero-dimensional system $(Y, S, \mathbb{S})$ and a factor map $\pi : Y \to X$ such that $h(\nu|\mu) = 0$ for every invariant measure $\nu \in \mathcal{M}_S(Y)$ and $\mu = \pi\nu$.*

Remark 7.6.2 By the Conditional Variational Principle (Theorem 6.8.8), such an extension is principal, i.e., the topological conditional entropy $\mathbf{h}(S|T)$ is zero. This fact does not follow directly from the proof below, as we do not examine covers of X.

Proof of Theorem 7.6.1 We begin by explaining why it suffices to prove the theorem for homeomorphisms T only. In case T is surjective we simply replace the system by its natural extension. In the not surjective case we need to go back to the construction of the natural extension (preceding Definition 6.8.11). In the intermediate system $(X_1', T_1', \mathbb{N}_0)$ the map is a homeomorphism and this system factors onto $(X_1, T_1, \mathbb{N}_0)$ containing $(X, T, \mathbb{N}_0)$ as a subsystem. If we prove that $(X_1', T_1', \mathbb{N}_0)$ has a principal zero-dimensional extension, then this extension is also a principal zero-dimensional extension of $(X_1, T_1, \mathbb{N}_0)$. The preimage of the original space X becomes a principal zero-dimensional extension of $(X, T, \mathbb{S})$.

So, it suffices to work with systems $(X, T, \mathbb{S})$ where T is a homeomorphism. The initial part of the proof consists of two lemmas, the first one being quite general and fairly well known. By $\mathbb{T}$ we denote the torus (circle) obtained from the interval $[0, 1]$ by the identification $0 = 1$. By λ we denote the normalized Lebesgue measure on $\mathbb{T}$.

Lemma 7.6.3 *For every invariant measure $\mu \in \mathcal{M}_T(X)$ there exists an irrational rotation R on $(\mathbb{T}, \lambda)$ which is disjoint of μ (in the sense of Furstenberg, i.e., such that the only $(T \times R)$-invariant joining of μ and λ is $\mu \times \lambda$).*

Proof This follows from three elementary facts in ergodic theory:

1. For an invariant measure μ, in order not to be disjoint of some ergodic measure ν, it is necessary that the ergodic measures not disjoint of ν contribute a positive mass in the ergodic decomposition of μ.
2. An ergodic μ is not disjoint of the rotation by ϱ if and only if $e^{n\pi i\varrho}$ is an eigenvalue for μ (under T) for some natural n.
3. Every ergodic measure has at most countably many eigenvalues.

We leave the easy deduction of the assertion from these facts to the reader. □

For each $\mu \in \mathcal{M}_T(X)$ we fix one rotation disjoint of μ and denote it by R_μ. Now recall the notation introduced near the statement of Fact 6.6.1: if $\mu' \in \mathcal{M}(X)$ (i.e., μ' is a probability measure on X, usually not invariant under T), then $\mathrm{M}_n(\mu')$ denotes the average $\frac{1}{n}\sum_{i=0}^{n-1} T^i(\mu')$. Fact 6.6.1 states that, regardless of the starting measure, long enough averages are close to the set of invariant measures. We will need the following:

Lemma 7.6.4 *Let $\mu \in \mathcal{M}_T(X)$ be not supported by a finite set. Then, for any neighborhood W of $\mu \times \lambda$ in $\mathcal{M}(X \times \mathbb{T})$, there exists a neighborhood U_μ of μ in $\mathcal{M}(X)$ such that for any $(x,t) \in X \times I$ and any $n \in \mathbb{N}$ the condition $\mathrm{M}_n(\boldsymbol{\delta}_x) \in U_\mu$ implies that $\mathrm{M}_n(\boldsymbol{\delta}_{(x,t)}) \in W$, where the averaging in the product is with respect to the map $T \times R_\mu$.*

Proof Suppose the statement of the lemma is not true. Then there exists a sequence of measures of the form $\mathrm{M}_{n_k}(\boldsymbol{\delta}_{(x_k,t_k)})$ such that the measures $\mathrm{M}_{n_k}(\boldsymbol{\delta}_{x_k})$ converge to μ yet all the averages $\mathrm{M}_{n_k}(\boldsymbol{\delta}_{(x_k,t_k)})$ lie outside U. Because μ is not supported by a finite set, the parameters n_k in this sequence must grow to infinity. Any accumulation point of the sequence $\mathrm{M}_{n_k}(\boldsymbol{\delta}_{(x_k,t_k)})$ is a $T \times R_\mu$-invariant measure which is outside U and whose marginals are μ (being the limit of $\mathrm{M}_{n_k}(\boldsymbol{\delta}_{x_k})$) and λ (being the only R_μ-invariant measure on I). But the only $T \times R_\mu$-invariant measure with marginals μ and λ is $\mu \times \lambda$, which is in U, a contradiction. □

We return to the main proof, which we start by extending the system $(X, T, \mathbb{S})$ to its direct product with an odometer to a base $(p_k)_{k\in\mathbb{N}}$ (see Definition A.3.1; for each k, p_{k+1} is a multiple of p_k). Each measure $\mu \in \mathcal{M}_T(X)$ lifts to (possibly many) joinings of μ with the unique invariant measure on the odometer. Since the odometer has entropy zero, all such lifts have zero conditional entropy given μ (see (4.4.3)). So, it suffices to construct the extension (with properties as in the hypothesis) of this product. From now on $(X, T, \mathbb{S})$ denotes the product. Note that the system has no periodic orbits (and hence no invariant measures supported on finite sets). Lifting from the odometer the cylinder corresponding to the symbol 1 in row k, we obtain in X a clopen marker set F_k such that every orbit visits F_k in equal distances p_k of the time. The sets F_k decrease in k.

In the next step we consider the product space $X \times \mathbb{T}$. We now describe the construction of a partition $\mathcal{A}_{\mathcal{F}}$ of the product space $X \times \mathbb{T}$, associated with a finite family $\mathcal{F}$ of functions $f : X \to [0,1]$. The same construction will be used later, in the section devoted to entropy structures and in Part III, so it might be worthwhile to memorize it.

Definition 7.6.5 Let $f : X \to [0,1]$ be a continuous function. With f we associate the two-set partition $\mathcal{A}_f$ of $X \times (0,1]$ into the sets "below" and "above" the graph of f:

$$\mathcal{A}_f = \big\{\{(x,t) : 0 < t \leq f(x)\}, \{(x,t) : 1 \geq t > f(x)\}\big\}.$$

If $\mathcal{F}$ is a finite family of continuous functions $f : X \to [0,1]$, then we let

$$\mathcal{A}_{\mathcal{F}} = \bigvee_{f \in \mathcal{F}} \mathcal{A}_f.$$

By the identification $0 = 1$, the above partition can be applied to $X \times \mathbb{T}$, as well. Clearly, we have

$$\mathcal{A}_{\mathcal{F}_1 \cup \mathcal{F}_2} = \mathcal{A}_{\mathcal{F}_1} \vee \mathcal{A}_{\mathcal{F}_2} \quad \text{and} \quad \mathcal{F}_2 \supset \mathcal{F}_1 \implies \mathcal{A}_{\mathcal{F}_2} \succcurlyeq \mathcal{A}_{\mathcal{F}_1}.$$

We are in a position to establish the alphabets in the consequent rows of the symbolic array space of our future zero-dimensional extension. We pick an increasing (with respect to inclusion) sequence $(\mathcal{F}_k)_{k \in \mathbb{N}}$ of families of continuous functions $f : X \to [0,1]$. We can easily arrange that the corresponding partitions refine in the product $X \times \mathbb{T}$, i.e., the diameters of $\mathcal{A}_{\mathcal{F}_k}$ decrease to zero. We also make sure that $\mathcal{F}_k$ contains the characteristic function of the marker set F_k; this function is continuous since F_k is clopen. This implies that the set $F_k \times \mathbb{T}$ is measurable (i.e., is a union of cells) with respect to the partition $\mathcal{A}_{\mathcal{F}_k}$. In practice, this means that in any dynamics on $X \times \mathbb{T}$ lifting T, the p_k-markers in the $\mathcal{A}_{\mathcal{F}_k}$-names are "built in" single symbols. We let Λ_k be a set of labels assigned to the partition $\mathcal{A}_{\mathcal{F}_k}$ and we imagine the symbols corresponding to the cells contained in $F_k \times \mathbb{T}$ to look like "$|a$" (while others just look like "a"); the vertical bar visualizes the marker.

Our symbolic array space will consist of arrays $(y_{k,n})_{k \in \mathbb{N}, n \in \mathbb{Z}}$ which, for every $k \in \mathbb{N}$, in row number k have symbols from Λ_k. Moreover, we will restrict the symbolic space to the arrays $y = (y_{k,n})$ whose columns satisfy a certain *column condition*:

- For every n the projections onto X of the cells of $\mathcal{A}_{\mathfrak{F}_k}$ labeled $y_{k,n}$ have nonempty intersection over $k \in \mathbb{N}$.

Since the diameters of these cells decrease to zero with k (for n fixed), the above intersection is a single point in X. We denote this point by $\pi_{X,n}(y)$. Clearly, the map $\pi_{X,n}$ is continuous on its domain. We let Y_{c} be the collection of all arrays y satisfying the column condition and such that $T(\pi_{X,n}(y)) = \pi_{X,n+1}(y)$ for every $n \in \mathbb{Z}$. It is obvious that Y_{c} is a closed set of arrays, invariant under the horizontal shift S, and that the map $\pi_X = \pi_{X,0}$ is a topological factor from $(Y_{\mathsf{c}}, S, \mathbb{S})$ onto $(X, T, \mathbb{S})$. This system will be our "working environment" inside of which we will select much more delicate extensions of $(X, T, \mathbb{S})$. Since every subsystem of $(Y_{\mathsf{c}}, S, \mathbb{S})$ will have the same action (the iterates $n \in \mathbb{S}$ of the horizontal shift S) we will denote it by one letter only, (for instance Y_0, Y_1, etc.) skipping the transformation S and the semigroup $\mathbb{S}$. Below we will consistently denote the measures on X by the letter μ (with subscripts, superscripts, etc.), similarly, the letter ν is reserved for the measures on the symbolic array system Y_{c}, while the letter ξ is used for the measures on the product space $X \times \mathbb{T}$. Exception: $\mu \times \lambda$ also denotes a measure on this product.

In the product space $X \times \mathbb{T}$ we introduce (temporarily) the "trivial" dynamics of $T \times \mathrm{Id}$, where Id denotes the identity map; this is a principal extension of $(X, T, \mathbb{S})$. Every point (x, t) in the product space has its *array-name* with respect to the sequence of partitions $\mathcal{A}_{\mathfrak{F}_k}$: in row number k we see the (bilateral) name of (x, t) with respect to $\mathcal{A}_{\mathfrak{F}_k}$. This array-name obviously belongs to Y_{c}. Recall that we have $\mathfrak{F}_k \subset \mathfrak{F}_{k+1}$ which translates to $\mathcal{A}_{\mathfrak{F}_k} \preccurlyeq \mathcal{A}_{\mathfrak{F}_{k+1}}$, which in turn means that every symbol in the array-name of (x, t) determines all the symbols above it in the same column (i.e., with the same n and smaller k). Also notice that every such array is marked to base (p_k) (see Definition A.3.2): in row number k we have periodically repeated p_k-markers visualized as vertical bars. The k-rectangles (see Definition A.3.3) in these arrays stand in 1-1 correspondence with the cells of $\mathcal{A}_{\mathfrak{F}_k}^{p_k}$ (the "power" is with respect to $T \times \mathrm{Id}$) contained in the marker set $F_k \times \mathbb{T}$. They will be called *fundamental k-cells*. Indeed, whenever a pair (x, t) belongs to a fundamental k-cell $C \in \mathcal{A}_{\mathfrak{F}_k}^{p_k}$, its array-name has a specific k-rectangle at positions 0 through $p_k - 1$ and rows 1 through k. We will denote this k-rectangle by $\hat{C}$.

Because the cells of the partitions $\mathcal{A}_{\mathfrak{F}_k}$ are not closed, the collection of all array-names is not closed, either. We take the closure of this collection (in the symbolic array space) and denote it Y_0. Notice that Y_0 maintains the following three properties: it is a subset of Y_{c}, it consists of arrays marked to base (p_k), and every symbol determines the symbols above it. The system Y_0 is not only an extension of $(X,T,\mathbb{S})$ (via π_X) but also of the product system $(X \times \mathbb{T}, T \times \mathrm{Id}, \mathbb{S})$ (via a map which we denote by π_0). The situation is pictured on the commuting diagram:

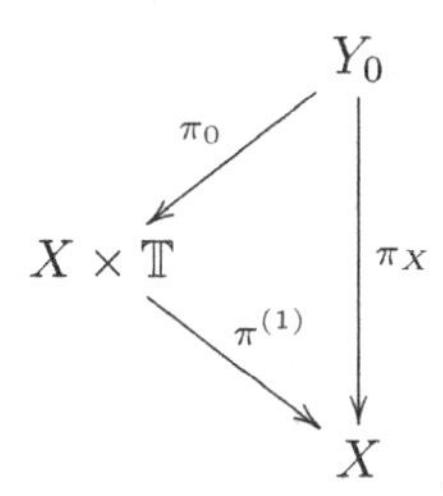

(here $\pi^{(1)}$ is simply the projection on the first axis).

The symbolic array extension via π_0 is 1-1 except on points in $X \times \mathbb{T}$, whose orbits visit the boundaries of the cells, (i.e., the graphs of the involved functions f). In general we have no guarantee that Y_0 is a principal extension (of the product system or, equivalently, of $(X,T,\mathbb{S})$) because (some of) the graphs of the functions in the families $\mathcal{F}_k$ may happen to have positive measure for some $(T \times \mathrm{Id})$-invariant measures, and then we have no control over the entropy of the lifted measures. Among measures in the product system, whose *any* lift has guaranteed conditional entropy zero are the product measures $\mu \times \lambda$. It is so, because the graphs of the involved functions f have measure zero for every such measure, so the extension is 1-1 almost everywhere for such measures. We denote by $\mathcal{M}_0$ the set of lifts against π_0 of the product measures.

We will construct the desired extension Y as a subsystem of Y_c which will be (in some sense) a limit of an auxiliary sequence of mutually conjugate subsystems Y_k. We will define Y_k inductively, by constructing the maps $\Phi_k : Y_{k-1} \to Y_k$ (these maps will be block codes defined on the $(k+1)$-rectangles). The main goal is to ensure that for any k the set $\mathcal{M}_S(Y_k)$ is contained within an open set $V_k \subset \mathcal{M}_S(Y_\mathsf{c})$, where the sequence V_k (which we will also define inductively) is decreasing and satisfies the following property:

$$\text{For any } k > 0 \text{ and any measure } \nu \in V_k\,,\ h(\nu, \Lambda_{k-1}|\mathfrak{A}_X) < \varepsilon_k\,, \tag{7.6.6}$$

where Λ_{k-1} means the partition of the symbolic array space corresponding to the alphabet in the row number $k-1$ (for $k = 1$ that will be the trivial partition), $\mathfrak{A}_X$ is the sigma-algebra lifted from X and ε_k is a preassigned decreasing to zero sequence of positive numbers.

We have already constructed Y_0. Suppose, for some $k \geq 1$ we have constructed the systems $Y_0, Y_1, \ldots, Y_{k-1}$ (all contained in Y_c) related as in the next diagram.

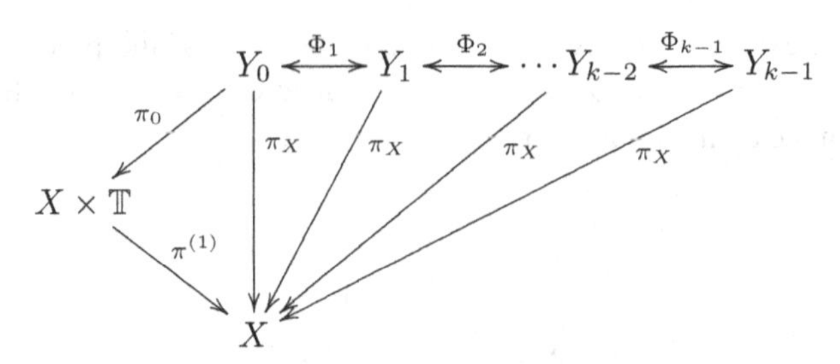

We assume that the maps Φ_i $(i = 1, \ldots, k-1)$ are topological conjugacies by block-codes replacing the i-rectangles occurring in Y_{i-1} by other i-rectangles (also occurring in Y_{i-1}) in a way depending only on the $(i+1)$-rectangle, and leaving the rows $i+1, i+2, \ldots$ intact. Such codes lose the property that each symbol determines the symbols above it, but still, the last (unchanged) row of a $(i+1)$-rectangle completely determines its image. We will soon define the code Φ_k on the $(k+1)$-rectangles.

On $X \times \mathbb{T}$ we select the product measures $\mu \times \lambda$ where $\mu \in \mathcal{M}_T(X)$, and we let $\mathcal{M}_{k-1}$ denote the set of their lifts to Y_{k-1} (against the composition of maps that leads from Y_{k-1} to $X \times \mathbb{T}$). As we said, for any $\nu \in \mathcal{M}_0$ we have $h(\nu|\mathfrak{A}_X) = 0$. Since each Φ_i is a conjugacy, we also have $h(\nu|\mathfrak{A}_X) = 0$ for any $\nu \in \mathcal{M}_{k-1}$. In particular, for such measures,

$$h(\nu, \Lambda_{k-1}|\mathfrak{A}_X) = 0.$$

Since the partition Λ_{k-1} is clopen, the function $\nu \mapsto h(\nu, \Lambda_{k-1}|\mathfrak{A}_X)$ is upper semicontinuous on invariant measures (see Fact 6.6.7 item (d) with X in the role of the factor), and thus there is an open set V_k in $\mathcal{M}_S(Y_\mathsf{c})$ containing $\mathcal{M}_{k-1}$ and contained in V_{k-1} (we are assuming inductively that $\mathcal{M}_S(Y_{k-1}) \subset V_{k-1}$), on which $h(\nu, \Lambda_{k-1}|\mathfrak{A}_X) < \varepsilon_k$. We can assume that V_k is a ball around $\mathcal{M}_{k-1}$ of some positive radius ρ. Since the metric between measures in zero-dimensional spaces (see (7.3.1)) is convex and $\mathcal{M}_{k-1}$ is convex, so is V_k. It is easy to see that there exists some $j \geq k$ and $\delta > 0$ such that two measures are less than ρ apart whenever their values on every j-rectangle differ by less than 3δ. Without loss of generality we can assume that $j = k$ (we can achieve this by skipping the families $\mathcal{F}_k, \mathcal{F}_{k+1}, \ldots, \mathcal{F}_{j-1}$ in the construction; they are contained in $\mathcal{F}_j$ anyway). Summarizing, we have arranged that if $\nu_0 \in \mathcal{M}_{k-1}$ and $\nu \in \mathcal{M}_S(Y_\mathsf{c})$, then

$$|\nu(\hat{C}) - \nu_0(\hat{C})| < 3\delta \text{ for all } k\text{-rectangles } \hat{C} \implies \nu \in V_k. \qquad (7.6.7)$$

Now we get back to the product system $X \times \mathbb{T}$. For every measure $\mu \in \mathcal{M}_T(X)$ the product measure $\mu \times \lambda$ gives the value zero to the boundaries of the cells C of the partition $\mathcal{A}_{\mathcal{F}_k}^{p_k}$. This implies that $\mu \times \lambda$ is a continuity point of the map $\xi \mapsto \xi(C)$ for measures ξ on the product system. So, the condition

$$|\xi(C) - (\mu \times \lambda)(C)| < \delta \quad \text{for all} \quad C \in \mathcal{A}_{\mathcal{F}_k}^{p_k} \tag{7.6.8}$$

holds on some neighborhood W of $\mu \times \lambda$ in $\mathcal{M}(X \times \mathbb{T})$. By Lemma 7.6.4, there is a neighborhood U_μ of μ in $\mathcal{M}(X)$ such that whenever $\mathrm{M}_n(\boldsymbol{\delta}_x) \in U_\mu$, then $\mathrm{M}_n(\boldsymbol{\delta}_{(x,t)})$ satisfies (7.6.8) in the role of ξ. Out of the sets U_μ, we select a finite cover, say $U_{\mu_1}, \ldots, U_{\mu_r}$, of $\mathcal{M}_T(X)$ in $\mathcal{M}(X)$. The union of this cover is an open set U around $\mathcal{M}_T(X)$.

Without loss of generality we can assume that p_{k+1} is so large that Fact 6.6.1 holds for any $n \geq p_{k+1}$. That is to say, every measure of the form $\mathrm{M}_{p_{k+1}}(\boldsymbol{\delta}_x)$ is already contained in U, so it is contained in some U_{μ_i}, hence $\mathrm{M}_{p_{k+1}}(\boldsymbol{\delta}_{(x,t)})$ satisfies (7.6.8) in the role of ξ and with μ_i in the role of μ.

Now we go to the symbolic array space again. Let $\hat{D}$ be a $(k+1)$-rectangle in Y_{k-1}. The last row of this rectangle is unchanged (the same as in its preimage in Y_0) so it corresponds to a fundamental $(k+1)$-cell (denoted by D). Choose a point (x_D, t_D) from this fundamental $(k+1)$-cell and a measure $\mu_D = \mu_i$ such that $\mathrm{M}_{p_{k+1}}(\boldsymbol{\delta}_{x_D})$ belongs to U_{μ_i}. Recall that the fundamental k-cells partition the set $F_k \times \mathbb{T}$, which is invariant under the map $(T \times R_{\mu_D})^{p_k}$. Thus (x_D, t_D) has a name under the action of $(T \times R_{\mu_D})^{p_k}$ on $F_k \times \mathbb{T}$ with respect to the partition into the fundamental k-cells. Take the initial block of length $q = p_{k+1}/p_k$ of this name. It is an ordered list of q fundamental k-cells, each associated with a unique k-rectangle in Y_{k-1}, so we have a sequence of k-rectangles $\hat{C}_1, \ldots, \hat{C}_q$. Observe that the number of times a k-rectangle $\hat{C}$ occurs in this list equals the number of times (x_D, t_D) visits the corresponding fundamental k-cell C under the action of $(T \times R_{\mu_D})^{p_k}$. This number equals $p_{k+1}\mathrm{M}_{p_{k+1}}(\boldsymbol{\delta}_{(x_D,t_D)})(C)$. As we know, $\mathrm{M}_{p_{k+1}}(\boldsymbol{\delta}_{(x_D,t)_D})$ satisfies (7.6.8) along with μ_D, i.e.,

$$|\mathrm{M}_{p_{k+1}}(\boldsymbol{\delta}_{(x_D,t_D)})(C) - (\mu_D \times \lambda)(C)| < \delta,$$

hence

$$\text{the number of occurrences of } \hat{C} \text{ on the list is } p_{k+1}\big((\mu_D \times \lambda)(C) \pm \delta\big). \tag{7.6.9}$$

At this point we define the image of $\hat{D}$ under Φ_k as follows: In rows 1 through k it has the ordered list of k-rectangles $\hat{C}_1, \ldots, \hat{C}_q$ described above and in the row $k+1$ it retains the original contents of $\hat{D}$. Observe that $\Phi_k(\hat{D})$ satisfies the column condition; the projections of the cells corresponding to the symbols in column n of this rectangle all contain the point $T^n x_D$.

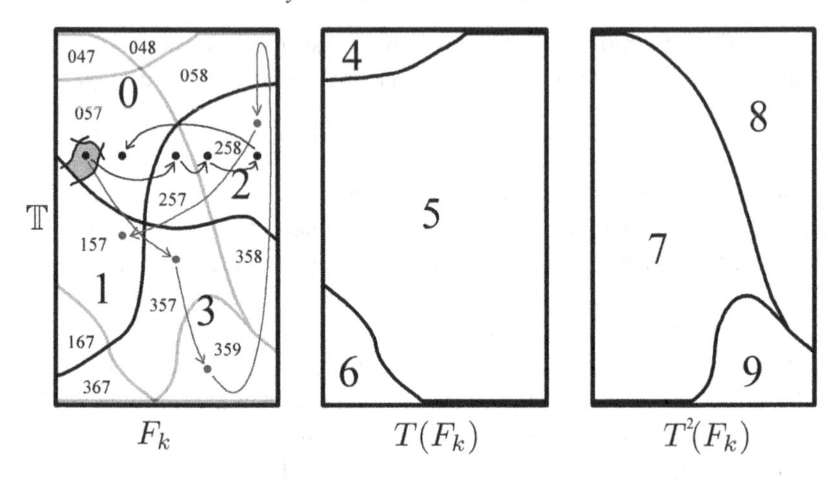

Figure 7.1 The code Φ_k.

The idea of the construction of the code Φ_k is shown in Figure 7.1. The bases of the three large rectangles are the sets $F_k, T(F_k), \ldots, T^{p_j-1}(F_k)$. For simplicity, we imagine the transformation T as the rigid translation between these sets, except on the last one, which is mapped somehow back to F_k. The large rectangles are Cartesian products with $\mathbb{T}$ (shown as the interval). The family $\mathcal{F}_k$ consists, in this example, of the characteristic functions of $T^i(F_k)$ ($i = 0, 1, 2$) and of two more functions (the black curves). The partition $\mathcal{A}_{\mathcal{F}_k}$ of $X \times \mathbb{T}$ is labeled $\{0, 1, \ldots, 9\}$. The resulting fundamental k-cells (enclosed by black and grey curves) are labeled $047, 048, \ldots, 359$ (these are our k-rectangles). The fundamental $(k+1)$-cell D in $X \times \mathbb{T}$ corresponding to the selected $(k+1)$-rectangle $\hat{D}$ is shown in grey (with pieces of the enclosing functions from $\mathcal{F}_{k+1}^{p_{k+1}}$) and the point (x_D, t_D) is inside. The kth row of $\hat{D}$ is obtained by reading the labels of the fundamental k-cells along the trajectory of (x_D, t_D) for q iterates of $(T \times \mathrm{Id})^{p_k}$ (the black dots). In this example it starts with $|057|257|258|258|057|\ldots$ The code Φ_k changes this row (and the ones above) by following the orbit of (x_D, t_D) under the action of $(T \times R_{\mu_D})^{p_k}$ (the grey dots). In this example the kth row of $\Phi_k(\hat{D})$ begins with $|057|357|359|258|157|\ldots$ Notice that the projection of the nth symbol in both names contains the point $T^n x_D$.

For any point $y \in Y_{k-1}$ we define its image, $\Phi_k(y)$, by replacing every $(k+1)$-rectangle $\hat{D}$ in y by $\Phi_k(\hat{D})$. We let $Y_k = \Phi_k(Y_{k-1})$. The column condition and the fact that Φ_k preserves rows from $k+1$ onwards ensure that

$Y_k \subset Y_c$, and that Φ_k is a conjugacy. The crucial thing in this construction is that the applied rotation changes from $(k+1)$-rectangle to $(k+1)$-rectangle and is usually inconsistent with the rotation applied in the earlier rows. The arrays in Y_k could not be obtained as array-names for any fixed dynamics on the product space.

We will now verify that $\mathcal{M}_S(Y_k) \subset V_k$. The statement (7.6.9) can now be interpreted as an approximation of the number of occurrences of $\hat{C}$ in $\Phi_k(\hat{D})$ (because of the markers, $\hat{C}$ cannot occur in $\Phi_k(\hat{D})$ at a position not divisible by p_k). Further, since p_{k+1} is very large, the corresponding frequency (the number of occurrences divided by the length p_{k+1}) is nearly equal to the value $\mu_{(\Phi_k(\hat{D}))}(\hat{C})$ (recall that $\mu_{(\Phi_k(\hat{D}))}$ is the periodic measure supported by the orbit of the sequence $\dots\Phi_k(\hat{D})\Phi_k(\hat{D})\Phi_k(\hat{D})\dots$). So, (7.6.9) yields

$$|\mu_{(\Phi_k(\hat{D}))}(\hat{C}) - (\mu_D \times \lambda)(C)| < 2\delta,$$

for all k-rectangles $\hat{C}$. Let ν_D denote the lift of $\mu_D \times \lambda$ to Y_k. Of course $\nu_D \in \mathcal{M}_k$. Each C has null boundary for $\mu_D \times \lambda$, hence the cell C lifts to $\hat{C}$ up to the measure ν_D and thus $(\mu_D \times \lambda)(C) = \nu_D(\hat{C})$. Eventually, we have

$$|\mu_{(\Phi_k(\hat{D}))}(\hat{C}) - \nu_D(\hat{C})| < 2\delta.$$

The rest of the argument is easier. Let $\nu \in \mathcal{M}_S(Y_k)$. In order to decide whether $\nu \in V_k$ we only need to observe the values of ν on k-rectangles (see (7.6.7)). It suffices to examine the projection of ν onto the symbolic factor of Y_k constituted by the first $k+1$ rows. Since V_k is convex, it suffices to examine ergodic ν only. Every such projected ergodic measure can be approximated by periodic measures $\mu_{(B)}$ where B is a long rectangle occurring in the first $k+1$ rows of Y_k. We can choose B as a concatenation of the $(k+1)$-rectangles. The $(k+1)$-rectangles occurring in Y_k have the form $\Phi_k(\hat{D})$. By Fact 7.3.2, the periodic measure $\mu_{(B)}$ can be very accurately replaced by the convex combination of the measures $\mu_{(\Phi_k(\hat{D}))}$. By making p_{k+1} large enough, we can make all the inaccuracies affect the values on k-rectangles by no more than another δ. In the end we get

$$|\nu(\hat{C}) - \nu_0(\hat{C})| < 3\delta,$$

for all k-rectangles $\hat{C}$, where ν_0 is a convex combination of the measures ν_D and still belongs to the (convex) set $\mathcal{M}_k$. By the criterion (7.6.7), we have shown that $\nu \in V_k$.

Having defined the sequence of systems Y_k (subsets of Y_c), we let

$$Y = \bigcap_{m=1}^{\infty} \overline{\bigcup_{k=m}^{\infty} Y_k}.$$

In other words, Y is the set of all arrays y such that $y = \lim_k y_k$, $y_k \in Y_k$. It is easy to see that this is a closed subsystem of Y_c and an extension of X. The important observation is that any invariant measure ν on Y is in every V_k. This follows by the same argument as the one used to show that the invariant measures on Y_k are in V_k; that argument depended only on the properties of the $(k+1)$-rectangles occurring in Y_k, and Y has the same $(k+1)$-rectangles.

To show that Y is a principal extension of X we need to show that the conditional entropy of Y given $\mathfrak{A}_X$ is 0 for every invariant measure $\nu \in \mathcal{M}_S(Y)$. For any $0 < k < k'$ we have $h(\nu, \Lambda_k|\mathfrak{A}_X) \le h(\nu, \Lambda_{k'-1}|\mathfrak{A}_X)$, (because $\Lambda_{k'-1} \succcurlyeq \Lambda_k$). On the other hand, since $\nu \in V_{k'}$, we know that $h(\nu, \Lambda_{k'-1}|\mathfrak{A}_X) < \varepsilon_{k'}$ (this is (7.6.6)). It follows that $h(\nu, \Lambda_k|\mathfrak{A}_X) < \varepsilon_{k'}$, and since k' is arbitrary, $h(\nu, \Lambda_k|\mathfrak{A}_X) = 0$. Taking the supremum over k, we conclude that $h(\nu|\mathfrak{A}_X) = 0$. □

Remark 7.6.10 Notice that the elements of the extension Y are (regardless of surjectivity of T) marked symbolic arrays with bilateral rows. This fact will be used in the construction of symbolic extensions.

As an immediate application of Theorem 7.6.1 we can generalize the second statement of Fact 7.2.4 to all systems:

Corollary 7.6.11 *The entropy function $\mu \mapsto h(\mu)$ on $\mathcal{M}_T(X)$ is of Young class* LU, *for any dynamical system $(X, T, \mathbb{S})$.*

Proof We pass to the principal zero-dimensional extension. Here, the second statement of Fact 7.2.4 holds. Since the extension is principal, the entropy function on $\mathcal{M}_T(X)$ equals the push-down of the entropy function on the extension (see Definition A.1.25 for the meaning of "pushing down"). It is elementary to see that push-down preserves Young class LU. □

Remark 7.6.12 Theorem 7.6.1 and the elementary proof of the Variational Principle in dimension zero provide an alternative way of proving the "harder" direction of the Variational Principle for general systems, as follows: Take any system $(X, T, \mathbb{S})$ and its extension $(Y, S, \mathbb{S})$ as in Theorem 7.6.1. Then $\mathbf{h}(S) \ge \mathbf{h}(T)$. By the Variational Principle in dimension zero, there exists an invariant measure $\nu \in \mathcal{M}_S(Y)$ such that $h(\nu, S) \ge \mathbf{h}(S) - \varepsilon$. The image $\mu = \pi\nu$ then satisfies $h(\mu, T) = h(\nu, S) \ge \mathbf{h}(S) - \varepsilon \ge \mathbf{h}(T) - \varepsilon$.

We cannot deduce the other direction this way, because without the Variational Principle we do not know whether the extension built in the proof of Theorem 7.6.1 preserves topological entropy (unless we reexamine the construction for that).

We close this chapter by providing the missing proofs of Theorem 6.9.9 and Theorem 6.9.10 from the preceding chapter.

Proof of Theorem 6.9.9 The task is to show that a system $(X, T, \mathbb{S})$ whose topological entropy is zero has an extension $(Y, S, \mathbb{S})$, in form of a bilateral subshift, via a map that collapses asymptotic pairs. We also need to show that the topological entropy of $(Y, S, \mathbb{S})$ is zero. The construction of symbolic extensions is in general a difficult task, discussed at length in Chapter 9. This proof is an exercise on constructing symbolic extensions in the easiest case of zero-entropy systems. By Theorem 7.6.1, Remark 7.6.10 and Lemma 6.9.4, we can immediately assume that $(X, T, \mathbb{S})$ is zero-dimensional and its elements are bilateral marked symbolic arrays. Let $\mathcal{R}_{k,n}$ denote the family of all rectangles of height k and length n appearing in the first k rows of X. Because the system has entropy zero, the cardinalities of these families grow subexponentially with n, in particular for each k there exists n_k such that $\log \#\mathcal{R}_{k,n_k} < n_k 2^{-k}$ (and the right-hand side is an integer). By dropping some of the markers we can easily arrange that $n_k = p_k$, the length of the k-rectangles. Now we let $\mathcal{R}_k$ denote the family of all k-rectangles, and we have $\#\mathcal{R}_k < 2^{p_k 2^{-k}}$. This implies that there exists an injective function (code) Φ_k from all k-rectangles into the family of all binary (i.e., over $\{0, 1\}$) blocks of length $p_k 2^{-k}$. We can now create the symbolic extension. Initially it will be not precisely symbolic, as its elements will consist of a pair: an element of the odometer and a symbolic row. For each $x \in X$ we create its "preimage," y, as follows: we take the same element of the odometer as is represented by the markers in x. The symbolic row of y is filled inductively: Above the left half of each 1-rectangle R of x we put in y the image of R via the code Φ_1 (this image has length exactly half of the length of R). After this step "half" of y is filled with zeros and ones, leaving the rest to be filled in the steps to come. In the following steps we apply an additional twist: the image $\Phi_k(R)$ of each k-rectangle R in x is placed not above R but, instead, above the neighboring k-rectangle (to the right). The contents of $\Phi_k(R)$ are written there into the consecutive free slots in that sector (starting from the left). This will use only half of the free slots available, leaving the rest to be used in the steps to come (see Figure 7.2).

It is easy to see that after all steps are completed, depending on the positioning of the markers, the symbolic row of y is either completely filled, or there

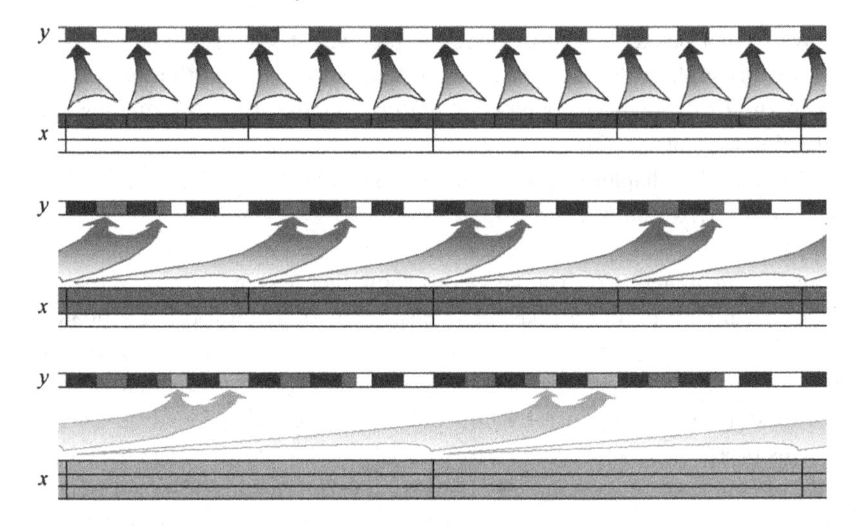

Figure 7.2 First three steps of the construction of the symbolic preimage y of x. The arrows show where the information is stored.

remain some unfilled slots. We fill these slots in every possible way, producing multiple preimages for x; for this reason $(Y, S, \mathbb{S})$ is not conjugate to $(X, T, \mathbb{S})$, only a topological extension. We skip the standard description of the factor map from Y to X, which relies on simply uncoding all the k-rectangles from the contents of appropriate places in y, located with the help of the odometer part of y.

Now consider two different arrays x and x' in X. If they have different positioning of the markers, then they factor to two distinct elements of the odometer. Any pair of their preimages y and y' contains the same distinct pair of elements of the odometer. Since the action on the odometer is an isometry in an appropriate metric, such elements are never asymptotic. Now suppose x and x' have the same structure of markers. They must still differ at at least one position (k, n). This implies that for every $k' \geq k$ the k'-rectangles covering this position in x and x' are different. Then they have different images by $\Phi_{k'}$. It is thus clear that the symbolic preimages y of x and y' of x' (any choice of a pair of such preimages) will differ at infinitely many positions tending toward infinity. In other words, y and y' cannot be (forward) asymptotic. We have proved that asymptotic pairs y, y' must be collapsed in X.

Now we calculate the topological entropy of the extension $(Y, S, \mathbb{S})$. Clearly the odometer has entropy zero, so we only need to compute the entropy of the symbolic row, and here it suffices to count the blocks occurring between neighboring pairs of p_k-markers (see Exercise 7.1). Take a block B in some

$y \in Y$ lying between two consecutive p_k-markers. Its contents is almost completely determined by the k-rectangle of the image x positioned directly below B. This rectangle determines all but $p_k 2^{-k-1}$ entries in B. This implies that the number of all such blocks B is at most the number of all k-rectangles in X (which is not larger than $2^{p_k 2^{-k}}$) times $2^{p_k 2^{-k-1}}$. The logarithm of this product divided by the length p_k goes to zero with k. Thus $\mathbf{h}(S) = 0$.

To complete the construction we must replace the odometer in the construction of Y by a symbolic system. The standard method is to extend the odometer to a binary Toeplitz system of entropy zero, as described in Example A.3.4. Such an extension induces an extension of Y to a symbolic system with two rows; the element of the odometer will be now replaced by a symbolic row containing the elements of the Toeplitz system. By Lemma 6.9.4, after this modification of Y, the property that the factor map onto X collapses asymptotic pairs will be preserved. This ends the proof. □

Proof of Theorem 6.9.10 For an arbitary system $(X, T, \mathbb{S})$ with zero topological entropy we need to find its zero-dimensional extension which is NAP (possesses no asymptotic pairs). Let $(X, T, \mathbb{S})$ be the initial system (of entropy zero). By Theorem 6.9.9, there exists a bilateral subshift extension, also of entropy zero, say $(Y_1, S_1, \mathbb{S})$, via a map π_0 that collapses asymptotic pairs. Applying the same theorem again, $(Y_1, S_1, \mathbb{S})$ has a bilateral subshift extension $(Y_2, S_2, \mathbb{S})$ via a map π_1 that collapses asymptotic pairs. And so on. We obtain a sequence of bilateral subshifts $(Y_k, S_k, \mathbb{S})$ bound by factor maps π_k that collapse asymptotic pairs. The zero-dimensional extension is obtained as the corresponding inverse limit of subshifts (see Definition 6.5.11). Suppose this inverse limit has an asymptotic pair $y \neq y'$. By the definition of an inverse limit, there is k such that the images y_k and y'_k of y and y' in Y_k are different. Then analogously defined y_{k+1} and y'_{k+1} are also different in Y_{k+1}. On the other hand, the last pair is asymptotic, which implies that it is collapsed by π_k, i.e., that $y_k = y'_k$, a contradiction. □

Exercises

7.1 Consider a subshift which factors to the odometer to base (p_k). We can visualize this factor as a system of p_k-periodic markers. Let $\mathcal{B}_k$ denote the family of all blocks occurring between the p_k-markers (called *k-blocks*). Prove that the topological entropy of this subshift equals $\lim_k \frac{1}{p_k} \log \#\mathcal{B}_k$.

7.2 Describe explicitly the natural extension of a unilateral subshift. Consider separately the surjective and non-surjective cases.

7.3 Prove that every bilateral (i.e., injective) zero-dimensional dynamical system $(X, T, \mathbb{S})$ without periodic points is conjugate to a subshift over the countable alphabet $\mathbb{N}_0 \cup \{\infty\}$. (This exercise is harder than the average. The restriction on periodic points can be weakened, but not skipped.)

7.4 Using the marker technique prove that if $(X, T, \mathbb{S})$ is a bilateral subshift and has no periodic points, then it is conjugate to a subshift $(Y, S, \mathbb{S})$ over an alphabet of cardinality $\lfloor 2^{\mathbf{h}} \rfloor + 1$ (as in Thoerem 7.2.3).

7.5 Show by example that the assumption about periodic points in the preceding exercise cannot be skipped.

8

The entropy structure

The topological entropy of a dynamical system is a rather crude measurement of its complexity. In order to thoroughly understand the dynamics it is essential not only to replace the topological entropy by the entropy function $\mu \mapsto h(\mu, T)$ on invariant measures, but also to replace the topological entropy detectable in a given resolution, say $\mathbf{h}_1(T, \varepsilon)$ or $\mathbf{h}_2(T, \varepsilon)$, by some function, say $\mu \mapsto h(\mu, T, \varepsilon)$, reflecting the dynamical entropy of each measure detectable in that resolution. Such a function has not been presented in this book yet. As we shall see, there is an essential difference between entropy of a measure with respect to a *measurable* resolution, i.e., a partition (even if we require that the cells have diameters bounded by ε) and the entropy of a measure with respect to a *topological* resolution, which we are about to define. The difference is in semicontinuity properties and in the "type of convergence" to the entropy function as the resolution refines. This type of convergence turns out to be the most important feature in the part of entropy theory leading to digitalization and data compression of topological dynamical systems.

The material in this section is based mainly on the paper [Downarowicz, 2005a].

8.1 The type of convergence

In this section we will try to understand what it means, for two monotone nets of real-valued functions converging to the same limit function, to converge "the same way," or to represent the same "type of convergence." We will try to capture a kind of "multilayer defect of uniformity" in this convergence. The common domain of the functions will be denoted by $\mathfrak{X}$. In the application to the theory of entropy structure the role of $\mathfrak{X}$ will be played by the set $\mathcal{M}_T(X)$ of all T-invariant measures of a topological dynamical system $(X, T, \mathbb{S})$. In this section the letter x refers, exceptionally, to an element of $\mathfrak{X}$ and not of X. Let us recall that in order to avoid confusingly sounding words we use *increasing* and *decreasing* in the meaning *nondecreasing* and *nonincreasing*, respectively.

8.1.1 Introduction to defect theory

The notions discussed in this section do not belong to the standard apparatus in the theory of dynamical systems and may be new to many of the readers. We will hence provide more details, intuition and elementary examples, than in other sections.

The main subject here is an increasing net of nonnegative functions defined on some abstract domain $\mathfrak{X}$,

$$\mathcal{H} = (h_\kappa)_{\kappa \in \mathcal{K}},$$

where κ ranges over a directed family $\mathcal{K}$ (see Appendix A.1.3) and, for all $\iota, \kappa \in \mathcal{K}$,

$$\iota \geq \kappa \quad \Longrightarrow \quad h_\iota \geq h_\kappa .$$

If $\mathcal{K}$ is fixed, we will just write (h_κ). (We could have managed with sequences, but sometimes it will be more convenient to use nets.)

We also assume that this net converges at every point $x \in \mathfrak{X}$ to a finite limit, which we denote by $h(x)$. By the *limit function* we will understand the resulting function $h : \mathfrak{X} \to [0, \infty)$. The convergence $h_\kappa \nearrow h$ may happen to be uniform (i.e., such that for every ε there is κ such that $h_\kappa \geq h - \varepsilon$) but in general it is not. We are interested in classifying the nonuniform cases. Later we will restrict our attention exclusively to the case where the domain $\mathfrak{X}$ is compact and metric. However, many statements of this section are valid in a wider generality and, whenever it is possible without additional effort, we will give the more general version.

The net $\mathcal{H}$ determines both its limit function h and the associated *net of tails* $(\theta_\kappa)_{\kappa \in \mathcal{K}}$, where $\theta_\kappa = h - h_\kappa$. Note that the net of tails converges decreasingly to zero. Conversely, h and the net of tails determine $\mathcal{H}$. In what follows we will often switch between net $\mathcal{H}$ and the net of tails, and we will identify the "type of convergence" (whatever it means) of (h_κ) to h with that of (θ_κ) to the zero function.

And so, the nonuniformity of the convergence $h_\kappa \nearrow h$ (equivalently, of the convergence $\theta_\kappa \searrow 0$) is very crudely measured by the *global defect of uniformity*:

$$D_{\mathfrak{X}} = \lim_{\kappa \in \mathcal{K}} \downarrow \sup_{x \in \mathfrak{X}} \theta_\kappa(x).$$

If $\mathfrak{X}$ is a metric space, we can define the *local defect of uniformity* at a point:

$$D_x = \inf_{\varepsilon>0} \lim_{\kappa\in\mathcal{K}} \downarrow \sup_{y\in B(x,\varepsilon)} \theta_\kappa(y) \tag{8.1.1}$$

(recall that $B(x,\varepsilon)$ is the ε-ball around x). It is clear that $D_x \le D_{\mathfrak{X}}$ for every $x \in \mathfrak{X}$. If $D_x = 0$ for all x, then we say that the convergence is *locally uniform.* In general, this does not imply uniform convergence, but it does in the most interesting, for us, case of a compact domain (see Theorem 8.1.3 below).

Somewhat surprisingly, the local defect of uniformity has to do with upper semicontinuity. It suffices to confront the formula (8.1.1) (with the infima in reversed order) with the definition of the upper semicontinuous envelope (see Appendix, (A.1.18)) to see that in case when h is locally bounded (bounded on a neighborhood of every point), each θ_κ is also locally bounded, and we have, at each point,

$$D_x = \lim_{\kappa\in\mathcal{K}} \downarrow \widetilde{\theta}_\kappa(x). \tag{8.1.2}$$

This observation leads us to studying upper semicontinuity properties of the net of tails. These upper semicontinuity properties become the key tool in the classification of the "types of convergence." We begin by proving

Theorem 8.1.3 *If $\mathfrak{X}$ is compact, then*

$$D_{\mathfrak{X}} = \sup_{x\in\mathfrak{X}} D_x.$$

Proof We have already noted that $D_x \le D_{\mathfrak{X}}$ for every $x \in \mathfrak{X}$. If $D_x = \infty$ at some point, then the assertion holds. Otherwise for each x there is a neighborhood $U_x \ni x$ and an index κ_x such that θ_{κ_x} is bounded on U_x. By compactness, we can find a threshold index κ_0 above which the functions θ_κ are bounded. Then

$$D_{\mathfrak{X}} = \lim_{\kappa\ge\kappa_0} \downarrow \sup_{x\in\mathfrak{X}} \theta_\kappa(x) \le \lim_{\kappa\ge\kappa_0} \downarrow \sup_{x\in\mathfrak{X}} \widetilde{\theta}_\kappa(x) = \sup_{x\in\mathfrak{X}} \lim_{\kappa\ge\kappa_0} \downarrow \widetilde{\theta}_\kappa(x) = \sup_{x\in\mathfrak{X}} D_x \le D_{\mathfrak{X}},$$

where the change of the order of sup and lim $\downarrow$ is validated by the elementary property of decreasing nets of upper semicontinuous functions (see Appendix, Fact A.1.24). □

In some situations the local defect D_x and the global defect $D_{\mathfrak{X}}$ can be used to estimate the defect of upper semicontinuity of the limit function. Recall (see (A.1.20)) that the *defect of upper semicontinuity* of a function f is defined as the difference function $\dddot{f} = \widetilde{f} - f$.

Fact 8.1.4 *Let (h_κ) be an increasing net converging to a locally bounded limit function h on a metric space $\mathfrak{X}$. Assume that all functions h_κ are upper semicontinuous. Then, at every point $x \in \mathfrak{X}$ we have*

$$\dddot{h}(x) \le D_x,$$

in particular

$$\sup_x \dddot{h}(x) \le D_{\mathfrak{X}}.$$

Proof Fix some $x \in \mathfrak{X}$ and let U be an open neighborhood of x. Fix an $\varepsilon > 0$. For any κ we have $h_\kappa(x) \le h(x) < h(x) + \varepsilon$, so, by upper semicontinuity of h_κ, there is an open neighborhood $U_\kappa \subset U$ of x on which $h_\kappa < h(x) + \varepsilon$. This neighborhood contains a point x' for which $h(x') \ge \widetilde{h}(x) - \varepsilon = h(x) + \dddot{h}(x) - \varepsilon$. We have

$$\sup_{y \in U}(h - h_\kappa)(y) \ge h(x') - h_\kappa(x') > h(x) + \dddot{h}(x) - \varepsilon - (h(x) + \varepsilon) = \dddot{h}(x) - 2\varepsilon.$$

Because this is true for any κ, $U \ni x$ and $\varepsilon > 0$, we get $D_x \ge \dddot{h}(x)$, as needed. The last statement follows from the inequality $\sup_x D_x \le D_{\mathfrak{X}}$. □

The function D_x still captures only very roughly the nonuniformity of the convergence. In fact, it captures what we call the "defect of the first order." The following example (more precisely, class of examples) shows that there is something more, some "defect of higher order," to which the function D_x is insensitive. This makes necessary the creation of further, more delicate, tools for our classification. The understanding of this class of examples is crucial in the theory of entropy structures. Although it shows only an easy sample of how complicated the defect of uniformity can be, we recommend digesting this example thoroughly before proceeding further.

Example 8.1.5 (The pick-up sticks game) A class of examples is created in the following manner: The space $\mathfrak{X}$ is countable and compact. We choose an infinite subset $A \subset \mathfrak{X}$ (it can be the whole space) and order it (completely arbitrarily) $A = \{x_1, x_2, \dots\}$. The net $\mathcal{H}$ is, in this class of examples, the sequence (h_k) defined as $h_k = \mathbb{1}_{\{x_1, x_2, \dots, x_k\}}$. Notice that each of these functions is upper semicontinuous, each difference function $h_k - h_{k-1}$ (equal to $\mathbb{1}_{\{x_k\}}$) is upper semicontinuous, and the sequence converges to the function $h = \mathbb{1}_A$. The peculiar name of the example comes from observing the tails $\theta_k = \mathbb{1}_A - \mathbb{1}_{\{x_1, x_2, \dots, x_k\}}$. If we draw the "solid graph" of h by attaching to each point of A a vertical "stick" of unit height, then the functions θ_k are obtained by removing the sticks one after another, following the ordering. This resembles the pick-up sticks game.

In this class, the variety of different types of defects depends solely on the structure of the set A and how it "sits" in $\mathfrak{X}$ (it does not depend on the ordering of the space). We present three cases, of increasing intricacy.

Game 1. In the first game, the space $\mathfrak{X}$ consists of a sequence $A = \{x_1, x_2, \dots\}$ (grey dots in Figure 8.1) converging to a limit point $x_0 \in \mathfrak{X} \setminus A$ (the black dot). The function θ_k is obtained by removing the first k sticks.

Figure 8.1 The simplest pick-up sticks game. The top picture shows the function h (the starting position for the game). The bottom picture shows the defect function D_x.

The defect D_x is zero except at x_0 where it equals 1. Notice that the same defect function will be obtained if we include x_0 in the set A (and assign to it some position in the enumeration). This will change the limit function h; there will be one more stick attached to the black dot, but it will be removed in a finite move and from that moment the game will be the same as before.

Game 2. The second game is this:

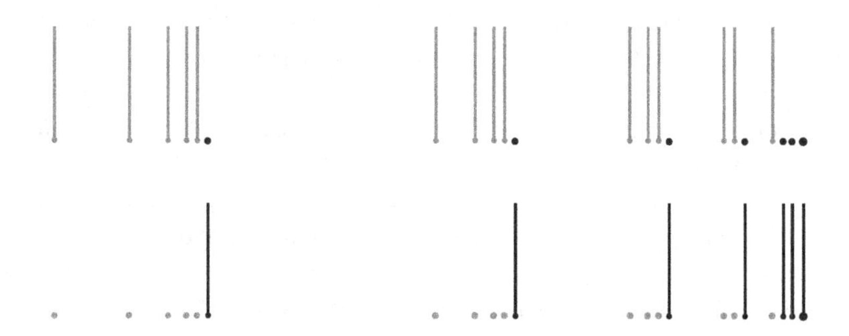

Figure 8.2 The second game. The function h and the defect function.

The space is as before plus to each point x_n ($n \geq 1$) we attach a sequence $x_{n,k}$ converging to x_n. We do it so that no other accumulation points are created. The set A now consists of all the newly added points (the grey dots in Figure 8.2). For each black dot (except the rightmost, x_0), locally the game is a copy of that in the preceding example, hence the defect occurs at each x_n

and equals 1. Since there are infinitely many grey sticks also near the rightmost point x_0, the defect at that point also occurs and equals 1.

Game 3. The third game is played on the same space $\mathfrak{X}$ as the preceding one, but we change the set A, including in it all points except the rightmost point x_0. In Figure 8.3, the game is played with both dark-grey and light-grey sticks (in any order).

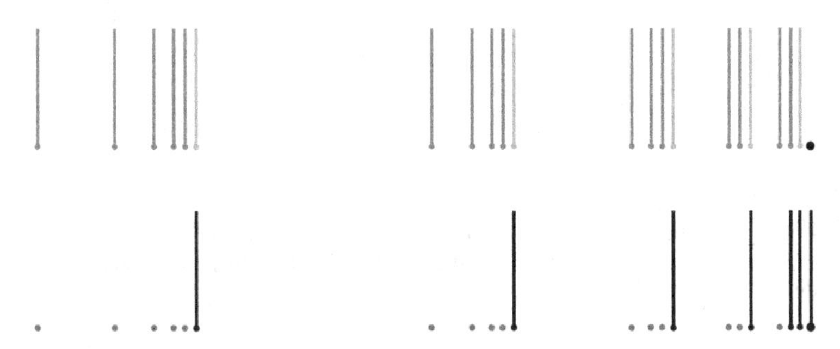

Figure 8.3 The third pick-up sticks game has a different function h, but the same defect function as the preceding game.

Each light-grey stick is removed in a finite step, and from this step onward the game looks locally the same as in the preceding example. So, the defect function is the same as in the preceding example. This argument does not apply to x_0, near which the game never looks the same as before, because now there are infinitely many light-grey sticks. Nevertheless, the defect at x_0 also occurs and equals 1. So, the entire defect function is identical as in the preceding example.

Although the function D_x is the same in the second and third games, there is an essential difference in the "type of nonuniformity," undetected by D_x. Namely, if we look at the support of the defect function and observe the game on this set only (without seeing what happens outside), then in the second game we see nothing happening. The function h is zero here. This means that all the defect "comes from outside." If we do the same in the third game, the function h will be nonzero. In fact, we will see exactly the same situation as in the first game, which generates the defect at x_0. So, in addition to the defect "coming from outside," there will be also a defect "coming from inside," a feature not present in the preceding two games. We will call it *the defect of the second order*. In this game, this defect (as a function) is the characteristic function of $\{x_0\}$.

The defect of uniformity (of a convergence) of higher orders is best understood in terms of defects of upper semicontinuity. Recall that a function f is upper semicontinuous if and only if its defect $\dddot{f}$ equals zero at every point. By analogy, we introduce a new term:

Definition 8.1.6 A net of functions f_κ is said to be *asymptotically upper semicontinuous* if

$$\lim_{\kappa \in \mathcal{K}} \dddot{f}_\kappa(x) = 0$$

at every point $x \in \mathfrak{X}$.

Observe that in case of h locally bounded, by (8.1.2) we have

$$D_x = \lim_{\kappa \in \mathcal{K}} \downarrow \widetilde{\theta}_\kappa(x) = \lim_{\kappa \in \mathcal{K}} \downarrow \widetilde{\theta}_\kappa(x) - \lim_{\kappa \in \mathcal{K}} \downarrow \theta_\kappa(x) = \lim_{\kappa \in \mathcal{K}} \downarrow \dddot{\theta}_\kappa(x).$$

So, we can interpret D_x also as the "defect of asymptotic upper semicontinuity" of the net of tails.

The following paragraphs contain an intuitive discussion of terms defined formally in the next subsection.

Consider a locally bounded function f which is not upper semicontiunuous. We can "repair it," i.e., "make it upper semicontinuous" by adding to it its defect function, because $f + \dddot{f} = \widetilde{f}$ is already upper semicontinuous. In fact, $\dddot{f}$ is the smallest nonnegative such "repair function." By analogy, we will say that a nonnegative function u "repairs" a net (f_κ) if the net $(u + f_\kappa)$ is asymptotically upper semicontinuous. Sometimes there is no such finite function u and the only way to "repair" a net is to add the infinity function. By repairing a net we cannot hope to make it uniformly convergent, but at least we make it "better" from the point of view of semicontinuity properties. The pointwise infimum of all repair functions will be shown to be a repair function as well. This *smallest repair function* (which is either finite or infinite everywhere) will be the object of our highest interest.

Now we can illuminate the difference between the second and third pick-up sticks games. In the second game the defect function D_x repairs the sequence of tails (θ_k): all the functions $D_x + \theta_k$ are easily seen to be upper semicontinuous (not only asymptotically, but immediately). The function D_x is in fact the smallest repair function for this net. All this is easily seen by consulting Figure 8.2: since all defects "come from outside" the defect function fills in all the "bad jumps", without creating new ones. In the third game it does not work that way. The function D_x repairs all the defects "coming from outside," but it cannot repair the "defect coming from inside," because it only shifts up the bad jump, which remains a bad jump. The smallest repair function is, in this case, the sum of D_x and "defect of the second order," the characteristic function of $\{x_0\}$. Consult Figure 8.4 below to see this.

Figure 8.4 The function D_x (black) plus $\mathbb{1}_{\{x_0\}}$ (dashed) repairs the net in game 3.

8.1.2 Repair functions and superenvelopes

A superenvelope is simply a repair function plus the limit function. We will frequently switch between one and another, depending on the convenience. Let us verbalize the notions.

Definition 8.1.7 A function u such that $(u + \theta_\kappa)$ is asymptotically upper semicontinuous will be called a *repair function* for the net (θ_κ), and $E = u + h$ will be called a *superenvelope* of $\mathcal{H}$. By convention, the constant infinity function is both a repair function and a superenvelope.

Except the infinity, every other repair function (and superenvelope) is finite.

Lemma 8.1.8 *Any finite repair function u is upper semicontinuous.*

Proof $0 = \lim_{\kappa \in \mathcal{K}} \left(\widetilde{(u + \theta_\kappa)} - (u + \theta_\kappa) \right) = \lim_{\kappa \in \mathcal{K}} \widetilde{(u + \theta_\kappa)} - u \geq \widetilde{u} - u \geq 0,$
hence $\ddot{u} = 0$. □

Definition 8.1.9 Let $u_{\mathcal{H}}$ denote the infimum of all repair functions. We let $\mathrm{E}\mathcal{H} = u_{\mathcal{H}} + h$.

Lemma 8.1.10 *The function $u_{\mathcal{H}}$ is a (the smallest) repair function (i.e., $\mathrm{E}\mathcal{H}$ is the smallest superenvelope).*

Proof The statement holds if $u_{\mathcal{H}} \equiv \infty$. If not, fix an $x \in \mathfrak{X}$ and let u be a repair function such that $(u - u_{\mathcal{H}})(x) < \varepsilon$. Next, let κ be such that

$\overset{\cdots\cdots}{(u + \theta_\kappa)}(x) < \varepsilon$. Then

$$\overset{\cdots\cdots}{(u_{\mathcal{H}} + \theta_\kappa)}(x) = (\widetilde{u_{\mathcal{H}} + \theta_\kappa})(x) - (u_{\mathcal{H}} + \theta_\kappa)(x) \leq$$
$$(\widetilde{u + \theta_\kappa})(x) - (u + \theta_\kappa)(x) + (u - u_{\mathcal{H}})(x) < 2\varepsilon.$$

□

Among the monotone nets $\mathcal{H}$ we distinguish those which satisfy a special additional semicontinuity condition.

Definition 8.1.11 An increasing net $\mathcal{H} = (h_\kappa)$ of nonnegative functions is said to have *upper semicontinuous differences* if for every pair of indices $\kappa_2 > \kappa_1$ the difference function $h_{\kappa_2} - h_{\kappa_1}$ (equivalently $\theta_{\kappa_1} - \theta_{\kappa_2}$) is upper semicontinuous.

In the special case of upper semicontinuous differences we have the following easy characterization of finite repair functions (and superenvelopes):

Lemma 8.1.12 *Let $\mathcal{H} = (h_\kappa)$ have upper semicontinuous differences. Then $u \geq 0$ is a finite repair function for (θ_κ) if and only if $u + \theta_\kappa$ is upper semicontinuous for every $\kappa \in \mathcal{K}$. The above translates that $E \geq h$ is a finite superenvelope if and only if $E - h_\kappa$ is upper semicontinuous for each κ.*

Proof If the latter condition is satisfied, then u is obviously a repair function. Conversely, let u be a finite repair function of (θ_κ). Fix some $\kappa \in \mathcal{K}$. For every $\kappa' > \kappa$ we have (using A.1.21)

$$\overset{\cdots\cdots}{(u + \theta_\kappa)} = \overset{\cdots\cdots\cdots\cdots\cdots}{((u + \theta_{\kappa'}) + (\theta_\kappa - \theta_{\kappa'}))} \leq \overset{\cdots\cdots}{(u + \theta_{\kappa'})} + \overset{\cdots\cdots}{(\theta_\kappa - \theta_{\kappa'})} = \overset{\cdots\cdots}{(u + \theta_{\kappa'})}.$$

The right-hand side converges to zero pointwise with κ', so $\overset{\cdots\cdots}{(u + \theta_\kappa)} = 0$. □

Corollary 8.1.13 *If, in addition, at least one function h_κ is upper semicontinuous, then every finite superenvelope E (including $\mathrm{E}\mathcal{H}$) can be represented as a sum of upper semicontinuous functions, $E = (E - h_\kappa) + h_\kappa$, so it is itself an upper semicontinuous function.*

Although $u_{\mathcal{H}}$ is usually not equal to 0 everywhere (equality holds only for uniformly convergent nets, see Corollary 8.1.20 below), on complete metric spaces the equality $u_{\mathcal{H}}(x) = 0$ holds on a residual set:

Theorem 8.1.14 *Let $\mathcal{H} = (h_\kappa)$ be an increasing net of nonnegative functions on a metric domain, with a finite limit function h. If the function $u_{\mathcal{H}}$ is finite, then the set $\{x : u_{\mathcal{H}}(x) > 0\}$ is of first category in $\mathfrak{X}$.*

Proof Recall that $u_{\mathcal{H}}$ is upper semicontinuous, so, for any $\varepsilon > 0$, the set $\{x : u_{\mathcal{H}}(x) \geq \varepsilon\}$ is closed. We will show that this set does not contain any open set. This clearly implies the hypothesis. Suppose $u_{\mathcal{H}}(x) \geq \varepsilon$ on an open set U. There exists a nonzero continuous function $f \leq \varepsilon$ which is zero outside U. Let $u = u_{\mathcal{H}} - f$. We have $u \geq 0$ and

$$\widetilde{(u + \theta_\kappa)} = \widetilde{(u_{\mathcal{H}} + \theta_\kappa - f)} \leq \widetilde{(u_{\mathcal{H}} + \theta_\kappa)} + \widetilde{(-f)} = \widetilde{(u_{\mathcal{H}} + \theta_\kappa)}.$$

The last term converges to zero with κ, so u is a repair function, smaller than $u_{\mathcal{H}}$, a contradiction. □

We discuss one very special case of $\mathcal{H}$, when the computation of $\mathrm{E}\mathcal{H}$ is very easy. It is the continuous case. It justifies the name "superenvelope."

Fact 8.1.15 *If all functions h_κ in $\mathcal{H}$ are continuous, then $\mathrm{E}\mathcal{H} = \widetilde{h}$. If there exists $\kappa_0 \in \mathcal{K}$ such that $h_\kappa - h_{\kappa_0}$ is continuous for all $\kappa \geq \kappa_0$, then $\mathrm{E}\mathcal{H} = h_{\kappa_0} + \widetilde{\theta}_{\kappa_0}$.*

Proof In the first case Corollary 8.1.13 applies, hence $\mathrm{E}\mathcal{H}$ is upper semicontinuous. It majorates h, so, it also majorates $\widetilde{h}$. On the other hand, for every κ, $\widetilde{h} - h_\kappa$ is upper semicontinuous, so $\widetilde{h}$ is a superenvelope. The latter case is seen by considering the net $\mathcal{H}' = (h'_\kappa)_{\kappa \geq \kappa_0}$ where $h'_\kappa = h_\kappa - h_{\kappa_0}$, which falls into the first case. So, we have $\mathrm{E}\mathcal{H}' = \widetilde{h - h_{\kappa_0}} = \widetilde{\theta}_{\kappa_0}$. Now we add back the function h_{κ_0} to both h'_κ and $\mathrm{E}\mathcal{H}'$. Since the differences do not change, we easily see that $\mathrm{E}\mathcal{H}' + h_{\kappa_0}$ is the minimal superenvelope of $\mathcal{H}$. □

8.1.3 The transfinite sequence

This section describes a tool allowing to determine $u_{\mathcal{H}}$ (and thus $\mathrm{E}\mathcal{H}$) more effectively, using transfinite induction. It will be widely used in the forthcoming applications. The domain is some metric space $\mathfrak{X}$.

Definition 8.1.16 Let $\mathcal{H} = (h_\kappa)_{\kappa \in \mathcal{K}}$ be an increasing net of nonnegative functions converging to a finite limit function h. Recall that the net of tails $\theta_\kappa = h - h_\kappa$ is decreasing and consists of nonnegative functions. We define the *transfinite sequence* associated with $\mathcal{H}$ by setting

$$u_0 = u_0^{\mathcal{H}} \equiv 0,$$

then, for an ordinal α such that u_β is already defined for all $\beta < \alpha$, we let

$$u_\alpha = u_\alpha^{\mathcal{H}} = \lim_{\kappa \in \mathcal{K}} \downarrow \widetilde{(\sup_{\beta < \alpha} u_\beta + \theta_\kappa)}.$$

Notice that for each α the function u_α is either the constant infinity function, or it is finite everywhere. Clearly, the transfinite sequence (u_α) is increasing. If α has a predecessor, then in the above definition we can replace $\sup_{\beta<\alpha} u_\beta$ by $u_{\alpha-1}$.

It is obvious that if $u_{\alpha+1} = u_\alpha$ on an open set U, then $u_\beta = u_\alpha$ on U for all ordinals $\beta \geq \alpha$. It is a trivial observation that the sequence (u_α) must stop growing at some ordinal (for instance, the cardinality of all real functions cannot be exceeded).

Definition 8.1.17 The smallest ordinal α such that $u_{\alpha+1} = u_\alpha$ on $\mathfrak{X}$ will be denoted by $\alpha_0^{\mathcal{H}}$ or simply α_0 and called *the order of accumulation* of $\mathcal{H}$.

Notice that if $u_\alpha \equiv \infty$ for some α, then α_0 is the smallest ordinal for which this happens. The most interesting case occurs when u_{α_0} is finite.

Remark 8.1.18 In the papers [Boyle and Downarowicz, 2004; Downarowicz, 2005a] the definition of u_α was different for limit ordinals α:

$$u_\alpha = \widetilde{\sup_{\beta<\alpha} u_\beta}.$$

The new definition has two advantages: it is consistent for all ordinals, and it gives a nicer estimate in Theorem 8.1.29 below.

We will now try to interpret the transfinite sequence. First observe that the function u_1 equals $\lim\limits_{\kappa\in\mathcal{K}} \downarrow \widetilde{\theta}_\kappa$, which is the same as the defect function D_x (see (8.1.2)). We can hope that (like in our first two examples of the pick-up sticks game) this will be the repair function for the net of tails. If not, then we look at the "unsuccessfully repaired" net $(u_1 + \theta_\kappa)_\kappa$ and find its "defect of asymptotic upper semicontinuity,"

$$\overset{\cdots\cdots\cdots}{\lim_{\kappa\in\mathcal{K}}}(u_1 + \theta_\kappa) = \lim_{\kappa\in\mathcal{K}} \widetilde{(u_1 + \theta_\kappa)} - \lim_{\kappa\in\mathcal{K}} (u_1 + \theta_\kappa) = u_2 - u_1.$$

We can try to repair the net of tails by adding both u_1 plus this last function, i.e., jointly by adding u_2. In other words, u_2 is the "second attempted repair function." In our third example this was the right function. If this fails, we can continue in the same manner. The functions u_α can be interpreted as consecutive "attempted repair functions." The key theorem below says that eventually, in step number α_0, we will succeed!

Theorem 8.1.19 *Let $\mathcal{H}$ be an increasing net of nonnegative functions on a metric space $\mathfrak{X}$, converging to a finite limit h. Then*

$$u_{\alpha_0} = u_{\mathcal{H}} \quad \textit{i.e.,} \quad \mathrm{E}\mathcal{H} = u_{\alpha_0} + h.$$

Proof At first we show that if u is any repair function, then $u_\alpha \le u$ for every α. This inequality holds for $\alpha = 0$. Suppose it holds for all $\beta < \alpha$. Then we have

$$u_\alpha = \lim_{\kappa\in\mathcal{K}} \widetilde{(\sup_{\beta<\alpha} u_\beta + \theta_\kappa)} \le \lim_{\kappa\in\mathcal{K}} \widetilde{(u + \theta_\kappa)} = \lim_{\kappa\in\mathcal{K}} \overset{\cdots}{(u + \theta_\kappa)} + \lim_{\kappa\in\mathcal{K}} (u+\theta_\kappa) = u.$$

Next we will show that u_{α_0} is a repair function. Indeed, we can assume that u_{α_0} is finite everywhere (otherwise it is infinity everywhere, and what we need is obvious), and then

$$\overset{\cdots}{(u_{\alpha_0} + \theta_\kappa)} = \widetilde{(u_{\alpha_0} + \theta_\kappa)} - u_{\alpha_0} - \theta_\kappa,$$

which converges pointwise to $u_{\alpha_0+1} - u_{\alpha_0} = 0$. □

The now obvious statement below concerns a very special case:

Corollary 8.1.20 *The following conditions are equivalent for an increasing net $\mathcal{H} = (h_\kappa)$ of nonnegative functions on a compact metric domain, with a limit function h:*

1. $h_\kappa \to h$ *uniformly,*
2. $u_\mathcal{H} \equiv 0$,
3. $u_1 \equiv 0$,
4. $\alpha_0 = 0$,
5. $\mathrm{E}\mathcal{H} \equiv h$. □

8.1.4 Uniform equivalence

The transfinite sequence effectively captures the defects of asymptotic upper semicontinuity, or, if one prefers, the defects of uniformity of all orders. There is a natural equivalence relation among increasing nets of real-valued functions $\mathcal{H} = (h_\kappa)_{\kappa\in\mathcal{K}}$, which preserves the above mentions defects. The equivalence classes can be interpreted as "types of convergence." The definition is so general that it does not even require the space $\mathfrak{X}$ to possess any topological structure.

Definition 8.1.21 Let $\mathcal{H} = (h_\kappa)_{\kappa\in\mathcal{K}}$ and $\mathcal{H}' = (h'_\iota)_{\iota\in J}$ be two increasing nets of functions on a common domain $\mathfrak{X}$. We say that $\mathcal{H}$ is *uniformly equivalent* to $\mathcal{H}'$ if for every $\gamma > 0$, every $\kappa \in \mathcal{K}$ and every $\iota \in J$ there exist indices $\kappa_1 \in \mathcal{K}$ and $\iota_1 \in J$ such that $h_{\kappa_1} > h'_\iota - \gamma$ and $h'_{\iota_1} > h_\kappa - \gamma$ at all points of the domain. For *decreasing* nets of nonnegative functions (θ_κ), uniform equivalence is defined analogously, with reversed inequalities: $\theta_{\kappa_1} < \theta'_\iota + \gamma$ and $\theta'_{\iota_1} < \theta_\kappa + \gamma$.

Roughly speaking, two increasing (decreasing) nets are uniformly equivalent if each element of one of them is nearly majorated (minorated) by an element of the other. Note that we admit the two nets to be indexed by different sets of indices. It is immediate to see that uniform equivalence is in fact an equivalence relation separately among increasing and among decreasing nets of functions on some domain $\mathfrak{X}$. We leave the easy verification to the reader (Exercise 8.6).

Recall that any increasing or decreasing net of functions converges pointwise to its pointwise supremum or infimum, respectively, called the limit function (which *a priori* admits infinite values). It is clear that uniformly equivalent nets share a common limit function. The limit function is hence an *invariant of the uniform equivalence relation*. Soon we will provide a wider list of such invariants when the domain is a metric space.

The example below justifies our choice of the name for the uniform equivalence relation: uniformly convergent nets constitute an equivalence class, the most fundamental one, in fact. The verification is immediate.

Fact 8.1.22 *An increasing net $\mathcal{H} = (h_\kappa)$ converges uniformly to its limit h if and only if it is uniformly equivalent to the constant net (h_ι) where the indices ι belong to an arbitrary directed family, and $h_\iota = h$ for every ι.* □

It follows directly from the definition of a subnet (see (A.1.6)) that any subnet of a monotone net of functions is uniformly equivalent to the whole net. However, if we take a sub-net (i.e., a "false subnet") obtained by restricting the net to a subfamily of indices which is internally directed (i.e., satisfies (A.1.5)) but does not satisfy the condition (A.1.6), then the resulting net need not be uniformly equivalent to the original net.

Example 8.1.23 Fix some nonzero function $h_0 \geq 0$ and let $\mathcal{H}$ be the "self-indexed" net of all functions $0 \leq h \leq h_0$ not equal to h_0, ordered by the usual inequality. It is easy to see that this net converges uniformly to h_0, however, it contains many sub-nets converging to other limit functions, as well as sub-nets converging to h_0, but not uniformly. None of these sub-nets is a genuine subnet.

The key fact, which we use later to establish uniform equivalence between some specific nets of functions on a compact domain, is the following one:

Lemma 8.1.24 *Let $\mathcal{H} = (h_\kappa)_{\kappa \in \mathcal{K}}$ be an increasing net of nonnegative functions defined on a compact domain $\mathfrak{X}$, having a finite limit h and upper semicontinuous differences (Definition 8.1.11). Let $\mathcal{K}' \subset \mathcal{K}$ be a directed subfamily of $\mathcal{K}$ not necessarily indexing a subnet. Then the corresponding sub-net $\mathcal{H}' = (h_{\kappa'})_{\kappa' \in \mathcal{K}'}$ is uniformly equivalent to $\mathcal{H}$ if and only if it converges to h.*

Proof One implication is obvious since uniformly equivalent nets have the same limit function. It is also obvious that every element of the net $\mathcal{H}'$ is dominated by one from the net $\mathcal{H}$, namely by itself. For the converse, we fix some $\kappa \in \mathcal{K}$ and for each $\kappa' \in \mathcal{K}'$ we choose in $\mathcal{K}$ a successor of both κ and κ', denoted $\kappa \vee \kappa'$. By assumption, $h_{\kappa'}$ converge with κ' to h. The function $h_{\kappa\vee\kappa'}$ lies between $h_{\kappa'}$ and h, so the net $(h_{\kappa\vee\kappa'})_{\kappa'\in\mathcal{K}'}$ also converges to h. Thus the functions $h_{\kappa\vee\kappa'} - h_{\kappa'}$ converge to 0 with κ'. Also, by assumption, they are upper semicontinuous. This convergence, however, need not be monotone. We let

$$g_{\kappa'} = \inf_{\kappa''\leq\kappa'} (h_{\kappa\vee\kappa''} - h_{\kappa''}).$$

This is already a decreasing to zero net of nonnegative upper semicontinuous functions on a compact domain. By compactness and Fact A.1.14, it converges uniformly, which implies that for every $\gamma > 0$ there exists some $\kappa' \in \mathcal{K}'$ with $g_{\kappa'} < \gamma$. Since $h_{\kappa\vee\kappa''} \geq h_\kappa$ and $h_{\kappa''} \leq h_{\kappa'}$ for every $\kappa'' \leq \kappa'$, we have $g_{\kappa'} \geq h_\kappa - h_{\kappa'}$ (the right-hand side no longer needs to be nonnegative, but it does not matter). This implies that $h_\kappa - h_{\kappa'} < \gamma$, i.e., $h_{\kappa'} > h_\kappa - \gamma$. □

Now comes the feature of the uniform equivalence relation that is most important for us: preservation of superenvelopes and of the entire transfinite sequence. We gather here all other invariants (that we know) of the uniform equivalence relation. Easy examples show that this set of invariants is not complete.

Theorem 8.1.25 *Let $\mathcal{H} = (h_\kappa)_{\kappa\in\mathcal{K}}$ and $\mathcal{H}' = (h'_\iota)_{\iota\in J}$ be a pair of uniformly equivalent increasing nets of nonnegative functions defined on a metric space $\mathfrak{X}$, with finite limit functions h and h', respectively. Then*

1. *$h = h'$,*
2. *$\mathcal{H}$ and $\mathcal{H}'$ have the same superenvelopes (and repair functions),*
3. *$u_{\boldsymbol{\alpha}}^{\mathcal{H}} = u_{\boldsymbol{\alpha}}^{\mathcal{H}'}$ for every ordinal $\boldsymbol{\alpha}$,*
4. *$\boldsymbol{\alpha}_0^{\mathcal{H}} = \boldsymbol{\alpha}_0^{\mathcal{H}'}$,*
5. *$\mathrm{E}\mathcal{H} = \mathrm{E}\mathcal{H}'$.*

Proof The common limit function statement is obvious. The infinite repair function is common to all nets. Consider a finite repair function u for the net of tails of $\mathcal{H}$. Fix a point $x \in \mathfrak{X}$ and a $\gamma > 0$. Let κ be such that both $h_\kappa(x) > h(x) - \gamma$ and $(u + \theta_\kappa)(x) < \gamma$. Let ι be such that $h'_\iota > h_\kappa - \gamma$, i.e., $h_\kappa - h'_\iota < \gamma$ (at all points). Then $(h_\kappa - h'_\iota) \leq \gamma$. Since $h'_\iota(x) \leq h(x)$, we also have

$h'_\iota(x) - h_\kappa(x) < \gamma$. Letting $\theta'_\iota = h - h'_\iota$ we write

$$\overline{(u + \theta'_\iota)}(x) \le \overline{(u + \theta_\kappa)}(x) + \overline{(h_\kappa - h'_\iota)}(x) =$$

$$\overline{(u + \theta_\kappa)}(x) + (\widetilde{h_\kappa - h'_\iota})(x) - (h_\kappa(x) - h'_\iota(x)) \le 3\gamma.$$

We have proved that u is a repair function for the net of tails (θ'_ι) of $\mathcal{H}'$.

To see that $\mathcal{H}$ and $\mathcal{H}'$ produce the same transfinite sequence first note that uniform equivalence of $\mathcal{H}$ and $\mathcal{H}'$ implies uniform equivalence of their corresponding nets of tails, (θ_κ) and (θ'_ι), which in turn implies the uniform equivalence of the nets $(\widetilde{\theta}_\kappa)$ and $(\widetilde{\theta}'_\iota)$. Since uniformly equivalent nets have the same limit, the function u_1 is common. Next we proceed by an obvious induction, using one more trivial observation, that if we add a common function to two uniformly equivalent nets, we obtain two uniformly equivalent nets.

The last two statements are direct consequences of the preceding ones. □

8.1.5 The order of accumulation*

Although the order of accumulation α_0 does not play a crucial role in the following sections, when applied to the entropy structure it becomes a new invariant of topological conjugacy (as we will show in the sequel), allowing us to classify all topological dynamical systems into ω_1 classes. Such classification was completely unknown until recently, and hence the order of accumulation may have some independent interest of its own [see e.g. Burguet and McGoff, in print; McGoff, in print].

Given an increasing net $\mathcal{H}$ of nonnegative functions on a metric space $\mathfrak{X}$, we can refine the notion of the order of accumulation α_0 of $\mathcal{H}$, so that it becomes a function on $\mathfrak{X}$:

Definition 8.1.26 For each point $x \in \mathfrak{X}$ we define the *order of accumulation of $\mathcal{H}$ at x*, denoted $\alpha_0(x)$, as the smallest ordinal α such that $u_\alpha(x) = u_{\alpha_0}(x)$.

It is obvious that $\alpha_0 = \sup_{x\in\mathfrak{X}} \alpha_0(x)$.

The following theorem motivates the terminology "order of accumulation of $\mathcal{H}$." We begin by a slight modification of a definition from general topology (see e.g. *Cantor–Bendixson level* in [Hart *et al.*, 2004, section g-2 Scattered Spaces]):

Definition 8.1.27 A point x in a topological space $\mathfrak{X}$ has *topological order of accumulation* 0 when it is an isolated point, i.e., when $\{x\}$ is open. Inductively, for an ordinal α, x is (an accumulation point) of topological order α if it is an isolated point relatively in the (closed) set of all points which are not of order smaller than α. The topological order of accumulation of a point x will be denoted by $\mathsf{ord}(x)$.

The above induction starts only in spaces with isolated points, otherwise no point has a defined order of accumulation. Among compact spaces, only countable spaces have this order defined everywhere. The function $x \mapsto \mathsf{ord}(x)$ is then upper semicontinuous, hence attains its maximum (this is true also for functions with ordinal values) denoted $\mathsf{ord}(\mathfrak{X})$ and called the *topological order of accumulation of* $\mathfrak{X}$, which is a countable ordinal (we skip the proofs of these purely topological facts).

Remark 8.1.28 The above can be expressed in terms of the *Cantor–Bendixson derivative* [see e.g. Sierpinski, 1952], as follows. Given a set $A \subset \mathfrak{X}$, the *derived set* A' is A minus the isolated (relatively in A) points of A. We let $\mathfrak{X}^0 = \mathfrak{X}$ and if $\mathfrak{X}^\alpha$ is defined for an ordinal α, then we let $\mathfrak{X}^{\alpha+1} = (\mathfrak{X}^\alpha)'$. For a limit ordinal α, $\mathfrak{X}^\alpha$ is defined as the intersection $\bigcap_{\beta<\alpha} \mathfrak{X}^\beta$. Clearly, there exists an ordinal α_0 (called the *Cantor–Bendixson rank* of $\mathfrak{X}$) such that $\mathfrak{X}^{\alpha_0} = \mathfrak{X}^{\alpha_0+1}$, i.e., $\mathfrak{X}^{\alpha_0}$ is a *perfect set* (which, of course, can be empty). The topological order of accumulation of a point x is defined outside this perfect set and coincides with the largest ordinal α such that $x \in \mathfrak{X}^\alpha$. The Cantor–Bendixson Theorem [see e.g. Sierpinski, 1952] asserts that in a separable metric space (in particular, compact metric), the perfect part is either empty or uncountable, while its complement is at most countable. If X is countable, then its perfect part is empty, and then $\mathsf{ord}(\mathfrak{X})$ coincides either with the Cantor–Bendixson rank of $\mathfrak{X}$ (if the latter is a limit ordinal), or else with that rank minus one.

Theorem 8.1.29 *Let $\mathfrak{X}$ be a compact metric space. Let $\mathcal{H}$ be an increasing net of nonnegative functions on $\mathfrak{X}$. Then at each point x, such that* $\mathsf{ord}(x)$ *is defined, the order of accumulation of $\mathcal{H}$ at x does not exceed the topological order of accumulation of x: $\alpha_0(x) \leq \mathsf{ord}(x)$. In particular, if $\mathfrak{X}$ is countable, we have $\alpha_0^{\mathcal{H}} \leq \mathsf{ord}(\mathfrak{X})$.*

Proof We will prove that every point x, at which $\mathsf{ord}(x)$ is defined, has an open neighborhood U_x where $u_{\mathsf{ord}(x)+1} = u_{\mathsf{ord}(x)}$. This clearly implies that the transfinite sequence at x does not grow above $u_{\mathsf{ord}(x)}$, as claimed.

If x is an isolated point, then $U_x = \{x\}$ is a neighborhood of x and, since $\widetilde{\theta}_\kappa(x) = \theta_\kappa(x)$ for every κ, we have $u_1(x) = 0$, i.e., $u_{\mathsf{ord}(x)+1} = u_{\mathsf{ord}(x)}$ on U_x.

Take an ordinal α and assume we have proved that every point y with $\mathsf{ord}(y) < \alpha$ has a neighborhood on which $u_{\mathsf{ord}(y)+1} = u_{\mathsf{ord}(y)}$. As we have noted, then the transfinite sequence does not grow at y above $u_{\mathsf{ord}(y)}$, in particular, $u_{\alpha+1}(y) = u_\alpha(y)$. Now take a point x with $\mathsf{ord}(x) = \alpha$. There is a neighborhood U_x of x which contains only x and points y of topological order of accumulation smaller than α, for which $u_{\alpha+1}(y) = u_\alpha(y)$. Thus, it remains to check that $u_{\alpha+1} = u_\alpha$ at the point x. We have

$$u_{\alpha+1}(x) = \lim_{\kappa\in\mathcal{K}} \downarrow \widetilde{(u_\alpha + \theta_\kappa)}(x) = \lim_{\kappa\in\mathcal{K}} \downarrow \inf_V \sup_{y\in V}(u_\alpha + \theta_\kappa)(y) =$$
$$\inf_V \lim_{\kappa\in\mathcal{K}} \downarrow \sup_{y\in V}(u_\alpha + \theta_\kappa)(y),$$

where V ranges over all neighborhoods of x. It suffices to take V such that $\overline{V} \subset U_x$.

Consider two cases: (a) For some V there is a subnet such that for each index κ in this subnet, the supremum over y is attained at x. Restricting to this V and this subnet, we can write

$$u_{\alpha+1}(x) \le \lim_{\kappa\in\mathcal{K}} \downarrow (u_\alpha + \theta_\kappa)(x) = u_\alpha(x).$$

In the complementary case (b), for every V and all sufficiently large κ, we can replace the supremum over V by the (perhaps larger) supremum over $\overline{V} \setminus \{x\}$. On the latter set $u_\alpha(y) = u_{\mathrm{ord}(y)}(y)$ where $\mathrm{ord}(y) < \alpha$, so $u_\alpha = \sup_{\beta<\alpha} u_\beta$. Thus

$$u_{\alpha+1}(x) = \lim_{\kappa\in\mathcal{K}} \downarrow \inf_V \sup_{y\in V}(u_\alpha + \theta_\kappa) \le \inf_V \lim_{\kappa\in\mathcal{K}} \downarrow \sup_{y\in\overline{V}\setminus\{x\}} (\sup_{\beta<\alpha} u_\beta + \theta_\kappa)(y) \le$$
$$\le \inf_V \lim_{\kappa\in\mathcal{K}} \downarrow \sup_{y\in\overline{V}} \widetilde{(\sup_{\beta<\alpha} u_\beta + \theta_\kappa)}(y).$$

Since $\overline{V}$ is compact, we can apply Fact A.1.24 to the last expression. This yields

$$u_{\alpha+1}(x) \le \inf_V \sup_{y\in\overline{V}} \lim_{\kappa\in\mathcal{K}} \downarrow \widetilde{(\sup_{\beta<\alpha} u_\beta + \theta_\kappa)}(y) = \inf_V \sup_{y\in\overline{V}} u_\alpha(y) =$$
$$\widetilde{u_\alpha}(x) = u_\alpha(x),$$

where the last equality follows from upper semicontinuity of u_α. □

The reason we are mostly interested in countable ordinals is in the theorem below:

Theorem 8.1.30 *If $\mathcal{H}$ is an increasing net of nonnegative functions on a compact metric space $\mathfrak{X}$, then $\alpha_0^{\mathcal{H}}$ is countable.*

Proof Let $\mathfrak{X}_\alpha = \{(x,t) \in \mathfrak{X} \times [0,\infty] : 0 \le t \le u_\alpha(x)\}$. Since u_α is upper semicontinuous, this is a compact set. The space of all compact subsets of $\mathfrak{X} \times [0,\infty]$ is compact (in the Hausdorff metric dist), and $\alpha < \beta < \gamma$ implies

$\mathrm{dist}(\mathfrak{X}_\alpha, \mathfrak{X}_\beta) \le \mathrm{dist}(\mathfrak{X}_\alpha, \mathfrak{X}_\gamma)$ (by inclusion). Therefore, for any positive ε, there can be only finitely many α such that $\mathrm{dist}(\mathfrak{X}_\alpha, \mathfrak{X}_{\alpha+1}) > \varepsilon$. Together, there are at most countably many α such that $\mathrm{dist}(\mathfrak{X}_\alpha, \mathfrak{X}_{\alpha_+1}) > 0$, hence $\mathrm{dist}(\mathfrak{X}_{\alpha_0}, \mathfrak{X}_{\alpha_0+1}) = 0$ for some countable ordinal α_0 (which is equivalent to $u_{\alpha_0} = u_{\alpha_0+1}$). □

We end this section with a statement proved in [Burguet and McGoff, in print], whose proof is too long and too complicated to be included in this book.

Fact 8.1.31 *For every countable ordinal α there exists a countable compact space $\mathfrak{X}$ with $\mathrm{ord}(\mathfrak{X}) = \alpha$, and an increasing sequence $\mathcal{H} = (h_k)_{k\in\mathbb{N}}$ of upper semicontinuous nonnegative functions with upper semicontinuous differences, converging to a bounded limit, such that $\alpha_0(x) = \mathrm{ord}(x)$ at every point. In particular, $\alpha_0^{\mathcal{H}} = \alpha$.* □

8.2 U.s.d.a.-sequences on simplices

This section is devoted to the special case of so-called u.s.d.a.-sequences defined on a Choquet simplex $\mathcal{K}$. The abbreviation stands for *upper semicontinuous differences* and *affine*. In fact, we assume also that the functions h_k alone are upper semicontinuous. This additional assumption can be replaced by the usual upper semicontinuous differences condition for the sequence $(h_k)_{k\ge0}$ enhanced by $h_0 \equiv 0$. We neglect to articulate this additional requirement in the abbreviation. Below we prove a number of further facts concerning such sequences, which will be extensively used later for entropy structures.

Definition 8.2.1 By a *u.s.d.a.-sequence* we shall mean an increasing sequence $\mathcal{H} = (h_k)_{k\in\mathbb{N}}$ of upper semicontinuous affine functions with upper semicontinuous differences, defined on a Choquet simplex $\mathcal{K}$.

8.2.1 Alternative computation of $\mathrm{E}\mathcal{H}$

The first fact does not use affinity assumption. Recall that the other assumptions alone imply that any finite function $E \ge h$ is a superenvelope if and only if all difference functions $E - h_k$ are upper semicontinuous, and E is upper semicontinuous and hence bounded (see Lemma 8.1.13 and Corollary 8.1.12).

Definition 8.2.2 Let $\mathcal{H} = (h_k)_{k\ge0}$ be the sequence enhanced by $h_0 \equiv 0$. By a *topping* of $\mathcal{H}$ we shall mean any sequence $\mathcal{G} = (g_k)_{k\ge1}$ of continuous functions satisfying $g_k \ge h_k - h_{k-1}$. By $\Sigma\mathcal{G}$ we will mean the sum $\sum_{k=1}^{\infty} g_k$.

Fact 8.2.3 *If $\mathcal{H}$ is an increasing sequence of nonnegative upper semicontinuous functions with upper semicontinuous differences defined on a compact domain $\mathfrak{X}$, then*

$$\mathrm{E}\mathcal{H} = \inf_{\mathcal{G}} \widetilde{\Sigma\mathcal{G}},$$

where the infimum is taken over all toppings $\mathcal{G}$ of $\mathcal{H}$. There exists a topping $\mathcal{G}$ such that

$$\sup_{x \in \mathfrak{X}} \mathrm{E}\mathcal{H}(x) = \sup_{x \in \mathfrak{X}} \Sigma\mathcal{G}(x).$$

Proof Notice that the set of all toppings of $\mathcal{H}$ is a directed family: $\mathcal{G} = (g_k)$ is a successor of $\mathcal{G}' = (g'_k)$ when $g_k \leq g'_k$ for every k, and the common successor of two toppings $\mathcal{G}, \mathcal{G}'$ is obtained as the sequence of the pointwise infima of g_k and g'_k. Thus the infimum in the first assertion of the theorem is in fact the limit of a decreasing net (indexed by $\mathcal{G}$). Fix a k. We have

$$\lim_{\mathcal{G}} \downarrow \widetilde{\Sigma\mathcal{G}} = \lim_{\mathcal{G}} \downarrow \Big(\sum_{i \leq k} g_i + \widetilde{\sum_{i>k} g_i}\Big) = \lim_{\mathcal{G}} \downarrow \Big(\sum_{i \leq k} g_i\Big) + \lim_{\mathcal{G}} \downarrow \Big(\widetilde{\sum_{i>k} g_i}\Big)$$

(we are using the easy fact that a continuous function can be pulled off the upper semicontinuous envelope). For a fixed i, g_i decreases (with $\mathcal{G}$) to the upper semicontinuous function $h_i - h_{i-1}$, hence the first limit equals $\sum_{i \leq k}(h_i - h_{i_1}) = h_k$. So, the displayed formula shows that $\lim_{\mathcal{G}} \widetilde{\Sigma\mathcal{G}} - h_k$ is upper semicontinuous (the last displayed limit is either finite and upper semicontinuous or both sides are constant infinite). Thus, the function $\lim_{\mathcal{G}} \widetilde{\Sigma\mathcal{G}}$ is a superenvelope of $\mathcal{H}$, and hence it is not smaller than $\mathrm{E}\mathcal{H}$. We need to show the reversed inequality. Suppose $\mathrm{E}\mathcal{H}$ is finite (otherwise the case is trivial). Let $g \geq \mathrm{E}\mathcal{H}$ be continuous. We will construct a topping $\mathcal{G}$ with $\Sigma\mathcal{G} \leq g$. This will end the proof because then also $\widetilde{\Sigma\mathcal{G}} \leq g$ and, taking infimum over g, $\widetilde{\Sigma\mathcal{G}} \leq \mathrm{E}\mathcal{H}$ (since $\mathrm{E}\mathcal{H}$ is upper semicontinuous, it equals the infimum of all such functions g).

We proceed inductively: It is clear that

$$h_1 - h_0 = h_1 \leq g - (\mathrm{E}\mathcal{H} - h_1),$$

and the assumptions of the Sandwich Theorem A.2.26 hold for the inequality. Thus there is a continuous g_1 in between. Suppose we have found continuous functions $g_1, g_2, \ldots, g_k$ such that $g_i \geq h_i - h_{i-1}$ ($i = 1, 2, \ldots, k$), and

$$\sum_{i \leq k} g_i \leq g - (\mathrm{E}\mathcal{H} - h_k)$$

(this was true for $k = 1$). This can be rewritten as

$$h_{k+1} - h_k \le g - \sum_{i \le k} g_i - (\mathrm{E}\mathcal{H} - h_{k+1}),$$

and again, the assumptions of the Sandwich Theorem are fulfilled. Thus, a continuous g_{k+1} exists between the sides. The inductive assumption is now satisfied for $k + 1$. In this manner a topping $\mathcal{G} = (g_k)$ of $\mathcal{H}$ has been constructed. Since each $\mathrm{E}\mathcal{H} - h_k$ is nonnegative, $\sum_{i \le k} g_i \le g$ holds for each k, and so $\Sigma\mathcal{G} \le g$, as desired.

The proof of the second assertion of the theorem consists in applying the last argument to the constant function $g = \sup_{x \in \mathfrak{X}} \mathrm{E}\mathcal{H}(x)$. □

In the sequel we shall use the full strength of the u.s.d.a.-sequences. We say a topping $\mathcal{G}$ of $\mathcal{H}$ is *offset* if $g_k > h_k - h_{k-1}$ (at each point) for every k. A topping is *affine* simply when it consists of affine functions. Recall that $\widehat{f}$ denotes the *upper semicontinuous concave envelope* of f (see Definition A.2.2)

Fact 8.2.4 *Let $\mathcal{H}$ be a u.s.d.a.-sequence defined on a Choquet simplex $\mathcal{K}$. The following equalities hold:*

$$\mathrm{E}\mathcal{H} = \inf_{\mathcal{G}_\mathsf{A}} \widetilde{\Sigma\mathcal{G}_\mathsf{A}} = \inf_{\mathcal{G}_\mathsf{A}} \widehat{\Sigma\mathcal{G}_\mathsf{A}} = \inf_{\mathcal{G}'_\mathsf{A}} \widehat{\Sigma\mathcal{G}'_\mathsf{A}} = \lim_{\mathcal{G}'_\mathsf{A}} \downarrow \widehat{\Sigma\mathcal{G}'_\mathsf{A}},$$

where $\mathcal{G}_\mathsf{A}$ ranges over all affine toppings of $\mathcal{H}$, and $\mathcal{G}'_\mathsf{A}$ ranges over all affine offset toppings of $\mathcal{H}$. The function $\mathrm{E}\mathcal{H}$ is concave.

Proof Obviously, in the formula $\mathrm{E}\mathcal{H} = \inf_{\mathcal{G}} \widetilde{\Sigma\mathcal{G}}$ of Fact 8.2.3, we can take the infimum only over offset toppings. Then (using the Combined Separation Theorem A.2.29), every such topping dominates an affine topping $\mathcal{G}_\mathsf{A}$. This proves the first equality. The second equality holds because the function $\Sigma\mathcal{G}_\mathsf{A}$ is either finite and then affine or it attains infinity at some point, hence, in both cases, the applications of $\widetilde{\ }$ and of $\widehat{\ }$ produce the same function (see Fact A.2.5). At this point we note that $\mathrm{E}\mathcal{H}$ is concave (if finite), and then by upper semicontinuity, supharmonic. The third equality is obvious and the last one follows because (by Theorem A.2.29 again) the affine offset toppings form a directed family. □

Another function of interest to us in this case is

$$\mathrm{E}_\mathsf{A}\mathcal{H} = \inf_{E_\mathsf{A}} E_\mathsf{A},$$

the pointwise infimum of all affine superenvelopes of $\mathcal{H}$. This is also a concave nonnegative upper semicontinuous function. But this is in fact not a new object.

Theorem 8.2.5 *For a u.s.d.a.-sequence on a Choquet simplex $\mathcal{K}$ we have*

$$\mathrm{E}_{\mathsf{A}}\mathcal{H} = \mathrm{E}\mathcal{H}, \quad \inf_{E_{\mathsf{A}}} \sup_{x \in \mathcal{K}} E_{\mathsf{A}}(x) = \sup_{x \in \mathcal{K}} \mathrm{E}\mathcal{H}(x).$$

Remark 8.2.6 Note that since the affine superenvelopes usually do not form a directed family (see Example 8.2.17), the second equality is not a direct consequence of the first by simply exchanging suprema and infima (Fact A.1.24).

Proof of Theorem 8.2.5 The inequalities "$\geq$" are obvious and equalities hold if $\mathrm{E}\mathcal{H} \equiv \infty$. Suppose $\mathrm{E}\mathcal{H}$ is finite (hence bounded) and let g be an affine continuous function above the (finite) sum of some fixed affine topping: $g \geq \Sigma\mathcal{G}_{\mathsf{A}}$, $\mathcal{G}_{\mathsf{A}} = (g_k)_{k \geq 1}$. We will find an affine superenvelope E_{A} below g. Set

$$E_{\mathsf{A}} = g - (\Sigma\mathcal{G}_{\mathsf{A}} - h).$$

This E_{A} is an affine function not larger than g. Moreover, for each $k \in \mathbb{N}$,

$$E_{\mathsf{A}} - h_k = (g - \Sigma\mathcal{G}_{\mathsf{A}}) + \sum_{i=k+1}^{\infty} (h_i - h_{i-1}) =$$

$$g - \sum_{i=1}^{k} g_i - \sum_{i=k+1}^{\infty} (g_i - (h_i - h_{i-1})),$$

which is nonnegative (see middle expression) and upper semicontinuous (each $g_i - (h_i - h_{i-1})$ is lower semicontinuous and so is the sum as a nondecreasing limit). So, E_{A} is an affine superenvelope of $\mathcal{H}$. Thus $\mathrm{E}_{\mathsf{A}}\mathcal{H} \leq \widehat{\Sigma\mathcal{G}_{\mathsf{A}}}$. But this is true for any affine topping $\mathcal{G}_{\mathsf{A}}$, so, by Fact 8.2.4, $\mathrm{E}_{\mathsf{A}}\mathcal{H} \leq \mathrm{E}\mathcal{H}$.

For the second statement let $g \equiv \sup_{x \in \mathcal{K}} \mathrm{E}\mathcal{H}(x) + \varepsilon$. Using the elementary Fact A.1.14 and the fact that the family of all affine offset toppings is directed, we can find an affine offset topping $\mathcal{G}'_{\mathsf{A}}$ with $\widetilde{\Sigma\mathcal{G}'_{\mathsf{A}}} < g$. The above argument now produces an affine superenvelope E_{A} below $\sup_{x \in \mathcal{K}} \mathrm{E}\mathcal{H}(x) + \varepsilon$, which ends the proof. □

8.2.2 The case of a Bauer simplex

Recall that a Bauer simplex is a Choquet simplex whose set of extreme points is compact. We will use the harmonic prolongation and the barycenter map. All these notions are explained in the Appendices A.2.4 and A.2.5.

Fact 8.2.7 *If $\mathcal{H}$ is a u.s.d.a.-sequence defined an a Bauer simplex $\mathcal{K}$, then the restriction map is a 1-1 correspondence between all affine superenvelopes of $\mathcal{H}$ and all superenvelopes of $\mathcal{H}|_{\mathsf{ex}\mathcal{K}}$.*

Proof We begin with any superenvelope E (defined on ex$\mathcal{K}$) of $\mathcal{H}|_{\mathrm{ex}\mathcal{K}}$. Applying the harmonic prolongation and since $h_k = (h_k|_{\mathrm{ex}\mathcal{K}})^{\mathsf{har}}$ for every k, we get $E^{\mathsf{har}} - h_k = (E - h_k|_{\mathrm{ex}\mathcal{K}})^{\mathsf{har}}$, which is nonnegative and upper semicontinuous. So, E^{har} is a superenvelope of $\mathcal{H}$ and it is clearly affine. Conversely, if E' is an affine superenvelope of $\mathcal{H}$ on $\mathcal{K}$, then, by restricting the upper semicontinuous differences $E' - h_k$ to ex$\mathcal{K}$, we obtain upper semicontinuous functions $E'|_{\mathrm{ex}\mathcal{K}} - h|_{\mathrm{ex}\mathcal{K}}$. So, $E'|_{\mathrm{ex}\mathcal{K}}$ is a superenvelope of $\mathcal{H}|_{\mathrm{ex}\mathcal{K}}$. If E' is obtained from the initial E by harmonic prolongation, then, by the 1-1 correspondence on upper semicontinuous functions, $E'|_{\mathrm{ex}\mathcal{K}} = E$. □

The above implies, in particular, that among affine superenvelopes of $\mathcal{H}$ there exists the smallest one (it is the harmonic prolongation of the smallest superenvelope of $\mathcal{H}|_{\mathrm{ex}\mathcal{K}}$). Combining this with Theorem 8.2.5 we conclude:

Theorem 8.2.8 *If $\mathcal{H}$ is a u.s.d.a.-sequence defined on a Bauer simplex $\mathcal{K}$, then*

$$\mathrm{E}\mathcal{H} = (\mathrm{E}(\mathcal{H}|_{\mathrm{ex}\mathcal{K}}))^{\mathsf{har}}.$$

In particular, E$\mathcal{H}$ *is affine.* □

In Example 8.2.17 we will show that for general Choquet simplices the situation may be significantly different: the smallest affine superenvelope need not exist, even more: every affine superenvelope may exceed the pointwise supremum of E$\mathcal{H}$. This example will also show that affine superenvelopes need not form a directed family (because the limit would equal the smallest superenvelope and it would be affine).

Now, we take a look at the transfinite sequence and the order of accumulation of a sequence $\mathcal{H}$ defined on a Bauer simplex $\mathcal{K}$. If f is a measurable, bounded from at least one side, function defined on ex$\mathcal{K}$, then $(\widetilde{f})^{\mathsf{har}} = \widetilde{f^{\mathsf{har}}}$ (this follows by the 1-1 correspondence of Fact A.2.25). It is thus very easy to see (tracing the transfinite construction) that:

Fact 8.2.9 *If $\mathcal{K}$ is a Bauer simplex, $\mathcal{H}$ is a sequence of nonnegative measurable functions on* ex$\mathcal{K}$ *and $\mathcal{H}'$ is defined on $\mathcal{K}$ as the harmonic prolongation of $\mathcal{H}$ (or, we can start with a sequence $\mathcal{H}'$ of harmonic functions on $\mathcal{K}$ and define $\mathcal{H}$ by restriction to* ex$\mathcal{K}$*, see Fact A.2.19), then, for every ordinal α, we have $u_\alpha^{\mathcal{H}'} = (u_\alpha^{\mathcal{H}})^{\mathsf{har}}$. In particular, the order of accumulation of $\mathcal{H}'$ is the same as that of $\mathcal{H}$.* □

8.2.3 Lifting and push-down techniques

We return to u.s.d.a.-sequences defined on a general Choquet simplex $\mathcal{K}$. There is a natural Bauer simplex $\mathcal{M}$ associated with $\mathcal{K}$ namely the simplex of all probability measures carried by the compact set $\overline{\mathsf{ex}\mathcal{K}}$. The barycenter map $\mathsf{bar}(\cdot)$ is an affine continuous map from $\mathcal{M}$ onto $\mathcal{K}$. In the statements provided in this section, we will use this (larger) simplex and take advantage of the simple form of the smallest superenvelope on a Bauer simplex shown in the preceding section to prove facts about the smallest superenvelope on $\mathcal{K}$.

For a function f on $\mathcal{K}$, by $f^{\mathcal{M}}$ we denote the lift of f to $\mathcal{M}$ against the barycenter map, i.e., the composition $f \circ \mathsf{bar}$. For a function f defined on $\mathcal{M}$, on $\mathcal{K}$ we define the *push-down* $f^{[\mathcal{K}]}$ of f by the formula

$$f^{[\mathcal{K}]}(x) = \sup\{f(\xi) : \xi \in \mathcal{M}, \mathsf{bar}(\xi) = x\}$$

(see also Appendix A.1.6).

Theorem 8.2.10 *Let $\mathcal{H}$ be a u.s.d.a.-sequence on a simplex $\mathcal{K}$. Then*

$$\sup_{x\in\mathcal{K}} \mathrm{E}\mathcal{H}(x) = \sup_{x\in\overline{\mathsf{ex}\mathcal{K}}} \mathrm{E}\mathcal{H}(x).$$

From the proof we isolate a lemma, which we will use once more, later.

Lemma 8.2.11 *We have*

$$\mathrm{E}\mathcal{H} = (\mathrm{E}(\mathcal{H}^{\mathcal{M}}))^{[\mathcal{K}]}.$$

Proof If E is a superenvelope of $\mathcal{H}^{\mathcal{M}}$, then the push-down function $E^{[\mathcal{K}]}$ is a superenvelope of $\mathcal{H}$. Indeed, since each of the functions $h_k^{\mathcal{M}}$ in the lift $\mathcal{H}^{\mathcal{M}}$ is constant on the fibers by the barycenter map, $(E-h_k)^{[\mathcal{K}]}$ equals $E^{[\mathcal{K}]}-h_k$. On the other hand, as a push-down of an upper semicontinuous function it is upper semicontinuous (see Fact A.1.26). So, $E^{[\mathcal{K}]}$ is a superenvelope of $\mathcal{H}$. Applying this to $\mathrm{E}(\mathcal{H}^{\mathcal{M}})$ we get that the right-hand side function in the assertion of the lemma is a superenvelope of $\mathcal{H}$. Also, it must be the minimal one since if there were a smaller one, its lift to $\mathcal{M}$ would become a superenvelope of $\mathcal{H}^{\mathcal{M}}$, strictly smaller, at some point, than $\mathrm{E}(\mathcal{H}^{\mathcal{M}})$, which is impossible. □

Proof of Theorem 8.2.10 By Theorem 8.2.8, the assertion holds on Bauer simplices. For a general simplex $\mathcal{K}$ we write

$$\begin{aligned}\sup_{x\in\mathcal{K}} \mathrm{E}\mathcal{H}(x) &\geq \sup_{x\in\overline{\mathsf{ex}\mathcal{K}}} \mathrm{E}\mathcal{H}(x) = \sup_{x\in\overline{\mathsf{ex}\mathcal{K}}} (\mathrm{E}(\mathcal{H}^{\mathcal{M}}))^{[\mathcal{K}]}(x) \geq \sup_{y\in\mathsf{ex}\mathcal{M}} \mathrm{E}(\mathcal{H}^{\mathcal{M}})(y)\\ &= \sup_{y\in\mathcal{M}} \mathrm{E}(\mathcal{H}^{\mathcal{M}})(y) = \sup_{x\in\mathcal{K}}(\mathrm{E}(\mathcal{H}^{\mathcal{M}}))^{[\mathcal{K}]}(x) = \sup_{x\in\mathcal{K}} \mathrm{E}\mathcal{H}(x),\end{aligned}$$

applying Lemma 8.2.11, the theorem already proved for the Bauer simplex $\mathfrak{M}$, and Lemma 8.2.11 again. □

In Example 8.2.18 we will show that the closure mark over ex$\mathfrak{K}$ in the formulation of the theorem cannot be dropped (unless $\mathfrak{K}$ is Bauer, of course).

We can further refine the above statements (Theorem 8.2.10 and Lemma 8.2.11), by providing a formula allowing us to fully retrieve E$\mathcal{H}$ from the smallest superenvelope of $\mathcal{H}$ restricted to $\overline{\mathsf{ex}\mathfrak{K}}$. The formula combines all the techniques used by us so far in the context of simplices (restriction, harmonic prolongation and push-down; lifting is also used implicitly by treating the points in $\overline{\mathsf{ex}\mathfrak{K}}$ as the extreme points of $\mathfrak{M}$, and will be used explicitly in the proof). Despite their unpleasant graphic appearance, this formula, and that in the next lemma, are very useful.

Theorem 8.2.12 *If $\mathcal{H}$ is a u.s.d.a.-sequence on a simplex $\mathfrak{K}$, then*

$$\mathrm{E}\mathcal{H} = \left(\left(\mathrm{E}(\mathcal{H}|_{\overline{\mathsf{ex}\mathfrak{K}}})\right)^{\mathsf{har}_{\mathfrak{M}}}\right)^{[\mathfrak{K}]},$$

where $\cdot^{\mathsf{har}_{\mathfrak{M}}}$ denotes the harmonic prolongation of a function defined on $\overline{\mathsf{ex}\mathfrak{K}}$ onto the Bauer simplex $\mathfrak{M}$, and $\cdot^{[\mathfrak{K}]}$ is the push-down onto $\mathfrak{K}$ via the barycenter map from $\mathfrak{M}$ to $\mathfrak{K}$.

Proof By Theorem 8.2.8, on $\mathfrak{M}$, the smallest superenvelope $\mathrm{E}(\mathcal{H}^{\mathfrak{M}})$ equals $(\mathrm{E}(\mathcal{H}^{\mathfrak{M}}|_{\mathsf{ex}\mathfrak{M}}))^{\mathsf{har}_{\mathfrak{M}}}$. But ex$\mathfrak{M}$ $=$ $\overline{\mathsf{ex}\mathfrak{K}}$ and $\mathcal{H}^{\mathfrak{M}}|_{\mathsf{ex}\mathfrak{M}}$ equals $\mathcal{H}|_{\overline{\mathsf{ex}\mathfrak{K}}}$, so we can write

$$\mathrm{E}(\mathcal{H}^{\mathfrak{M}}) = \left(\mathrm{E}(\mathcal{H}|_{\overline{\mathsf{ex}\mathfrak{K}}})\right)^{\mathsf{har}_{\mathfrak{M}}}.$$

Now we apply the push-down operation $\cdot^{[\mathfrak{K}]}$ to both sides, and, using Lemma 8.2.11, replace the left-hand side by E$\mathcal{H}$. □

We continue with the technical lemmas.

Lemma 8.2.13 *Let h be a harmonic function on a simplex $\mathfrak{K}$. Let h_0 be defined as $h \cdot 1\!\!1_{\mathsf{ex}\mathfrak{K}}$. Then, with the notation established above,*

$$\widetilde{h} = \left(\left(\widetilde{h|_{\overline{\mathsf{ex}\mathfrak{K}}}}\right)^{\mathsf{har}_{\mathfrak{M}}}\right)^{[\mathfrak{K}]} = \left(\left(\widetilde{h_0|_{\overline{\mathsf{ex}\mathfrak{K}}}}\right)^{\mathsf{har}_{\mathfrak{M}}}\right)^{[\mathfrak{K}]},$$

where the upper semicontinuous envelope is relative on the set $\overline{\mathsf{ex}\mathfrak{K}}$.

Proof Let $x \in \mathfrak{K}$. One of the measures supported by $\overline{\mathsf{ex}\mathfrak{K}}$ with barycenter at x is ξ^x (the unique measure supported by ex$\mathfrak{K}$ with barycenter at x), so the right-hand term evaluated at x is not smaller than $\left(\widetilde{h_0|_{\overline{\mathsf{ex}\mathfrak{K}}}}\right)^{\mathsf{har}_{\mathfrak{M}}}(\xi^x)$, which is not smaller than $\left(h_0|_{\overline{\mathsf{ex}\mathfrak{K}}}\right)^{\mathsf{har}_{\mathfrak{M}}}(\xi^x)$ which equals $\int h_0 \, d\xi^x$, which is the same as $\int h\,d\xi^x$. Because h is harmonic, this is $h(x)$. So the right-hand side function is not smaller than h and since it is upper semicontinuous (use Fact A.2.25 and Fact A.1.26) it is not smaller than $\widetilde{h}$. Clearly, the right-hand side is not larger than the middle expression, which, evaluated at x equals $\left(\widetilde{h|_{\overline{\mathsf{ex}\mathcal{K}}}}\right)^{\mathsf{har}_{\mathcal{M}}}(\xi)$ for some ξ supported by $\overline{\mathsf{ex}\mathcal{K}}$ with barycenter at x (the attainment follows from upper semicontinuity and since the fiber of x by the barycenter map is compact). The latter equals

$$\int \widetilde{h|_{\overline{\mathsf{ex}\mathcal{K}}}}\,d\xi.$$

Notice that the meaning of $\widetilde{\ }$ here is different than when we write $\widetilde{h}$. Here, one has to consider majorating continuous functions defined only on $\overline{\mathsf{ex}\mathcal{K}}$. Thus, at each point of $\overline{\mathsf{ex}\mathcal{K}}$, $\widetilde{h|_{\overline{\mathsf{ex}\mathcal{K}}}}$ is not larger than $\widetilde{h}$ (which is evaluated in the wider context of $\mathcal{K}$). Thus the last integral does not exceed $\int \widetilde{h}\,d\xi$, which, by the supharmonic property of $\widetilde{h}$ (see Fact A.2.5) does not exceed $\widetilde{h}(x)$. □

Below we give an effective way of determining that a particular function is a repair function. This applies not only to u.s.d.a.-sequences, but to any sequence defined on a simplex, for which the tails θ_k are harmonic. This lemma will become essential later, when dealing with symbolic extensions of smooth interval maps.

Lemma 8.2.14 *Let $(\theta_k)_{k\in\mathbb{N}}$ be a decreasing to zero (pointwise) sequence of nonnegative harmonic functions on a simplex $\mathcal{K}$. Let u be a nonnegative upper semicontinuous harmonic function on $\mathcal{K}$. Suppose that for every $x \in \overline{\mathsf{ex}\mathcal{K}}$ and every $\gamma > 0$ the inequality*

$$\theta_k(y) \le u(x) - u(y) + \gamma \tag{8.2.15}$$

holds for some k (depending on x) and all $y \in \mathsf{ex}\mathcal{K}$ belonging to some open neighborhood of x. Then u is a repair function for the sequence (θ_k).

Proof Denote $\theta_{k,0} = \theta_k \cdot 1\!\!1_{\mathsf{ex}\mathcal{K}}$. For $x \in \overline{\mathsf{ex}\mathcal{K}}$ the assumption (8.2.15) implies

$$(\theta_{k,0} + u)(y) \le u(x) + \gamma, \tag{8.2.16}$$

for all $y \in \mathsf{ex}\mathcal{K}$ near x. For non-extreme y near x, by the upper semicontinuity of u, we have $u(y) \le u(x) + \gamma$, which we can also write as (8.2.16), because now $\theta_{k,0}(y) = 0$. Altogether, on $\overline{\mathsf{ex}\mathcal{K}}$, we have obtained

$$\widetilde{(\theta_{k,0} + u)}(x) \le u(x) + \gamma.$$

The above is valid if $\widetilde{\ }$ is taken on $\overline{\mathsf{ex}\mathcal{K}}$ only. Passing to the limit, we get

$$\lim_k \widetilde{(\theta_{k,0} + u)} = u$$

pointwise on $\overline{\text{ex}\mathcal{K}}$. Via the Lebesgue Dominated Theorem, this implies

$$\lim_k \left(\widetilde{\theta_{k,0} + u}\right)^{\mathsf{har}_\mathcal{M}} - \left(u|_{\overline{\text{ex}\mathcal{K}}}\right)^{\mathsf{har}_\mathcal{M}} = u^\mathcal{M},$$

where the last equality is a simple consequence of u being harmonic. On each fiber (which is compact) of the barycenter map from $\mathcal{M}$ to $\mathcal{K}$, we have obtained a decreasing sequence of upper semicontinuous functions (the harmonic prolongation preserves upper semicontinuity on Bauer simplices), so, we can use the elementary Fact A.1.24 to get

$$\lim_k \left(\left(\widetilde{\theta_{k,0} + u}\right)^{\mathsf{har}_\mathcal{M}}\right)^{[\mathcal{K}]} = (u^\mathcal{M})^{[\mathcal{K}]} = u.$$

Now we invoke Lemma 8.2.13: Because $\theta_k + u$ is harmonic and

$$\theta_k + u \geq \theta_{k,0} + u \geq (\theta_k + u)_0$$

(where $(\theta_k + u)_0 = (\theta_k + u) \cdot 1\!\!1_{\text{ex}\mathcal{K}}$), we have

$$\left(\left(\widetilde{\theta_{k,0} + u}\right)^{\mathsf{har}_\mathcal{M}}\right)^{[\mathcal{K}]} = (\widetilde{\theta_k + u}),$$

with $\tilde{}$ on the right taken already on $\mathcal{K}$. We have proved the pointwise convergence on $\mathcal{K}$:

$$\lim_k (\widetilde{\theta_k + u}) = u.$$

Now we subtract $\theta_k + u$ from both sides and since $\theta_k \to 0$ pointwise, we get

$$\dot{(\theta_k + u)} \to 0,$$

i.e., u is a repair function for the sequence (θ_k). □

We end this section with examples of some pathologies.

Note that if $\mathcal{H}$ is u.s.d.a.-sequence on a simplex which is not Bauer, then $\mathrm{E}\mathcal{H}$ is a concave function, without necessarily being affine. This is the case where the minimal affine superenvelope does not exist. The next example shows an even worse scenario.[1]

Example 8.2.17 There exists a u.s.d.a.-sequence $\mathcal{H}$ for which

$$\sup_{x \in \mathcal{K}} E_\mathsf{A}(x) > \sup_{x \in \mathcal{K}} \mathrm{E}\mathcal{H}(x),$$

for every affine superenvelope E_A of $\mathcal{H}$. Moreover, in this example $\mathrm{E}\mathcal{H} = \widetilde{h}$, which implies that $\sup_{x\in\mathcal{K}} \mathrm{E}\mathcal{H}(x) = \sup_{x\in\mathcal{K}} h(x)$.

[1] The idea of this example was suggested by R. Phelps.

Let $\mathcal{K}$ be the Choquet simplex whose set of extreme points consists of a point b_1, a sequence $(a_n)_{n\geq 1}$ converging to b_1, and a sequence $(b_n)_{n\geq 2}$ converging to

$$b = \sum_{n=1}^{\infty} 2^{-n} a_n.$$

Define $\mathcal{H} = (h_k)$, by letting $h_k = (\mathbb{1}_{\{b_1,b_2,\ldots,b_k\}})^{\mathsf{har}}$ (see Figure 8.5).

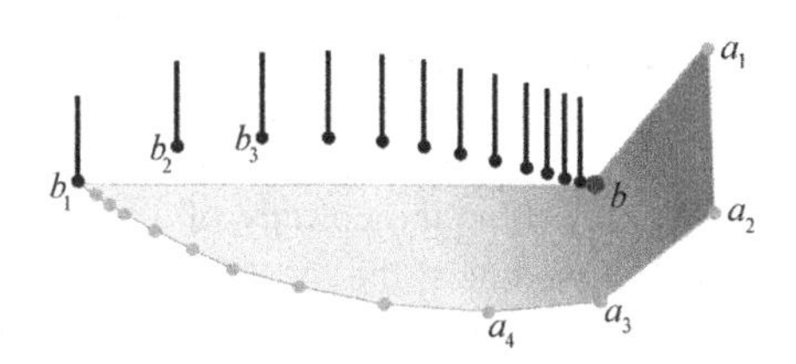

Figure 8.5 The extreme points of $\mathcal{K}$ and the point b. The shaded area symbolizes the "span" of the points a_n, to which b belongs. On the restriction to this set we play the pick-up sticks game with the black sticks.

Since both h_1 and the differences $h_k - h_{k-1}$ for $k \geq 2$ are harmonic prolongations of the upper semicontinuous and convex functions $\mathbb{1}_{\{b_k\}}$, they are, by Fact A.2.20, upper semicontinuous (and obviously affine). So $\mathcal{H}$ is a u.s.d.a.-sequence. To compute $\mathrm{E}\mathcal{H}$ we will use Theorem 8.2.12. The restriction of $\mathcal{H}$ to $\overline{\mathsf{ex}\mathcal{K}}$ is precisely the pick-up sticks game as in Example 8.1.5, Game 1, and its smallest superenvelope equals 1 at each b_n and at the limit point b. This is precisely $\widetilde{h|_{\overline{\mathsf{ex}\mathcal{K}}}}$. So, $\mathrm{E}\mathcal{H}$ equals $\left(\left(\widetilde{h|_{\overline{\mathsf{ex}\mathcal{K}}}}\right)^{\mathsf{har}_{\mathcal{M}}}\right)^{[\mathcal{K}]}$ which, by the first equation in Lemma 8.2.13, equals $\widetilde{h}$. In particular, the maximal value of $\mathrm{E}\mathcal{H}$ is 1.

Now suppose E_{A} is an affine superenvelope of $\mathcal{H}$ not exceeding 1. Clearly, $E_{\mathsf{A}}(b_n) = 1$ for each $n \geq 1$. By upper semicontinuity, $E_{\mathsf{A}}(b) = 1$. Because

$$1 = E_{\mathsf{A}}(b) = \sum_{n=1}^{\infty} 2^{-n} E_{\mathsf{A}}(a_n),$$

E_{A} must equal 1 at all points a_n $(n \geq 1)$. But now $E_{\mathsf{A}} - h_1$ is not upper semicontinuous, because it has a "bad" jump at b_1. We have proved that every affine superenvelope exceeds 1 at some point. This concludes the proof of the desired properties. (It is easy to see what an affine superenvelope looks like: it slightly exceeds 1 at some point a_{n_0}, which allows it to decrease to zero at the points a_n with large n.)

Example 8.2.18 Here is an easy example of a u.s.d.a.-sequence $\mathcal{H}$ such that

$$\sup_{x\in\mathcal{K}} \mathrm{E}\mathcal{H}(x) > \sup_{x\in\mathsf{ex}\mathcal{K}} \mathrm{E}\mathcal{H}(x).$$

Consider a simplex whose extreme points consist of two elements a_0, a_1 and a sequence $(a_n)_{n\geq 2}$ converging to $b = \frac{1}{2}(a_0 + a_1)$. Set $h_k = (\mathbb{1}_{\{a_0,a_1,a_2,\ldots,a_k\}})^{\mathsf{har}}$

(so that $h_1(a_0) = h_1(a_1) = h_1(b) = 1$). The restriction to $\overline{\mathrm{ex}\mathcal{K}}$ is a sequence whose smallest superenvelope equals 1 on $\mathrm{ex}\mathcal{K}$ and 2 at b (just like in Example 8.1.5, Game 1, the version with x_0 included in the set A). This is preserved by the harmonic prolongation to $\mathcal{M}$ and then by the push-down to $\mathcal{K}$. By Theorem 8.2.12 we have just proved that $\mathrm{E}\mathcal{H}$ equals 1 at all the extreme points of $\mathcal{K}$ and 2 at the (unique) nonextreme point b belonging to the closure of $\mathrm{ex}\mathcal{K}$.

8.3 Entropy of a measure with respect to a topological resolution

There are numerous ways to define the entropy of a measure with respect to a topological resolution. As we know, the resolution itself can be represented in many ways: either by a positive distance ε, or by an open cover, or even (as we shall explain) by a family of continuous functions. For some of these, there are still multiple ways to define the notion. But, as we will show, all of these notions, treated as nets indexed by the resolution, are mutually uniformly equivalent on the simplex of invariant measure.

8.3.1 Three definitions

We begin with *entropy of a measure with respect to an open cover* as defined by Pierre-Paul Romagnoli [Romagnoli, 2003]. Let $\mathcal{U}$ be an open cover of X. For an invariant measure μ, we define:

Definition 8.3.1

$$H(\mu, \mathcal{U}) = \inf\{H(\mu, \mathcal{P}) : \mathcal{P} \succcurlyeq \mathcal{U}\},$$
$$h(\mu, T, \mathcal{U}) = \lim_n \tfrac{1}{n} H(\mu, \mathcal{U}^n),$$

where $\mathcal{P}$ denotes a measurable partition and $\mathcal{P} \succcurlyeq \mathcal{U}$ means that each element of $\mathcal{P}$ is contained in some element of $\mathcal{U}$.

Note that the limit defining $h(\mu, T, \mathcal{U})$ is the same as the infimum, because the corresponding sequence is easily seen to be subadditive (Exercise 8.10).

Since the family of all open covers is directed, we have in fact a net of functions on $\mathcal{M}_T(X)$, indexed by $\mathcal{U}$. It is obvious that this net increases.

Remark 8.3.2 In the above definitions we can replace the topological resolution associated with the cover $\mathcal{U}$ by one associated with the distance $\varepsilon > 0$. For that, the condition $\mathcal{P} \succcurlyeq \mathcal{U}$ should be replaced by $\mathrm{diam}(\mathcal{P}) < \varepsilon$. In this manner one obtains a variant of the above notion, $h(\mu, T, \varepsilon)$, *the entropy of a measure with respect to* ε. This time the subadditivity need not hold, so the limit must be replaced by, say, lim sup. Now we have a net indexed by the directed family of positive numbers ε (ordered by the reversed inequality). It is easy to see that this net also increases. See also Exercise 8.11.

Assume that $\mathcal{U} = \{U_1, \ldots, U_l\}$ is a *finite* open cover of X. By $\mathcal{P}_\mathcal{U}$ we will denote the partition *generated* by $\mathcal{U}$, i.e., the partition by the sets of the form

$$A_F = \bigcap_{i \in F} U_i \cap \bigcap_{i \notin F} U_i^c, \tag{8.3.3}$$

where F is a finite subset of $\{1, 2, \ldots, l\}$. The following fact is important.

Fact 8.3.4 *If $\mathcal{U}$ is a finite cover, then the infimum in the definition of $H(\mu, \mathcal{U})$ is achieved on a partition $\mathcal{P}$ measurable with respect to $\mathcal{P}_\mathcal{U}$.*

Remark 8.3.5 [Romagnoli, 2003] contains a stronger fact: the infimum is achieved on a partition of the form

$$\mathcal{P} = \{U_1,\ U_2 \setminus U_1,\ U_3 \setminus (U_1 \cup U_2),\ \ldots,\ U_l \setminus (U_1 \cup \cdots \cup U_{l-1})\},$$

where $\{U_1, \ldots, U_l\}$ is some ordering of $\mathcal{U}$.

Proof of Fact 8.3.4 First notice that the infimum of the entropy is attained on partitions $\mathcal{P}$ having at most l elements, each element of the cover containing at most one element of the partition. Indeed, two elements of the partition contained in the same element of the cover can be added together, which produces a new partition inscribed in $\mathcal{U}$, and of lower entropy.

We order every such a partition $\mathcal{P} = \{A_1, \ldots, A_l\}$ (some of the sets A_i may be empty) in the unique way satisfying $A_i \subset U_i$ for each i. Notice that $\mathcal{P}$ induces a partition of every set A_F (defined in (8.3.3)) by the sets $A_{F,i} = A_F \cap A_i$ such that $i \in F$. In fact, this is a 1-1 correspondence; if we partition each of the sets A_F arbitrarily into some sets $A_{F,i}$ indexed by $i \in F$, then the (disjoint) unions $A_i = \bigcup_{F \ni i} A_{F,i}$ form a partition $\mathcal{P}$ of X as above. The vectors of values assigned by the measure μ to all partitions $\{A_{F,i} : i \in F\}$ of A_F form a compact convex set (simplex) whose extreme points correspond to trivial partitions, i.e., such that $A_F = A_{F,i}$ up to measure zero, for some $i \in F$. It is now clear that the probability vectors corresponding to all partitions $\mathcal{P}$ form a compact convex set (but not a simplex any more), whose extreme points correspond to partitions which are trivial on each set A_F, i.e., measurable with respect to $\mathcal{P}_\mathcal{U}$. (In fact, if the measure has atoms, then we will obtain only a subset of this convex set, but it always contains all the extreme partitions.) The last thing to notice is that the entropy, being continuous and concave, attains its infimum over a compact convex set at (at least) one of its extreme points (here this is completely trivial, because the set of extreme points is finite; the quoted fact holds more generally, see Exercise 8.8). □

It is worth mentioning (although we will not use it) that for homeomorphisms Romagnoli proved the following variational principle [Romagnoli, 2003, Theorem 2]:

Theorem 8.3.6

$$\sup_{\mu \in \mathcal{M}_T(X)} h(\mu, T, \mathcal{U}) = \mathbf{h}(T, \mathcal{U}). \qquad \square$$

Our next notion is *the local entropy* of a measure. It was originally introduced by S. Newhouse [Newhouse, 1989, 1990] as a refinement of Misiurewicz's "conditional entropy given an epsilon" and provided an upper bound for the defect of upper semicontinuity of the entropy function individually at each measure. The dependence on the ergodic measure μ is achieved by running an infimum over sets F of large measure μ. Contrary to Romagnoli's notion, this one reflects the entropy *not detected* in a given resolution, i.e., a kind of conditional entropy *given* the resolution. In the finite entropy case, one can think of local entropy as the difference between the entropy and entropy with respect to the given resolution. The definition below differs slightly from the original; we change the notation a little, we rid it of some unnecessary technicalities, and we give priority to the version with covers.

Definition 8.3.7 Let $\mathcal{U}$ and $\mathcal{V}$ be open covers. Let $F \subset X$ be a measurable set. The item (a) coincides with the Definition 6.3.1, we repeat it for convenience. We define successively:

(a) $\mathbf{H}(\mathcal{U}|F, \mathcal{V}) =$
$\log \max\{\min\{\#\mathcal{U}_{F\cap V} : \mathcal{U}_{F\cap V} \subset \mathcal{U}, F \cap V \in \bigcup \mathcal{U}_{F\cap V}\} : V \in \mathcal{V}\}$;
(b) $\mathbf{h}(T, \mathcal{U}|F, \mathcal{V}) = \limsup_n \frac{1}{n}\mathbf{H}(\mathcal{U}^n|F, \mathcal{V}^n)$;
(c) $\mathbf{h}(T|F, \mathcal{V}) = \sup_{\mathcal{U}} \mathbf{h}(T, \mathcal{U}|F, \mathcal{V})$;
(d) $h(T|\mu, \mathcal{V}) = \lim_{\sigma \to 1} \inf\{\mathbf{h}(T|F, \mathcal{V}) : \mu(F) > \sigma\}$.

The last term is called *the local entropy of μ given $\mathcal{V}$*. We apply the formula (d) only to ergodic measures μ and for other invariant measures we use the harmonic prolongation (see (A.2.15)).

Remark 8.3.8 The notion (d) has bad affinity properties on nonergodic measures. For example, on a combination of two ergodic measures it assumes the maximum of the two values, nothing close to their corresponding combination. This is why we employ the harmonic prolongation.

Remark 8.3.9 The reason why μ occurs in (d) on the condition's side is purely accidental; it replaces F (which clearly is a conditioning object). As we shall see in zero-dimensional spaces, local entropy corresponds to conditional entropy which is denoted with μ on the other side of the bar. The notation as in (d) has been used already in several papers, so we decided to keep it.

Remark 8.3.10 Local entropy can be adapted to the other type of topological resolution: the cover $\mathcal{V}$ can be replaced by the scale defined by the distance ε – one needs to replace the cover $\mathcal{V}^n$ in (b) by the cover created by the (n, ε)-balls. This leads in (d) to a notion $h(T|\mu, \varepsilon)$.

We can also replace the resolution $\mathcal{U}$ by a distance δ. This time the final notion will not change. We leave the details to the reader as Exercise 8.13.

The last type of entropy of a measure with respect to a topological resolution presented in this book is based on the same approach that was already used in the construction of principal zero-dimensional extensions. In this definition we are dealing with a somewhat artificial type of topological resolution determined by the family of continuous functions, nevertheless, as we shall soon see, this entropy notion enjoys the best topological properties among all notions introduced in this section. It will play a crucial role in proving many asymptotic (as the resolution refines) properties of the other notions as well.

Definition 8.3.11 Let $\mathcal{F}$ be a finite family of continuous functions $f : X \to [0, 1]$. For an invariant measure μ, we define the *entropy of μ with respect to $\mathcal{F}$* by

$$h(\mu, T, \mathcal{F}) = h(\mu \times \lambda, T \times \mathrm{Id}, \mathcal{A}_{\mathcal{F}}),$$

where $\mathcal{A}_{\mathcal{F}}$ is the partition of $X \times [0, 1]$ introduced in Definition 7.6.5, and λ denotes the Lebesgue measure on $[0, 1]$.

The collection of all finite families $\mathcal{F}$ as above is clearly a directed family (with respect to inclusion), so, once again, we are dealing with a net of functions. Obviously, this net is also increasing.

There are more ways to compute the entropy of a measure with respect to a topological resolution. Anatole Katok introduced such a notion based on counting the minimal number of (n, ε)-balls whose union achieves certain positive measure value σ [Katok, 1980]. Another definition, by Misha Brin and A. Katok uses an analog of the Shannon–McMillan–Breiman Theorem: for each ergodic measure μ, $\frac{1}{n}$th of the logarithm of the measure of the (n, ε)-ball around x converges almost surely in n, and this limit is used to define $h(\mu, \varepsilon)$ [Brin and Katok, 1983]. We refer the reader to [Downarowicz,

2005a] for a detailed discussion of these and other notions and the inequalities between them.

In this book, we will use one more notion of this kind. It computes the (unconditional) entropy with respect to a cover (just like Romagnoli's notion in Definition 8.3.1), but in a manner somewhat analogous to the way Newhouse local entropy computes the conditional entropy given a cover: the dependence on the ergodic measure is achieved by running an infimum over sets F of positive measure.

Definition 8.3.12 Let $(X, T, \mathbb{S})$ be a topological dynamical system and let $\mu \in \mathcal{M}_T(X)$ be ergodic. We let

$$\mathbf{h}(\mu, T, \mathcal{U}) = \inf\{\mathbf{h}(T, \mathcal{U}|F) : \mu(F) > 0\},$$

where $\mathbf{h}(T, \mathcal{U}|F)$ is $\mathbf{h}(T, \mathcal{U}|F, \mathcal{V})$ of Definition 8.3.7 for trivial $\mathcal{V}$. For nonergodic measures we prolong $\mathbf{h}(\mu, T, \mathcal{U})$ using the harmonic prolongation. This notion will be called *variant entropy of μ with respect to $\mathcal{U}$*.

Question 8.3.13 As we shall soon see (Lemma 8.4.13), the above notion converges over refining covers to the same limit (the entropy of μ) as Romagnoli's notion. We do not know whether these notions are identical before taking the limit, i.e., whether $\mathbf{h}(\mu, T, \mathcal{U}) = h(\mu, T, \mathcal{U})$ for any $\mu \in \mathcal{M}_T(X)$ and any open cover $\mathcal{U}$.

8.3.2 Properties of the above notions

For each of the first three notions of entropy with respect to a topological scale (treated as functions on invariant measures) we will take a brief look at the semicontinuity and affinity properties. Then we will examine how these notions behave under topological extensions. Throughout this section $(X', T', \mathbb{S})$ is a topological extension of $(X, T, \mathbb{S})$ via a map π.

Lemma 8.3.14 *Let $\mathcal{U}$ be a finite cover of X. The function $\mu \mapsto h(\mu, T, \mathcal{U})$ is upper semicontinuous and concave.*

Proof Concavity is obvious, since we are dealing with an infimum (applied two times) of the concave functions $H(\mu, \mathcal{P})$ (see (1.3.5)). We sketch the proof of the semicontinuity. We fix a measure μ and a partition $\mathcal{P}$ measurable with respect to the sigma-algebra generated by $\mathcal{U}$, which realizes the infimum in the definition of $H(\mu, \mathcal{U})$ (use Fact 8.3.4). Now, using regularity of the measure, we can modify the cover $\mathcal{U}$ to $\mathcal{U}'$ inscribed in $\mathcal{U}$, by making each element of $\mathcal{U}$ slightly smaller (so that the measure of the difference is small), and with boundary of measure zero. The partition $\mathcal{P}'$ obtained from $\mathcal{U}'$ by the same set operations as $\mathcal{P}$ is obtained from $\mathcal{U}$ is inscribed in $\mathcal{U}'$ (hence in $\mathcal{U}$), and its ℓ^1-distance from $\mathcal{P}$ is small. So, $H(\mu, \mathcal{P}')$ is smaller than $H(\mu, \mathcal{U}) + \varepsilon$, while the function $H(\cdot, \mathcal{P}')$ is continuous at μ (see Lemma 6.6.7). This proves that $H(\mu, \mathcal{U})$ is upper semicontinuous at μ. Upper semicontinuity of the function $h(\cdot, T, \mathcal{U})$ is now clear, since, by subadditivity, this is the infimum of the upper semicontinuous functions $\frac{1}{n}H(\cdot, \mathcal{U}^n)$. □

It seems rather hard to check whether the difference $h(\cdot, T, \mathcal{U}) - h(\cdot, T, \mathcal{V})$, where $\mathcal{U} \succcurlyeq \mathcal{V}$, is upper semicontinuous, and whether the function $h(\cdot, T, \mathcal{U})$ is affine. Fortunately, as it will soon turn out, we can afford to skip this verification. We pass to studying the behavior of $h(\mu, T, \mathcal{U})$ under extensions. The next lemma originates from [Romagnoli, 2003].

Lemma 8.3.15 *If $\mathcal{U}$ is a finite cover of X and $\mathcal{U}'$ is the cover of X' obtained by lifting $\mathcal{U}$ against π, then*

$$h(\mu, T, \mathcal{U}) = h(\mu', T', \mathcal{U}')$$

for any $\mu \in \mathcal{M}_T(X)$ and $\mu' \in \mathcal{M}_{T'}(X')$ such that $\pi\mu' = \mu$.

Proof Since $\mathcal{U}'$ is lifted from X, so is $\mathcal{U}'^n$ (where the iterates are with respect to the action of T'), so is $\mathcal{P}_{\mathcal{U}'^n}$, the partition generated by it, and any partition measurable with respect to the last one. By Fact 8.3.4, only such partitions are needed to compute $h(\mu', T', \mathcal{U}')$ which then obviously equals $h(\mu, T, \mathcal{U})$. □

Lemma 8.3.16 *If $\mathcal{U}$ and $\mathcal{V}$ are two covers of X, then, for any $\mu \in \mathcal{M}_T(X)$,*

$$h(\mu, T, \mathcal{U}) \leq h(\mu, T, \mathcal{V}) + \mathbf{h}(T, \mathcal{U}|\mathcal{V}).$$

Proof It suffices to prove that $H(\mu, \mathcal{U}) \leq H(\mu, \mathcal{V}) + \mathbf{H}(\mathcal{U}|\mathcal{V})$ and then apply this to the covers $\mathcal{U}^n$ and $\mathcal{V}^n$, divide both sides by n and pass to the limit over n. Let $\mathcal{P}$ be a finite partition inscribed in $\mathcal{V}$. For each $A \in \mathcal{P}$ choose a $V \in \mathcal{V}$ containing A. The set V is covered by a family of at most $2^{\mathbf{H}(\mathcal{U}|\mathcal{V})}$ sets of the form $U \cap V$, where $U \in \mathcal{U}$. Thus, we can partition A into at most that many disjoint sets, each contained entirely in one of the above sets $U \cap V$. In this manner we have defined a refinement of $\mathcal{P}$, say $\mathcal{P}'$, inscribed in $\mathcal{U}$ and such that $H(\mu, \mathcal{P}'|\mathcal{P}) \leq \mathbf{H}(\mathcal{U}|\mathcal{V})$. Because $H(\mu, \mathcal{P}'|\mathcal{P}) = H(\mu, \mathcal{P}') - H(\mu, \mathcal{P})$, we have obtained

$$H(\mu, \mathcal{P}') \leq H(\mu, \mathcal{P}) + \mathbf{H}(\mathcal{U}|\mathcal{V}).$$

The left-hand side can be replaced by the not larger term $H(\mu, \mathcal{U})$. Now, on the right, we can apply the infimum over all partitions $\mathcal{P}$ inscribed in $\mathcal{V}$. □

The next fact is completely obvious, by the elementary monotonicity (1.6.6) (or, alternatively, by Fact 8.3.4):

Lemma 8.3.17 *If $\mathcal{U}$ is a finite cover of X by disjoint clopen sets, then $h(\mu, T, \mathcal{U}) = h(\mu, T, \mathcal{P})$, where $\mathcal{P}$ is $\mathcal{U}$ treated as a partition.* □

Next we pass to the local entropy. Directly by the last step of Definition 8.3.7 (and using Fact A.2.15), the corresponding function $h(T|\mu, \mathcal{V})$ on invariant measures is harmonic, and hence affine. Semicontinuity properties of this function seem rather difficult to check. Moreover, it would be much more important to check upper semicontinuity of the difference function $h(T|\mu, \mathcal{V}) - h(T|\mu, \mathcal{V}')$, where $\mathcal{V}' \succcurlyeq \mathcal{V}$, which seems even more difficult. Fortunately, as it will turn out soon, we will manage to proceed without verifying these properties for this particular notion. We pass immediately to the behavior under extensions.

Lemma 8.3.18 *If $\mathcal{V}$ is a finite cover of X and $\mathcal{V}'$ is the cover of X' obtained by lifting $\mathcal{V}$, then*

$$h(T|\mu, \mathcal{V}) \leq h(T'|\mu', \mathcal{V}') \leq h(T|\mu, \mathcal{V}) + \mathbf{h}(T'|T), \tag{8.3.19}$$

for any $\mu \in \mathcal{M}_T(X)$ and any of its lifts $\mu' \in \mathcal{M}_{T'}(X')$. If $(X', T', \mathcal{S})$ is a principal extension of $(X, T, \mathcal{S})$, then $h(T|\mu, \mathcal{V}) = h(T'|\mu', \mathcal{V}')$. □

Proof First notice that we can restrict the infimum in Definition 8.3.7 (d) to closed sets F only. On one hand, such infimum is obviously not smaller, on the other, by regularity, in each measurable set F there is a closed subset G with nearly the same measure as F, and then $h(T|F, \mathcal{V}) \geq h(T|G, \mathcal{V})$, so the infimum over the closed sets is not larger.

Consider a closed set $F' \in X'$ and let $G = \pi(F')$ and $G' = \pi^{-1}(G)$. We say that G' is a *fiber saturation* of F. Since F' is compact and π is continuous, G and G' are compact too, in particular, measurable. Now it is clear that if $\mathcal{U}, \mathcal{U}'$ denote some cover of X and its lift, respectively, then $\mathbf{h}(T, \mathcal{U}|G, \mathcal{V}) = \mathbf{h}(T', \mathcal{U}'|G', \mathcal{V}')$. Also, it is clear that $\mathbf{h}(T', \mathcal{U}'|F', \mathcal{V}') = \mathbf{h}(T', \mathcal{U}'|G', \mathcal{V}')$ since any family of lifted sets covering some set also covers its fiber saturation. This implies the first inequality in (8.3.19), because the term $h(T'|\mu', \mathcal{V}')$ is a supremum over all covers $\mathcal{U}'$ of X', including the lifted ones.

For the second inequality, let $\mathcal{U}'$ now denote an arbitrary cover of X', and let $\mathcal{W}$ be a cover of X with a lift $\mathcal{W}'$. For any closed set $F' \in X'$ and its fiber saturation G' we have (using (6.3.10) with permuted covers)

$$\mathbf{H}(\mathcal{U}'|F', \mathcal{V}') \leq \mathbf{H}(\mathcal{U}'|G', \mathcal{V}') \leq \mathbf{H}(\mathcal{W}'|G', \mathcal{V}') + \mathbf{H}(\mathcal{U}'|\mathcal{W}').$$

Note that $\mathbf{H}(\mathcal{W}'|G', \mathcal{V}') = \mathbf{H}(\mathcal{W}'|F', \mathcal{V}')$. The proof is completed by applying the above to the covers raised to "power" n, dividing by n and passing to lim sup over n (note that the sum of lim sup's is not smaller than lim sup of the sums), refining $\mathcal{W}$ and next $\mathcal{U}'$, and finally letting the set F vary as instructed in Definition 8.3.7.

The last statement in the lemma now becomes obvious. □

Lemma 8.3.20 *If $\mathcal{V}$ and $\mathcal{W}$ are two covers of X, then, for any $\mu \in \mathcal{M}_T(X)$,*

$$h(T, |\mu, \mathcal{W}) \leq h(T|\mu, \mathcal{V}) + \mathbf{h}(T, \mathcal{V}|\mathcal{W}).$$

Proof By (6.3.10), for any open cover $\mathcal{U}$ and a measurable set F we have $\mathbf{H}(\mathcal{U}|F, \mathcal{W}) \leq \mathbf{H}(\mathcal{U}|F, \mathcal{V}) + \mathbf{H}(\mathcal{V}|\mathcal{W})$. We apply the above to $\mathcal{U}^n$, $\mathcal{V}^n$ and $\mathcal{V}^n$, divide by n, pass to lim sup over n, take the supremum over all covers $\mathcal{U}$ and the infimum over all sets F with large measure, to obtain the hypothesis. □

Lemma 8.3.21 *Let X be zero-dimensional. If $\mathcal{V}$ is a cover by disjoint clopen sets, then*

$$h(T|\mu, \mathcal{V}) = h(\mu, T|\mathcal{Q}),$$

where $\mathcal{Q}$ is $\mathcal{V}$ treated as a partition.

Proof Since both sides are harmonic functions of μ, it suffices to work with ergodic measures. The proof uses the Shannon–McMillan–Breiman Theorem for both directions. We choose an $\varepsilon > 0$. We fix a cover $\mathcal{U}$ by disjoint clopen sets, finer than $\mathcal{V}$, such that $h(\mu, T, \mathcal{P}) \geq h(\mu, T) - \varepsilon$, where $\mathcal{P}$ is $\mathcal{U}$ treated as a partition (if μ has infinite entropy, this requirement is replaced by $h(\mu, T, \mathcal{P}) > 1/\varepsilon$; the rest of the proof should be modified accordingly). Now we choose $1/2 < \sigma < 1$, and we let n_σ be so large that the set G exceeds σ in measure, where G consists of the points which satisfy, up to ε and for every $n \geq n_\sigma$, the hypothesis of the Shannon–McMillan–Breiman Theorem for the processes generated by the partitions $\mathcal{P}$ and $\mathcal{Q}$. Consider a cell $V \in \mathcal{V}^n$ not disjoint of G. Its measure ranges within $2^{-n(h(\mu,T,\mathcal{Q})\pm\varepsilon)}$. For at least one of such cells, say for V_0, the measure of its intersection with G is at least $\sigma 2^{-n(h(\mu,T,\mathcal{Q})+\varepsilon)}$, hence at least $2^{-n(h(\mu,T,\mathcal{Q})+2\varepsilon)}$. Any cell of $\mathcal{U}^n$ containing a point from G has measure in the range $2^{-n(h(\mu,T,\mathcal{P})\pm\varepsilon)}$, a cell intersecting V is contained in it, and different cells of $\mathcal{U}^n$ are disjoint. This implies that the number of such cells needed to cover any $V \cap G$ is at most $2^{n(h(\mu,T,\mathcal{P})-h(\mu,T,\mathcal{Q})+2\varepsilon)}$, while for $V_0 \cap G$ it is at least $2^{n(h(\mu,T,\mathcal{P})-h(\mu,T,\mathcal{Q})-3\varepsilon)}$. In both estimates we can replace $h(\mu, T, \mathcal{P})$ by $h(\mu, T)$ at a cost of another ε. Because $\mathcal{Q}$ is finite, the difference $h(\mu, T) - h(\mu, T, \mathcal{Q})$ equals the conditional entropy $h(\mu, T|\mathcal{Q})$ (recall Fact 4.1.6). We have proved that $\mathbf{h}(T, \mathcal{U}|G, \mathcal{V}) = h(\mu, T|\mathcal{Q}) \pm 4\varepsilon$. Since the covers $\mathcal{U}$ of the considered type generate, we obtain that $\mathbf{h}(T|G, \mathcal{V})$ ranges within $h(\mu, T|\mathcal{Q}) \pm 4\varepsilon$. The infimum over all sets F (in place of G) with measure at least σ is not larger. This clearly implies the inequality "$\leq$"

in the assertion. For the reversed inequality notice that any measurable set F of measure sufficiently close to 1 contains a subset G' of the set G defined above, of measure slightly smaller than σ. For G' the above estimate still holds: $\mathbf{h}(T|G', \mathcal{V}) \geq h(\mu, T|\mathcal{Q}) - 4\varepsilon$, and clearly, $\mathbf{h}(T|F, \mathcal{V})$ is not smaller. Thus the infimum over all large enough sets F is not smaller than this estimate. □

Let us now address the notion $h(\mu, T, \mathfrak{F})$. First of all, this is the usual dynamical entropy (under the action of $T \times \mathrm{Id}$) of an invariant measure (in this case $\mu \times \lambda$) with respect to a finite partition of the space (here it is $\mathcal{A}_{\mathfrak{F}}$, a partition of $X \times [0,1]$). So, all formulae valid for the dynamical entropy apply, with the join of partitions and the relation $\preccurlyeq$ replaced by the ordinary union and inclusion of the families of continuous functions, respectively. In particular, denoting

$$h(\mu, T, \mathfrak{F}|\mathfrak{G}) = h(\mu, T, \mathfrak{F} \cup \mathfrak{G}) - h(\mu, T, \mathfrak{G}) \qquad (8.3.22)$$

the monotonicity and subadditivity formulae gathered in Fact 2.4.2 hold. Both $h(\mu, T, \mathfrak{F})$ and $h(\mu, T, \mathfrak{F}|\mathfrak{G})$ are affine functions of the measure (Theorem 2.5.1). Now, we pass to semicontinuity.

Lemma 8.3.23 *Let $\mathfrak{F} \subset \mathfrak{G}$ be two finite families of continuous functions $f : X \to [0,1]$. Then the functions $\mu \mapsto h(\mu, T, \mathfrak{F})$ and $\mu \mapsto h(\mu, T, \mathfrak{G}) - h(\mu, T, \mathfrak{F})$ are upper semicontinuous on $\mathcal{M}_T(X)$.*

Proof The first assertion has already been noted and used in the construction of the principal zero-dimensional extension. We recall the idea. By definition, $h(\mu, T, \mathfrak{F}) = h(\mu \times \lambda, T \times \mathrm{Id}, \mathcal{A}_{\mathfrak{F}})$. The boundaries of the cells of $\mathcal{A}_{\mathfrak{F}}$ are contained in the graphs of the functions in $\mathfrak{F}$, and these graphs have measure zero for any measure of the form $\mu \times \lambda$. Thus Lemma 6.6.7 applies to every such measure. The second assertion follows by noticing that

$$H(\mu \times \lambda, \mathcal{A}_{\mathfrak{G}}^n) - H(\mu \times \lambda, \mathcal{A}_{\mathfrak{F}}^n) = H(\mu \times \lambda, \mathcal{A}_{\mathfrak{G}}^n | \mathcal{A}_{\mathfrak{F}}^n).$$

The left-hand side divided by n is a sequence of continuous functions of μ converging to $h(\mu, T, \mathfrak{G}) - h(\mu, T, \mathfrak{F})$. The right-hand side shows that this sequence is decreasing (see Fact 2.3.1) hence upper semicontinuity follows. □

We also note the completely obvious fact concerning extensions:

Lemma 8.3.24 *Let $\mathfrak{F}$ be a finite family of continuous functions $f : X \to [0,1]$ and let $\mathfrak{F}'$ be its lift to X'. Then $h(\mu, T, \mathfrak{F}) = h(\mu', T', \mathfrak{F}')$ for any $\mu \in \mathcal{M}_T(X)$ and any of its lifts $\mu' \in \mathcal{M}_{T'}(X')$.* □

Lemma 8.3.25 *If $\mathcal{P}$ is a finite partition of X into clopen sets and $\mathcal{F}$ is the family of the characteristic functions of the cells of $\mathcal{P}$ (such functions are continuous), then, for every $\mu \in \mathcal{M}_T(X)$, we have*

$$h(\mu, T, \mathcal{F}) = h(\mu, T, \mathcal{P}).$$

Proof This is immediate, since in this case $\mathcal{A}_{\mathcal{F}}$ is the partition of $X \times [0,1]$ obtained by lifting $\mathcal{P}$ against the projection map, which sends $\mu \times \lambda$ to μ. □

8.4 Entropy structure

From now on we will assume that the system $(X, T, \mathbb{S})$ has finite topological entropy. Entropy structure, although it can be also defined for systems with infinite topological entropy, plays the most important role in the theory of symbolic extensions, and this theory is void for systems with infinite entropy.

Before we say what the entropy structure is, we prove the key fact concerning the various notions of entropy of a measure with respect to a topological resolution introduced in the preceding section.

Theorem 8.4.1 *Let $(X, T, \mathbb{S})$ be a topological dynamical system with finite topological entropy. The following three increasing nets of nonnegative functions, defined on the simplex $\mathcal{M}_T(X)$ of all T-invariant measures on X, are unifomly equivalent: $\mathcal{H}^{\mathrm{Rom}} = \big(h(\mu, T, \mathcal{U})\big)_{\mathcal{U}}$, $\mathcal{H}^{\mathrm{New}} = \big(h(\mu, T) - h(T|\mu, \mathcal{V})\big)_{\mathcal{V}}$ and $\mathcal{H}^{\mathrm{fun}} = \big(h(\mu, T, \mathcal{F})\big)_{\mathcal{F}}$.*

Remark 8.4.2 By virtue of Exercises 8.11 and 8.12, the nets $\big(h(\mu, T, \varepsilon)\big)_{\varepsilon}$ and $\big(h(\mu, T) - h(T|\mu, \varepsilon)\big)_{\varepsilon}$ can be added to the above collection.

Proof of Theorem 8.4.1 At first assume that the space X is zero-dimensional. Then the family of finite disjoint covers by clopen sets is a subnet of the net of all covers (it fulfills the condition (A.1.6)). Now, by Lemmas 8.3.17, 8.3.21 and 8.3.25, for such covers the notions $h(\mu, T, \mathcal{U})$, $h(\mu, T) - h(T|\mu, \mathcal{U})$ and $h(\mu, T, \mathcal{F}_{\mathcal{U}})$, coincide (here $\mathcal{F}_{\mathcal{U}}$ is the family of the characteristic functions of the cells of $\mathcal{U}$). Because subnets of monotone nets are always uniformly equivalent to the whole net, we obtain that the first two (of the three discussed) full nets are uniformly equivalent. We cannot claim it yet for the full net $h(\mu, T, \mathcal{F})$ because the subfamily $\mathcal{F}_{\mathcal{U}}$ does not satisfy (A.1.6) among all finite families of continuous functions ordered by inclusion, so, in this case, we are not dealing with a subnet (only a sub-net). Here, however, we can use Lemmas 8.3.23 and 8.1.24: the net indexed by $\mathcal{F}_{\mathcal{U}}$ converges to the finite entropy function, the same as the full net $h(\mu, T, \mathcal{F})$, the full net has upper semicontinuous differences, and the domain is compact. Thus, these nets are uniformly equivalent.

The general case (of not necessarily zero-dimensional systems) of the theorem will follow immediately from the existence of zero-dimensional principal extensions (Theorem 7.6.1) and Lemma 8.4.3 provided below. □

Lemma 8.4.3 *Let $(X', T', \mathbb{S})$ be a principal extension of a system $(X, T, \mathbb{S})$ with finite topological entropy. Then the three above discussed nets: $\mathcal{H}^{\text{Rom}}$, $\mathcal{H}^{\text{New}}$ and $\mathcal{H}^{\text{fun}}$ defined on $\mathcal{M}_T(X)$ and lifted to $\mathcal{M}_{T'}(X')$ are uniformly equivalent to the corresponding nets defined directly on $\mathcal{M}_{T'}(X')$.*

Proof We begin with the first two nets. By Lemma 8.3.15, the net $h(\mu, T, \mathcal{U})$ lifted to $\mathcal{M}_{T'}(X')$ becomes $h(\mu', T', \mathcal{U}')$, where $\mathcal{U}'$ is the lift of $\mathcal{U}$. Similarly, by Lemma 8.3.18, $h(T|\mu, \mathcal{U})$ becomes $h(T'|\mu', \mathcal{U}')$. In the extension, the family of all covers lifted from X is a sub-net usually without being a subnet, so, we only have one (trivial) direction of the inequality needed to prove uniform equivalence: every function $h(\mu', T', \mathcal{U}')$ (or $h(\mu', T') - h(T'|\mu', \mathcal{U}')$) in the net indexed by the lifted covers is dominated by a member of the corresponding full net (namely by itself). We need to show the converse (up to some fixed $\varepsilon > 0$). So, choose an arbitrary open cover $\mathcal{W}'$ of X' (not necessarily lifted from X). Because we are dealing with a principal extension, there is a cover $\mathcal{U}$ of X such that $\mathbf{h}(T', \mathcal{W}'|\mathcal{U}') < \varepsilon$. Then, by Lemmas 8.3.16 and 8.3.20 (both with accordingly permuted covers), we get the inequalities $h(\mu', T', \mathcal{W}') \le h(\mu', T', \mathcal{U}') + \varepsilon$ and $h(T'|\mu', \mathcal{U}') \le h(T'|\mu', \mathcal{W}') + \varepsilon$. Using finite topological entropy we can rewrite the last inequality as $h(\mu', T') - h(T'|\mu', \mathcal{W}') \le h(\mu', T') - h(T'|\mu', \mathcal{U}') + \varepsilon$. This concludes the proof of uniform equivalence for the first two nets.

The uniform equivalence of the net $h(\mu, T, \mathcal{F})$ and the analogous net in the extension is proved as follows: If $\mathcal{F}'$ denotes the family $\mathcal{F}$ lifted to X', then, by the trivial Lemma 8.3.24, $h(\mu, T, \mathcal{F}) = h(\mu', T', \mathcal{F}')$. We have obtained, as before, only a sub-net (not subnet) of the net on $\mathcal{M}_{T'}(X')$ indexed by all finite families of continuous functions on X'. But we know that before lifting this net converges to the finite entropy function on $\mathcal{M}_T(X)$, and since the extension is principal, the lift of this function coincides with the entropy function on $\mathcal{M}_{T'}(X')$. Now Lemmas 8.3.23 and 8.1.24 imply that the sub-net is uniformly equivalent to the whole net, because it has the same finite limit function, the whole net has upper semicontinuous differences, and the domain is compact. □

The three nets of Theorem 8.4.1 determine one and the same class of uniform equivalence. This class becomes our master entropy invariant in topological dynamics.

Definition 8.4.4 Let $(X, T, \mathbb{S})$ be a topological dynamical system with finite topological entropy. The *entropy structure* is defined as the class of the uniform equivalence between increasing nets of nonnegative functions defined on $\mathcal{M}_T(X)$, containing the three above discussed nets: $\mathcal{H}^{\mathrm{Rom}}, \mathcal{H}^{\mathrm{New}}$ and $\mathcal{H}^{\mathrm{fun}}$.

By convention, any member of this class is also called an entropy structure (instead of saying that it *belongs to* we say that it *is* an entropy structure). This causes no ambiguity. The entropy structure will be denoted by $\mathcal{H}^T$ or simply by $\mathcal{H}$, if the considered system is fixed.

The statement below shows the importance of the notion of entropy of a measure with respect to a family of functions. It explains why we could neglect the verification of affinity or upper semicontinuity of the differences for the other two notions. Observe that on any compact space there exists a sequence $\mathcal{F}_k$ of finite families of continuous functions from X to $[0,1]$ such that the associated partitions $\mathcal{A}_{\mathcal{F}_k}$ refine in the product space $X \times [0,1]$. Using such a sequence and the properties proved in Lemma 8.3.23 we easily deduce the first part of Fact 8.4.5 below. Corollary 8.1.13 and Lemma 8.1.8, combined with Theorem 8.1.25 (which says that $\mathrm{E}\mathcal{H}$ does not depend on the choice of the representative sequence as the entropy structure), yield the following:

Fact 8.4.5 *Let $(X,T,\mathbb{S})$ be a topological dynamical system with finite topological entropy. Then the entropy structure $\mathcal{H}$ of this system contains a u.s.d.a.-sequence. If $\mathrm{E}\mathcal{H}$ is finite, then $\mathrm{E}\mathcal{H}$ and the smallest repair function $u_{\mathcal{H}} = \mathrm{E}\mathcal{H} - h$ are upper semicontinuous.* □

Although the realization theorem quoted below is of high importance in the theory of entropy structures, we have decided not to rewrite its lengthy and technical proof in this book. We refer the reader to the original paper [Downarowicz and Serafin, 2003].

Theorem 8.4.6 *Every uniform equivalence class defined on a Choquet simplex containing a u.s.d.a.-sequence represents (up to affine homeomorphism) the entropy structure of some topological dynamical system. Moreover, the system can be chosen zero-dimensional, minimal and invertible.* □

The entropy structure carries the information about the "type of convergence" (to the entropy function) of the net of functions representing the entropy of invariant measures with respect to topological resolutions, as the resolution refines. It is an invariant of topological conjugacy in the sense of the following theorem, which is an immediate consequence of Lemma 8.4.3 and the fact that topological conjugacy is a case of a principal extension.

Theorem 8.4.7 *Let $(X,T,\mathbb{S})$ and $(Y,S,\mathbb{S})$ be two topologically conjugate systems with finite topological entropy, where $\pi : Y \to X$ is the conjugating* *homeomorphism. Let* $\mathcal{H} = (h_\kappa)$ *be an entropy structure of the system* $(X, T, \mathbb{S})$. *Then the net* $\mathcal{H}' = (h'_\kappa)$ *on* $\mathcal{M}_S(Y)$, *where* $h'_\kappa(\nu) = h_\kappa(\pi\nu)$, *is an entropy structure of the system* $(Y, S, \mathbb{S})$. □

Invariants of the uniform equivalence relation applied to the entropy structure become invariants of topological conjugacy. Some of them have very important interpretation in terms of symbolic extensions, which will be discussed in the next chapter. And so, we obtain *superenvelopes of the entropy structure*, the *smallest superenvelope of the entropy structure* denoted by $\mathrm{E}\mathcal{H}^T$ or just $\mathrm{E}\mathcal{H}$, *the smallest repair function* $u_\mathcal{H}$, *the transfinite sequence associated with the entropy structure* whose elements will be denoted by u^T_α or just u_α, the order of accumulation of this sequence, called for short *the order of accumulation of entropy* and denoted just by α_0, and its refinement, the function on invariant measures representing for each $\mu \in \mathcal{M}_T(X)$ *the order of accumulation of entropy at* μ, denoted by $\alpha_0(\mu)$. The superenvelopes and specifically the function $\mathrm{E}\mathcal{H}^T$ (or $h + u_\mathcal{H}$) will be given a special meaning as the *symbolic extension entropy function*, and its supremum over all invariant measures equals the parameter called the *topological symbolic extension entropy*, an extremely important invariant of topological conjugacy, responsible for the "vertical data compression," the subject of the next chapter.

We already know that the entropy structure of a principal extension coincides with the lift of the entropy structure of the original system. This clearly implies that any lifted superenvelope is a superenvelope in the extension. However, there may be other superenvelopes in the extension (not constant on the fibers of invariant measures). The next lemma explains the relation.

Lemma 8.4.8 *Let* $(X', T', \mathbb{S})$ *be a principal extension of* $(X, T, \mathbb{S})$. *Let* E *be a superenvelope of the entropy structure* $\mathcal{H} = (h_k)$ *of* $(X, T, \mathbb{S})$, *and let* E' *be a superenvelope of the entropy structure* $\mathcal{H}' = (h'_k)$ *of* $(X', T', \mathbb{S})$. *Then the lift of* E *to* $\mathcal{M}_{T'}(X')$ *is a superenvelope of* $\mathcal{H}'$, *while the push-down* $E'^{[\mathcal{M}_T(X)]}$ *is a superenvelope of* $\mathcal{H}$. *The smallest superenvelope* $\mathrm{E}\mathcal{H}'$ *equals the lift of* $\mathrm{E}\mathcal{H}$.

Proof We know that $\mathcal{H}'$ is the lift of $\mathcal{H}$ (see Lemma 8.4.3). We also know that the entropy structures in both systems can be chosen to be u.s.d.a.-sequences, hence, by Lemma 8.1.12, the superenvelopes can be tested by upper semicontinuity of $E - h_k$ (and analogously for the extension). Now the first statement in the assertion is immediate as the lift of $E - h_k$ is, on one hand, upper semicontinuous, on the other, it equals the lift of E minus h'_k (because the latter equals the lift of h_k). The second statement holds by symmetric reasons: the push-down $(E' - h'_k)^{[\mathcal{M}_T(X)]}$ is, on one hand, upper semicontinuous (see Fact

A.1.26), on the other it equals $E'^{[\mathcal{M}_T(X)]} - h_k$. The last claim of the lemma follows, for example, from the transfinite characterization of $\mathrm{E}\mathcal{H}$ (Theorem 8.1.19); for a sequence $\mathcal{H}'$ which is a lift (via a continuous map) of $\mathcal{H}$, all the functions in the transfinite sequence $u_\alpha^{T'}$ are equal to the lifted functions u_α^T, correspondingly. So, $\mathrm{E}\mathcal{H}'$ equals the lift of $\mathrm{E}\mathcal{H}$. □

The entropy structure $\mathcal{H}^T$ also allows several familiar entropy invariants to be recovered: the limit function h coincides with the entropy function on invariant measures, the supremum of this function over all invariant measures equals the topological entropy of the system. In the next theorem we will also show that the topological tail entropy $\mathbf{h}^*(T)$ can be recovered as the maximum over all invariant measures of the function u_1^T, the first one in the transfinite sequence associated with the entropy structure.

Theorem 8.4.9 (Tail Entropy Variational Principle[2]) *In a topological dynamical system $(X,T,\mathbb{S})$ with finite topological entropy we have*

$$\mathbf{h}^*(T) = \sup_{\mu\in\mathcal{M}_T(X)} u_1^T(\mu).$$

Proof As in many cases before, it suffices to prove this for zero-dimensional systems. Then, for other systems we apply the principal zero-dimensional extension. Since both $\mathbf{h}^*(T)$ (see Corollary 6.4.15) and the entropy structure with its invariants, in particular the function u_1, are preserved by principal extensions, the proof will be complete.

In zero-dimensional systems $\mathbf{h}^*(T)$ is characterized by Equation (7.5.2):

$$\mathbf{h}^*(T) = \lim_k \downarrow \mathbf{h}(T|T_k),$$

where T_k is the subshift factor generated by a finite partition $\mathcal{P}_{\Lambda_k}$. By the conditional variational principle (Theorem 6.8.8), we can write

$$\mathbf{h}^*(T) = \inf_k \sup_{\mu\in\mathcal{M}_T(X)} h(\mu|\mu_k),$$

which, by the finite entropy assumption can be rewritten as

$$\mathbf{h}^*(T) = \inf_k \sup_{\mu\in\mathcal{M}_T(X)} (h(\mu) - h(\mu,T,\mathcal{P}_{\Lambda_k})).$$

The partitions $\mathcal{P}_{\Lambda_k}$ are clopen and they can be chosen to refine in X, the sequence of functions $\mathcal{H} = (h_k)$ on $\mathcal{M}_T(X)$, where $h_k(\mu) = h(\mu,T,\mathcal{P}_{\Lambda_k})$, is an entropy structure for $(X,T,\mathbb{S})$, and the functions $h(\mu) - h(\mu,T,\mathcal{P}_{\Lambda_k})$

[2] This theorem was first proved in [Downarowicz, 2005a] for homeomorphisms, and generalized to continuous maps in [Burguet, 2009]. Here, the general version is derived directly, using zero-dimensional extensions.

are the tails θ_k. So, we have $\mathbf{h}^*(T) = \inf_k \sup_\mu \theta_k(\mu)$. Because the pointwise supremum of a function is the same as that of its upper semicontinuous envelope, we can write as well $\mathbf{h}^*(T) = \inf_k \sup_\mu \widetilde{\theta}_k(\mu)$. Now, we have a decreasing sequence of upper semicontinuous functions on a compact domain, so the elementary Fact A.1.24 allows to switch the supremum and infimum, to obtain $\mathbf{h}^*(T) = \sup_\mu \inf_k \widetilde{\theta}_k(\mu)$, where the right-hand side equals $\sup_\mu u_1^T(\mu)$, as claimed. □

Corollary 8.4.10 *Combining the above theorem with the interpretation of the function $u_1 = D_x$, and with Theorem 8.1.3, we obtain the following interpretation of the topological tail entropy: $\mathbf{h}^*(T)$ equals the global defect of uniformity of the convergence of the entropy structure to the entropy function.*

Corollary 8.4.11 *Since the entropy structure contains a net consisting of upper semicontinuous functions (namely the u.s.d.a.-sequence $h(\mu, T, \mathcal{F}_k)$), we can apply Fact 8.1.4 to obtain that $u_1(\mu)$ estimates from above the defect of upper semicontinuity of the entropy function at the point μ. Using the second part of that fact and Corollary 8.4.10 we recover the well-known fact that $\mathbf{h}^*(T)$ estimates from above the global defect of upper semicontinuity of the entropy function [Misiurewicz, 1976].*

Corollary 8.4.12 *The notion of an asymptotically h-expansive system receives a new interpretation: A system $(X, T, \mathbb{S})$ is asymptotically h-expansive (this condition implies finite topological entropy) if and only if the entropy structure converges uniformly to the entropy function, i.e., when any of the equivalent conditions of Corollary 8.1.20 holds for the entropy structure. We recover the well-known fact that in such case the entropy function is upper semicontinuous.*

We end this section with a somewhat incomplete statement concerning the fourth notion introduced in the preceding section, the variant entropy $\mathbf{h}(\mu, T, \mathcal{U})$ (recall Definition 8.3.12). It is true that the corresponding net (indexed by $\mathcal{U}$) is uniformly equivalent to the other three nets discussed here, hence it provides yet another explicit example of an entropy structure. Because we will not use the full strength of this uniform equivalence, we will just prove that the net converges to the entropy function, giving up the verification of the type of convergence (which is rather technical, [see Downarowicz, 2005a]).

Lemma 8.4.13 *For every invariant measure μ, $\sup_\mathcal{U} \mathbf{h}(\mu, T, \mathcal{U}) = h(\mu, T)$.*

Proof One inequality is easy. Given a cover $\mathcal{U}$ there exists a finite partition $\mathcal{P}$ inscribed in $\mathcal{U}$. By the Shannon–McMillan–Breiman Theorem, for each $\varepsilon > 0$

there is a set F of positive measure which, for every large enough n, is covered by no more than $2^{n(h(\mu,T,\mathcal{P})+\varepsilon)}$ cylinders from $\mathcal{P}^n$. Obviously, F can be covered by the same amount of elements of $\mathcal{U}^n$. This implies

$$\mathbf{h}(T,\mathcal{U}|F) \leq h(\mu,T,\mathcal{P}) + \varepsilon.$$

The left-hand side is larger than or equal to $\mathbf{h}(\mu,T,\mathcal{U})$ (see Definition 8.3.12), the right-hand side is not larger than $h(\mu,T) + \varepsilon$. Since $\mathcal{U}$ and ε are arbitrary, we get $\sup_{\mathcal{U}} \mathbf{h}(\mu,T,\mathcal{U}) \leq h(\mu,T)$.

For the converse inequality we will show that for any finite partition $\mathcal{P}$ with boundaries of measure μ zero, and any $\delta > 0$, there exists a cover $\mathcal{U}$ such that for any set F of positive measure,

$$\mathbf{h}(T,\mathcal{U}|F) > h(\mu,T,\mathcal{P}) - \delta. \tag{8.4.14}$$

Since the entropy $h(\mu,T)$ can be approximated by the entropies $h(\mu,T,\mathcal{P})$ of the above kind, this will imply the hypothesis.

By regularity of μ, there exists an open set U of measure smaller than $\gamma \geq 0$ (how small we will decide in a moment), which contains the boundaries of all the cells of $\mathcal{P}$. We define the open cover as

$$\mathcal{U} = \{A \cup U : A \in \mathcal{P}\}.$$

Given a set F of positive measure, by making it a bit smaller, we can assume that all points in F satisfy the Ergodic Theorem for the set U, that is, for each sufficiently large n, every n-orbit starting from F visits U at most $n\gamma$ times. Also, by Theorem 2.8.9, we know that, for large enough n, F intersects at least $2^{n(h(\mu,T,\mathcal{P})-\gamma)}$ different cylinders from $\mathcal{P}^n$, in other words, at least that many blocks $B \in \mathcal{P}^n$ occur as $x[0,n-1]$ for some $x \in F$. For every $x \in F$, let us put a marker (say, a $*$) above these coordinates $i \in [0,n-1]$ in the block $x[0,n-1]$ for which $T^i x \in U$. Now observe that two points belong to a common set of the cover $\mathcal{U}^n$ if and only if their blocks $x[0,n-1]$ disagree only at places where at least one of them has a star. This implies that each cylinder B (from the family intersecting F) intersects (within F) at most as many sets in $\mathcal{U}^n$ as there are blocks that can be obtained from B by altering not more than $2n\gamma$ of its symbols. The logarithm of this number is approximately $n(H(2\gamma,1-2\gamma)+2\gamma\log l)$, where l is the cardinality of $\mathcal{P}$. This amount estimates the difference between the logarithms of the cardinality of all blocks B intersecting F and of the smallest cardinality of a subfamily of $\mathcal{U}^n$ needed to cover F. The logarithm of the latter is hence at least $n(h(\mu,T,\mathcal{P}) - \gamma - H(2\gamma,1-2\gamma) + 2\gamma\log l)$, which, by an appropriate choice of γ can be made larger than $n(h(\mu,T,\mathcal{P})-\delta)$. Dividing by n and passing with n to infinity we get the inequality (8.4.14). □

Exercises

8.1 Let $\mathfrak{X} = [0,1]$ and set $h_k = \mathbb{1}_{\{x_1,x_2,\dots,x_k\}}$, where (x_i) is a sequence dense in $[0,1]$. What is $\mathrm{E}\mathcal{H}$? What is α_0?

8.2 Prove that $u_{\mathcal{H}}(x) \le \mathsf{ord}(x)u_1(x)$, at every point of finite topological order of accumulation.

8.3 Let $\mathfrak{X} = \{x_1, x_2, \dots\}$ be a countable compact space. Assume that $\mathsf{ord}(x)$ is finite at every point. Fix an arbitrary sequence of positive real numbers $a_0, a_1, \dots$ and let $h(x) = a_{\mathsf{ord}(x)}$. Let $h_k = h\mathbb{1}_{\{x_1,x_2,\dots,x_k\}}$. Verify that the values of $u_\alpha(x)$ depend on $\mathsf{ord}(x)$ as in the table below (we skipped the function u_0 which is zero everywhere and the isolated points, where all u_α equal zero). This is a tedious exercise. You can try it just for $\mathsf{ord}(x) = 2$.

$\mathsf{ord}(x)$	1	2	3	$\dots$	n
$u_1(x)$	a_0	$\max\{a_0,a_1\}$	$\max\{a_0,a_1,a_2\}$	$\dots$	$\max\{a_0,a_1,\dots,a_{n-1}\}$
$u_2(x)$	a_0	a_0+a_1	$\max\{a_0+a_1,a_0+a_1,a_1+a_2\}$	$\dots$	$\max\{a_0+a_1,\dots,a_{n-2}+a_{n-1}\}$
$u_3(x)$	a_0	a_0+a_1	$a_0+a_1+a_2$	$\dots$	$\max\{a_0+a_1+a_2,\dots,a_{n-3}+a_{n-2}+a_{n-1}\}$
$\vdots$	$\vdots$	$\vdots$	$\vdots$	$\vdots$	$\vdots$
$u_n(x)$	a_0	a_0+a_1	$a_0+a_1+a_2$	$\dots$	$a_0+a_1+\dots+a_{n-1}$

8.4 Assume that $\mathcal{H}$ consists of upper semicontinuous functions (on a metric space). Prove that $\mathrm{E}\mathcal{H} = \widetilde{h}$ implies $\alpha_0 \le 1$.

8.5 (David Burguet) Check that the superenvelopes of $\mathcal{H}$ are precisely the fixpoints of the monotone operator $f \mapsto \lim_\kappa \downarrow \widetilde{(f+\theta_\kappa)}$ defined on the complete lattice of nonnegative upper semicontinuous functions on $\mathfrak{X}$ (including the constant infinity function), and use the Tarski–Knaster Theorem [Tarski, 1955] to deduce the existence of the smallest superenvelope.

8.6 Check that uniform equivalence is indeed an equivalence relation among increasing (decreasing) nets of nonnegative functions.

8.7 Check that if (θ_κ) is uniformly equivalent to (θ'_ι) (both defined on a metric space), then $(\widetilde{\theta}_\kappa)$ is uniformly equivalent to $(\widetilde{\theta}'_\iota)$. Does the converse hold?

8.8 Prove that an upper semicontinuous convex function defined on a compact convex set attains its maximum at an extreme point. Hint: Use the Choquet Theorem.

8.9 Deduce, using Lemma 8.2.13, that if $\mathcal{H}$ is defined on a Choquet simplex $\mathcal{K}$ and consists of harmonic functions, then for each ordinal α the function $u_\alpha^{\mathcal{H}}$ is determined by its restriction to $\overline{\mathsf{ex}\mathcal{K}}$, which equals $u_\alpha^{\mathcal{H}|_{\overline{\mathsf{ex}\mathcal{K}}}}$. In particular, the order of accumulation of $\mathcal{H}$ is the same as that of $\mathcal{H}|_{\overline{\mathsf{ex}\mathcal{K}}}$.

8.10 Show that the sequence $H(\mu, \mathcal{U}^n)$ (see Definition 8.3.1) is subadditive.

8.11 Verify that for every invariant measure μ and any cover $\mathcal{U}$, two versions of Romagnoli's entropy (see Definition 8.3.1 and Remark 8.3.2) satisfy

$$h(\mu, T, \mathsf{diam}(\mathcal{U})) \leq h(\mu, T, \mathcal{U}) \leq h(\mu, T, \mathsf{Leb}(\mathcal{U})).$$

8.12 Similarly, check that two versions of the Newhouse local entropy (see Definition 8.3.7 and Remark 8.3.10) satisfy

$$h(T|\mu, \mathsf{diam}(\mathcal{V})) \geq h(T|\mu, \mathcal{V}) \geq h(T|\mu, \mathsf{Leb}(\mathcal{V})).$$

8.13 For $\delta > 0$ let $\mathbf{h}(T, \delta | F, \mathcal{V}) =$

$$\limsup_{n\to\infty} \frac{1}{n} \log \max\{\#E : E \text{ is } (n, \delta)\text{-separated}, E \subset F \cap V, V \in \mathcal{V}^n\}.$$

Show that $\sup_\delta \mathbf{h}(T, \delta | F, \mathcal{V}) = \mathbf{h}(T|F, \mathcal{V})$ (see Definition 8.3.7).

8.14 Verify that the function $\mu \mapsto h(T|\mu, \mathcal{V})$ in Definition 8.3.7(d) is measurable (in fact, it is of Young class LU).

9
Symbolic extensions

Given a topological dynamical system $(X, T, \mathbb{S})$ of finite entropy, we are interested in computing the "amount of information" per unit of time transferred in this system. Suppose we want to compute *verbatim* the vertical data compression, as described in the Introduction (Section 0.3.5): the logarithm of the minimal cardinality of the alphabet allowing the system to be losslessly encoded in real time. Since we work with topological dynamical systems, we want the coding to respect not only the measurable, but also the topological, structure. There is nothing like the Krieger Generator Theorem in topological dynamics. A topological dynamical system of finite entropy (even invertible) need not be conjugate to a subshift over a finite alphabet. Thus, we must first create its *lossless digitalization*, which has only one possible form: a subshift, in which the original system occurs as a factor. In other words, we are looking for a *symbolic extension*. Only then can we try to optimize the alphabet. It turns out that such a vertical data compression is not governed by the topological entropy only by a different (possibly much larger) parameter.

9.1 What are symbolic extensions?

First of all, notice that if a system has infinite topological entropy, it simply does not have any symbolic extensions, as these have finite topological entropy. We do not accept extensions in form of subshifts over infinite alphabets, even countable, like $\mathbb{N}_0 \cup \infty$. It follows from Theorem 7.6.1 and Exercise 7.3 that every system has such an extension which is principal. Alas, these extensions are useless in terms of vertical data compression. So, in all we are about to say, we assume that $(X, T, \mathbb{S})$ has finite topological entropy and a symbolic extension always means a subshift over a finite alphabet.

Next, we want to explain that in either case of $\mathbb{S}$ we are always going to look for a symbolic extension in the form of a *bilateral* subshift, i.e., a subset of $\Lambda^{\mathbb{Z}}$, never of $\Lambda^{\mathbb{N}_0}$. Without any assumptions, the system may happen to possess an invertible factor of positive topological entropy, which eliminates the existence of unilateral symbolic extensions. To be specific, a unilateral subshift clearly has a unilateral measure-theoretic generator for every invariant measure. Any invertible factor must be measurable with respect to $\mathfrak{A}^\infty$, which, in subshifts, equals the Pinsker sigma-algebra. Thus the factor cannot have positive zero.

So, if we want a unified theory of symbolic extensions, we should agree on the definition of a symbolic extension given below:

Definition 9.1.1 Let $(X, T, \mathbb{S})$ be a topological dynamical system. By a *symbolic extension* of $(X, T, \mathbb{S})$ we understand a bilateral subshift $(Y, S, \mathbb{S})$, where $Y \subset \Lambda^{\mathbb{Z}}$ (Λ - finite), together with a topological factor map $\phi : Y \to X$.

The construction of symbolic extensions with minimized topological entropy relies on controlling the measure-theoretic entropies of the measures in the extension. This leads to the following notion:

Definition 9.1.2 Let $(Y, S, \mathbb{S})$ be an extension of $(X, T, \mathbb{S})$ via a map ϕ : $Y \to X$. The *extension entropy function* is defined on $\mathcal{M}_T(X)$ as

$$h^{\phi}_{\mathsf{ext}}(\mu) = \sup\{h(\nu, S) : \nu \in \mathcal{M}_S(Y),\ \phi\nu = \mu\} = (h(\cdot, \nu))^{[\mathcal{M}_T(X)]}(\mu),$$

where h denotes the entropy function on $\mathcal{M}_S(Y)$ and the push-down is with respect to $\phi : \mathcal{M}_S(Y) \to \mathcal{M}_T(X)$.

An alternative formula uses topological fiber entropy:

$$h^{\phi}_{\mathsf{ext}}(\mu) = h(\mu, T) + \mathbf{h}(S|\mu), \tag{9.1.3}$$

which is obtained by the elementary Fact (4.1.6) (and applying the appropriate supremum on both sides) and the Inner Variational Principle (Theorem 6.8.4) with the roles of $(X, T, \mathbb{S})$ and $(Y, S, \mathbb{S})$ reversed.

We now introduce the key notions of this chapter:

Definition 9.1.4 Let $(X, T, \mathbb{S})$ be a topological dynamical system. The *symbolic extension entropy function* is defined on $\mathcal{M}_T(X)$ as

$$h_{\mathsf{sex}}(\mu) = h_{\mathsf{sex}}(\mu, T) = \inf\{h^{\phi}_{\mathsf{ext}}(\mu) : \phi \text{ is a symbolic extension of } (X, T, \mathbb{S})\}.$$

The *topological symbolic extension entropy* of $(X, T, \mathbb{S})$ is

$$\mathbf{h}_{\mathsf{sex}}(T) = \inf\{\mathbf{h}(S) : (Y, S, \mathbb{S}) \text{ is a symbolic extension of } (X, T, \mathbb{S})\}.$$

In both cases, the supremum over the empty set is ∞.

A system has no symbolic extensions if and only if $\mathbf{h}_{\mathsf{sex}}(T) = \infty$ if and only if $h_{\mathsf{sex}}(\mu, T) = \infty$ for all (equivalently, some) invariant measures.

If $\mathbf{h}_{\mathsf{sex}}(T)$ is finite, we can create a symbolic extension whose topological entropy is only a bit larger, for example, smaller than $\log(\lfloor 2^{\mathbf{h}_{\mathsf{sex}}(T)} \rfloor + 1)$. By Theorem 7.2.3, the alphabet in this symbolic extension can be optimized to contain $\lfloor 2^{\mathbf{h}_{\mathsf{sex}}(T)} \rfloor + 1$ elements. On the other hand, there is no way to do better than that. In this manner the symbolic extension entropy controls the vertical data compression in the topological sense.

Historically, in the first attempt to capture the "entropy jump" when passing a to symbolic extension Mike Boyle defined *topological residual entropy* which, in our notation, is the difference

$$\mathbf{h}_{\mathsf{res}}(T) = \mathbf{h}_{\mathsf{sex}}(T) - \mathbf{h}(T), \tag{9.1.5}$$

[see Boyle, 1991; Boyle *et al.*, 2002]. Following this line, one defines the *residual entropy function* on invariant measures as

$$h_{\mathsf{res}}(\mu) = h_{\mathsf{sex}}(\mu) - h(\mu). \tag{9.1.6}$$

Using (9.1.3), the latter notion can be expressed as the infimum of fiber entropies over symbolic extensions. By the Symbolic Extension Entropy Variational Principle (see the next section), the residual entropy is the difference of suprema of two functions, and there are easy examples (see Exercise 9.1) that it need not equal the supremum of the difference functions. Hence there is no such thing as "residual entropy variational principle." As we shall see later, residual entropy function corresponds precisely to the smallest repair function of the net of tails of the entropy structure. To reduce the multitude of notions, we will try to avoid using residual entropy in the sequel.

We remark that searching for symbolic extensions is, in terms of entropy, equivalent to searching for *expansive extensions* (where expansiveness is understood under the action of $\mathbb{Z}$) or asymptotically h-expansive extensions. On one hand, every bilateral subshift is expansive, on the other, every asymptotically h-expansive system has a principal symbolic extension – this will follow from Theorem 9.3.3.

9.2 The Symbolic Extension Entropy Theorem

This section contains the main theorem about symbolic extension entropy [Boyle and Downarowicz, 2004]. It connects this notion with the entropy structure, in which it strongly refers to invariant measures.

Theorem 9.2.1 (Symbolic Extension Entropy Theorem) *Let $(X, T, \mathbb{S})$ denote a topological dynamical system with finite entropy $\mathbf{h}(T) < \infty$ and with entropy structure $\mathcal{H} = (h_k)$. For a function $E\colon \mathcal{M}_T(X) \to [0, +\infty)$ the following conditions are equivalent:*

1. *E is a bounded affine superenvelope of $\mathcal{H}$.*
2. *There exists a symbolic extension $\phi\colon (Y, S, \mathbb{S}) \to (X, T, \mathbb{S})$ with $h_{\mathsf{ext}}^{\phi} = E$.*

In particular, $h_{\mathsf{sex}} \equiv \mathrm{E}\mathcal{H}$, or, equivalently, $h_{\mathsf{res}} \equiv u_{\mathcal{H}}$ (the smallest repair function).

As a direct consequence of the last statement and Theorem 8.2.5, we get

Corollary 9.2.2 (The Symbolic Extension Entropy Variational Principle)

$$\mathbf{h}_{\mathsf{sex}}(T) = \sup\{h_{\mathsf{sex}}(\mu) : \mu \in \mathcal{M}_T(X)\} = \sup\{\mathrm{E}\mathcal{H}(\mu) : \mu \in \mathcal{M}_T(X)\}.$$

Question 9.2.3 We do not know whether and how it is possible to compute $\mathbf{h}_{\mathsf{sex}}(T)$ in purely topological terms, i.e., without using invariant measures (not counting the purely topological, but useless, articulation of the definition).

Proof of Theorem 9.2.1 One direction of the proof is fairly easy and we start with it. Choose the entropy structure of $(X, T, \mathbb{S})$ to be the sequence $\mathcal{H}^{\mathrm{fun}} = (h_k)$, with $h_k = h(\cdot, T, \mathcal{F}_k)$, where $\mathcal{F}_k$ are such families of continuous functions that the associated partitions $\mathcal{A}_{\mathcal{F}_k}$ refine in $X \times [0, 1]$. It follows from Fact 8.3.23 that our particular entropy structure is a u.s.d.a.-sequence. This allows us to identify all superenvelopes E by the criterion that $E - h_k$ is upper semicontinuous for each k (see Lemma 8.1.12).

Let $\phi\colon (Y, S, \mathbb{S}) \to (X, T, \mathbb{S})$ be a symbolic extension and let Λ denote the alphabet of Y. Since the zero-coordinate partition $\mathcal{P}_\Lambda$ generates the sigma-algebra for every invariant measure $\nu \in \mathcal{M}_S(Y)$, on $\mathcal{M}_S(Y)$ the entropy function $h(\cdot, S)$ coincides with $h(\cdot, S, \mathcal{P}_\Lambda) = h(\cdot, S, \mathcal{F}_\Lambda)$, where $\mathcal{F}_\Lambda$ is the family of the (continuous) characteristic functions of the symbols in Λ (i.e., of the cylinders of length 1). Moreover, for every k we can lift the family $\mathcal{F}_k$ from X to Y, and then $h(\cdot, S) = h(\cdot, S, \mathcal{F}_\Lambda \cup \mathcal{F}_k)$. Combining this with Lemma 8.3.23 we get that the difference $h(\cdot, S) - h(\cdot, S, \mathcal{F}_k)$ is upper semicontinuous (and affine) on $\mathcal{M}_S(Y)$. By pushing down these functions to $\mathcal{M}_T(X)$ and applying the elementary Fact A.2.22 (note that the factor map on measures preserves ergodicity) we obtain that for each k the function

$$(h(\cdot, S) - h(\cdot, S, \mathcal{F}_k))^{[\mathcal{M}_T(X)]} = (h(\cdot, S))^{[\mathcal{M}_T(X)]} - h(\cdot, T, \mathcal{F}_k) = h_{\mathsf{ext}}^{\phi} - h_k$$

is upper semicontinuous and affine. So, h_{ext}^{ϕ} is an affine superenvelope of $\mathcal{H}$ and the easy direction is done.

The proof of the other direction is complex. It will be interrupted by numerous auxiliary definitions and lemmas.

Given a bounded affine superenvelope E_{A} of the entropy structure $\mathcal{H}$, we need to construct a symbolic extension $\phi : (Y, S, \mathbb{S}) \to (X, T, \mathbb{S})$ such that $h^{\phi}_{\mathsf{ext}} = E_{\mathsf{A}}$. We will construct a symbolic extension $(Y, S, \mathbb{S})$ of a principal zero-dimensional extension $(X', T', \mathbb{S})$ of $(X, T, \mathbb{S})$. As we know, the entropy structure $\mathcal{H}'$ of the principal extension is obtained by lifting the original entropy structure $\mathcal{H}$, and the lift of the superenvelope E_{A} is a superenvelope of $\mathcal{H}'$. If we match the extension entropy function of the extension ϕ' from Y to X' with the (lifted) superenvelope E_{A}, then the extension entropy function of the composition $\phi = \pi \circ \phi'$ (here π denotes the factor map from X' onto X) will match E_{A} on $\mathcal{M}_T(X)$.

Using Theorem 7.6.1 we can replace our system $(X, T, \mathbb{S})$ by its zero-dimensional principal extension whose elements are bilateral symbolic marked arrays (see Remark 7.6.10). In other words, we can assume that our system $(X, T, \mathbb{S})$ is already a marked symbolic array system (see Definition A.3.2): $(X, T, \mathbb{S})$ is represented as the shift transformation on a closed shift-invariant family of arrays

$$x = (x_{k,n})_{k \in \mathbb{N}, n \in \mathbb{Z}}.$$

The alphabet in row number k is Δ_k and each element has in this row p_k-periodically repeated p_k-markers visualized as vertical bars. We are free to choose the base (p_k) of the odometer according to our needs, and we will specify it in Stage 2. We let

$$x_k = (x_{s,n})_{s \le k, n \in \mathbb{Z}}$$

denote the projection of x onto the first k rows. In this setting, $(X_k, T_k, \mathbb{S})$ is the (bilateral) symbolic system obtained from $(X, T, \mathbb{S})$ by the projection π_k onto the first k rows. The alphabet of X_k is the product $\Lambda_k = \Delta_1 \times \cdots \times \Delta_k$.

The zero-coordinate partitions $\mathcal{P}_{\Lambda_k}$ are in fact clopen covers and together, under the action of T, they generate the sigma-algebra for every invariant measure supported by X. So, the sequence of functions $\mu \mapsto h_k(\mu, T, \mathcal{P}_{\Lambda_k})$ is an entropy structure for $(X, T, \mathbb{Z})$. Moreover, it is a u.s.d.a.-sequence and E_{A} is its affine superenvelope, so we know that $E_{\mathsf{A}} - h_k$ is affine and upper semicontinuous for every k. This fact will be extensively exploited in the proof.

For given k, the blocks B of length p_k over the alphabet Δ_k, ending with a marker, will be called k-blocks and their collection will be denoted by $\mathcal{B}_k$. The k-rectangles (see Definition A.3.3) occur in rows 1 through k between pairs of consecutive markers in row k. A 1-rectangle is synonymous with 1-block, while each $(k+1)$-rectangle is a concatenation $D = R^{(1)} R^{(2)} \ldots R^{(q_k)}$ of

$q_k = p_{k+1}/p_k$ k-rectangles with a $(k+1)$-block added as a new row, which we will write as

$$R = \begin{pmatrix} R^{(1)} R^{(2)} \dots R^{(q_k)} \\ B \end{pmatrix} = \begin{pmatrix} D \\ B \end{pmatrix}.$$

Note that a k-rectangle is in fact a block over Λ_k. We let $\mathcal{R}_k$ denote the collection of all possible (formal) k-rectangles over the alphabet Λ_k, while $\mathcal{R}_k(X)$ stands for the collection of all k-rectangles that actually occur in X.

The proof is organized in three major *stages*. In stage 1, given an affine superenvelope E_{A}, we construct a certain integer-valued function on k-rectangles called the *oracle* (see below for the definition). In stage 2, given an oracle, we construct the symbolic extension. Stage 3 is the verification that the extension entropy function indeed matches E_{A}.

Definition 9.2.4 An *oracle* is a sequence of functions $\mathcal{O}_k \colon \mathcal{R}_k \mapsto \mathbb{N}$ such that $\mathcal{O}_k(R) = 0$ if and only if $R \notin \mathcal{R}_k(X)$, and

$$\mathcal{O}_k(R^{(1)})\mathcal{O}_k(R^{(2)}) \cdots \mathcal{O}_k(R^{(q_k)}) \geq \sum_{B \in \mathcal{B}_{k+1}} \mathcal{O}_{k+1} \begin{pmatrix} R^{(1)} R^{(2)} \dots R^{(q_k)} \\ B \end{pmatrix}.$$

STAGE 1. *Constructing an oracle from an affine superenvelope.*

From an affine superenvelope of the particular entropy structure $\mathcal{H} = (h_k)$, where $h_k(\mu) = h(\mu, T, \mathcal{P}_{\Lambda_k})$, we are going to derive an oracle. To simplify the writing we introduce some shortcut notation:

$$\mathcal{K} = \mathcal{M}_T(X), \;\; \mathcal{K}_k = \mathcal{M}_{T_k}(X_k), \;\; \mu_k = \pi_k \mu.$$

Notice that the projections π_k (understood as maps on invariant measures) send $\mathcal{K}$ onto $\mathcal{K}_k$, and since each measure μ on X depends on its values on cylinders corresponding to rectangular blocks of finite size, each measure μ is determined by the sequence of its images μ_k. This is to say that the fibers $\pi_k^{-1}(\pi_k \mu)$ decrease to the one-point intersection $\{\mu\}$.

In this context, we make the following general observation. Although the lemma is applicable in a much wider context, for easier reference, we denote the space by $\mathcal{K}$ and its elements by μ.

Lemma 9.2.5 *Let $\mathcal{K}$ and $\mathcal{K}_k$ ($k \in \mathbb{N}$) be compact metric spaces and assume that $\pi_k : \mathcal{K} \to \mathcal{K}_k$ are continuous surjections such that for every $\mu \in \mathcal{K}$ the fibers $\pi_k^{-1}(\pi_k \mu)$ decrease to the one-point intersection $\{\mu\}$. Suppose that*

$f : \mathcal{K} \to [-\infty, \infty)$ is an upper semicontinuous function. Set $f_k = f^{[\mathcal{K}_k]}$ regarded (by lifting) as a function on $\mathcal{K}$ (formally, $f_k(\mu) = \sup\{f(\nu) : \pi_k \nu = \pi_k \mu\}$). Then $f = \lim_k \downarrow f_k$.

Proof It is clear from the definition of the push-down, that the functions f_k are above f and decrease. Since compact sets decreasing to a one-point set must shrink in diameter to zero, it follows from the definition of upper semicontinuity (Definition A.1.7) that, at each μ, $f(\mu) \geq \lim_k f_k(\mu)$. □

We return to our specific context of the entropy structure $\mathcal{H}$ and its affine superenvelope E_{A}. Note that for each k we have $h_k(\mu) = h(\mu_k)$, so h_k is in fact defined on $\mathcal{K}_k$ from where it is lifted to $\mathcal{K}$. Moreover, for each k, the map π_k (which originates as the projection onto the first k rows of the arrays) can be applied not only to $\mathcal{K}$ but also to $\mathcal{K}_j$ for each $j > k$. In this manner, by the appropriate lifting, h_k is also defined on $\mathcal{K}_j$. These remarks will help us approximate the upper semicontinuous function $E_{\mathsf{A}} - h$ by a decreasing sequence of rather carefully chosen functions:

Lemma 9.2.6 *There exists a sequence g_k of affine continuous functions, each defined on $\mathcal{K}_k$, such that regarded as functions on $\mathcal{K}$ they satisfy the following conditions:*

$$\lim_k \downarrow g_k = E_{\mathsf{A}} - h, \tag{9.2.7}$$

$$\forall_k \; g_k > E_{\mathsf{A}} - h_k, \tag{9.2.8}$$

$$\forall_k \; g_k - g_{k+1} > h_{k+1} - h_k. \tag{9.2.9}$$

Proof Recall that E_{A} and all $E_{\mathsf{A}} - h_k$ are upper semicontinuous. By Lemma 9.2.5, the functions

$$E_k = E_{\mathsf{A}}^{[\mathcal{K}_k]}$$

decrease to E_{A} on $\mathcal{K}$. Thus $E_k - h_k$ decrease to $E_{\mathsf{A}} - h$. Since h_k is lifted from $\mathcal{K}_k$, the difference $E_k - h_k$ can be written as $(E_{\mathsf{A}} - h_k)^{[\mathcal{K}_k]}$, which, by the properties of pushing-down and lifting (Fact A.1.26 and Fact A.2.22) is upper semicontinuous and affine on $\mathcal{K}$.

By the elementary Fact A.1.11, we can represent each function $E_k - h_k$ as a decreasing limit (in i) of some continuous functions $f_{k,i}$ defined on $\mathcal{K}_k$. By adding small constants, we can assume that the sequence decreases strictly. Since, for $i \leq k$, the function $f_{i,k}$ (the exchange of indices is not a mistake) can be lifted to $\mathcal{K}_k$, on $\mathcal{K}_k$ we can define $f_k = \inf\{f_{1,k}, f_{2,k}, \ldots, f_{k,k}\}$. Lifting each f_k to $\mathcal{K}$, we obtain a strictly decreasing sequence of continuous functions. Since, for $i \leq k$, $f_{i,k} > E_i - h_i \geq E_k - h_k$, we have $f_k > E_k - h_k$. On the other hand, for each i, $\lim_k f_k \le \lim_k f_{i,k} = E_i - h_i$, hence $\lim_k f_k \le \lim_k E_k - h_k$, so we have equality, i.e., $\lim_k f_k = E_{\mathsf{A}} - h$. The functions f_k nearly satisfy our requirements, but we still need to make them affine and take care of (9.2.9). We will do that inductively.

By the Combined Separation Theorem A.2.29 (on $\mathcal{K}_1$), we can replace f_1 by a smaller continuous and affine function g_1 still strictly larger than $E_1 - h_1$. The condition (9.2.9) refers to g_2, so we will take care of it in the following steps. Once we have defined g_k (on $\mathcal{K}_k$), we lift it to $\mathcal{K}_{k+1}$, where $g_k + h_k \ge E_k \ge E_{k+1}$, and thus $g_k - (h_{k+1} - h_k) > E_{k+1} - h_{k+1}$. Also $f_{k+1} > E_{k+1} - h_{k+1}$, so

$$\min\{g_k - (h_{k+1} - h_k), f_{k+1}\} > E_{k+1} - h_{k+1}.$$

On the left-hand side we have a lower semicontinuous function and on the right an affine upper semicontinuous function. Using Theorem A.2.29 one more time, we can find an affine continuous function g_{k+1}, defined on $\mathcal{K}_{k+1}$, which satisfies $g_{k+1} > E_{k+1} - h_{k+1}$ (and thus the inductive assumption (9.2.8)) and $g_{k+1} < g_k - (h_{k+1} - h_k)$ (this is (9.2.9); in particular this condition implies that the sequence g_k strictly decreases). Since also $g_{k+1} \le f_{k+1}$, the functions g_k (lifted to $\mathcal{K}$) converge to $E_{\mathsf{A}} - h$, as required in (9.2.7). □

Since the left-hand side of (9.2.9) is continuous and the right-hand side is upper semicontinuous, there exists a positive ε_k such that

$$g_k - g_{k+1} - 3\varepsilon_k > h_{k+1} - h_k.$$

We can arrange the numbers ε_k so that they decrease to zero.

Now, we will choose the base (p_k) of the odometer (i.e., a sequence satisfying $p_{k+1} = q_k p_k$) growing rapidly enough to satisfy several conditions. First, note that the functions $\mu \mapsto H_n(\mu_{k+1}|\mu_k) = \frac{1}{n} H(\mu, \mathcal{P}^n_{\Lambda_{k+1}}|\mathcal{P}^n_{\Lambda_k})$ are continuous and decrease (in n) to $h_{k+1} - h_k$ (this is Fact 2.3.1). By the elementary Fact A.1.14 this sequence eventually (for indices above some n_k) lies strictly below $g_k - g_{k+1} - 3\varepsilon_k$. We can choose p_k larger than n_k, and then

$$g_k(\mu) - g_{k+1}(\mu) - 3\varepsilon_k > H_{p_k}(\mu_{k+1}|\mu_k), \tag{9.2.10}$$

for all k and $\mu \in \mathcal{K}$ (in fact all terms depend on the projection of μ denoted $\mu_{k+1} \in \mathcal{K}_{k+1}$). We suspend the construction for a while, as the sequel requires a technicality. (We are still imposing conditions upon the base (p_k).)

The next lemma supplements our calculations of Section 2.8. We will prove a direct consequence of the conditional data compression Lemma 2.8.7. For this, we return to the setup and use the notation introduced in Definition 2.8.6 and above it.

Lemma 9.2.11 *Let $\Lambda = \Lambda_1 \times \Lambda_2$ be a product alphabet. Then for every $n \in \mathbb{N}$ and $\varepsilon > 0$ there exists an $m_{(n,\varepsilon)} \in \mathbb{N}$ such that for every $m \geq m_{(n,\varepsilon)}$ the following holds*

$$\sup_{D \in \Lambda_1^m} \sum_{B \in \Lambda^m : B_1 = D} 2^{-mH_n(B|B_1)} \leq 2^{m\varepsilon},$$

where $B_1 \in \Lambda_1^m$ denotes the block appearing in the first row of B.

Proof Fix an $m \in \mathbb{N}$ and $D \in \Lambda_1^m$. Consider the family

$$\mathcal{B}_D = \{B \in \Lambda^m : B_1 = D\}$$

as a probability space with the uniform probability $\mathsf{Prob}(B) = 1/\#\mathcal{B}_D = p$ (note that $p \geq l^{-m}$, where l is the cardinality of Λ) with defined on it random variable $\mathcal{X}(B) = H_n(B|B_1)$. The expression to estimate equals $\frac{1}{p}\mathsf{E}(2^{-m\mathcal{X}})$ (where E denotes the expected value). By Lemma 2.8.7, we have for every $t \geq 0$ that $\mathbf{C}_{\text{cond}}[n, m, t] \leq 2^{m(t+\varepsilon)}$, i.e., that

$$\#\{B \in \mathcal{B}_D \colon H_n(B|B_1) \leq t\} \leq 2^{m(t+\varepsilon)}$$

for m sufficiently large and every D of length m. Clearly, the above cardinality does not exceed $\#\mathcal{B}_D$, and it is zero for $t < 0$. Eventually, we can write

$$p \cdot \#\{B \in \mathcal{B}_D \colon H_n(B|B_1) \leq t\} \leq \begin{cases} 0\,, & t < 0; \\ \min(p2^{m(t+\varepsilon)}, 1)\,, & t \geq 0. \end{cases}$$

This can be interpreted as saying that the distribution function of $\mathcal{X}$, given by $F_{\mathcal{X}}(t) = \mathsf{Prob}\{\mathcal{X} \leq t\}$, is below the distribution function $F_{\mathcal{Y}}$ of $\mathcal{Y}$, where $\mathcal{Y}$ is a random variable with density

$$f(t) = pm \ln 2 \cdot 2^{m(t+\varepsilon)}$$

on the interval $[0, \frac{-\log p}{m} - \varepsilon]$, and with an atom of mass $p2^{m\varepsilon}$ at zero. This means that the mass of probability of the distribution of $\mathcal{X}$ is moved to the right compared to that of $\mathcal{Y}$. Because the function 2^{-mx} is decreasing, the expected value $\mathsf{E}(2^{-m\mathcal{X}})$ is not larger than $\mathsf{E}(2^{-m\mathcal{Y}})$, while

$$\begin{aligned} \mathsf{E}(2^{-m\mathcal{Y}}) &= pm \ln 2 \cdot 2^{m\varepsilon}(\tfrac{-\log p}{m} - \varepsilon) + 1 \cdot p2^{m\varepsilon} \\ &\leq pm \ln 2 \cdot 2^{m\varepsilon}(\log l - \varepsilon + \tfrac{1}{m \ln 2}). \end{aligned}$$

For large m, by slightly enlarging ε in the exponent, we can ignore all terms which depend subexponentially on m. What remains is just $p2^{m\varepsilon}$. Dividing by p we obtain the required estimate $\frac{1}{p}\mathsf{E}(2^{-m\mathcal{X}}) \leq 2^{m\varepsilon}$. □

We get back to the construction, where we are still making assumptions on the speed of growth of the base (p_k) of the odometer. Here are the next (and not last) two requirements:

$$p_{k+1} \geq m_{(p_k,\varepsilon_k)} \tag{9.2.12}$$

as defined in Lemma 9.2.11 applied to the product $\Lambda_k \times \Delta_{k+1}$ $(= \Lambda_{k+1})$,

$$2^{p_{k+1}(b+\varepsilon_k)} \geq \lceil 2^{p_{k+1}b} \rceil \text{ for every } b \geq 0, \tag{9.2.13}$$

where $\lceil\cdot\rceil$ denotes the integer ceiling. (For the last condition it suffices that $2^{p_{k+1}\varepsilon_k} \geq 2$.)

Observe that $\mathcal{K}_{k+1}$ is a subset of the set of all shift-invariant probability measures on the symbolic space over the alphabet Λ_{k+1}. The function $\mu \mapsto H_{p_k}(\mu_{k+1}|\mu_k)$ is well defined (as entropy) and continuous on this larger set. Using the Hahn–Banach Theorem [see e.g. Rudin, 1991], for each k we find a continuous and affine prolongation of g_k (denoted by the same letter) to the set of all invariant measures on the symbolic space over the alphabet Λ_k, which, by lifting, is defined also on invariant measures on the symbolic space over the alphabet Λ_{k+1}, where we also have a prolongation of g_{k+1}. Here the inequality (9.2.10) holds on some open neighborhood U_{k+1} of $\mathcal{K}_{k+1}$.

Recall that the periodic measure $\mu_{(R)}$ carried by the orbit of ...RRR..., where R is any sufficiently long block appearing in X_{k+1} (i.e., a rectangle), is in U_{k+1} (this follows easily from Fact 6.6.1 and Fact 7.3.2 item 3). Replacing a rectangle by its periodic measure, we can apply the functions H_{p_k}, g_k and g_{k+1} directly to such rectangles. We may specify two more requirements on the speed of growth of the numbers p_k. First, take the length p_{k+1} of a $(k+1)$-rectangle so large that

$$\text{if } R \text{ is a } (k+1)\text{-rectangle, then } \mu_{(R)} \text{ is in } U_{k+1} \tag{9.2.14}$$

(and then (9.2.10) applies to $\mu_{(R)}$). Finally, by appeal to g_k being uniformly continuous (their domain is compact) and affine, we take p_k large enough to imply that the value of g_k on any concatenation $R^{(1)}R^{(2)}\dots R^{(q)}$ of k-rectangles is close to the corresponding convex combination of values (see Fact 7.3.2 item 1):

$$\left| g_k(R^{(1)}R^{(2)}\dots R^{(q)}) - \tfrac{1}{q}\sum_{i=1}^{q} g_k(R^{(i)}) \right| < \varepsilon_k \,. \tag{9.2.15}$$

We can now define the desired oracle. For a k-rectangle R appearing in X we let

$$\mathcal{O}_k(R) = \lceil 2^{p_k g_k(R)} \rceil \,. \tag{9.2.16}$$

We need to verify the condition in Definition 9.2.4 of an oracle. Let $D = R^{(1)}R^{(2)}\dots R^{(q_k)}$ be a concatenation of k-rectangles appearing in X. If D does not occur in X, then the right side of the condition in Definition 9.2.4 is zero and the inequality holds trivially; so suppose D occurs in X. Then, using (9.2.13), (9.2.10) (which we can, by (9.2.14)), and (9.2.15), for the consecutive inequalities, we can write

$$\sum_{B\in\mathcal{B}_{k+1}} \mathcal{O}_{k+1}\binom{D}{B} = \sum_{B\in\mathcal{B}_{k+1}} \left\lceil 2^{p_{k+1}g_{k+1}\binom{D}{B}}\right\rceil \leq$$
$$\sum_{B\in\mathcal{B}_{k+1}} 2^{p_{k+1}\left(g_{k+1}\binom{D}{B}+\varepsilon_k\right)} \leq \sum_{B\in\mathcal{B}_{k+1}} 2^{p_{k+1}\left(g_k(D)-H_{p_k}\left(\binom{D}{B}|D\right)-2\varepsilon_k\right)}$$
$$= 2^{p_{k+1}(g_k(D)-2\varepsilon_k)} \sum_{B\in\mathcal{B}_{k+1}} 2^{-p_{k+1}H_{p_k}\left(\binom{D}{B}|D\right)} \leq$$
$$2^{p_k\sum_{i=1}^{q_k} g_k(R^{(i)})}\cdot 2^{-p_{k+1}\varepsilon_k} \sum_{B\in\mathcal{B}_{k+1}} 2^{-p_{k+1}H_{p_k}\left(\binom{D}{B}|D\right)}.$$

The first term in the last expression above is not larger than

$$\mathcal{O}_k(R^{(1)})\mathcal{O}_k(R^{(2)})\cdots\mathcal{O}_k(R^{(q_k)}).$$

By (9.2.12) and Lemma 9.2.11, the remaining part of that expression is not larger than $2^{-p_{k+1}\varepsilon_k}\cdot 2^{p_{k+1}\varepsilon_k} = 1$. This verifies the condition in Definition 9.2.4.

STAGE 2. *Constructing a symbolic extension from an oracle.*

We now describe the construction of a symbolic extension $(Y,S,\mathbb{S})$ of $(X,T,\mathbb{S})$ given an oracle. Initially, the extension Y will be not exactly symbolic, as it will have the form of a joining of a symbolic system over some finite alphabet Λ with the odometer to the base (p_k) represented by the symbolic marked array system with "empty" rows, i.e., containing only the markers and empty cells (like on the Figure A.2 with zeros representing the empty cells and ones being the markers). Each element y of Y will be pictured as having the symbolic sequence in row number 0, and the element of the odometer in rows numbered $1,2,\dots$. In Y, by *k-blocks* we will understand the blocks of length p_k occurring in the zero row between the coordinates where the markers occur in the row number k. Later the odometer will be replaced by another symbolic system of entropy zero.

The factor map from Y to X will preserve the odometer, i.e., the configuration of the markers in all rows numbered $1,2,\dots$ The Λ-contents of the zero row will be constructed in the inductive steps below. Roughly speaking, we will gradually restrict the freedom as to what can appear in the preimage y of a point x "above" (i.e., at the same coordinates as) a k-rectangle in x. In each step we will construct an extension $(Y_k, S_k, \mathbb{S})$ of a system slightly larger than X_k, namely of the space $\overline{X}_k$ of all bi-infinite concatenations of all k-rectangles occurring in X_k. The intersection of the systems Y_k will turn out to be an extension of X.

Step 1.
Note that $\mathcal{R}_1 = \mathcal{B}_1$. Let $l_Y \in \mathbb{N}$ be such that $l_Y^{p_1} \geq \sum_{R \in \mathcal{R}_1} \mathcal{O}_1(R)$, and let Λ be an alphabet with cardinality l_Y.

For each $R \in \mathcal{R}_1$ pick a family $\mathcal{E}_1(R)$ of $\mathcal{O}_1(R)$ different blocks C of length p_1 over Λ (for R not appearing in X this family is empty). The cardinality of Λ allows it to be done so that for different blocks R the families $\mathcal{E}_1(R)$ are disjoint. We let $\mathcal{E}_1$ be the union of all families $\mathcal{E}_1(R)$. The set Y_1 is defined as the collection of all marked arrays such that all 1-blocks in the zero row belong to $\mathcal{E}_1$ (and the other rows contain just the odometer). The map ϕ_1 is first defined on the 1-blocks and it sends each 1-block B to the unique 1-rectangle R such that $B \in \mathcal{E}_1(R)$, then ϕ_1 is defined on Y_1 as the block code by replacing the 1-blocks by their image 1-rectangles. We skip the standard verification that so defined Y_1 is closed, shift invariant, and that ϕ_1 is continuous onto $\overline{X}_1$. We emphasize that, while constructing a preimage of some x_1, above each 1-block R we are free to choose any out of $\mathcal{O}_1(R)$ elements of $\mathcal{E}_1(R)$, independently of what is chosen to the left and right.

Step $k+1$.
Suppose the task has been completed for some k. That means, for all k-rectangles R we have selected disjoint collections $\mathcal{E}_k(R)$, each of exactly $\mathcal{O}_k(R)$ blocks of length p_k over Λ. The union of $\mathcal{E}_k(R)$ over all k-rectangles R is denoted by $\mathcal{E}_k$. The set Y_k consists of all such marked arrays y (with the zero row containing the symbols from Λ and other rows containing just the markers) that all the k-blocks in the zero row of that array belong to $\mathcal{E}_k$. The map ϕ_k acts on Y_k onto $\overline{X}_k$ as the code replacing each k-block B by the unique k-rectangle R such that $B \in \mathcal{E}_k(R)$. We need to go one step further.

Consider a concatenation $D = R^{(1)} R^{(2)} \ldots R^{(q_k)}$ of k-rectangles. Assume that this concatenation occurs in X_k between the $(k+1)$-markers. Above it, in Y_k, there occur exactly as many as

$$\mathcal{O}_k(R^{(1)}) \mathcal{O}_k(R^{(2)}) \cdots \mathcal{O}_k(R^{(q_k)})$$

blocks, namely all possible concatenations of blocks: first from $\mathcal{E}_k(R^{(1)})$, next from $\mathcal{E}_k(R^{(2)})$, and so on. By the inequality in Definition 9.2.4 of an oracle, it is thus possible, for each $(k+1)$-rectangle $R = \binom{D}{B}$ occurring in X_{k+1}, to select from these concatenations a family $\mathcal{E}_{k+1}(R)$ of cardinality $\mathcal{O}_{k+1}(R)$ in such a way that the families corresponding to different blocks B (with the same contents of the first k rows) are disjoint. Note that if two $(k{+}1)$-rectangles differ already in the first k rows, then their families are chosen from disjoint collections of concatenations, so obviously they are disjoint. We let $\mathcal{E}_{k+1}$ be the union of the families $\mathcal{E}_{k+1}(R)$ over all $(k{+}1)$-rectangles R. As before, we let Y_{k+1} be the collection of all arrays whose all $(k{+}1)$-blocks in the zero row belong to $\mathcal{E}_{k+1}$. The map ϕ_{k+1} acts on Y_{k+1} onto $\overline{X}_{k+1}$ as the code replacing each $(k{+}1)$-block B by the unique $(k{+}1)$-rectangle R such that $B \in \mathcal{E}_{k+1}(R)$. The task has been now completed for $k{+}1$.

Notice the obvious fact that the sets Y_k decrease with k and the maps ϕ_k are consistent: for $y \in Y_{k+1}$, $\phi_k(y) = \pi_k \circ \phi_{k+1}(y)$ (here π_k denotes the projection to the first k rows).

When this induction is completed, we define Y as the intersection $\bigcap_k Y_k$, and the map ϕ by the rule, that $\phi(y)$ is the marked array x for which $\pi_k(x) = \phi_k(y)$, for every k. This is a factor map from $(Y, S, \mathbb{S})$ onto a system $(\overline{X}, \overline{T}, \mathbb{S})$ containing $(X, T, \mathbb{S})$ as a subsystem. (To be precise, $\overline{X}$ may contain some arrays whose left "half" is taken from one element of X and the right "half" from another.) Of course, this presents no problem, as we can always restrict ϕ to the preimage of X. In the calculations of the extension entropy, in the end, we care only about the measures μ supported by X, and any lift of such a μ to Y is automatically supported by this preimage.

The last step in order to obtain a genuine symbolic extension of $(\overline{X}, \overline{T}, \mathbb{S})$ (and of $(X, T, \mathbb{S})$) is replacing the odometer in the representation of $(Y, S, \mathbb{S})$ by its symbolic extension of entropy zero. As in the proof of Theorem 6.9.9, we can use a regular Toeplitz system (see Example A.3.4). This completes the construction of a symbolic extension of $(X, T, \mathbb{S})$ using a given oracle.

STAGE 3. *Entropy calculation*

It remains to verify that for $\mu \in \mathcal{M}_T(X)$ we have the equality $h^{\phi}_{\mathsf{ext}}(\mu) = E_{\mathsf{A}}(\mu)$, or, equivalently (see (9.1.3)), that $\mathbf{h}(S|\mu) = (E_{\mathsf{A}} - h)(\mu)$, where $\mathbf{h}(S|\mu)$ is the topological fiber entropy with respect to the extension ϕ. We begin with a lemma.

Lemma 9.2.17 *With the notation as used throughout the above construction, let $\mu \in \mathcal{M}_{\overline{T}}(\overline{X})$ and let μ_k denote $\pi_k\mu$. Then*

$$\mathbf{h}(S|\mu) = \lim_k \downarrow \mathbf{h}(S_k|\mu_k)$$

(recall that S_k is the shift map on Y_k).

Proof The proof reduces to a simple exchange of suprema and infima for upper semicontinuous functions. Fix $\mu \in \mathcal{M}_{\overline{T}}(\overline{X})$ (this fixes all the measures μ_k). Let M_k be the set of all invariant measures ν supported by Y_k and mapped by ϕ_k to μ_k. The function $\nu \mapsto h(\nu|\mu_k)$ is upper semicontinuous on M_k. If $\nu \in M_{k+1}$, then also $\nu \in M_k$ (because $\phi_k = \pi_k \circ \phi_{k+1}$) and $h(\nu|\mu_{k+1}) \leq h(\nu|\mu_k)$ (because $h(\nu|\mu_k) = h(\nu) - h(\mu_k)$, and analogously for $k+1$, while μ_k is a factor of μ_{k+1}). The sets M_k are compact and their (decreasing) intersection is precisely the set M of these invariant measures supported by Y which map via ϕ to μ. Moroever, for $\nu \in M$ we have $h(\nu|\mu) = \lim\limits_k \downarrow h(\nu|\mu_k)$ (because $h(\mu) = \lim\limits_k \uparrow h(\mu_k)$).

Summarizing, we have a decreasing sequence of compact sets M_k, and on each of them a nonnegative upper semicontinuous function, say f_k, so that $f_{k+1} \leq f_k$ on M_{k+1}. On the intersection M of M_k all these functions are well defined and their (decreasing) limit is some upper semicontinuous function f. What we want, is that the suprema over M_k of f_k converge to the supremum over M of f. But this follows immediately from the exchange of suprema and infima Fact A.1.24. All we need, is to unify the domains of the functions f_k. This can easily be done by prolonging each f_k and the limit function f to the largest domain M_1 by simply assigning the value 0 where the functions were not defined. These functions remain upper semicontinuous and now the sequence f_k decreases to f on M_1, so Fact A.1.24 applies. Of course, the added value zero does not affect the pointwise suprema of the functions, so we obtain the desired equality. □

In view of the preceding lemma, all we need is to compute $\mathbf{h}(S_k|\mu_k)$ (for the factor map $\phi_k : Y_k \to \overline{X}_k$) for any invariant measure μ_k supported by $\overline{X}_k$. Following the Definition 6.7.1 we must begin with $\mathbf{H}(\mathcal{P}_\Lambda^n|x_k)$, where $x_k \in \overline{X}_k$. By definition, this is the logarithm of the cardinality of the family of blocks $y[0, n-1]$ as y ranges over the fiber $\phi_k^{-1}(x_k)$. If n is much larger than p_k, then the block $x_k[0, n-1]$ can be broken as $A_1 R^{(1)} R^{(2)} \ldots R^{(q)} A_2$, where each $R^{(i)}$ is a k-rectangle, while A_1, A_2 are some incomplete k-rectangles, i.e., blocks over Λ_k each not exceeding p_k in length. By the construction of ϕ_k, in the preimages of x_k, above each $R^{(i)}$ there are admitted precisely $\mathcal{O}_k(R^{(i)})$ different blocks, independently from what occurs above the neighboring blocks, so together this amounts to $\mathcal{O}_k(R^{(1)})\mathcal{O}_k(R^{(2)}) \cdots \mathcal{O}_k(R^{(q)})$ possibilities above $R^{(1)} R^{(2)} \ldots R^{(q)}$. The number of blocks above A_1 and A_2 does not exceed $l_Y^{2p_k}$. Taking the logarithm and dividing by n, we get

$$\frac{1}{n}\mathbf{H}(\mathcal{P}_\Lambda^n|x_k) \approx \frac{1}{n}\sum_{i=1}^{q} \log(\mathcal{O}_k(R^{(i)})),$$

where the error is at most $2\frac{p_k}{n}\log l_Y$ and decreases to zero with n. By the formula (9.2.16) defining our particular oracle and by (9.2.13) we can write $\log(\mathcal{O}_k(R^{(i)})) = p_k(g_k(R^{(i)})+\delta_i)$, where each δ_i is at most ε_k. Consequently,

$$\frac{1}{n}\mathbf{H}(\mathcal{P}_\Lambda^n|x_k) \approx \frac{qp_k}{n}\cdot\frac{1}{q}\sum_{i=1}^{q}(g_k(R^{(i)})+\delta_i).$$

Since $qp_k \approx n$ and the function g_k (hence the above average) is bounded, we can ignore qp_k/n at a cost of another error vanishing as n grows. We have

$$\frac{1}{n}\mathbf{H}(\mathcal{P}_\Lambda^n|x_k) = \Big(\frac{1}{q}\sum_{i=1}^{q} g_k(R^{(i)})\Big)+\delta,$$

where δ is the average of δ_i (plus the last inaccuracy) and does not exceed $2\varepsilon_k$. Now we apply (9.2.15), and get

$$\tfrac{1}{n}\mathbf{H}(\mathcal{P}_\Lambda^n|x_k) = g_k(R_{(1)}R_{(2)}\dots R_{(q)})+\delta$$

(now the term δ is at most $3\varepsilon_k$). Recall that g_k applies to measures, and $R_{(1)}R_{(2)}\dots R_{(q)}$ stands for the periodic measure $\mu_{(R_{(1)}R_{(2)}\dots R_{(q)})}$. By the elementary Fact 7.3.2 (items (1) and (2)) this measure is, for large n, very close to $\frac{1}{n}\sum_{i=0}^{n-1}\boldsymbol{\delta}_{T^ix}$, so close that under the uniformly continuous function g_k (which we have prolonged to all invariant measures on the symbolic space $\Lambda_k^{\mathbb{S}}$) the inaccuracy is, once again, negligible:

$$\frac{1}{n}\mathbf{H}(\mathcal{P}_\Lambda^n|x_k) = g_k\Big(\frac{1}{n}\sum_{i=0}^{n-1}\boldsymbol{\delta}_{T^ix}\Big)+\delta.$$

Now it is time to integrate with respect to μ_k:

$$\frac{1}{n}\mathbf{H}(\mathcal{P}_\Lambda^n|\mu_k) = \frac{1}{n}\int\mathbf{H}(\mathcal{P}_\Lambda^n|x_k)\,d\mu_k = \int g_k\Big(\frac{1}{n}\sum_{i=0}^{n-1}\boldsymbol{\delta}_{T^ix}\Big)\,d\mu_k+\delta.$$

(this time δ is the average error term, still not exceeding $3\varepsilon_k$). The function g_k is affine and continuous, hence harmonic, so we can pull it outside the integral. Inside, we are left with the average of μ_k and its $n-1$ images by T, which, by invariance, is μ_k again. We conclude that

$$\tfrac{1}{n}\mathbf{H}(\mathcal{P}_\Lambda^n|\mu_k) = g_k(\mu_k)+\delta,$$

and, passing with n to infinity, that

$$\mathbf{h}(\mathcal{P}_\Lambda|\mu_k) = g_k(\mu_k)+\delta. \tag{9.2.18}$$

The partition (and cover) $\mathcal{P}_\Lambda$ generates only the zero row in Y_k. In order to compute $\mathbf{h}(S_k|\mu_k)$ we should refine $\mathcal{P}_\Lambda$ with a sequence of clopen covers lifted from the other rows, i.e., lifted from the odometer. However, using (6.3.7), and since the odometer has topological entropy zero, we can easily see that $\mathbf{h}(S_k|\mu_k) = \mathbf{h}(\mathcal{P}_\Lambda|\mu_k) = g_k(\mu_k) + \delta$.

The proof of the equivalence between 1 and 2 in the formulation of the theorem is concluded by the application of Lemma 9.2.17. W take $\mu \in \mathcal{M}_T(X)$ (in particular it belongs to $\mathcal{M}_{\overline{T}}(\overline{X})$). By the lemma, $\mathbf{h}(S|\mu) = \lim_k \downarrow \mathbf{h}(S_k|\mu_k)$. Since the error term δ in (9.2.18) vanishes as k grows, and since for measures $\mu \in \mathcal{M}_T(X)$ we have (9.2.7), we can write (using (9.1.3)):

$$h^\phi_{\mathsf{ext}}(\mu) = h(\mu) + \mathbf{h}(S|\mu) = h(\mu) + \lim_k \downarrow \mathbf{h}(S_k|\mu_k) =$$

$$h(\mu) + \lim_k g_k(\mu_k) = h(\mu) + \lim_k g_k(\mu) = h(\mu) + (E_{\mathsf{A}} - h)(\mu) = E_{\mathsf{A}}(\mu).$$

The claim $h_{\mathsf{sex}} \equiv \mathrm{E}\mathcal{H}$ now becomes obvious, by the appropriate definitions. □

9.3 Properties of symbolic extension entropy

We discuss some fundamental properties of the function h_{sex} and the parameter $\mathbf{h}_{\mathsf{sex}}(T)$. We begin with the statements concerning attainability. The first fact is a direct consequence of Theorem 9.2.1. The last statement then follows from Theorem 8.2.8.

Theorem 9.3.1 *Let $(X, T, \mathbb{S})$ be a topological dynamical system with finite symbolic extension entropy. There exists a symbolic extension $\phi : Y \to X$ such that $h^\phi_{\mathsf{ext}} \equiv h_{\mathsf{sex}}$ if and only if the smallest superenvelope $\mathrm{E}\mathcal{H}$ of the entropy structure is affine on $\mathcal{M}_T(X)$. In particular, it is always so, when $\mathcal{M}_T(X)$ is a Bauer simplex.* □

(We remark, however, that we do not know of any general dynamical criterion to check whether $\mathcal{M}_T(X)$ is a Bauer simplex, except by individual investigation or in obvious cases when there are finitely many ergodic measures.)

Attainability of the topological symbolic extension entropy $\mathbf{h}_{\mathsf{sex}}(T)$, although it might be considered more important in terms of applications, does not translate to any clear condition concerning the superenvelopes of the entropy structure. Of course, $\mathbf{h}_{\mathsf{sex}}(T)$ is attained as topological entropy of a symbolic extension whenever the attainability of Theorem 9.3.1 holds, but there is a range of less restrictive examples as well. Direct application of Theorem 9.2.1 yields the (somewhat vague) criterion below.

Theorem 9.3.2 *Let $(X,T,\mathbb{S})$ be a topological dynamical system with finite symbolic extension entropy. There exists a symbolic extension $(Y,S,\mathbb{S})$ of $(X,T,\mathbb{S})$ such that $\mathbf{h}(S) = \mathbf{h}_{\mathsf{sex}}(T)$ if and only if there exists an affine superenvelope E_{A} of the entropy structure such that*

$$\sup_{\mu\in\mathcal{M}_T(X)} E_{\mathsf{A}}(\mu) = \sup_{\mu\in\mathcal{M}_T(X)} \mathrm{E}\mathcal{H}(\mu). \qquad \square$$

The attainability Theorem 9.3.1 has another, very important consequence. Suppose $h_{\mathsf{sex}} \equiv h < \infty$ for some system. By the Symbolic Extension Entropy Theorem 9.2.1 this is the same as $\mathrm{E}\mathcal{H} \equiv h$. We already know, that this is equivalent to asymptotic h-expansiveness (see Corollaries 8.4.12 and 8.1.20). On the other hand, since h is affine on invariant measures, so is $\mathrm{E}\mathcal{H}$, and then, by Theorem 9.3.1, it is attained, i.e., equals h^{ϕ}_{ext} for some symbolic extension. An extension satisfies $h^{\phi}_{\mathsf{ext}} \equiv h$ if and only if it is a principal extension. Conversely, the existence of a principal symbolic extension implies (directly, without invoking any theorems) that $h_{\mathsf{sex}} \equiv h$, which is equivalent to asymptotic h-expansiveness. In this manner, we have given asymptotic h-expansiveness a new, spectacular meaning, in terms of symbolic extensions.

Theorem 9.3.3 *A topological dynamical system $(X,T,\mathbb{S})$ is asymptotically h-expansive if and only if it has a principal symbolic extension.* $\square$

The above fact in full generality was first established in [Boyle *et al.*, 2002] (by a slightly different method, without using entropy structures).

We remark that principal symbolic extensions are particularly nice in terms of vertical data compression. They realize a digitalization which is not only lossless, but also "*gainless*," at least as far as entropy of invariant measures is concerned. In every other case, the encoding involves some "unwanted" extra complexity (in form of entropy) for orbits representing at least one invariant measure.

The next statement is a direct translation of Theorem 8.2.10 and relaxes the Symbolic Extension Entropy Variation Principle.

Theorem 9.3.4 *Let $(X,T,\mathbb{S})$ be a topological dynamical system. Then*

$$\mathbf{h}_{\mathsf{sex}}(T) = \sup\{h_{\mathsf{sex}}(\mu) : \mu \in \overline{\mathsf{ex}\mathcal{M}_T(X)}\}. \qquad \square$$

Immediate, though ineffective, examples of systems with h_{sex} not affine (even with its supremum over ergodic measures smaller than the global supremum), or such that every affine superenvelope exceeds $\mathbf{h}_{\mathsf{sex}}(T)$, are provided via the realization Theorem 8.4.6 by the Examples 8.2.17 and 8.2.18 of abstract u.s.d.a.-sequences on simplices. We remark also that all examples of increasing sequences $\mathcal{H}$ of nonnegative functions on a compact domain $\mathfrak{X}$ (without any convex structure; see Section 8.1) also produce examples of entropy structures in dynamical systems. It suffices to take the Bauer simplex $\mathcal{K} = \mathcal{M}(\mathfrak{X})$ of all Borel probability measures on $\mathfrak{X}$ and prolong the sequence $\mathcal{H}$ via the harmonic prolongation onto $\mathcal{K}$. Such procedure preserves, for instance, the order of accumulation (see Fact 8.2.9). Then we can apply the realization theorem to the extended $\mathcal{H}$ on $\mathcal{K}$. Of course, this method produces only systems $(X, T, \mathbb{S})$ for which $\mathcal{M}_T(X)$ is a Bauer simplex, so that the attainability holds.

In spite of this indirect technique, it might be of interest to see directly some examples of dynamical systems with various types of behavior of the entropy structure. This is why we replicate from [Boyle and Downarowicz, 2004] two explicit zero-dimensional examples in the form of symbolic array systems. In each of them, by default, we take the entropy structure $\mathcal{H} = (h_k)$, where $h_k(\mu) = h(\mu_k)$ with μ_k being the image of μ by the projection onto X_k (the subshift in the first k rows). The construction of a third example of this kind, where the supremum of $h_{\mathsf{sex}}(\mu)$ over all invariant measures is strictly larger than the supremum over all ergodic measures, is left to the reader as Exercise 9.4. We emphasize that this last pathology makes the symbolic extension entropy function h_{sex} exceptional among entropy-related functions on invariant measures. For most of other known entropy-like functions the supremum equals the supremum over ergodic measures.

Example 9.3.5 This is a very simple example with $\mathbf{h}_{\mathsf{sex}}(T)$ strictly larger than $\mathbf{h}(T)$. The strategy is simply to construct a system with $\mathcal{H}$ essentially as in the abstract Example 8.1.5, Game 1, (the version with x_0 included in A).

Let $(X_0, S, \mathbb{Z})$ denote an ergodic bilateral subshift with a unique invariant measure μ_0 and entropy 1 (this eliminates periodic systems). Let B_k ($k \geq 2$) be a sequence of blocks occurring in X_0 of lengths increasing with k and such that $\mu_{(B_k)} \to \mu_0$ (in a uniquely ergodic system such convergence is in fact automatic). Let T denote the shift map on the set X of all symbolic arrays $x = (x_{k,n})_{k\in\mathbb{N}, n\in\mathbb{Z}}$ satisfying the following conditions:

1 The first row x_1 of x either belongs to X_0 or it is $x_{(B_k)} = \ldots B_k B_k B_k \ldots$ for some $k \geq 2$. By taking closure we must also admit that x_1 may have the form $y(-\infty, m]y'[m+1, \infty)$ for some $y, y' \in X_0$, $m \in \mathbb{Z}$.
2 If the first row is $x_{(B_k)}$, then the kth row of x is an element of X_0.
3 All other rows are filled with zeros.

The set of arrays constructed in this way is closed and invariant under the horizontal shift. The structure of invariant measures is as follows: there is one measure supported by matrices having nonperiodic first row and all other rows filled with zeros; this measure is isomorphic to the original μ_0 (hence we denote it also by μ_0), its entropy is 1 and $h_k(\mu_0) = 1$ for $k \geq 1$. Moreover, for each $k > 1$, there are finitely many measures $\mu_{k,i}$ supported by arrays having nonzero kth row and periodic first row ($\mu_{k,i}$ is a joining of $\mu_{(B_k)}$ and μ_0, and there are finitely many such joinings). Again, each of these measures has entropy 1, with $h_j(\mu_{k,i}) = 1$ for $j \geq k$ and $h_j(\mu_{k,i}) = 0$ for $j < k$. With increasing k the measures $\mu_{k,i}$ accumulate at μ_0.

The entropy function h is 1 on all measures, so $\mathbf{h}(T) = 1$, while $\mathrm{E}\mathcal{H} = 1 + 1^{\mathsf{har}}_{\{\mu_0\}}$, hence $\mathbf{h}_{\mathsf{sex}}(T) = 2$.

Example 9.3.6 In this example $\mathbf{h}_{\mathrm{top}}(S) > \mathbf{h}_{\mathrm{sex}}(T)$ for every symbolic extension $(Y, S, \mathbb{S})$ of $(X, T, \mathbb{S})$. This is achieved while $\mathrm{E}\mathcal{H} = \widetilde{h}$ (which implies order of accumulation of entropy 1, see Exercise 8.4). The idea is to obtain a system for which the entropy structure behaves essentially as in the abstract Example 8.2.17.

Let B_k be as in the previous example: blocks of increasing lengths occurring in a strictly ergodic subshift X_0. Let C_k ($k \geq 2$; we skip $k = 1$ only in order to better match the enumeration of Example 8.2.17) be the block with the following structure:

$$C_k = B_2 B_2 \ldots\ldots\ldots\ldots\ldots\ldots B_2, B_3 B_3 \ldots B_3, \cdots, B_{k-1} B_{k-1}, B_k,$$

where the repetitions of B_i occupy roughly 2^{-i-1} of the length of C_k, and the precision of these proportions improves with k (the number of repetitions for each i increases with k). As before, $x_{(B_k)}, x_{(C_k)}, \mu_{(B_k)}, \mu_{(C_k)}$ denote respective periodic sequences and the corresponding periodic measures. It is seen that the measures $\mu_{(C_k)}$ converge weakly-star to $\sum_{k=1}^{\infty} 2^{-k} \mu_{(B_{k+1})}$. Let T be the shift map on the space X of all symbolic arrays $x = (x_{k,n})_{k \in \mathbb{N}, n \in \mathbb{Z}}$ satisfying the conditions:

1. The first row x_1 of x either belongs to X_0 or it has the form $x_{(B_k)}$ or $x_{(C_k)}$ for some $k \geq 2$. By taking closure we must additionally admit sequences of the forms $y_1(-\infty, m] y_2[m+1, \infty)$ with $y_1, y_2 \in X_0$, $x_{(B_k)}(-\infty, m] x_{(B_{k+1})}[m+1, \infty)$, and $y_1(-\infty, m] x_{(B_2)}[m+1, \infty)$.
2. If the first row is $x_{(C_k)}$, then the kth row of x is an element of X_0.
3. All other rows are filled with zeros.

The structure of invariant measures is now the following: the measure μ_0 is as in the previous example, and h_1 is the (harmonic prolongation of) the characteristic function at μ_0 (μ_0 corresponds to the point b_1 in Example 8.2.17). This measure is approached by periodic measures μ_k supported by arrays with $x_{(B_k)}$ in the first row and zeros otherwise (these measures have entropy zero; they correspond to the points a_k in Example 8.2.17). In addition, for each $k \geq 2$, we have finitely many measures $\nu_{k,j}$ supported by arrays with $x_{(C_k)}$ in the first row and a nonperiodic kth row. Then $h_k - h_{k-1}$ is the (harmonic prolongation of the) characteristic function of the set of measures $\nu_{k,j}$. With increasing k, these latest measures approach the combination $\sum_{k=1}^{\infty} 2^{-k} \mu_{k+1}$ (like the points b_k in Example 8.2.17, which are now replaced by "groups of points"). The behavior of $\mathcal{H}$ of the example in Example 8.2.17 is hence copied.

The rest of this section is devoted to examining how the function h_{sex} and the parameter $\mathbf{h}_{\mathsf{sex}}$ behave under joinings (including products) and inverse limits. We obtain formulae similar to those for the usual entropy. The results come from [Boyle and Downarowicz, 2006].

Theorem 9.3.7 (Joinings and Products) *Suppose $(Z, R, \mathbb{S})$ is the direct product of $(X, T, \mathbb{S})$ and $(Y, S, \mathbb{S})$. Let $\xi \in \mathcal{M}_R(Z)$ and let $\mu \in \mathcal{M}_T(X)$, $\nu \in \mathcal{M}_S(Y)$ be the respective projections of ξ (ξ is a joining of μ and ν). Then*

$$h_{\mathsf{sex}}(\xi, R) \leq h_{\mathsf{sex}}(\mu, T) + h_{\mathsf{sex}}(\nu, S). \tag{9.3.8}$$

If $\xi = \mu \times \nu$, then

$$h_{\mathsf{sex}}(\xi, R) = h_{\mathsf{sex}}(\mu, T) + h_{\mathsf{sex}}(\nu, S). \tag{9.3.9}$$

As a result,

$$\mathbf{h}_{\mathsf{sex}}(R) = \mathbf{h}_{\mathsf{sex}}(T) + \mathbf{h}_{\mathsf{sex}}(S). \tag{9.3.10}$$

Proof We fix an $\varepsilon > 0$ and let ξ, μ and ν be as in the formulation of the theorem. Let $(X^o, T^o, \mathbb{S})$ and $(Y^o, S^o, \mathbb{S})$ be symbolic extensions of $(X, T, \mathbb{S})$ and $(Y, S, \mathbb{S})$ via maps ϕ and ψ, respectively, such that

$$h^{\phi}_{\mathsf{ext}}(\mu) < h_{\mathsf{sex}}(\mu, T) + \tfrac{\varepsilon}{2},$$

and analogously for ν. Then the product system on $X^o \times Y^o$ is a symbolic extension of Z via $\pi = \phi \times \psi$, and any lift of ξ is a joining of a lift of μ with a lift of ν. By subadditivity of entropy for joinings (see (4.1.11)), and by applying supremum over all such lifts, we obtain $h^{\pi}_{\mathsf{ext}}(\xi) < h_{\mathsf{sex}}(\mu, T) + h_{\mathsf{sex}}(\nu, S) + \varepsilon$. Since ε is arbitrary, we arrive at (9.3.8).

The equality for the product measure $\xi = \mu \times \nu$ is much less trivial. We need to use the entropy structures. Let $(\mathcal{F}_k)$ and $(\mathcal{G}_k)$ be sequences of families of continuous functions giving rise to entropy structures $\mathcal{H}^{\text{fun}}$ on $(X, T, \mathbb{S})$ and $(Y, S, \mathbb{S})$, respectively. Then the families $(\mathcal{F}_k \cup \mathcal{G}_k)$ (the union of lifted families) give rise to an entropy structure in the product system $(Z, R, \mathbb{S})$.

We have simply $h(\xi) = h(\mu) + h(\nu)$ and $h_k(\xi) = h_k(\mu) + h_k(\nu)$, hence also $\theta_k(\xi) = \theta_k(\mu) + \theta_k(\nu)$. We will complete the proof by showing, using transfinite induction, that for every ordinal $\boldsymbol{\alpha}$

$$u_{\boldsymbol{\alpha}}(\xi) \geq u_{\boldsymbol{\alpha}}(\mu) + u_{\boldsymbol{\alpha}}(\nu).$$

This inequality is trivial for $\boldsymbol{\alpha} = 0$. Next we suppose it holds for all $\boldsymbol{\beta} < \boldsymbol{\alpha}$ for some ordinal $\boldsymbol{\alpha} \geq 1$. Since the supremum $\sup_{\boldsymbol{\beta}<\boldsymbol{\alpha}} u_{\boldsymbol{\beta}}$ is in fact an increasing limit (over the ordinals $\boldsymbol{\beta} < \boldsymbol{\alpha}$), we have

$$\sup_{\boldsymbol{\beta}<\boldsymbol{\alpha}} u_{\boldsymbol{\beta}}(\xi) \geq \sup_{\boldsymbol{\beta}<\boldsymbol{\alpha}} u_{\boldsymbol{\beta}}(\mu) + \sup_{\boldsymbol{\beta}<\boldsymbol{\alpha}} u_{\boldsymbol{\beta}}(\nu).$$

Fix $k \in \mathbb{N}$. Pick sequences $\mu_i \to \mu$ and $\nu_i \to \nu$ ($i \in \mathbb{N}$) such that

$$\lim_i (\sup_{\boldsymbol{\beta}<\boldsymbol{\alpha}} u_{\boldsymbol{\beta}} + \theta_k)(\mu_i) = \widetilde{(\sup_{\boldsymbol{\beta}<\boldsymbol{\alpha}} u_{\boldsymbol{\beta}} + \theta_k)}(\mu), \quad \text{and}$$

$$\lim_i (\sup_{\boldsymbol{\beta}<\boldsymbol{\alpha}} u_{\boldsymbol{\beta}} + \theta_k)(\nu_i) = \widetilde{(\sup_{\boldsymbol{\beta}<\boldsymbol{\alpha}} u_{\boldsymbol{\beta}} + \theta_k)}(\nu).$$

Then

$$\widetilde{(\sup_{\beta<\alpha} u_\beta + \theta_k)}(\xi) \geq \limsup_{i\to\infty}(\sup_{\beta<\alpha} u_\beta + \theta_k)(\mu_i \times \nu_i) \geq$$

$$\limsup_{i\to\infty}\Big(\sup_{\beta<\alpha} u_\beta(\mu_i) + \sup_{\beta<\alpha} u_\beta(\nu_i) + \theta_k(\mu_i) + \theta_k(\nu_i)\Big) =$$

$$\widetilde{(\sup_{\beta<\alpha} u_\beta + \theta_k)}(\mu) + \widetilde{(\sup_{\beta<\alpha} u_\beta + \theta_k)}(\nu).$$

We conclude that

$$u_\alpha(\xi) \geq u_\alpha(\mu) + u_\alpha(\nu).$$

This completes the transfinite induction. By the product rule for measure-theoretic entropy (4.4.5) and the transfinite characterization of $\mathrm{E}\mathcal{H}$ (Theorem 8.1.19), we get

$$\mathrm{E}\mathcal{H}^R(\xi) \geq \mathrm{E}\mathcal{H}^T(\mu) + \mathrm{E}\mathcal{H}^S(\nu).$$

The identity $h_{\mathsf{sex}} = \mathrm{E}\mathcal{H}$ (Theorem 9.2.1) and the Symbolic Extension Entropy Variational Principle (Corollary 9.2.2) imply the missing inequalities in (9.3.9) and (9.3.10), respectively. □

As to inverse limits, we only comment that in general the limit passage analogous to that for the usual entropy (Fact 6.5.12) does not hold. For instance, every zero-dimensional system is an inverse limit of symbolic systems, for which the symbolic extension entropy function obviously equals the entropy function. But there are zero-dimensional systems with symbolic extension entropy function strictly larger than the entropy function.

We pass to power systems. A subtlety here is that the power system usually supports more invariant measures than the original. The comparison of the symbolic extension entropy can only be made for measures invariant under T, and for the topological notion. Below are the relevant *power rules*.

Theorem 9.3.11 (Powers) *Fix some $n \in \mathbb{S}$. For any system $(X, T, \mathbb{S})$ the restriction of $h_{\mathsf{sex}}(\,\cdot\,, T^n)$ to $\mathcal{M}_T(X)$ equals $|n| h_{\mathsf{sex}}(\,\cdot\,, T)$ and*

$$\mathbf{h}_{\mathsf{sex}}(T^n) = |n|\mathbf{h}_{\mathsf{sex}}(T).$$

Proof If $\mathbb{S} = \mathbb{Z}$, then the systems $(X, T^n, \mathbb{Z})$ and $(X, T^{-n}, \mathbb{Z})$ have the same invariant measures and the same symbolic extensions (with the left shift S and right shift S^{-1}, respectively) and since for each invariant measure ν on the extension we have $h(\nu, S, \mathbb{Z}) = h(\nu, S^{-1}, \mathbb{Z})$, the systems $(X, T^n, \mathbb{Z})$ and $(X, T^{-n}, \mathbb{Z})$ determine the same symbolic extension entropy function. As a consequence, we need to consider nonnegative n only (in either case of $\mathbb{S}$). The case $n = 0$ is trivial; systems with entropy zero have zero symbolic extension entropy as well.

Now assume $n > 0$ and suppose that $\phi : (Y, S, \mathbb{S}) \to (X, T, \mathbb{S})$ is a symbolic extension. Let ϕ_n denote the same map ϕ regarded as a factor map from $(Y, S^n, \mathbb{S})$ onto $(X, T^n, \mathbb{S})$. Suppose $\nu \in \mathcal{M}_S(Y)$ and $\phi\nu = \mu$. Then $\nu \in \mathcal{M}_{S^n}(Y)$ and $h(\nu, S^n) = nh(\nu, S)$. Therefore $nh^{\phi}_{\text{ext}}(\mu) \le h^{\phi_n}_{\text{ext}}(\mu)$ (we cannot claim equality, as μ may have other S^n-invariant lifts, which are not S-invariant). It follows that $nh_{\text{sex}}(\mu, T) \le h_{\text{sex}}(\mu, T^n)$.

Conversely, suppose $\psi : (Y, S, \mathbb{S}) \to (X, T^n, \mathbb{S})$ is a symbolic extension. Define $Y' = Y \times \{0, 1, \dots, n-1\}$ and define $S' : Y' \to Y'$ by the following rules: $(y, i) \mapsto (y, i+1)$ if $0 \le i < n-1$, and $(y, n-1) \mapsto (Sy, 0)$. Then $(Y', S', \mathbb{S})$ is a symbolic system (it is an easy exercise to find its symbolic representation) and it is elementary to verify that the map $\psi' : Y' \to X$ given by $(y, i) \mapsto (T^i \circ \psi)(y)$ is a topological extension (a symbolic extension of T). Let $\mu \in \mathcal{M}_T(X) \subset \mathcal{M}_{T^n}(X)$ and let ν be a lift of μ via ψ. Let ν' be the product of $\nu \in \mathcal{M}_S(Y)$ with the equidistributed probability measure on $\{0, 1, \dots, n-1\}$. Then $\nu' \in \mathcal{M}_{S'}(Y')$, $\psi'\nu' = \mu$ and $h(\nu', S') = \frac{1}{n}h(\nu, S)$. Therefore $h^{\psi}_{\text{ext}}(\mu) \le nh^{\psi'}_{\text{ext}}(\mu)$. It follows that $h_{\text{sex}}(\mu, T^n) \le nh_{\text{sex}}(\mu, T)$. This completes the proof of power rule in Theorem 9.3.11 concerning the symbolic extension entropy function. Next, this power rule and the Symbolic Extension Entropy Variational Principle together yield that

$$n\mathbf{h}_{\text{sex}}(T) = \sup\{h_{\text{sex}}(\mu, T^n) : \mu \in \mathcal{M}_T(X)\} \le \mathbf{h}_{\text{sex}}(T^n).$$

On the other hand, for any $\varepsilon > 0$ there exists a symbolic extension $(Y, S, \mathbb{S})$ of $(X, T, \mathbb{S})$ with $\mathbf{h}(S) < \mathbf{h}_{\text{sex}}(T) + \varepsilon$. Then $(Y, S^n, \mathbb{S})$ is a symbolic extension of $(X, T^n, \mathbb{S})$, whose topological entropy is $n\mathbf{h}(S) < n\mathbf{h}_{\text{sex}}(T) + n\varepsilon$. Since ε is arbitrary (and n fixed), it follows that $\mathbf{h}_{\text{sex}}(T^n) \le n\mathbf{h}_{\text{sex}}(T)$. □

9.4 Symbolic extensions of interval maps

In order to fully appreciate the theory of entropy structures we need an example of an important class of systems for which this theory allows us to compute (or at least estimate) the symbolic extension entropy and hence the vertical data compression, while explicit construction of symbolic extensions, in the same generality, seems inaccessible. A spectacular such example is the class of smooth transformations of the interval (or of the circle). We can not only estimate the symbolic extension entropy function in terms of more familiar parameters, such as the integral of the derivative, but even indicate a concrete function that is realized as the entropy function in a symbolic extension.

All this is done purely theoretically, without actually building any symbolic extensions.

Let f be a C^1 transformation of the interval or of the circle X. In either case, the derivative f' is a real-valued continuous function defined on the interval (in case of the circle we mean the derivative f' of the covering function $f : \mathbb{R} \to \mathbb{R}$; although the covering function is given only up to an integer constant, the derivative is unique and satisfies $f'(x+1) = f'(x)$, hence it is determined by its restriction to the interval). Let $\mu \in \mathcal{M}(X)$. We denote

$$\chi(\mu) = \int \log |f'(x)| \, d\mu.$$

Clearly, χ is a harmonic and upper semicontinuous function of the measure. It need not be continuous, as $\log |f'|$ may assume $-\infty$ (see also Remark A.1.10). In fact, if f has a *critical point* (i.e., such that $f'(x) = 0$), then $\chi(\mu) = -\infty$ on a dense set of probability measures (all, not only invariant). For ergodic measures, $\chi(\mu)$ is called the *Lyapunov exponent* of μ. We let $\chi^+ = \max\{0, \chi\}$ (the positive part of χ), and by $\overline{\chi}^+$ we denote the harmonic prolongation of χ^+ onto the simplex $\mathcal{M}_f(X)$ of f-invariant measures. (Note that in general $\overline{\chi}^+(\mu)$ is not the same as the integral of $\log^+ |f'(x)|$; the harmonic prolongation is with respect to the ergodic decomposition, and such equality fails already for ergodic measures.) In the one-dimensional case, the Margulis–Ruelle inequality asserts that

$$h(\mu) \leq \overline{\chi}^+(\mu), \tag{9.4.1}$$

for every f-invariant measure. We will use this fact in this section; the proof will be provided in the next chapter.

We also denote $\mathsf{L}(f) = \sup_{x \in X} \log^+ |f'(x)| < \infty$. (For C^1-maps this number coincides with the maximum of zero and the logarithm of the Lipschitz constant.) It is easy to see, by the chain rule, that the sequence $\mathsf{L}(f^n)$ (where f^n denotes the composition power of f) is subadditive. Thus the limit

$$\mathsf{R}(f) = \lim_{n \to \infty} \tfrac{1}{n} \mathsf{L}(f^n)$$

exists and equals the infimum. It is obvious (by the ergodic theorem and harmonicity) that the function χ (hence also $\overline{\chi}^+$ and the entropy function h) restricted to invariant measures is bounded from above by the constant $\mathsf{R}(f)$. The advantage of $\mathsf{R}(f)$ over $\mathsf{L}(f)$ is that while $\mathsf{L}(f)$ clearly depends on the choice of the metric on X, both $\mathsf{R}(f)$ and the function χ are invariants of C^1 conjugacy (we leave the easy verification of this statement to the reader).

We define the *degree of smoothness* r of f inductively: f is of class C^r for $r \leq 1$ if it is r-Hölder, i.e., there exists a constant c such that $|f(x) - f(y)| \leq$

$c|x-y|^r$. For $r > 1$ we require that f is differentiable, and f' is of class C^{r-1}. Our definition is slightly weaker than usual, as for integer r we do not require differentiability r times, only $r-1$ times, with the last derivative Lipschitz. We make one exception: by C^1 we will understand the class of differentiable functions with continuous derivative.

We are in a position to formulate the main theorem of this section [Downarowicz and Maass, 2009].

Theorem 9.4.2 *Let f be a C^r transformation of the interval or of the circle X, where $r > 1$. Then, the function*

$$E_{\mathsf{A}} = h + \frac{\overline{\chi}^+}{r-1}$$

is an affine superenvelope of the entropy structure, hence it is realized as the extension entropy function for some symbolic extension.

Corollary 9.4.3 *As an immediate consequence of the above we get the first inequality below, and combining it with the Variational Principle, the estimate of $\chi(\mu)$ by $\mathsf{R}(f)$, and the Margulis–Ruelle inequality* (9.4.1), *we get the second one:*

$$h_{\mathsf{sex}}(\mu) \le h(\mu) + \frac{\overline{\chi}^+(\mu)}{r-1} \quad \textit{for } \mu \in \mathcal{M}_f(X), \quad \textit{and} \tag{9.4.4}$$

$$\mathbf{h}_{\mathsf{sex}}(f) \le \mathbf{h}(f) + \frac{\mathsf{R}(f)}{r-1} \le \frac{r\mathsf{R}(f)}{r-1}. \tag{9.4.5}$$

□

The proof of Theorem 9.4.2 will be performed by showing that the function $u = \frac{\overline{\chi}^+}{r-1}$ fulfills the assumption of Lemma 8.2.14, where (θ_k) is the sequence of tails in the entropy structure determined by the Newhouse local entropy (see Definitions 8.4.4 and 8.3.7). That lemma will imply that u is a repair function, that is, $h + u$ is a superenvelope of the entropy structure. Clearly, it is also affine.

We start with a counting lemma.[1]

Lemma 9.4.6 *Let $g : [0,1] \to \mathbb{R}$ be a C^r function, where $r > 0$. Then there exists a constant $c > 0$ such that for every $0 < s < 1$ the number of the connected components of the set $\{x : g(x) \neq 0\}$ on which $|g|$ reaches or exceeds the value s is at most $c \cdot s^{-\frac{1}{r}}$.*

[1] D. Burget generalized it to higher dimensions [Burguet, 2010].

Proof For $0 < r \leq 1$, g is Hölder, i.e., there exists a constant $c_1 > 0$ such that $|g(x) - g(y)| \leq c_1|x-y|^r$. If $|g(x)| > s$ and y is a zero point for g, then

$$|x-y| > c_1{}^{-\frac{1}{r}} \cdot s^{\frac{1}{r}}.$$

The component containing x is at least that long and the number of such components is at most $c \cdot s^{-\frac{1}{r}}$, where $c = c_1{}^{\frac{1}{r}}$.

For larger r we proceed inductively: suppose that the lemma holds for $r-1$. Let g be of class C^r. Consider only such components $I = (a_I, b_I)$ of the set $\{x : g(x) \neq 0\}$, on which $|g|$ reaches or exceeds s. By elementary considerations of the graph of g, with every such component we can disjointly associate an interval (x_I, y_I), so that $x_I \in I$ is the rightmost point at which $|g|$ attains its maximum on I, while y_I is the leftmost critical point lying to the right of I (see Figure 9.1).

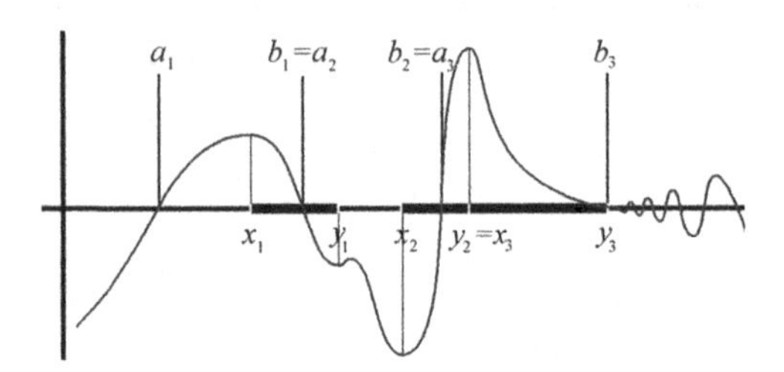

Figure 9.1 The intervals (x_I, y_I) for $I = 1, 2, 3$. It may happen that $y_I = b_I$ or that $y_I = x_{I+1}$, but the intervals (x_I, y_I) remain disjoint.

There are two possible cases: either

$$\text{(a) } y_I - x_I > s^{\frac{1}{r}}, \quad \text{or} \quad \text{(b) } y_I - x_I \leq s^{\frac{1}{r}}.$$

Clearly, the number of components I satisfying (a) is smaller than $s^{-\frac{1}{r}}$. If a component satisfies (b), then, by the mean value theorem, $|g'|$ attains on (x_I, b_I) a value at least $s \cdot s^{-\frac{1}{r}} = s^{\frac{r-1}{r}}$. Because g' is of class C^{r-1}, by the inductive assumption, the number of such intervals (x_I, y_I) (hence of components I) does not exceed $c \cdot (s^{\frac{r-1}{r}})^{-\frac{1}{r-1}} = c \cdot s^{-\frac{1}{r}}$. Jointly, the number of the considered components I is at most $3 + (c+1) \cdot s^{-\frac{1}{r}} \leq c_1 \cdot s^{-\frac{1}{r}}$ (recall that $s < 1$). The number 3 is added because the above argument need not apply to the first, last and last but one component (if there are such); the first and last component need not contain a critical point. □

For $g = f'$ we obtain the following

Corollary 9.4.7 *Let $f : [0,1] \to [0,1]$ be a C^r function, where $r > 1$. Then there exists a constant $c > 0$ such that for every $0 < s < 1$ the number of branches of monotonicity of f on which $|f'|$ reaches or exceeds s is at most $c \cdot s^{-\frac{1}{r-1}}$.* □

Next we apply the above to counting possible ways by which a point (with a specific derivative for the composition power of f) may traverse the branches of monotonicity. We make a formal definition.

Definition 9.4.8 Let $f : [0,1] \to [0,1]$ be a function of class C^1. Let $\mathcal{I} = (I_1, I_2, \ldots, I_n)$ be a finite sequence of branches of monotonicity of f, (i.e., any formal finite sequence whose elements belong to the countable set of branches, admitting repetitions). Denote

$$a_i = \min\{-1, \max\{\log|f'(x)| : x \in I_i\}\}. \tag{9.4.9}$$

Choose $S \leq -1$. We say that $\mathcal{I}$ *admits* the value S if

$$\frac{1}{n}\sum_{i=1}^{n} a_i \geq S.$$

Notice that if there exist points $y_i \in I_i$ with $\log|f'(y_i)| \leq -1$ for each i satisfying $\frac{1}{n}\sum_{i=1}^{n}\log|f'(y_i)| \geq S$, then $\mathcal{I}$ admits the value S.

Lemma 9.4.10 *Let $f : [0,1] \to [0,1]$ be a C^r function, where $r > 1$. Fix $\varepsilon > 0$. Then there exists $S_\varepsilon \leq -1$ such that for every n and $S < S_\varepsilon$ the logarithm of the number of sequences $\mathcal{I}$ of length n which admit the value S is at most*

$$n\frac{-S}{r-1}(1+\varepsilon).$$

Proof Without loss of generality assume that S is a negative integer. Consider a sequence of n branches of monotonicity which admits the value S. Such a sequence determines the sequence of numbers (a_i) defined in (9.4.9). Denote $k_i = \lfloor a_i \rfloor$. Then $(-k_i)$ is a sequence of n positive integers with sum at most $n(1-S)$. Now, in a given sequence (k_i), each value k_i may be realized by any branch of monotonicity on which $\max\log|f'|$ lies between k_i and k_i+1 (or just exceeds -1 if $k_i = -1$). From Corollary 9.4.7 it follows that there are no more than $c2^{\frac{-k_i}{r-1}}$ such branches for each k_i. Jointly the logarithm of the number of sequences of branches of monotonicity corresponding to one sequence (k_i) is at most

$$n\log c - \frac{1}{r-1}\sum_{i=1}^{n} k_i \leq n\log c + \frac{n}{r-1}(1-S).$$

For large values of $-S$, the first term and the last constant term 1 can be skipped at a cost of multiplying the remaining term $-\frac{n}{r-1}S$ by $(1+\varepsilon/2)$. The sequences $(-k_i)$ with sum not exceeding $n(1-S)$ stand in a 1-1 correspondence with the increasing sequences of their partial sums, which are n-element subsets of $\{1, 2, \dots, n(1-S)\}$. Thus their cardinality is at most the binomial coefficient $\binom{n(1-S)}{n}$, whose logarithm is approximately $n(1-S)H(\frac{1}{1-S}, \frac{-S}{1-S})$. For large enough $-S$, the latter expression equals $-nS$ times a small number, say smaller than $\frac{\varepsilon}{2(r-1)}$. So, the logarithm of the cardinality of all sequences of branches of monotonicity which admit the value S is, regardless of n, estimated from above as in the assertion. □

As we have explained earlier, regardless of whether f is a transformation of the interval or of the circle X, the derivative f' can be regarded as a function defined on the interval $[0, 1]$. Let $C = \{x : f'(x) = 0\}$ be the set of the critical points of f. Fix $\varepsilon > 0$. Fix some open neighborhood V of C on which $\log |f'| < S_\varepsilon$. Then V^c can be covered by finitely many open intervals on which f is monotone. Let $\mathcal{V}$ be the cover consisting of V and these intervals.

Lemma 9.4.11 *Let f be a C^r transformation of the interval or of the circle X, where $r > 1$. Let V and $\mathcal{V}$ be as described above. Let ν be an ergodic measure and let*

$$S(\nu) = \int_V \log |f'| \, d\nu. \tag{9.4.12}$$

Then the local entropy (recall Definition 8.3.7) satisfies

$$h(X|\nu, \mathcal{V}) \leq \frac{-S(\nu)}{r-1}(1+\varepsilon). \tag{9.4.13}$$

Proof Let F be the set of points on which the nth Cesaro means of the function $\mathbb{I}_V \log |f'|$ are close to $S(\nu)$, (we need the relative error to be small) for n larger than some threshold integer (we are using the ergodic theorem; such a set F can have measure larger than $1-\sigma$). To estimate the local entropy we will use the variant of its definition based on counting (n, δ)-separated sets in elements of the cover $\mathcal{V}^n$ intersected with F (see Exercise 8.13).

And so, for $x \in F$ and large n consider a set

$$V_x^n = V_0 \cap T^{-1}(V_1) \cap \dots \cap T^{-n+1}(V_{n-1})$$

containing x, with $V_i \in \mathcal{V}$ (so that $V_x^n \in \mathcal{V}^n$). Consider the finite subsequence of times $0 \leq i_j \leq n-1$ when $V_{i_j} = V$. Let $n\zeta$ denote the length of this subsequence and assume $\zeta > 0$. For a fixed δ, let E be an (n, δ)-separated set in $V_x^n \cap F$ and let $y \in E$. The sequence (i_j) contains only (usually not all) times i when $f^i(y) \in V$. Thus, since $y \in F$, we have,

$$S(\nu) \leq \frac{1}{n}\Big(\sum_j \log |f'(T^{i_j}(y))| + A(y)\Big) + \beta,$$

where $A(y)$ is the similar sum over the times of visits to V not included in the sequence (i_j), while β is a small error term associated with the choice of the set F. Clearly $A(y) \leq 0$, so it can be skipped. Dividing by ζ we obtain

$$\frac{S(\nu) - \beta}{\zeta} \leq \frac{1}{n\zeta}\sum_j \log |f'(T^{i_j}(y))|.$$

The right-hand side above is smaller than S_ε (defined in Lemma 9.4.10). This implies that along the subsequence (i_j) the trajectory of y traverses a sequence $\mathcal{I}$ (of length $n\zeta$) of branches of monotonicity of f admitting the value $\frac{S(\nu)-\beta}{\zeta}$ smaller than S_ε. By Lemma 9.4.10, the logarithm of the number of such sequences $\mathcal{I}$ is dominated by

$$n\frac{-S(\nu) + \beta}{r - 1}(1 + \varepsilon). \tag{9.4.14}$$

By a small adjustment of ε, we can skip β in the numerator.

For indices i other than i_j the set V_i contains only one branch of monotonicity, so if two points from $V_x^n \cap F$ traverse the same sequence of branches during the times (i_j), then they traverse the same full sequence of branches during all times $i = 0, 1, \ldots, n - 1$. The number of (n, δ)-separated points which, during all times $i = 0, 1, \ldots, n - 1$, traverse the same given sequence of branches of monotonicity is at most $n/\delta + 1$. This follows from the easy observation that any collection of n monotone functions on the interval can δ-separate at most $n/\delta + 1$ points: look at the intervals between neighboring separated points. Each of them must be stretched to at least length δ by one of the functions. By monotonicity, each function can strech at most $1/\delta$ of these intervals, because their images are disjoint. Together, at most n/δ intervals can be stretched, which limits the cardinality of the points to $n/\delta + 1$. The logarithm of $n/\delta + 1$ should be added to the estimate (9.4.14). The proof is concluded by dividing by n, and letting $n \to \infty$ and then formally passing with δ to zero. □

Proof of Theorem 9.4.2 Fix an invariant measure μ and some $\gamma > 0$. In order to use Lemma 8.2.14 we only need to consider ergodic measures ν close to μ. If $\chi(\mu) < 0$, then, by upper semicontinuity of the function χ, for ν sufficiently close to μ, $\chi(\nu) < 0$, in which case ν has entropy zero (which follows e.g. from (9.4.1)), and hence $h(X|\nu, \mathcal{V}) = 0$ for any open cover $\mathcal{V}$.

What remains is the case $\chi(\mu) \geq 0$ and then we only need to consider ergodic ν with $\chi(\nu) > 0$. Clearly, then $\mu(C) = 0$. Since $\log|f'|$ is μ-integrable, the open neighborhood V of C (on which $\log|f'| < S_\varepsilon$) can be made so small that the (negative) integral of $\log|f'|$ over the closure of V is very close to zero (say, closer than ε). Then

$$\int_{\overline{V}^c} \log|f'(x)|\, d\mu < \chi(\mu) + \varepsilon. \tag{9.4.15}$$

The above integral is an upper semicontinuous function of the measure (it is so because $\overline{V}^c$ is an open set on which $\log|f'|$ is finite and continuous and negative on the boundary), hence (9.4.15) remains valid when the integration on the left-hand side is with respect to any invariant measure ν contained in a sufficiently small neighborhood of μ. All the more

$$\int_{V^c} \log|f'(x)|\, d\nu < \chi(\mu) + \varepsilon$$

(we have included the boundary in the set of integration, and the function is negative on that boundary). Then, with the notation of (9.4.12), we have

$$-S(\nu) = \int_{V^c} \log|f'(x)|\, d\nu - \chi(\nu) \leq \chi(\mu) - \chi(\nu) + \varepsilon. \tag{9.4.16}$$

We define the cover $\mathcal{V}$ with the above choice of the set V (recall that $\mathcal{V}$ consists of V and some intervals on which f is monotone). We can now apply Lemma 9.4.11. Substituting (9.4.16) into (9.4.13) we get

$$h(X|\nu, \mathcal{V}) \leq \frac{\chi(\mu) - \chi(\nu) + \varepsilon}{r-1}(1+\varepsilon).$$

Because the function $\frac{\chi^+}{r-1}$ is bounded, the contribution of the error terms ε can be made smaller than any preassigned additive term γ (as it appears in the next displayed formula).

Since χ^+ is obviously convex and upper semicontinuous, it is subharmonic (see (A.2.10)), hence $\chi(\mu) = \chi^+(\mu) \leq \overline{\chi}^+(\mu)$, while at an ergodic ν we have equality. We can thus write.[2]

$$h(X|\nu, \mathcal{V}) \leq \frac{\overline{\chi}^+(\mu) - \overline{\chi}^+(\nu)}{r-1} + \gamma. \tag{9.4.17}$$

[2] This inequality is of independent interest; it estimates local entropy in terms of Lyapunov exponents. It is called *The Antarctic Theorem* [Downarowicz and Maass, 2009]

We have proved that the harmonic function $u = \frac{\overline{\chi}^+}{r-1}$ satisfies the assumptions of Lemma 8.2.14, where $\theta_k = h(X|\nu, \mathcal{V}_k)$ is the tail function of the entropy structure given by the local entropy and a refining sequence of open covers (the diameter of $\mathcal{V}_k$ will eventually, for large k, become smaller than the Lebesgue number of $\mathcal{V}$, so (9.4.17) will apply to $\mathcal{V}_k$). The lemma now implies that u is a repair function, i.e., that $h + u$ is a superenvelope. □

Corollary 9.4.3 implies that for C^∞ maps, $h_{\mathsf{sex}} \equiv h$ (i.e., the system is asymptotically h-expansive). This is a special case of a theorem proved by J. Buzzi for all C^∞ transformations of compact Riemannian manifolds of any dimension [Buzzi, 1997], which will be discussed in slightly more detail in the following section.

Returning to the interval, a natural question arises: Are the estimates of Corollary 9.4.3 optimal? David Burguet provided a partial positive answer, by proving the following facts (the constructions are by far too complicated to be included in this book, see the original paper [Burguet, 2010]):

Theorem 9.4.18 *For any integer $r \geq 1$ there exists a C^r interval map f_r such that $h_{\mathsf{sex}}(\mu, f_r) = h(\mu) + \frac{\mathsf{L}(f_r)}{r-1}$ at at least one ergodic invariant measure μ. Moreover, for every $\varepsilon > 0$ there exists a C^r map $f_{r,\varepsilon}$ with $\mathsf{L}(f_{r,\varepsilon}) \geq \log 2$, satisfying $\mathbf{h}_{\mathsf{sex}}(\mu, f_{r,\varepsilon}) \geq \mathbf{h}(f_{r,\varepsilon}) + \frac{\mathsf{L}(f_r)}{r-1} - \varepsilon$.* □

Because $\mathsf{L}(f_r) \geq \overline{\chi}^+(\mu)$, the first inequality proves sharpness of the estimate (9.4.4) at at least one measure. The last inequality proves sharpness of the estimate (9.4.5). For $r = 1$, using either statement, we obtain an example of a C^1 map without any symbolic extension. The question, in its most radical form, remains still open:

Question 9.4.19 For $r > 1$, does there exist a C^r interval map f with positive entropy, such that the repair function $\frac{\overline{\chi}^+(\mu)}{r-1}$ is the smallest?

Exercises

9.1 Among the examples provided in this section find one for which the topological residual entropy $\mathbf{h}_{\mathsf{res}}(T)$ is strictly smaller than the residual entropy $h_{\mathsf{res}}(\mu)$ of some invariant measure.

9.2 Construct a principal symbolic extension of an irrational rotation of the circle.

9.3 Construct a principal symbolic extension of the identity map on the unit interval. Note that this is already a highly nontrivial subshift.

9.4 Build an explicit example of a system on marked symbolic arrays with entropy structure copying the behavior of Example 8.2.18, so that the supremum of $h_{\mathsf{sex}}(\mu)$ over ergodic measures is strictly smaller than $\mathbf{h}_{\mathsf{sex}}(T)$.

9.5 Build an explicit example of a finite entropy system without symbolic extensions.

10

A touch of smooth dynamics*

Smooth dynamics is concerned with smooth transformations of Riemannian manifolds. It is one of the most exploited areas of dynamical systems, and many papers and books are devoted to this branch. We refer the reader to the book [Katok and Hasselblatt, 1995] as a primary reference. These studies require background in smooth geometry, hyperbolic dynamics, foliation theory, and many more. Also here the entropy is one of the most important subjects.

As this book is designed to be self-contained, and there is obviously no room to provide all that background, deprived of the basic tools, we will actually be able to do very little. In fact we will prove only one rather elementary fact: an estimate of the measure-theoretic entropy in terms of characteristic exponents, a weaker version of the Margulis–Ruelle estimate of entropy for ergodic measures. Besides that, we will only state several results without a proof: the Pesin Entropy Formula, the Buzzi–Yomdin estimate of the topological tail entropy for C^r maps, and some results and questions concerning symbolic extensions.

10.1 Margulis–Ruelle Inequality and Pesin Entropy Formula

Let $T : \mathfrak{M} \to \mathfrak{M}$ be a C^1 transformation of a compact Riemannian manifold $\mathfrak{M}$ of dimension dim. We refrain from providing the detailed definition of the derivative D_xT of T at x, which is a transformation defined on the tangent bundle of $\mathfrak{M}$. The details can be found in any textbook on differential topology [see e.g. Hirsch, 1994]. For our purposes an intuitive and approximative understanding should suffice. Roughly speaking, the transformation T between a small neighborhood of x and a small neighborhood of Tx in appropriate local coordinate systems (centered at x and Tx, respectively) is nearly a linear map, which equals D_xT. By $\|D_xT\|$ we will understand the norm (the maximal length of the image of a unit vector) of D_xT. For C^1 maps this parameter depends continuously on x. Just like for interval maps, we let $L(T) = \max\{0, \log \|D_x T\| : x \in \mathfrak{M}\}$. This number is finite and equals the maximum of zero and the logarithm of the Lipschitz constant, for an appropriately chosen metric. What we need to know is that the cocycle

$$(x, n) \mapsto \log \|D_x T^n\|$$

is subadditive, i.e., $\log \|D_x T^{n+m}\| \leq \log \|D_x T^n\| + \log \|D_{T^n x} T^m\|$. Thus, the limit

$$R(T) = \lim_{n\to\infty} \tfrac{1}{n} L(T^n)$$

exists and equals the infimum. Further, applying the Subadditive Ergodic Theorem (see Theorem 2.1.4), for every (T-invariant Borel probability) ergodic measure μ on $\mathfrak{M}$, we have the almost everywhere convergence

$$\lim_{n\to\infty} \tfrac{1}{n} \log \|D_x T^n\| = \inf_n \tfrac{1}{n} \int \log \|D_x T^n\| \, d\mu = \chi(\mu). \qquad (10.1.1)$$

The parameter $\chi(\mu)$ is called the *maximal Lyapunov* (or *characteristic*) *exponent* of the measure μ. Clearly, $\chi(\mu) \leq R(T)$. In fact, the celebrated *Oseledets Theorem* [Oseledets, 1968] provides as many as dim Lyapunov exponents (counting with multiplicities) of which $\chi(\mu)$ is the largest; here we will restrict ourselves to the applications of the maximal exponent only. Using it, we can prove a simplified version of the Margulis–Ruelle Inequality, giving an estimate for the entropy of each ergodic measure. We abbreviate $\chi^+(\mu) = \max\{0, \chi(\mu)\}$.

Theorem 10.1.2 *Let $T : \mathfrak{M} \to \mathfrak{M}$ be a C^1 transformation of a compact Riemannian manifold $\mathfrak{M}$ of dimension* dim. *Let μ be an ergodic measure. Then*

$$h(\mu, T) \leq \mathsf{dim} \cdot \chi^+(\mu).$$

In particular, $\mathrm{h}(T) \leq \mathsf{dim} \cdot R(T) < \infty$.

Proof The last statement follows directly from the first one and the Variational Principle.

Since we fix μ, we can abbreviate $\chi^+(\mu)$ as χ^+. In view of the almost everywhere convergence (10.1.1), for $\delta > 0$ there exists a set $M \subset \mathfrak{M}$ and $n \in \mathbb{N}$ such that $\mu(M) > 1 - \gamma$ (where γ depends on δ and some constants, and will be specified later) and $\frac{1}{n} \log \|D_x T^n\| \leq \chi^+ + \delta$ for all $x \in M$. Because the derivative of T^n is uniformly continuous (T and hence T^n are of class C^1), there is a positive ε_0 such that $\log \|D_y T^n\| < n(\chi^+ + 2\delta)$ for all y in the ε_0-ball around each point $x \in M$.

We will observe $S = T^n$ rather than T. Now, we must resolve the standard obstacle: μ need not be ergodic under S. But, under S, μ has at most n ergodic components $\mu^{(i)}$ with the same entropy $h(\mu^{(i)}, S) = nh(\mu, T)$, and for at least one of them the set M has measure larger than $1 - \gamma$. By μ_0 we will denote a component $\mu^{(i)}$ which satisfies the last condition. Suppose we prove that $h(\mu_0, S) \le n \,\mathsf{dim} \cdot (\chi^+ + 3\delta)$. This will imply $h(\mu, T) \le \mathsf{dim} \cdot (\chi^+ + 3\delta)$, and since δ is arbitrary, we will get $h(\mu, T) \le \mathsf{dim} \cdot \chi^+$, as in the assertion. In virtue of Lemma 8.4.13, it now suffices to prove that for any open cover $\mathcal{U}$

$$\mathbf{h}(\mu_0, S, \mathcal{U}) \le n \,\mathsf{dim} \cdot (\chi^+ + 3\delta) \tag{10.1.3}$$

(we are using the variant entropy of a measure with respect to a cover, see Definition 8.3.12). Let F be a set of positive measure, such that every m-orbit (under S) starting in F visits the complement of M no more than $m\gamma$ times, for every m greater than some threshold integer m_0. Fix an open cover $\mathcal{U}$ and let ε denote the Lebesgue number of $\mathcal{U}$. We may assume that $\varepsilon < \varepsilon_0$. Now take an m larger than m_0 (how much larger, we will say in a moment), and let

$$\varepsilon_m = \varepsilon \cdot 2^{-mn(\chi^+ + 3\delta)}.$$

The whole manifold can be covered by a constant times $(1/\varepsilon_m)^{\mathsf{dim}}$ balls of radius $\varepsilon_m/2$ (locally the manifold looks like $\mathbb{R}^n$). Let B be one such ball containing a point $x \in F$. Note that the diameter of B does not exceed ε_m.

Observe the images of this ball under S^k for $k = 0, 1, \ldots, m-1$. Each time when $S^k x \in M$, and as long as $\mathsf{diam}(S^k(B)) < \varepsilon_0$, the norm of the derivative of S is bounded by $2^{n(\chi^+ + 2\delta)}$ on the entire set $S^k(B)$, hence the following image, $S^{k+1}(B)$, has diameter at most $2^{n(\chi^+ + 2\delta)}$ times $\mathsf{diam}(S^k(B))$. Otherwise, the analogous proportion can be estimated by $2^{nL(T)}$. By making m large enough (this makes ε_m very small), we can arrange that even applying the second (larger) proportion throughout the initial m_0 steps, the diameters of the images $S^k(B)$ remain smaller than ε (which is smaller than ε_0). For larger k we already control the proportion of times, when the first and second growth rates are applied, and then we can estimate the diameter of $S^k(B)$ more precisely:

$$\mathsf{diam}(S^k(B)) \le \mathsf{diam}(B) \cdot 2^{k(1-\gamma)n(\chi^+ + 2\delta)} \cdot 2^{k\gamma n L(T)}.$$

This estimate holds inductively, as long as the right-hand side does not exceed ε_0. We have

$$\log \mathsf{diam}(S^k(B)) \le \log \varepsilon - mn(\chi^+ + 3\delta) + kn((1-\gamma)(\chi^+ + 2\delta) + \gamma L(T)).$$

Now is the moment we specify γ: We require that the convex combination $(1-\gamma)(\chi^+ + 2\delta) + \gamma L(T)$ does not exceed $\chi^+ + 3\delta$. In such case

$$\log \mathsf{diam}(S^k(B)) \le \log \varepsilon - (m-k)n(\chi^+ + 3\delta),$$

hence $\mathsf{diam}(S^k(B)) < \varepsilon$ (which is indeed smaller than ε_0) for all observed times $k \le m$. This last estimate implies that every such image $S^k(B)$ is contained in an element of the cover $\mathfrak{U}$, hence B is fully contained in some element of $\mathfrak{U}^m$ (the "power" refers to the action of S). Thus, F can be covered by as many elements of $\mathfrak{U}^m$ as there are balls B that it intersects, i.e., at most a constant times $\varepsilon^{-\mathsf{dim}} \cdot 2^{mn\mathsf{dim}\cdot(\chi^+ + 3\delta)}$. Taking the logarithm, dividing by m and letting m grow to infinity, we obtain the desired estimate (10.1.3). □

For the sake of completeness, below we state the Margulis–Ruelle Inequality in its most general form [see e.g. Qian *et al.*, 2009].

Theorem 10.1.4 (Margulis–Ruelle Inequality) *Let $\mathfrak{M}$ be a compact Riemannian manifold and T a C^r-map ($r > 1$). Let μ be an invariant measure. Then*

$$h(\mu, T) \le \int \sum k_i(x)\chi_i^+(x)\, d\mu$$

(the sum includes all positive Lyapunov exponents $\chi_i^+(x)$ with multiplicities $k_i(x)$). □

We do not explain precisely what *Lyapunov exponents at points* $\chi_i(x)$ are, this is part of the Oseledets Theorem [Oseledets, 1968]. For an ergodic measure μ, these exponents (and multiplicities) are constant μ-almost everywhere, so the expression on the right can be written as $\sum k_i(\mu)\chi_i^+(\mu)$. This sum for an ergodic μ is obviously not larger than dim times the maximal Lyapunov exponent (or zero) $\chi^+(\mu)$, appearing in Theorem 10.1.2, hence that theorem is a weakening of the Margulis–Ruelle Inequality. The assertions are equivalent in dimension 1, where $\chi(\mu) = \int |f'(x)|\, d\mu(x)$ (in dimension 1 we denote the map by f and the derivative by f').

In certain situations the Margulis–Ruelle Inequality becomes an equality, providing an alternative way to compute the entropy. This is the case of *smooth measures*, i.e., measures which are absolutely continuous with respect to the Lebesgue measure with a C^2 density function. The corresponding theorem is known as the Pesin Entropy Formula [Pesin, 1977] (for automorphisms) and [Liu, 1998] (for endomorphisms):

Theorem 10.1.5 (Pesin Entropy Formula) *Let T be a C^r map ($r > 1$) of a compact Riemannian manifold $\mathfrak{M}$ and let μ be a smooth invariant measure on $\mathfrak{M}$. Then*

$$h(\mu, T) = \int \sum k_i(x)\chi_i^+(x)\, d\mu.$$

□

10.2 Tail entropy estimate

We state here a result proved by Jerome Buzzi estimating the topological tail entropy $\mathbf{h}^*(T)$ in case of a C^r map on a compact manifold [Buzzi, 1997]. The proof relies on an estimate of volume growth proved by Y. Yomdin [Yomdin, 1987].

Theorem 10.2.1 *Let T be a C^r map ($r > 1$) of a compact Riemannian manifold $\mathfrak{M}$ of dimension* dim. *Then*

$$\mathbf{h}^*(T) \leq \frac{\mathsf{dim} \cdot R(T)}{r}. \qquad \square$$

A similar estimate was obtained much earlier by Sheldon Newhouse [Newhouse, 1989], with $\mathbf{h}^*(T)$ replaced by another term $\mathbf{h}_{\mathrm{loc}}(T)$, defined as the supremum over all invariant measures of the Newhouse local entropy. Nowadays, using the entropy structure theory (the Tail Entropy Variational Principle and the fact that local entropy gives rise to an entropy structure) we know that $\mathbf{h}_{\mathrm{loc}}(T)$ equals $\mathbf{h}^*(T)$. This equality was not known in 1989.

A spectacular consequence of Theorem 10.2.1 is asymptotic h-expansiveness of C^∞ maps, which, combined with Theorem 9.3.3 gives a statement in terms of symbolic extensions. This was first observed in [Boyle *et al.*, 2002]. According to our leading interpretation, it says that C^∞-maps allow for a digitalization which is both "lossless and gainless."

Corollary 10.2.2 *Every C^∞ transformation of a compact Riemannian manifold $\mathfrak{M}$ is asymptotically h-expansive, i.e., admits a principal symbolic extension.* $\square$

Theorem 10.2.1 was refined by D. Burguet to the level of individual invariant measures [Burguet, 2008] (Theorem 10.2.1 follows from Theorem 10.2.3 given below via Theorem 8.4.9). Recall that u_1^T denotes the first function in the transfinite sequence associated with the entropy structure and equals $\lim_k \downarrow \widetilde{\theta}_k$, and its supremum is $\mathbf{h}^*(T)$.

Theorem 10.2.3 *Let T be a C^r map ($r > 1$) of a compact Riemannian manifold $\mathfrak{M}$ of dimension* dim. *Then for all invariant measures μ*

$$u_1^T(\mu) \leq \frac{\mathsf{dim} \cdot \overline{\chi}^+(\mu)}{r}. \qquad \square$$

Remark 10.2.4 On manifolds of dimension 1 (interval or circle), Theorem 10.2.3 and hence also Theorem 10.2.1 can be derived, without invoking Yomdin's algebraic machinery, from the Antarctic Theorem (9.4.17) and the (simplified) Margulis–Ruelle inequality (Theorem 10.1.2): by upper semicontinuity of $\overline{\chi}^+$, for measures ν in a neighborhood of μ, the function

$$\nu \mapsto \min\Big\{\frac{\overline{\chi}^+(\mu) - \overline{\chi}^+(\nu)}{r-1}, \chi^+(\mu)\Big\}$$

does not exceed $\frac{\chi^+(\mu)}{r}$ (plus a small error term). Next use the definition of u_1.

Except for C^∞ maps, the estimate of Theorem 10.2.1 or even that of Theorem 10.2.3 do not provide any information about symbolic extension entropy. This is due to the fact that whenever $\mathbf{h}^*(T)$ is positive, $\mathbf{h}_{\mathsf{sex}}(T)$ can be arbitrarily large, even infinite. The limitations on $\mathbf{h}_{\mathsf{sex}}(T)$ in smooth systems on manifolds of dimension higher than 1 are summarized in the next section.

10.3 Symbolic extensions of smooth systems

The one-dimensional case can be considered (almost) completely solved. Theorem 9.4.2 and the following Corollary 9.4.3 give an upper bound for the function h_{sex} and the number $\mathbf{h}_{\mathsf{sex}}(f)$, and by Burguet's example (Theorem 9.4.18) these estimates cannot be improved. In higher dimensions the situation, with a few exceptions, is open. One of the exceptions is the case of C^∞ maps, which have principal symbolic extensions, i.e., $h_{\mathsf{sex}} \equiv h$ (Theorem 10.2.1). It is believed that C^r transformations of compact Riemannian manifolds have symbolic extensions for any $r > 1$, and that an analog of Corollary 9.4.3 holds:

Conjecture 10.3.1 [Downarowicz and Newhouse, 2005] Let T be a C^r transformation of a compact Riemannian manifold $\mathfrak{M}$ of dimension dim, with $r > 1$. Then

$$\mathbf{h}_{\mathsf{sex}}(T) \le \mathbf{h}(T) + \frac{\mathsf{dim} \cdot R(T)}{r-1}.$$

We can refine this conjecture to the level of measures:

Conjecture 10.3.2 Let T and $\mathfrak{M}$ be as above. Then, on $\mathcal{M}_T(\mathfrak{M})$,

$$h_{\mathsf{sex}}(\mu) \le h(\mu) + \frac{\mathsf{dim} \cdot \overline{\chi}^+(\mu)}{r-1}.$$

Perhaps, this can be further improved by replacing the numerator $\mathsf{dim} \cdot \overline{\chi}^+(\mu)$ by the integral of the sum of the positive Lyapunov exponents with multiplicities, the same as in the Margulis–Ruelle Inequality or Pesin Entropy Formula. In case the above conjecture is true (in either version), one can ask whether the function on the right is, like in the one-dimensional case, a superenvelope.

At least we know that such estimates would be sharp. Chronologically, the first examples occur in [Downarowicz and Newhouse, 2005] and are more than just examples; the results show the behavior of a *typical* map in a class:

Theorem 10.3.3 *Let $\mathfrak{M}$ be a compact surface (Riemannian manifold of dimension 2). Among C^1 area-preserving non-Anosov diffeomorphisms $T : \mathfrak{M} \to \mathfrak{M}$ there is a residual set of mappings which do not have symbolic extensions.*

Theorem 10.3.4 *Let $\mathfrak{M}$ be a compact surface. Let $r \geq 2$ be an integer. In the set of all C^r diffeomorphisms $T : \mathfrak{M} \to \mathfrak{M}$ those which satisfy*

$$\mathbf{h}_{\mathsf{sex}}(T) \geq \mathbf{h}(T) + \frac{R(T)}{r-1}$$

contain an intersection of an open set and a residual set.

Notice that, compared to Conjecture 10.3.1, the factor dim is missing, and the theorems do not cover dimension 1. These faults are fixed by Burguet's example of a smooth interval map (Theorem 9.4.18). By a straightforward passage to product actions (and the product rule (9.3.10) for symbolic extension entropy), his example can be used to easily produce C^r mappings on higher-dimensional tori showing that the estimate in Conjecture 10.3.1 (if valid) is optimal in the class of all maps. Optimality for invertible maps is open. Optimality of the individual estimate in Conjecture 10.3.2 at each invariant measure (as well, of course, as the validity of the estimate) also remains an open question.

Let us mention one of the positive results considered a significant step forward. It is due to D. Burguet and it proves Conjecture 10.3.1 for C^2 surface diffeomorphisms [Burguet, in print].

Theorem 10.3.5 *Let $T : \mathfrak{M} \to \mathfrak{M}$ be a C^2 diffeomorphism of a compact surface. Then*

$$\mathbf{h}_{\mathsf{sex}}(T) \leq \mathbf{h}(T) + 2R(T).$$

In particular, such a T always has a symbolic extension. □

Remark 10.3.6 It is believed that the above estimate can be improved by removing the factor 2. This is motivated by the fact that for surface diffeomorphisms each ergodic measure has at most one positive Lyapunov exponent. Burguet's technique used in the proof does not allow this constant to be removed.

Part III

Entropy theory for operators

11

Measure-theoretic entropy of stochastic operators

11.1 A few words on operator dynamics

Definition 11.1.1 A continuous linear operator $\boldsymbol{T}$ on $L^1(\mu)$, where $(X, \mathfrak{A}, \mu)$ is a standard probability space, is called *doubly stochastic* when it satisfies three conditions

(i) $\boldsymbol{T}f$ is a nonnegative function whenever f is nonnegative,
(ii) $\boldsymbol{T}\mathbb{I} = \mathbb{I}$, where $\mathbb{I}$ is the constant function equal everywhere to 1,
(iii) $\int \boldsymbol{T}f \, d\mu = \int f \, d\mu$ for every $f \in L^1(\mu)$.

Definition 11.1.2 A continuous linear operator $\boldsymbol{T}$ on the space $C(X)$ of all continuous real functions on a compact metric space X which satisfies the above conditions (i) and (ii) is called a *Markov operator*.

It is elementary to see that every doubly stochastic operator is a contraction (i.e., Lipschitz with the constant 1) on $L^1(\mu)$, while every Markov operator is a contraction in the uniform norm. The reader will find more information on operators in the book [Neveu, 1965].

Every measure-preserving transfromation T of a probability space $(X, \mathfrak{A}, \mu)$ induces a doubly stochastic operator on $L^1(\mu)$ by composition: $\boldsymbol{T}f = f \circ T$. Similarly, every continuous transformation T of a compact metric space X defines by composition a Markov operator on $C(X)$. Thus, doubly stochastic operators are generalizations of measure-theoretic dynamical systems, while Markov operators generalize topological dynamical systems.

An important class of examples of operator dynamics is provided by stochastic processes generated by so-called transition probabilities. A *transition probability* is a map $x \mapsto P(x, \cdot)$ associating to each point a probability measure on the relevant sigma-algebra in such a way that for each measurable set A the function $x \mapsto P(x, A)$ is measurable. Then the operator given by

$$\boldsymbol{T}f(x) = \int f(y)P(x, dy) \tag{11.1.3}$$

is doubly stochastic on $L^1(\mu)$ for every $\boldsymbol{T}$-invariant measure μ, i.e., such that $\int P(x, A)\, d\mu = \mu(A)$ for each measurable set A. We can interpret the "dynamics" associated with such an operator so that the system that is currently in state x does not evolve to the predetermined state Tx (as happens in case of transformations) but chooses its next state randomly, according to the probability measure $P(x, \cdot)$.

It is known, for instance, that every Markov operator is associated with a transition probability, namely $P(x, \cdot) = \boldsymbol{T}^*\boldsymbol{\delta}_x$, where $\boldsymbol{T}^*$ denotes the dual operator acting on $C^*(X)$ identified (via the Riesz Theorem) with the space of all finite signed measures on X, and $\boldsymbol{\delta}_x$ is the point-mass at x. Every Markov operator (just like every continuous map on a compact space) has at least one invariant probability measure μ (i.e., such that $\boldsymbol{T}^*\mu = \mu$). For such a measure the formula (11.1.3) defines a doubly stochastic operator on $L^1(\mu)$. We will refer to this fact when discussing the variational principle for Markov operators. Transition probabilities will also occur in our examples.

Now we will try to say what we expect the entropy of operator dynamics to be. We focus on the measure-theoretic case. Suppose the phase space comes with a finite partition $\mathcal{P} = \{A_1, \dots, A_l\}$ and suppose the initial state of the system is some x whose "identity" is unknown to us. In a deterministic system (here we mean pointwise generated, i.e., associated with a transformation of the phase space) in each step we acquire the information in form of the label i of the cell A_i visited by $T^n x$, equivalently, the label of the cell of $T^{-n}(\mathcal{P})$ containing x. This allows us to locate the initial point in a specific cylinder, shrinking as n grows. The average "speed" (on the logarithmic scale) at which the cylinder shrinks is the dynamical entropy. In the operator case, in each step we only learn the values $\boldsymbol{T}^n \mathbb{1}_{A_i}(x)$ $(i = 1, 2, \dots, l)$. These values form a probability vector and each of them can be interpreted as the probability that the system starting from the state x will, after n steps, evolve to a state belonging to the cell A_i. Such information also helps us better identify the initial state. Usually we cannot locate it in a specific cylinder set, but our "uncertainty" about where the initial state might be should reduce, and the entropy should measure the exponential speed of this reduction.

For example, if the operator is such that the images of every function tend to some invariant function, the entropy of such an operator should be zero, because eventually we keep acquiring almost no new information. In this manner we are led to the following two postulates, which any reasonable notion of operator entropy should satisfy:

- If $\boldsymbol{T}$ is *pointwise generated*, then its operator entropy should coincide with the classical notion (either Kolmogorov–Sinai entropy, or topological entropy).
- If $\boldsymbol{T}$ has trivial behavior on functions, i.e., $\boldsymbol{T}^n f$ converges (to some function) for each f in the domain, then the operator entropy should be zero.

At this point we would like to illuminate the huge difference between so understood operator dynamics (and its entropy) and the dynamics of the associated transformation on trajectories. It is decisive for the correct understanding of operator entropy. Suppose that a doubly stochastic operator $\boldsymbol{T}$ is given by a transition probability $x \mapsto P(x, \cdot)$ and a $\boldsymbol{T}$-invariant measure μ on X. Then one can create a shift-invariant probability distribution Prob_μ on the space of all trajectories $X^{\mathbb{N}_0}$ (this is roughly the Ionescu-Tulcea Theorem, see [Ionescu-Tulcea, 1949]). In this manner, we obtain a classical dynamical system with a measure-preserving transformation (the shift). For a Markov operator $\boldsymbol{T}$ on $C(X)$ we can similarly pass to a continuous transformation (the shift) on a certain shift-invariant subset Y of $X^{\mathbb{N}_0}$ interpreted as the collection of all possible trajectories. Moreover, in both cases, the system on trajectories allows us to reconstruct the operator by the projection onto the coordinate zero. So, it seems that operator dynamics is just a particular case of pointwise dynamics. Well... yes, but not for our interpretation of information and entropy. In order to replace the operator dynamics by the dynamics of the shift transformation, we must essentially change the phase space, from X to $X^{\mathbb{N}_0}$. Unless $\boldsymbol{T}$ is a composition with a transformation, the initial state x does not determine its trajectory in $X^{\mathbb{N}_0}$, hence such a change replaces the states by something whose "identity" requires much more information. This is why the entropy of the operator dynamics has very little to do with the entropy of the associated shift map on trajectories.

Let us give a very simple example illustrating this difference. Consider a transition probability in which $P(x, \cdot) = \mu$ does not depend on x. It is clear that the associated stochastic operator $\boldsymbol{T}$ sends every function f to the constant function $g \equiv \int f \, d\mu$, invariant under $\boldsymbol{T}$. According to the second postulate, the operator entropy is immediately zero. On the other hand, in this example the Ionescu-Tulcea system on the space of trajectories is simply the independent process with the product measure $\mathsf{Prob}_\mu = \mu^{\mathbb{N}_0}$, and independent processes have positive Kolmogorov–Sinai entropy, infinite whenever μ is nonatomic.

In the literature one can find several attempts to define entropy for operators. In the measure-theoretic case, the most deeply investigated notion is the quantum dynamical entropy introduced by R. Alicki, J. Andries, M. Fannes and P. Tuyls (see [Alicki *et al.*, 1996] for the full course on this type of entropy) based on von Neumann's definition of entropy of a density matrix. A similar definition formulated for doubly stochastic operators (called there Markov operators) on the space of integrable functions was given by I. I. Makarov in [Makarov, 2000]. A quite different approach was presented by E. Ghys, R. Langevin and P. Walczak, in [Ghys *et al.*, 1986] and then studied by Kaminski and de Sam Lazaro [Kamiński and de Sam Lazaro, 2000].

The only topological entropy applicable to Markov operators on $C(X)$ known to us was defined by Langevin and Walczak in [Langevin and Walczak, 1994]. All these definitions satisfy the above two postulates. No other relations between them were ever established before [Downarowicz and Frej, 2005].

In this book we will exploit the most recent approach provided in [Downarowicz and Frej, 2005] to both measure-theoretic and topological entropies, allowing us, among other things, to establish equality between all of the preceding notions for doubly stochastic operators, and allowing us to prove a number of good properties not known earlier. The theory is under construction; we conclude Part III of this book by listing a number of open problems that await solution.

11.2 The axiomatic measure-theoretic definition

This section presents an axiomatic approach to measure-theoretic entropy. Accepting certain basic properties of entropy of measure-preserving transformations as indispensable, five construction steps and four axioms are formulated, which a general entropy of an action on functions should follow. Then we present the full proof of the fact that all entropy notions satisfying the axioms coincide on doubly stochastic operators. In particular, this establishes the equality, on such operators, between all the above-mentioned measure-theoretic entropies introduced by independent teams of authors.

11.2.1 The axioms

Given a standard probability space $(X, \mathfrak{A}, \mu)$, consider an operator (even not necessarily linear) $\boldsymbol{T} : \mathcal{L} \to \mathcal{L}$ on some collection $\mathcal{L}$ of measurable functions with range contained in [0,1] (of course, doubly stochastic operator fall in this category). Any reasonable way to define the *entropy* $h_\mu(\boldsymbol{T})$ of $\boldsymbol{T}$ with respect to μ would have to follow the major steps listed below:

(1°) One needs to specify $\mathbb{F}$, a collection of selected finite *families* $\mathcal{F}$ of functions belonging to $\mathcal{L}$. These families can be either ordered, admitting repetitions, or unordered i.e., treated as sets (hence without repetitions). The collection $\mathbb{F}$ should be $\boldsymbol{T}$-invariant, i.e., such that $\mathcal{F} \in \mathbb{F}$ implies $\boldsymbol{T}(\mathcal{F}) \in \mathbb{F}$, where $\boldsymbol{T}(\mathcal{F}) = \{\boldsymbol{T}f : f \in \mathcal{F}\}$. This collection should also contain a special *trivial family* $\mathcal{O}$ invariant under $\boldsymbol{T}$.

(2°) One has to specify an associative and commutative operation $\sqcup$ (called *join*) on these families, so that $\mathcal{F} \sqcup \mathcal{G} \in \mathbb{F}$ whenever $\mathcal{F} \in \mathbb{F}$ and $\mathcal{G} \in \mathbb{F}$, and with the cardinality of the joined family bounded by a number depending on the cardinalities of the components. For every $\mathcal{F} \in \mathbb{F}$ we should have $\mathcal{F} \sqcup \mathcal{O} = \mathcal{F}$. In case the families are ordered, all above statements are understood up to permutation.

(3°) One needs to define the *static entropy with respect to* μ on $\mathbb{F}$; a nonnegative (and finite) function $\mathcal{F} \mapsto H_\mu(\mathcal{F})$, which is zero for the trivial family.

(4°) Denoting

$$\mathcal{F}^n = \bigsqcup_{i=0}^{n-1} \boldsymbol{T}^i(\mathcal{F})$$

one then defines the *operator entropy* of $\mathcal{F}$ under the action of $\boldsymbol{T}$ as

$$h_\mu(\mathcal{F}) = h_\mu(\boldsymbol{T}, \mathcal{F}) = \limsup_{n\to\infty} \tfrac{1}{n} H_\mu(\mathcal{F}^n).$$

(5°) Eventually one sets

$$h_\mu(\boldsymbol{T}) = \sup_{\mathcal{F}\in\mathbb{F}} h_\mu(\boldsymbol{T}, \mathcal{F}).$$

For example, Kolmogorov–Sinai entropy for measurable maps uses $\mathbb{F}$ defined as families of characteristic functions corresponding to finite measurable partitions, and the join is obtained by pointwise multiplication "each by each" or equivalently by the application of pointwise infima. Some other, more general, definitions [e.g. Alicki *et al.*, 1996; Ghys *et al.*, 1986] use for $\mathbb{F}$ the measurable *partitions of unity*, $\mathcal{F} = \{f_1, \dots, f_r\}$ with each f_i nonnegative and with $\sum_i f_i = 1$ (actually $\sum_i f_i^2 = 1$ in [Alicki *et al.*, 1996]). In both cases the joins are performed via pointwise multiplication. In the definition in Section 11.3 we let $\mathbb{F}$ consist of all finite sets of measurable functions with range in $[0, 1]$, and the join is simply the union.

We remark that the construction of topological entropy (the version via covers) follows the same five steps. The only difference is the lack of reference to any measure. We leave matching the objects to the reader.

The construction consisting of steps (1°)–(5°) is usually accompanied by some definition of a conditional static entropy – a tool useful in verifying properties of entropy. At this level of generality we define this quantity by the formula analogous to (1.4.3)

$$H_\mu(\mathcal{F}|\mathcal{G}) = H_\mu(\mathcal{F} \sqcup \mathcal{G}) - H_\mu(\mathcal{G}). \tag{11.2.1}$$

The notion of static entropy should posses some elementary "nice" properties known for the Shannon static entropy of measurable partitions. This leads to formulation of several conditions which we call *axioms of static entropy*.

(A) **Monotonicity axiom** (compare (1.6.6), (1.6.7))
For $\mathcal{F}$, $\mathcal{G}$ and $\mathcal{H}$ belonging to $\mathbb{F}$ we require that

$$H_\mu(\mathcal{F}|\mathcal{H}) \le H_\mu(\mathcal{F} \sqcup \mathcal{G}|\mathcal{H}), \tag{11.2.2}$$

$$H_\mu(\mathcal{F}|\mathcal{G} \sqcup \mathcal{H}) \le H_\mu(\mathcal{F}|\mathcal{G}). \tag{11.2.3}$$

Remark 11.2.4 We are used to the situation where the monotonicities are with respect to some partial order among the objects on which the static entropy is defined: the relation of refining. To make our axiom (A) compatible with such an approach it suffices to agree that a family $\mathcal{F}$ is a *refinement* of $\mathcal{G}$ if and only if $\mathcal{G} = \mathcal{F} \sqcup \mathcal{H}$ for some $\mathcal{H}$.

(B) **Continuity axiom**
For two families $\mathcal{F} = \{f_1, \dots, f_r\}$ and $\mathcal{G} = \{g_1, \dots, g_r\}$ of the same length r we define their $L^1(\mu)$-distance as

$$\mathsf{dist}(\mathcal{F}, \mathcal{G}) = \min_\pi \Big\{ \max_{1 \le i \le r} \|f_i - g_{\pi(i)}\| \Big\}$$

($\|\cdot\|$ denotes the norm in $L^1(\mu)$), where the minimum ranges over all permutations π of the set $\{1, \dots, r\}$. If the lengths of the families are different, we enhance the smaller family by adding to it the appropriate number of zero functions. (The distance dist generalizes d_1 on partitions, see Section 1.7.) In this axiom we require that for every $r \ge 1$ and $\varepsilon > 0$ there is a $\delta > 0$ such that if $\mathcal{F}$, $\mathcal{G}$ and $\mathcal{H}$ have cardinalities at most r and $\mathsf{dist}(\mathcal{F}, \mathcal{G}) < \delta$, then

$$|H_\mu(\mathcal{F}|\mathcal{H}) - H_\mu(\mathcal{G}|\mathcal{H})| < \varepsilon \quad \text{and} \tag{11.2.5}$$

$$\big|H_\mu(\mathcal{H}|\mathcal{F}) - H_\mu(\mathcal{H}|\mathcal{G})\big| < \varepsilon. \tag{11.2.6}$$

(compare Facts 1.7.9 and 1.7.10).[1]

[1] In [Downarowicz and Frej, 2005] this axiom also requires continuity (in dist) of the operation $\sqcup$. The version given here is weaker while it has the same practical consequences.

(C) **Partitions axiom**
If $\mathcal{P}$ is a measurable partition of X, then $\mathbb{I}_{\mathcal{P}} = \{\mathbb{I}_A : A \in \mathcal{P}\}$ denotes the family of the corresponding characteristic functions. We require that characteristic functions of measurable sets belong to $\mathcal{L}$ and that $\mathbb{F}$ contains the families $\mathbb{I}_{\mathcal{P}}$ of all finite measurable partitions $\mathcal{P}$ of X. The entropy H_μ should coincide on partitions with the Shannon entropy:

$$H_\mu(\mathbb{I}_{\mathcal{P}}) = H(\mu, \mathcal{P}) \quad \Big(= - \sum_{A \in \mathcal{P}} \mu(A) \log \mu(A) \Big),$$

$$H_\mu\Big(\bigsqcup_{k=1}^{n} \mathbb{I}_{\mathcal{P}_k} \Big) = H\big(\mu, \bigvee_{k=1}^{n} \mathcal{P}_k\big).$$

The next (and last) axiom is rather technical. Roughly speaking, it asserts that the entropy of each family $\mathcal{F}$ must be majorated by the entropy of some finite partition.

(D) **Domination axiom**
Each family of functions $\mathcal{F} \in \mathbb{F}$ together with a partition $\varkappa$ of the unit interval determine the following partition of X:

$$\mathcal{F}^{-1}(\varkappa) = \bigvee_{f \in \mathcal{F}} f^{-1}(\varkappa). \tag{11.2.7}$$

This axiom requires that for every $r \geq 1$ and $\varepsilon > 0$ there exists $\gamma > 0$ such that every family $\mathcal{F}$ of cardinality not exceeding r and every partition $\varkappa$ of $[0,1]$ into finitely many subintervals of lengths not exceeding γ satisfy

$$H_\mu(\mathcal{F} | \mathbb{I}_{\mathcal{F}^{-1}(\varkappa)} \sqcup \mathcal{O}_\varkappa) < \varepsilon,$$

where $\mathcal{O}_\varkappa \in \mathbb{F}$ is some auxiliary family of functions, depending only on $\varkappa$, satisfying

$$\lim_n \frac{1}{n} H_\mu \left(\bigsqcup_{i=0}^{n-1} \mathcal{O}_\varkappa \right) = 0.$$

The family $\mathcal{O}_\varkappa$ is added in order to allow some of the definitions of operator entropy to comply with the axioms. Usually $\mathcal{O}_\varkappa$ is either trivial, which means it can be skipped, or it is a family of some constant functions.

By a brief inspection we verify that all the mentioned examples of notions of entropy indeed follow the construction steps (1°)–(5°), and that the associated notions of static entropy satisfy the axioms (see lemmas 11.1 and 11.4 in [Alicki *et al.*, 1996] for axiom (A) with regard to the quantum entropy, the axiom (D) is implicitly included in the proof of Theorem 11.2 there; see

[Kamiński and de Sam Lazaro, 2000] for the axiom (A) with regard to the entropy of [Ghys *et al.*, 1986] and the proof of the main theorem in [Ghys *et al.*, 1986] for axiom (D); the rest is either obvious or explicit).

11.2.2 Elementary properties of operator entropy

In spite of the fact that the axioms completely determine the final notion obtained in step (5°) (see Theorem 11.2.32), each of them, especially the first one, has its own set of consequences, which we now review.

Definition 11.2.1 of conditional entropy alone (without any help of the axioms) easily implies the analog of (1.6.3) (compare Exercise 1.3)

$$H_\mu(\mathcal{F} \sqcup \mathcal{G}|\mathcal{H}) = H_\mu(\mathcal{F}|\mathcal{G} \sqcup \mathcal{H}) + H_\mu(\mathcal{G}|\mathcal{H}). \tag{11.2.8}$$

This, and the monotonicities of axiom (A) (and in one case the existence of the trivial family), imply the familiar subadditivity formulae (compare (1.6.9) through (1.6.12)):

Fact 11.2.9

$$H_\mu(\mathcal{F} \sqcup \mathcal{G}|\mathcal{H}) \le H_\mu(\mathcal{F}|\mathcal{H}) + H_\mu(\mathcal{G}|\mathcal{H}), \tag{11.2.10}$$

$$H_\mu(\mathcal{F} \sqcup \mathcal{G}) \le H_\mu(\mathcal{F}) + H_\mu(\mathcal{G}), \tag{11.2.11}$$

$$H_\mu(\mathcal{F} \sqcup \mathcal{F}'|\mathcal{G} \sqcup \mathcal{G}') \le H_\mu(\mathcal{F}|\mathcal{G}) + H_\mu(\mathcal{F}'|\mathcal{G}'), \tag{11.2.12}$$

$$H_\mu(\mathcal{F}|\mathcal{H}) \le H_\mu(\mathcal{F}|\mathcal{G}) + H_\mu(\mathcal{G}|\mathcal{H}). \tag{11.2.13}$$

□

Remark 11.2.14 It is not hard to verify that (11.2.2) and (11.2.10) imply (11.2.3) (we leave it as Exercise 11.1). In [Downarowicz and Frej, 2005] the monotonicity axiom was given in an equivalent form consisting of (11.2.2) and (11.2.10). Here we have made a cosmetic change to make the axiom more homogeneous.

The formula (11.2.11) easily implies that for every $l \ge 1$

$$h_\mu(\boldsymbol{T}, \boldsymbol{T}^l\mathcal{F}) = h_\mu(\boldsymbol{T}, \mathcal{F}), \tag{11.2.15}$$

where $h_\mu(\boldsymbol{T}, \mathcal{F})$ is defined in step (4°). The verification is left to the reader as Exercise 11.2; we will use this equality in due time. The next immediate consequence of the sole axiom (A) is the *power rule*. Again, the proof is left as Exercise 11.3.

Fact 11.2.16 *For any doubly stochastic operator $\boldsymbol{T}$, $n \in \mathbb{N}$ and $\mathcal{F} \in \mathbb{F}$,*

$$h_\mu(\boldsymbol{T}^n, \mathcal{F}^n) = nh_\mu(\boldsymbol{T}, \mathcal{F}), \quad \textit{and}$$
$$h_\mu(\boldsymbol{T}^n) = nh_\mu(\boldsymbol{T}). \qquad \square$$

We remark that operators are usually not invertible, so we do not consider negative iterates (and actions of $\mathbb{Z}$). It can be easily proved (Exercise 11.4) that all invertible doubly stochastic operators or Markov operators are pointwise generated.

We proceed with other consequences of the axioms. The following fundamental fact establishes the first major postulate listed in the introduction to Part III.

Theorem 11.2.17 *If $\boldsymbol{T}$ is a composition with a measure-preserving transformation T on a probability space, then any notion of entropy following the construction steps* (1°)–(5°) *and with the associated static entropy satisfying axioms* (A)–(D) *coincides with the Kolmogorov–Sinai entropy.*

Proof Axiom (C) implies that $h(\mu, T) \leq h_\mu(\boldsymbol{T})$, while axioms (D) and (A) give the reversed inequality. The easy details are left to the reader. $\square$

Notice that we do not assume $H_\mu(\mathcal{F}) = H_\mu(\boldsymbol{T}\mathcal{F})$, hence, in general, we do not have subadditivity of the sequence $H_\mu(\mathcal{F}^n)$ and we do not have the alternative formula $h_\mu(\boldsymbol{T}, \mathcal{F}) = \lim_n H_\mu(\mathcal{F}|\mathcal{F}^{[1,n]})$. Neither do we assume that $\mathcal{F} \sqcup \mathcal{F} = \mathcal{F}$ (or even that $H_\mu(\mathcal{F}|\mathcal{F}) = 0$). This is the reason why axiom (A) alone is insufficient to even prove that $\boldsymbol{T}$-invariant families have dynamical entropy zero, and axioms (A) and (B) together do not imply the same for families whose orbits converge. In order to prove the second major postulate we need to involve the other axioms as well (actually (B) is not used).

Theorem 11.2.18 *If $\boldsymbol{T}$ is a doubly stochastic operator and $\mathcal{F}$ consists of functions f such that their orbits $\boldsymbol{T}^n f$ converge in $L^1(\mu)$, then $h_\mu(\mathcal{F}) = 0$. In particular, if the orbits of all functions converge, then $h_\mu(\boldsymbol{T}) = 0$.*

Proof Fix some $\varepsilon > 0$. Take a family $\mathcal{F}$ of cardinality r which fulfills the assumption of the theorem. Let γ be as specified in axiom (D) for r and ε and fix a finite partition $\varkappa$ of $[0,1]$ into subintervals of lengths smaller than γ. Choose a large integer l and set $\mathcal{F}_0 = \boldsymbol{T}^l\mathcal{F}$. We know that $h_\mu(\mathcal{F}) = h_\mu(\mathcal{F}_0)$. We will show the latter to be arbitrarily small. We will abbreviate $\varkappa_i = \mathbb{I}_{(\boldsymbol{T}^i\mathcal{F}_0)^{-1}(\varkappa)}$. Using (11.2.1), and then axiom (A) in the form of an iterated version of (11.2.12) and (11.2.11), we get

$$\frac{1}{n}H_\mu(\mathcal{F}_0{}^n) = \frac{1}{n}H_\mu\Big(\mathcal{F}_0{}^n\Big|\bigsqcup_{i=0}^{n-1}(\varkappa_i \sqcup \mathcal{O}_\varkappa)\Big) + \frac{1}{n}H_\mu\Big(\bigsqcup_{i=0}^{n-1}(\varkappa_i \sqcup \mathcal{O}_\varkappa)\Big) \leq$$
$$\frac{1}{n}\sum_{i=0}^{n-1} H_\mu(\boldsymbol{T}^i\mathcal{F}_0|\varkappa_i \sqcup \mathcal{O}_\varkappa) + \frac{1}{n}H_\mu\Big(\bigsqcup_{i=0}^{n-1}\varkappa_i\Big) + \frac{1}{n}H_\mu\Big(\bigsqcup_{i=0}^{n-1}\mathcal{O}_\varkappa\Big).$$

By axiom (D), the average in the last line above does not exceed ε and the last term converges to zero with n. The middle term involves partitions, so by axiom (C), it equals the Shannon entropy

$$\frac{1}{n}H\Big(\bigvee_{i=0}^{n-1}(\boldsymbol{T}^i\mathcal{F}_0)^{-1}(\varkappa)\Big).$$

By (1.4.3) and (1.6.9), this is not larger than

$$\frac{1}{n}H(\mathcal{F}_0{}^{-1}(\varkappa)) + \frac{1}{n}\sum_{i=1}^{n-1} H((\boldsymbol{T}^i\mathcal{F}_0)^{-1}(\varkappa)|\mathcal{F}_0{}^{-1}(\varkappa)).$$

If l is large, the families $\boldsymbol{T}^i\mathcal{F}_0$ are all close to $\mathcal{F}_0$ in the pseudometric dist, which easily implies that the partitions $(\boldsymbol{T}^i\mathcal{F}_0)^{-1}(\varkappa)$ are close to $\mathcal{F}_0^{-1}(\varkappa)$ in d_1. By the elementary property $H(\mathcal{P}|\mathcal{P}) = 0$ of the Shannon entropy (see (1.6.5); this is what we do not assume for the operator entropy) and its d_1-continuity of Fact 1.7.9, the last displayed expression is (for large l) smaller than ε. Now we let n grow to infinity, which yields $h_\mu(\mathcal{F}) = h_\mu(\mathcal{F}_0) < 2\varepsilon$. □

11.2.3 Asymptotic lattice stability

This subsection has nothing to do with entropy. It describes a general property of doubly stochastic operators on $L^1(\mu)$. Throughout the rest of this section $\boldsymbol{T}$ denotes a doubly stochastic operator on $L^1(\mu)$.

We will treat $L^1(\mu)$ as a lattice, that is, we will consider the lattice operations $\vee$ (maximum of two functions) and $\wedge$ (minimum of two functions). These operations are uniformly continuous in the $L^1(\mu)$-norm; it is elementary to see that

$$\|f_1 \vee f_2 - g_1 \vee g_2\| \leq \|f_1 - g_1\| + \|f_2 - g_2\| \tag{11.2.19}$$

(and similarly for $\wedge$). It is known that pointwise generated doubly stochastic operators are exactly those which preserve the lattice operations (see Exercise 11.5). We will show that doubly stochastic operators are, in a sense, not that different; they "eventually almost preserve" the lattice operations (we call this property *asymptotic lattice stability*).

Lemma 11.2.20 *Let f, g be two bounded measurable functions on X. For every $\delta > 0$ there exists $l_0 \in \mathbb{N}$ such that for every $l \geq l_0$ and any $n \in \mathbb{N}$,*

$$\|\boldsymbol{T}^n(\boldsymbol{T}^l f \vee \boldsymbol{T}^l g) - (\boldsymbol{T}^{n+l} f \vee \boldsymbol{T}^{n+l} g)\| < \delta$$

and

$$\|\boldsymbol{T}^n(\boldsymbol{T}^l f \wedge \boldsymbol{T}^l g) - (\boldsymbol{T}^{n+l} f \wedge \boldsymbol{T}^{n+l} g)\| < \delta.$$

Proof Clearly, we have $\boldsymbol{T}(f \vee g) \geq \boldsymbol{T}f$ and $\boldsymbol{T}(f \vee g) \geq \boldsymbol{T}g$. This and a symmetric argument imply

$$\boldsymbol{T}(f \vee g) \geq \boldsymbol{T}f \vee \boldsymbol{T}g \geq \boldsymbol{T}f \wedge \boldsymbol{T}g \geq \boldsymbol{T}(f \wedge g). \tag{11.2.21}$$

Since $\boldsymbol{T}$ preserves the measure, for each $n \in \mathbb{N}$ we obtain

$$\int \boldsymbol{T}^n f \vee \boldsymbol{T}^n g \, d\mu = \int \boldsymbol{T}(\boldsymbol{T}^n f \vee \boldsymbol{T}^n g) \, d\mu \geq \int \boldsymbol{T}^{n+1} f \vee \boldsymbol{T}^{n+1} g \, d\mu \geq$$
$$\int \boldsymbol{T}^{n+1} f \wedge \boldsymbol{T}^{n+1} g \, d\mu \geq \int \boldsymbol{T}^n f \wedge \boldsymbol{T}^n f \, d\mu,$$

producing two sequences of integrals: on the left a decreasing sequence and on the right an increasing one. Both sequences must converge. Given $\delta > 0$ one can find l_0 so large that for every $l \geq l_0$ and every $n \in \mathbb{N}$

$$0 \leq \int \boldsymbol{T}^l f \vee \boldsymbol{T}^l g \, d\mu - \int \boldsymbol{T}^{n+l} f \vee \boldsymbol{T}^{n+l} g \, d\mu \leq \delta.$$

Since $\boldsymbol{T}^n$ preserves the measure and the pointwise inequality (11.2.21) holds between $\boldsymbol{T}^n(\boldsymbol{T}^l f \vee \boldsymbol{T}^l g)$ and $\boldsymbol{T}^{n+l} f \vee \boldsymbol{T}^{n+l} g$, the above difference represents the desired L^1-distance. The proof for minima is symmetric. □

By an r-argument *lattice polynomial* we shall mean any finite formal expression involving a sequence of arguments $f_1, \ldots, f_r$ (not necessarily all of them), bound by the lattice operations (and brackets). An example of a 3-argument lattice polynomial is

$$\vartheta(f, g, h) = (f \vee g) \wedge (g \vee h) \wedge (f \vee h).$$

Note that at each point x the lattice polynomial chooses one of the values $f_i(x)$ depending only on the order of the numbers $f_1(x), \ldots, f_r(x)$. Thus there are only finitely many different r-argument lattice polynomials treated as functions $(L^1(\mu))^r \to L^1(\mu)$.

Definition 11.2.22 We say that a finite (ordered) family $\mathcal{F} = \{f_1, \ldots, f_r\}$ of functions in $L^1(\mu)$ is *lattice δ-stable* under $\boldsymbol{T}$ (for some $\delta > 0$) if for any r-argument lattice polynomial ϑ and any $n \geq 1$ the following holds

$$\|\boldsymbol{T}^n(\vartheta(\mathcal{F})) - \vartheta(\boldsymbol{T}^n(\mathcal{F}))\| < \delta.$$

Theorem 11.2.23 *For every finite family $\mathcal{F} = \{f_1, \dots, f_r\} \subset L^1(\mu)$, and any $\delta > 0$, there exists an integer $l_0 \geq 0$ such that for each $l \geq l_0$ the family $\mathcal{F}_l = \boldsymbol{T}^l(\mathcal{F})$ is lattice δ-stable under $\boldsymbol{T}$.*

Proof Since there are only finitely many r-argument lattice polynomials, it suffices to prove the stability under each lattice polynomial separately. The proof is by induction over the number of lattice operations in the polynomial. For lattice polynomials with zero operations the statement is trivial. For those with one operation ($\vee$ of $\wedge$), the assertion coincides with Lemma 11.2.20. Every lattice polynomial can be broken as

$$\vartheta(\mathcal{F}) = \vartheta_1(\mathcal{F}) \vee \vartheta_2(\mathcal{F}) \quad \text{or} \quad \vartheta_1(\mathcal{F}) \wedge \vartheta_2(\mathcal{F}),$$

where ϑ_1 and ϑ_2 have strictly less operations than ϑ. Since the proof in either case is identical, we consider the first option only. By the inductive assumption, we can find an integer l' such that the family $\mathcal{F}' = \boldsymbol{T}^{l'}(\mathcal{F})$ is $\delta/5$-stable under both ϑ_1 and ϑ_2. Lemma 11.2.20 applied to the functions $f = \vartheta_1(\mathcal{F}')$ and $g = \vartheta_2(\mathcal{F}')$ with the parameter $\delta/5$ produces an integer l_0'. We set $l_0 = l' + l_0'$. Now, for $l \geq l_0$ and any $n \geq 1$ we have

$$\vartheta(\boldsymbol{T}^n(\mathcal{F}_l)) = \vartheta_1(\boldsymbol{T}^n(\mathcal{F}_l)) \vee \vartheta_2(\boldsymbol{T}^n(\mathcal{F}_l)) \approx$$

$$\boldsymbol{T}^{n+l-l'}(\vartheta_1(\mathcal{F}')) \vee \boldsymbol{T}^{n+l-l'}(\vartheta_2(\mathcal{F}')) \approx \boldsymbol{T}^n\left(\boldsymbol{T}^{l-l'}(\vartheta_1(\mathcal{F}')) \vee \boldsymbol{T}^{l-l'}(\vartheta_2(\mathcal{F}'))\right) \approx$$

$$\boldsymbol{T}^n\left(\vartheta_1(\boldsymbol{T}^{l-l'}(\mathcal{F}')) \vee \vartheta_2(\boldsymbol{T}^{l-l'}(\mathcal{F}'))\right) = \boldsymbol{T}^n(\vartheta(\mathcal{F}_l)),$$

where the first approximation is up to $2\delta/5$ in the norm and follows from the $\delta/5$-stability of $\mathcal{F}'$ and (11.2.19), the second approximation is up to $\delta/5$ and follows from the inequality $l - l' \geq l_0'$ and the way we employed Lemma 11.2.20. The last approximation is up to $2\delta/5$ again, and follows by the same reasons as the first one, plus the fact that $\boldsymbol{T}^n$ is a contraction in the norm. Jointly, the approximation is up to δ, as we need. □

By an r-argument *lattice expression* we will understand a lattice polynomial with some finite number s of arguments applied to functions $g_1, \dots, g_s$ being linear combinations of the functions $1, f_1, \dots, f_r$ (with some fixed constant coefficients). An example of a 2-argument lattice expression is

$$\Theta(f, g) = ((2f - 1) \vee 3) \wedge (f + 2g + 1).$$

Since $\boldsymbol{T}$ is linear and preserves the constants, the application of $\boldsymbol{T}$ to the functions f_i is equivalent to the application of $\boldsymbol{T}$ to the functions g_j. Thus we derive the following consequence of Theorem 11.2.23:

Corollary 11.2.24 *For every finite family $\mathcal{F} = \{f_1, \dots, f_r\} \subset L^1(\mu)$, any r-argument lattice expression Θ, and any $\delta > 0$, there exists an integer $l \geq 0$ such that the family $\mathcal{F}_0 = \boldsymbol{T}^l(\mathcal{F})$ satisfies, for all $n \geq 1$,*

$$\|\boldsymbol{T}^n(\Theta(\mathcal{F}_0)) - \Theta(\boldsymbol{T}^n(\mathcal{F}_0))\| < \delta. \qquad \square$$

Since the number of r-argument lattice expressions is infinite, we cannot claim the existence of a common l for all such expressions.

Now we say how a characteristic function of the set $\{t < f < s\}$ (formally $\{x : t < f(x) < s\}$, where $t, s \in [0, 1]$) can be approximated by 1-argument lattice expressions. The fact below is completely obvious and the verification is left to the reader as Exercise 11.6:

Lemma 11.2.25 *Fix $f \in L^1(\mu)$ and a pair of real numbers $s > t$. For $m \in \mathbb{N}$ let*

$$\Theta_{m,t,s}(f) = (m((f - t) \wedge (s - f)) \vee 0) \wedge 1.$$

Then

$$\Theta_{m,t,s}(f) \leq 1\!\!1_{\{t<f<s\}} \quad \text{and} \tag{11.2.26}$$

$$\|1\!\!1_{\{t<f<s\}} - \Theta_{m,t,s}(f)\| \leq \mu\{f \in (t, t + \tfrac{1}{m}] \cup [s - \tfrac{1}{m}, s)\}. \tag{11.2.27}$$

$\square$

We now give the key technical tool to prove the uniqueness Theorem 11.2.32.

Lemma 11.2.28 *Fix a family $\mathcal{F} = \{f_1, \dots, f_r\}$ of bounded measurable functions, also fix $\gamma > 0$ and $\delta > 0$. Then there exists an $l \in \mathbb{N}$ and a finite partition $\varkappa$ of the range by intervals of lengths between $\gamma/2$ and γ such that for every $n \in \mathbb{N}$ and $\mathcal{F}_0 = \boldsymbol{T}^l(\mathcal{F})$ it holds that*

$$\mathrm{dist}(\boldsymbol{T}^n(1\!\!1_{\mathcal{F}_0^{-1}(\varkappa)}), 1\!\!1_{(\boldsymbol{T}^n(\mathcal{F}_0))^{-1}(\varkappa)}) < \delta.$$

Proof For simplicity assume that the range of all f_i is contained in $[0, 1]$. We fix some integer m (we will say later how large). Let $\varkappa_0$ be the partition of $[0, 1]$ into m equal intervals and denote by t_j ($j = 0, \dots, m$) their endpoints ordered increasingly. We can shift the endpoints by insignificant distances to assure that none of the functions $\boldsymbol{T}^n f_i$ ($n \in \mathbb{N}$, $i = 1, \dots, r$) assumes any of the values t_j with positive probability. We take all lattice polynomials with as many arguments as there are triples (i, j, k) with $i = 1, \dots, r$, $0 \leq j < k \leq m$, and we apply them to the functions $1\!\!1_{\{t_j<f_i<t_k\}}$. In this manner we obtain the characteristic functions of all sets in the algebra generated by $\mathcal{F}^{-1}(\varkappa_0)$. Next, in each of these polynomials, we replace $1\!\!1_{\{t_j<f_i<t_k\}}$ by $\Theta_{m,t_j,t_k}(f_i)$

(for all i, j, k). We have created a finite family of r-argument lattice expressions applied to $\mathcal{F}$. We pick an l for which the assertion of Corollary 11.2.24 is satisfied with $\delta' = 1/\sqrt{m}$ for all lattice expressions in this family (which includes the 1-argument lattice expressions $\Theta_{m,t_j,t_k}(f_i)$). We let $\mathcal{F}_0 = \boldsymbol{T}^l(\mathcal{F})$.

Recall that we have arranged that for each $f \in \mathcal{F}_0$ the measures of the sets $\{f \in (t_j, t_{j+1}]\}$ do not depend on including or excluding the endpoints. Clearly, with f fixed, only at most $\sqrt{m}$ of these sets can have measure equal to or larger than $1/\sqrt{m}$. So, all but (at most) $2r\sqrt{m}$ points t_j satisfy both

$$\mu\{f \in (t_j, t_{j+1}]\} < \tfrac{1}{\sqrt{m}} \quad \text{and} \quad \mu\{f \in [t_{j-1}, t_j)\} < \tfrac{1}{\sqrt{m}} \tag{11.2.29}$$

for all $f \in \mathcal{F}_0$. Choosing (at the start) m large enough we can guarantee that each interval of length $\gamma/2$ contains at least one such "good" point t_j. We can now easily select a subset of the "good" points t_j which partitions $[0, 1]$ into subintervals of lengths between $\gamma/2$ and γ. We denote these points by s_k ($k = 1, \ldots, r'$) and we let $\varkappa$ denote the corresponding partition. It now suffices to verify the assertion of the lemma.

Combining (11.2.27) and (11.2.29) we get

$$\|\mathbb{I}_{\{s_k < f \le s_{k+1}\}} - \Theta_{m,s_k,s_{k+1}}(f)\| < \tfrac{2}{\sqrt{m}} \tag{11.2.30}$$

for each k and $f \in \mathcal{F}_0$.

For a moment we fix $f \in \mathcal{F}_0$. The sum over k of the functions $\Theta_{m,s_k,s_{k+1}}(f)$ is majorated by the sum of the characteristic functions $\mathbb{I}_{\{s_k < f \le s_{k+1}\}}$, which is 1 everywhere. The integral of the first sum is at least $1 - \frac{2}{\gamma}\frac{2}{\sqrt{m}}$. This is preserved by $\boldsymbol{T}^n$, i.e, the integral of the sum of the functions $\boldsymbol{T}^n(\Theta_{m,s_k,s_{k+1}}(f))$ is also at least $1 - \frac{2}{\gamma}\frac{2}{\sqrt{m}}$. By the choice of l, if we apply $\boldsymbol{T}^n$ "inside" rather than "outside," the integral of the sum will change by at most $\frac{2}{\gamma}\frac{1}{\sqrt{m}}$, i.e., the integral of the sum of the functions $\Theta_{m,s_k,s_{k+1}}(\boldsymbol{T}^n f)$ is at least $1 - \frac{2}{\gamma}\frac{3}{\sqrt{m}}$. Each of the latter functions is majorated by $\mathbb{I}_{\{s_k < \boldsymbol{T}^n f \le s_{k+1}\}}$, and the sum of these last characteristic functions also equals 1 everywhere. Thus, for each k (and every f), we have proved

$$\|\mathbb{I}_{\{s_k < \boldsymbol{T}^n f \le s_{k+1}\}} - \Theta_{m,s_k,s_{k+1}}(\boldsymbol{T}^n f)\| \le \tfrac{2}{\gamma}\tfrac{3}{\sqrt{m}}. \tag{11.2.31}$$

Now fix a characteristic function $\mathbb{I}_A$ of a set $A \in \mathcal{F}_0^{-1}(\varkappa)$. This function equals a lattice polynomial ϑ applied to the functions $\mathbb{I}_{s_k < f \le s_{k+1}}$, which we abbreviate as $\vartheta\{\mathbb{I}_{\{s_k < f \le s_{k+1}\}} : f, k\}$. Then $\vartheta\{\mathbb{I}_{\{s_k < \boldsymbol{T}^n f \le s_{k+1}\}} : f, k\}$ is the characteristic function of a set $A' \in (\boldsymbol{T}^n(\mathcal{F}_0))^{-1}(\varkappa)$. We have

$$\begin{aligned}
&\|\boldsymbol{T}^n(\mathbb{I}_A) - \mathbb{I}_{A'}\| = \\
&\quad \|\boldsymbol{T}^n(\vartheta\{\mathbb{I}_{\{s_k<f\le s_{k+1}\}} : f,k\}) - \vartheta\{\mathbb{I}_{\{s_k<\boldsymbol{T}^n f\le s_{k+1}\}} : f,k\}\| \le \\
&\quad \|\boldsymbol{T}^n(\vartheta\{\mathbb{I}_{\{s_k<f\le s_{k+1}\}} : f,k\}) - \boldsymbol{T}^n(\vartheta\{\Theta_{m,s_k,s_{k+1}}(f) : f,k\})\| + \\
&\quad \|\boldsymbol{T}^n(\vartheta\{\Theta_{m,s_k,s_{k+1}}(f) : f,k\}) - \vartheta\{\Theta_{m,s_k,s_{k+1}}(\boldsymbol{T}^n f) : f,k\}\| + \\
&\quad \|\vartheta\{\Theta_{m,s_k,s_{k+1}}(\boldsymbol{T}^n f) : f,k\} - \vartheta\{\mathbb{I}_{\{s_k<\boldsymbol{T}^n f\le s_{k+1}\}} : f,k\}\|.
\end{aligned}$$

The first of the last three distances does not exceed $2/\sqrt{m}\cdot K$, where K is the maximal number of operations in all considered lattice polynomials ϑ (and depends only on r and γ), by using (11.2.30), (11.2.19) and the fact that $\boldsymbol{T}^n$ is a contraction. The second distance is at most $1/\sqrt{m}$ because the composition of ϑ with the expressions $\Theta_{m,s_k,s_{k+1}}$ is among the lattice expressions for which l was selected (using Corollary 11.2.24) for this very estimate to hold on $\mathcal{F}_0 = \boldsymbol{T}^l(\mathcal{F})$. The last distance does not exceed $\frac{2}{\gamma}\frac{3}{\sqrt{m}}\cdot K$, by (11.2.31) and again (11.2.19). Jointly the initial distance is at most $C/\sqrt{m}$, where C depends on r and γ. The number of functions $\mathbb{I}_A \in \mathcal{F}_0^{-1}(\varkappa)$ does not exceed another constant also depending only on r and γ. We conclude that with an appropriate *a priori* choice of m (and the resulting choice of l), the distance between the families as in the assertion of the lemma can be made smaller than δ, as required. □

11.2.4 Uniqueness of entropy for doubly stochastic operators

The next theorem says that there is in fact a unique reasonable definition of entropy for doubly stochastic operators. It implies that the quantum dynamical entropy introduced by R. Alicki, J. Andries, M. Fannes and P. Tuyls, the operator entropy defined by I. I. Makarov, and that by E. Ghys, R. Langevin and P. Walczak, coincide on doubly stochastic operators (because they are built along the steps (1°)–(5°) and satisfy the axioms). This collection is joined by the notion introduced in the next section. We remark that B. Frej proved that another notion (built not along the construction steps (1°)–(5°)) defined in [Maličký and Riečan, 1987] coincides (on doubly stochastic operators) with the other four [Frej, 2006].

Theorem 11.2.32 *If $\boldsymbol{T}$ is a doubly stochastic operator on $L^1(\mu)$, then axioms (A)–(D) (along with the construction steps (1°)–(5°)) completely determine the value of $h_\mu(\boldsymbol{T})$.*

Proof Consider an entropy definition following the construction steps (1°)–(5°) and satisfying axioms (A)–(D). Fix $\varepsilon > 0$ and find a family $\mathcal{F} \in \mathbb{F}$ (of some cardinality r) such that

$$h_\mu(\boldsymbol{T}) < h_\mu(\boldsymbol{T}, \mathfrak{F}) + \varepsilon \tag{11.2.33}$$

(see step (5°)). Let γ be as specified in the domination axiom (D) for the cardinality r of $\mathfrak{F}$ and ε. Let δ be as specified in the axiom (B) for the cardinality $r' = (2/\gamma)^r$ and ε. Find l so large that the assertion of Lemma 11.2.28 holds for $\mathfrak{F}_0 = \boldsymbol{T}^l(\mathfrak{F})$, γ and δ, and let $\varkappa$ be the corresponding partition of $[0,1]$. Recall that $\varkappa$ is a partition by intervals not longer than γ, hence it can be used in the domination axiom (D) with any family of cardinality r. Let $\mathcal{O}_\varkappa$ be the family (depending only on $\varkappa$) specified in that axiom. For each $i \in \mathbb{N}$ the axiom (D) applied to $\boldsymbol{T}^i(\mathfrak{F}_0)$ and $\varkappa$ reads

$$H_\mu(\boldsymbol{T}^i(\mathfrak{F}_0)|\mathbb{I}_{(\boldsymbol{T}^i(\mathfrak{F}_0))^{-1}(\varkappa)} \sqcup \mathcal{O}_\varkappa) < \varepsilon.$$

Using (11.2.12) (iterated $n-1$ times) we get, for each n,

$$H_\mu\left(\mathfrak{F}_0^n \middle| \bigsqcup_{i=0}^{n-1} (\mathbb{I}_{(\boldsymbol{T}^i(\mathfrak{F}_0))^{-1}(\varkappa)} \sqcup \mathcal{O}_\varkappa)\right) < n\varepsilon.$$

By (11.2.8) and the inequalities of Fact 11.2.9, the above implies

$$H_\mu(\mathfrak{F}_0^n) < H_\mu\left(\bigsqcup_{i=0}^{n-1} \mathbb{I}_{(\boldsymbol{T}^i(\mathfrak{F}_0))^{-1}(\varkappa)}\right) + H_\mu(\bigsqcup_{i=0}^{n-1} \mathcal{O}_\varkappa) + n\varepsilon.$$

Since $\lim_n \frac{1}{n} H_\mu(\bigsqcup_{i=0}^{n-1} \mathcal{O}_\varkappa) = 0$, applying step (4°), we obtain

$$h_\mu(\boldsymbol{T}, \mathfrak{F}_0) \leq \limsup_{n\to\infty} \frac{1}{n} H_\mu\left(\bigsqcup_{i=0}^{n-1} \mathbb{I}_{(\boldsymbol{T}^i(\mathfrak{F}_0))^{-1}(\varkappa)}\right) + \varepsilon. \tag{11.2.34}$$

Denoting the right-hand side expression (without "$+\varepsilon$") by $h_\mu(\boldsymbol{T}, \mathfrak{F}_0, \varkappa)$, and noting that, by (11.2.15), the family $\mathfrak{F}_0$ also satisfies (11.2.33), we have obtained

$$h_\mu(\boldsymbol{T}) \leq h_\mu(\boldsymbol{T}, \mathfrak{F}_0, \varkappa) + 2\varepsilon.$$

Notice that $h_\mu(\boldsymbol{T}, \mathfrak{F}_0, \varkappa)$ involves joinings and entropies exclusively of partitions hence is completely determined by the partitions axiom (C) (it also depends on the careful selection of the partition $\varkappa$, but this one was done independently of any notion of entropy).

On the other hand, Lemma 11.2.28 reads, for each $i \in \mathbb{N}$,

$$\mathsf{dist}(\boldsymbol{T}^i(\mathbb{I}_{\mathfrak{F}_0^{-1}(\varkappa)}), \mathbb{I}_{(\boldsymbol{T}^i(\mathfrak{F}_0))^{-1}(\varkappa)}) < \delta.$$

Notice that the cardinality of both families does not exceed $r' = (2/\gamma)^r$. By the choice of δ, the second inequality in the axiom (B) yields

$$H_\mu\big(\mathbb{I}_{(T^i(\mathcal{F}_0))^{-1}(\varkappa)}\big|T^i(\mathbb{I}_{\mathcal{F}_0^{-1}(\varkappa)})\big) < H_\mu\big(\mathbb{I}_{(T^i(\mathcal{F}_0))^{-1}(\varkappa)}\big|\mathbb{I}_{(T^i(\mathcal{F}_0))^{-1}(\varkappa)}\big) + \varepsilon.$$

By the axiom (C) and the elementary property (1.6.5) of the Shannon entropy the right-hand side is just ε. Using (11.2.12) and step (4°) again, we deduce that

$$h_\mu(T, \mathcal{F}_0, \varkappa) \le h_\mu(T, \mathbb{I}_{\mathcal{F}_0^{-1}(\varkappa)}) + \varepsilon \le h_\mu(T) + \varepsilon.$$

We have proved that

$$h_\mu(T, \mathcal{F}_0, \varkappa) - \varepsilon \le h_\mu(T) \le h_\mu(T, \mathcal{F}_0, \varkappa) + 2\varepsilon.$$

In this manner $h_\mu(T)$ is completely determined by the terms $h_\mu(T, \mathcal{F}_0, \varkappa)$, which are completely determined by the axioms. □

Remark 11.2.35 It might seem that the term $h_\mu(T, \mathcal{F}_0, \varkappa)$ allows one to define the operator entropy exclusively by means of partitions. Unfortunately, the approximation obtained at the end of the preceding proof is valid only for very carefully selected partitions $\varkappa$, depending on $\mathcal{F}_0$. There are examples that if we choose $\varkappa$ just to be any partition into intervals of appropriately bounded lengths, the term $h_\mu(T, \mathcal{F}_0, \varkappa)$ may be way too large (see Exercise 11.7 or [Downarowicz and Frej, 2005, Example 3.2]).

11.3 An explicit measure-theoretic definition

In this section we introduce a definition of operator entropy which follows the construction steps (1°)–(5°), and with static entropy fulfilling axioms (A)–(D). It uses a very natural and effective way of quantifying the exponential growth of "information content" in an evolution of a finite family of functions by tracing the partitions of $X \times [0, 1]$ determined by graphs of the functions. The same idea has been already used to define entropy of a transformation via families of functions (see Definition 8.3.11). In addition to the axioms, the static entropy used in this approach enjoys some other desirable properties.

Before we proceed, we advise to go back to Definition 7.6.5 of the partition $\mathcal{A}_\mathcal{F}$ associated with a finite family $\mathcal{F}$ of functions from X into $[0, 1]$; it will be used again in a moment. Obviously, the requirement that each $f \in \mathcal{F}$ is continuous can be dropped. We will assume the functions to be measurable. Notice that $\mathcal{A}_\mathcal{F}$ depends only on $\mathcal{F}$ treated as a set and is insensitive to both the order and possible repetitions. Observe that the procedure described below coincides with that used to define the entropy $h(\mu, T, \mathcal{F})$ of a family of continuous functions in case T is a continuous transformation in Definition 8.3.11. By $\mathcal{L}$ we denote all functions in $L^1(\mu)$ with range in $[0,1]$. The definition below applies to any (not even linear) transformation $\boldsymbol{T} : \mathcal{L} \to \mathcal{L}$.

(1°) We let $\mathbb{F}$ be the collection of all finite families $\mathcal{F} \subset \mathcal{L}$.
(2°) For $\mathcal{F}$ and $\mathcal{G}$, $\mathcal{F} \sqcup \mathcal{G}$ is defined as the union $\mathcal{F} \cup \mathcal{G}$.
(3°) We define $H_\mu(\mathcal{F}) = H(\mu \times \lambda, \mathcal{A}_\mathcal{F})$, where λ is the Lebesgue measure on the unit interval.
(4°) $h_\mu(\boldsymbol{T}, \mathcal{F}) = \limsup_{n\to\infty} \frac{1}{n} H_\mu(\mathcal{F}^n)$.
(5°) $h_\mu(\boldsymbol{T}) = \sup_{\mathcal{F}\in\mathbb{F}} h_\mu(\boldsymbol{T}, \mathcal{F})$.

The verification that so defined static entropy satisfies axioms (A)–(C) is immediate. At some point we must use the easy fact that the passage $\mathcal{F} \mapsto \mathcal{A}_\mathcal{F}$ is continuous with respect to dist and d_1. For (D), fix a partition $\varkappa$ of $[0,1]$ into intervals with small lengths and endpoints t_j. For $i = 1, \ldots, r$ define simple functions g_i by the rule $g_i(x) = t_j \iff t_j \le f_i(x) < t_{j+1}$. Denote $\mathcal{G} = \{g_1, \ldots, g_r\}$. Since $\mathsf{dist}(\mathcal{F}, \mathcal{G})$ is small, the (already proved) axiom (B) implies that $H_\mu(\mathcal{F}|\mathcal{G}) = H(\mu \times \lambda, \mathcal{A}_\mathcal{F}|\mathcal{A}_\mathcal{G})$ is also small, and, all the more, $H(\mu \times \lambda, \mathcal{A}_\mathcal{F}|\mathcal{F}^{-1}(\varkappa) \times \varkappa)$ is small (because $\mathcal{F}^{-1}(\varkappa) \times \varkappa \succcurlyeq \mathcal{A}_\mathcal{G}$). The last term equals $H_\mu(\mathcal{F}|\mathbb{I}_{\mathcal{F}^{-1}(\varkappa)} \sqcup \mathcal{O}_\varkappa)$, where $\mathcal{O}_\varkappa$ is the family of constant functions with values at the endpoints t_j of $\varkappa$ (this place is in fact the only reason why we have involved the family $\mathcal{O}_\varkappa$ in the axiom (D)). It is also obvious that whenever $\sqcup$ is the set union the condition $\lim_n \frac{1}{n} H_\mu(\bigsqcup_{k=0}^{n-1} \mathcal{O}_\varkappa) = 0$ holds not only for $\mathcal{O}_\varkappa$, but in fact for any finite family of functions.

Summarizing, we have provided a way to define particular notions of static and dynamical entropy of a family of functions $\mathcal{F}$, leading to the entropy $h_\mu(\boldsymbol{T})$ (by uniqueness, the same as using any other method within the scope). We now discuss some "good" properties of these notions.

First of all, we notice that whenever $\sqcup$ is the set union the static entropy enjoys the very natural, in view of the information origins of entropy, property

$$H_\mu(\mathcal{F}|\mathcal{F}) = 0.$$

It is not implied by the axioms, and it is satisfied neither for the entropy introduced in [Ghys *et al.*, 1986] [see Kamiński and de Sam Lazaro, 2000, corollary following Proposition 1] nor for the quantum entropy (try the two point space with the measure $\{\frac{1}{2}, \frac{1}{2}\}$, and the operational partition consisting of functions $(\frac{1}{\sqrt{2}}, 0)$ and $(\frac{1}{\sqrt{2}}, 1)$).

Since our particular static entropy of a family $\mathcal{F}$ is just the Shannon entropy of the associated partition, it is very easy to deduce its uniform continuity as a function of $\mathcal{F}$ of cardinality bounded by some m. This, combined with (11.2.12), and the fact that $\boldsymbol{T}$ is a contraction in the $L^1(\mu)$-norm, yields the continuity of our particular notion of the dynamical entropy as a function of $\mathcal{F}$, as well:

$$\mathsf{dist}(\mathcal{F},\mathcal{G}) < \delta \Longrightarrow |h_\mu(\boldsymbol{T},\mathcal{F}) - h_\mu(\boldsymbol{T},\mathcal{G})| < \varepsilon. \tag{11.3.1}$$

We pass to listing the properties of the operator entropy $h_\mu(\boldsymbol{T})$ analogous to those of the Kolmogorov–Sinai entropy, which are best derived with the help of the particular definition introduced in this section (but, by the uniqueness, enjoyed by the operator entropy regardless of the method of defining it). We start with the behavior of the entropy under factors.

Definition 11.3.2 Let $(X,\mathfrak{A},\mu)$ and $(Y,\mathfrak{B},\nu)$ be two probability spaces. Let $\boldsymbol{T}$ and $\boldsymbol{S}$ be doubly stochastic operators acting on $L^1(\mu)$ and $L^1(\nu)$, respectively. We say that

(i) $\boldsymbol{S}$ is a *factor* of $\boldsymbol{T}$ if there is a measurable map $\phi : X \to Y$ satisfying $\phi\mu = \nu$, and for every $f \in L^1(\nu)$, $(\boldsymbol{S}f)\circ\phi = \boldsymbol{T}(f\circ\phi)$.
(ii) $\boldsymbol{T}$ and $\boldsymbol{S}$ are *isomorphic* if the map ϕ defined above is invertible.

Fact 11.3.3 *Let $\boldsymbol{T}$ and $\boldsymbol{S}$ be doubly stochastic operators acting on $L^1(\mu)$ and $L^1(\nu)$, respectively.*

(i) If $\boldsymbol{S}$ is a factor of $\boldsymbol{T}$, then $h_\nu(\boldsymbol{S}) \le h_\mu(\boldsymbol{T})$.
(ii) If $\boldsymbol{T}$ and $\boldsymbol{S}$ are isomorphic, then their entropies are equal.

Proof Let $\phi : X \to Y$ be a factor map. Let $\mathcal{F} \subset L^1(\nu)$ be a finite family of functions with ranges in $[0,1]$ and let

$$\mathcal{F}^\phi = \{f\circ\phi : f \in \mathcal{F}\}$$

denote the lifted family of functions on X. By the definition of the factor we have $(\mathcal{F}^\phi)^n = (\mathcal{F}^n)^\phi$. The particular way of computing the entropy (via partitions $\mathcal{A}_\mathcal{F}$) yields that $H_\nu(\boldsymbol{S},\mathcal{F}) = H_\mu(\boldsymbol{T},\mathcal{F}^\phi)$, and similarly, $H_\nu(\boldsymbol{S},\mathcal{F}^n) = H_\mu(\boldsymbol{T},(\mathcal{F}^\phi)^n)$. Thus (i) is now a consequence of the limit step in (4°) and because in (5°) applied to $h_\mu(\boldsymbol{T})$ the supremum runs over all finite families of functions, which includes all the families lifted from Y.

Once (i) is proved, (ii) follows immediately. □

We state the next property without a proof. We refer the interested reader to the original paper [Frej and Frej, preprint]. The proof refers to another definition of entropy (satisfying the axioms) not discussed in this book. Given two doubly stochastic operators $\boldsymbol{T}$ and $\boldsymbol{S}$ acting on $L^1(\mu)$ and $L^1(\nu)$ (on two probability spaces $(X,\mathfrak{A},\mu)$ and $(Y,\mathfrak{B},\nu)$, respectively) we can create their product $T \times S$ acting on $L^1(\mu \times \nu)$ as follows: For a "product function" $h(x,y) = f(x)g(y)$ we let $(T \times S)h(x,y) = Tf(x)Sg(y)$. We prolong this operator linearly to all linear combinations of the product functions, and then, from this dense set, continuously to all of $L^1(\mu \times \nu)$. We skip the standard details. The theorem below generalizes the product rule for independent joinings of measure-preserving transformations (see (4.4.5)).

Theorem 11.3.4 $h_{\mu\times\nu}(T \times S) = h_\mu(T) + h_\nu(S)$. □

11.4 Not so bad properties of the operator entropy

Now we discuss some differences between the operator entropy of doubly stochastic operators as defined in the preceding section and the Kolmogorov–Sinai (or Shannon) entropy for measure-preserving transformations.

First of all, unlike measure-preserving transformations, doubly stochastic operators need not preserve the static entropy H_μ. Moreover, their application can increase the static entropy, as illustrated in the simple example below.

Example 11.4.1 Let T be the doubly stochastic operator on the unit interval (equipped with the Lebesgue measure) defined by $Tf(x) = \frac{1}{2}(f(x) + f(1-x))$. Take $\mathcal{F}$ consisting of the characteristic functions $\mathbb{I}_{[0,\frac{1}{4}]}$ and $\mathbb{I}_{(\frac{1}{4},1]}$. Then $H_\mu(\mathcal{F}) = 2\log 2 - \frac{3}{4}\log 3 < \log 2$, while $H_\mu(T\mathcal{F}) = \frac{3}{2}\log 2$. Note that the pathology cannot be removed by the sometimes useful trick of joining $\mathcal{F}$ with some set of constant functions.

Nevertheless, an asymptotic invariance of H_μ does hold:

Lemma 11.4.2 *Let T be a doubly stochastic operator. For every $\varepsilon > 0$ there exists $l \in \mathbb{N}$ such that for every n*

$$\left| H_\mu\left(T^{l+n}\mathcal{F}\right) - H_\mu\left(T^l\mathcal{F}\right)\right| < \varepsilon.$$

Proof Notice that the sets in $\mathcal{A}_\mathcal{F}$ are enclosed by graphs of functions obtained via lattice polynomials applied to $\mathcal{F}$. The assertion now follows from the asymptotic lattice stability (Theorem 11.2.23) and the continuity axiom (11.2.5). □

One of the key disadvantages of operator dynamics is the lack of subadditivity of the sequence $H_\mu(\mathcal{F}^n)$. By the axiom (A) (more precisely, by (11.2.11)), we do have

$$H_\mu(\mathcal{F}^{n+m}) \le H_\mu(\mathcal{F}^n) + H_\mu(T^n\mathcal{F}^m),$$

but we cannot drop T^n. The following result is a substitute of the subadditivity property valid for doubly stochastic operators. We call it *quasi-subadditivity*.

We skip the technical proof, which can be found in [Downarowicz and Frej, 2005].

Lemma 11.4.3 *Let $\boldsymbol{T}$ be a doubly stochastic operator. For every $\varepsilon > 0$ there exists $l \in \mathbb{N}$ and a constant c such that for every $n \in \mathbb{N}$ and $m \geq l$ we have*

$$H_\mu\left(\mathfrak{F}^{n+m}\right) \leq H_\mu\left(\mathfrak{F}^n\right) + H_\mu\left(\mathfrak{F}^m\right) + m\varepsilon + c. \qquad \square$$

Quasi-subadditivity allows us to prove another similarity to the classical notion, but only for the notion introduced in the preceding section. We include it here for the context, for the proof see [Downarowicz and Frej, 2005].

Theorem 11.4.4 *If $\boldsymbol{T}$ is a doubly stochastic operator, then the upper limit in step* ($4°$) *of the construction of the entropy is in fact a limit.* □

The limit no longer needs to coincide with the corresponding infimum.

We conclude this section (and chapter) with the interesting observation that the operator entropy generalizes the Kolmogorov–Sinai entropy in an essential way; in addition to the pointwise type dynamics, it captures also some pure "operator dynamics," not associated with any pointwise behavior. Recall that the Sinai Theorem (Theorem 4.5.1) asserts that all of the entropy of any measure-preserving system comes from a Bernoulli factor. The example below shows that operator entropy need not even come from any pointwise generated factor.

Example 11.4.5 There exists a doubly stochastic operator with positive entropy, which admits no nontrivial pointwise generated factors (i.e., factors which are isomorphic to some measure-preserving transformations).

Let $(X, \mathfrak{A}, \mu)$ be the set of one-sided 0-1 sequences $X = \{0,1\}^{\mathbb{N}}$ with the product sigma-algebra and with the uniform product measure $\mu = \{\frac{1}{2}, \frac{1}{2}\}^{\mathbb{N}}$. Let ν be the geometric distribution on natural numbers $\mathbb{N}$ given by $\nu(k) = 2^{-k}$. The element (x, k) of the product space $X \times \mathbb{N}$ can be visualized as the 0-1-valued sequence $(x_1, x_2, \ldots, \overset{*}{x_k}, x_{k+1}, \ldots)$ with a marker (star) over the position k. As usual, σ stands for the shift transformation on X. Moreover, for each finite block $B = (b_1, b_2, \ldots, b_k) \in \{0,1\}^k$ we define the map $\sigma_B : X \to X$ by

$$(\sigma_B x)_n = \begin{cases} b_n & \text{for } n \leq k \\ x_{n+1} & \text{for } n > k \end{cases}.$$

Next we define the operator $\boldsymbol{T}$ on $L^1(\mu \times \nu)$ as follows:

$$\boldsymbol{T} f((x,k)) = f((\sigma x, k-1)) \quad \text{if } k > 1,$$

$$\boldsymbol{T} f((x,1)) = \sum_{k=1}^{\infty} 2^{-k} \sum_{B \in \{0,1\}^k} 2^{-k} f((\sigma_B x, k)).$$

The corresponding transition probability $P((x,k), \cdot)$ can be described as the shift map (also shifting the position of the marker) on points with marker further to the right, while points with the marker over the first position are shifted, and then the initial block of length k (chosen according to the geometric distribution) is replaced by a random block of length k. A marker is added over the last position of the replaced block. Our first claim is that $\boldsymbol{T}$ is doubly stochastic with respect to the product measure $\mu \times \nu$. To see this consider the characteristic function $f = \mathbb{I}_C$ of a cylinder of the form

$$C = C(y_1, y_2, \ldots, y_k^*, y_{k+1}, \ldots y_n) = \{(x,k) : \forall_{i=1,\ldots,n}\, x_i = y_i\},$$

where $k \le n$ and $y_1, \ldots, y_n$ are fixed (it suffices to consider such cylinders because they generate in the product). Clearly, the integral of f is $2^{-(k+n)}$. All points in both cylinders $C(x_0, y_1, y_2, \ldots, y_k^*, y_{k+1}, \ldots y_n)$ (with $x_0 = 0$ or $x_0 = 1$) are sent deterministically into C. So, $\boldsymbol{T}f = 1$ on these two cylinders. The integral of $\boldsymbol{T}f$ over these cylinders thus equals $2 \cdot 2^{-(n+1)} \cdot 2^{-(k+1)} = 2^{-(k+n+1)}$ (the marker appears at the position $k+1$). Also, each point in any cylinder $C(x_0^*, x_1, \ldots, x_k, y_{k+1}, \ldots, y_n)$ (with any choice of $x_0, \ldots, x_k$) contributes with probability 2^{-2k} to C (via the map σ_B, where $B = (y_1, y_2, \ldots, y_k)$), so $\boldsymbol{T}f$ at such points is 2^{-2k}. The integral of $\boldsymbol{T}f$ over the union of such cylinders equals $2^{k+1} \cdot 2^{-2k} \cdot 2^{-(n+1)} \cdot 2^{-1} = 2^{-(k+n+1)}$. The sum of both parts equals the integral of f, so $\boldsymbol{T}$ is doubly stochastic with respect to $\mu \times \nu$.

Now, suppose $\boldsymbol{T}$ has a pointwise generated factor. This factor corresponds to a sub-sigma-algebra $\mathfrak{B}$ with the property $P((x,k), A) = 0$ or 1 for each $A \in \mathfrak{B}$ and almost every x. Consider a pair of cofinal points (x,k), (x',k'), i.e., such that $x_n = x'_n$ for n larger than some k''. We can assume $k'' > k$ and $k'' > k'$. Consider also a point (x'',k'') which coincides with both x and x' above the index k'' where it has the marker. Both points (x'',k'') and (x,k) are accessible with positive transition probabilities from the point $(0x, 1) = 0^*x$ (which has 0^* at the first position and then looks like x shifted to the right). This means (unless 0^*x falls in some zero measure set) that the factor identifies (x'',k'') with (x,k). Similarly, it identifies (x'',k'') with (x',k'), and hence (x,k) with (x',k'). We have shown that the factor identifies (almost surely) cofinal points regardless of the positioning of the markers in them. This implies that $\mathfrak{B}$ is (up to measure μ) contained in the tail sigma-algebra of $\{0,1\}^{\mathbb{Z}}$. By the Kolmogorov 0-1 Law [see e.g. Feller, 1968], such sigma-algebra is trivial with respect to the measure μ. So, $\boldsymbol{T}$ admits no nontrivial pointwise (measure-theoretic) factors.

Finally, we will show that $h_\mu(\boldsymbol{T}) \ge \log 2$. Consider the 0-1-valued function $f_i((x,k)) = x_i$. We will show that

$$\boldsymbol{T}^n f_i((x,k)) = \alpha \tfrac{1}{2} + (1-\alpha) x_{i+n}, \tag{11.4.6}$$

with $\alpha \le 2^{-(i-1)}$. This is obvious if $n < k$ (then $\alpha = 0$), because such $\boldsymbol{T}^n$ is deterministic at (x,k). Now consider $n = k$. This is the first time the point (x,k) is actually spread, and notice that with respect to $P^n((x,k),\cdot)$ the position of the marker has the geometric distribution ν. In the next step $n+1$, with probability $1/2$, the markers will be shifted (creating half of the same geometric distribution) and with probability $1/2$, spread again (with the same geometric distribution). As a result, the distribution of the marker's position with respect to $P^{n+1}((x,k),\cdot)$ remains the same. This applies to all further steps. We have proved that for $n \ge k$ the position of the marker has the geometric distribution with respect to $P^n((x,k),\cdot)$. Now, the points in the support of $P^n((x,k),\cdot)$ whose markers fall below the coordinate i have the value x_{i+n} at the position i because this position has never been altered (only shifted). This happens with probability

$$\sum_{j=1}^{i-1} 2^{-j} = 1 - 2^{-(i-1)},$$

contributing a factor $(1 - 2^{-(i-1)})x_{i+n}$ to $T^n f_i((x,k))$. Otherwise the value at the position i is 0 or 1 with equal chances, contributing the factor $2^{-(i-1)}\frac{1}{2}$. We have proved the formula (11.4.6) for $n \geq k$ with α equal to $2^{-(i-1)}$.

From what we have derived it is seen that the images of f_i behave almost as the functions f_{i+n}, except that instead of oscillating with amplitude 1 they oscillate with (nonconstant) amplitude not smaller than $1 - 2^{-(i-1)}$. For large i such functions generate entropy arbitrarily close to log 2.

Exercises

11.1 Prove that (11.2.2) and (11.2.10) imply (11.2.3).

11.2 Prove formula (11.2.15).

11.3 Prove the power rule of Fact 11.2.16.

11.4 Prove that any invertible doubly stochastic operator is pointwise generated.

11.5 Prove that a doubly stochastic operator is pointwise generated if and only if it preserves the lattice operations if and only if it sends characteristic functions to characteristic functions.

11.6 Prove Lemma 11.2.25.

11.7 Let $(X, \mathfrak{A}, \mu)$ be the set of one-sided 0-1 sequences $X = \{0,1\}^{\mathbb{N}}$ with the product σ-algebra and with the uniform product measure $\mu = \{\frac{1}{2}, \frac{1}{2}\}^{\mathbb{N}}$. Define the doubly stochastic operator T by

$$Tf(x) = \frac{1}{2}\left(f(\sigma x) + \int_X f(x)d\mu\right),$$

where σ denotes the shift $(x_n) \mapsto (x_{n+1})$ on X. Let $\mathfrak{F}$ contain only one function, namely $f(x) = x_0$, and consider the partition of the unit interval $\varkappa = \{[0, \frac{1}{2}], (\frac{1}{2}, 1]\}$. Prove that $h_\mu(T) = 0$ while $h_\mu(T, \mathfrak{F}, \varkappa)$ (the term appearing in (11.2.34)) equals log 2. (This example shows that without carefully specifying the partitions $\varkappa$ the terms $h_\mu(T, \mathfrak{F}, \varkappa)$ cannot be used to define the operator entropy).

12
Topological entropy of a Markov operator

12.1 Three definitions

For Markov operators we will explore three natural ways of defining topological entropy. All three of them lead to the same quantity $\mathbf{h}(\boldsymbol{T})$, called the *topological operator entropy*. As we will show, this notion satisfies both basic postulates: (1) For operators associated with continuous maps it coincides with the classical topological entropy of this map, and (2) for trivial actions (such that $\boldsymbol{T}^n f$ converges for every $f \in C(X)$) the topological operator entropy is zero.

Throughout this chapter X is a compact Hausdorff space and $\boldsymbol{T}$ denotes a Markov operator acting on $C(X)$. Assuming metrizability of X brings neither simplification nor strengthening of any results.

Our first definition of topological entropy of $\boldsymbol{T}$ takes after the notion of measure-theoretic entropy of a stochastic operator introduced in the preceding chapter. The covers $\mathfrak{U}^\varepsilon_{\mathcal{F}}$ are obtained by "thickening" the sets in $\mathcal{A}_{\mathcal{F}}$. The second definition uses continuity of functions in $\mathcal{F}$ to transport open covers from the unit interval to X. In the third one we make use of a certain pseudometric on X induced by a finite collection of functions. This leads us to the definition similar to Bowen's definition of entropy. Let us proceed.

For a continuous function $f : X \to [0, 1]$ we define

$$\begin{aligned} U^\varepsilon_{<f} &= \{(x, t) \in X \times [0, 1] : t < f(x) + \varepsilon\}, \\ U^\varepsilon_{>f} &= \{(x, t) \in X \times [0, 1] : t > f(x) - \varepsilon\}, \\ \mathfrak{U}^\varepsilon_f &= \{U^\varepsilon_{<f}, U^\varepsilon_{>f}\}. \end{aligned}$$

For a finite family $\mathcal{F} \subset C(X)$ of functions with range in $[0, 1]$, we create a finite open cover of the product space $X \times [0, 1]$ by the formula

$$\mathfrak{U}^\varepsilon_{\mathcal{F}} = \bigvee_{f \in \mathcal{F}} \mathfrak{U}^\varepsilon_f.$$

If $\mathcal{V}$ is a finite open cover of the unit interval, then we let

$$\mathcal{F}^{-1}(\mathcal{V}) = \bigvee_{f \in \mathcal{F}} f^{-1}(\mathcal{V}).$$

Recall that for any open cover $\mathcal{U}$ the symbol $N(\mathcal{U})$ denotes the minimal cardinality of a subcover of $\mathcal{U}$. As before, we denote $\mathcal{F}^n = \bigcup_{i=0}^n \boldsymbol{T}^i(\mathcal{F})$, where $\boldsymbol{T}^i(\mathcal{F}) = \{\boldsymbol{T}^i f : f \in \mathcal{F}\}$.

Definition 12.1.1 Let $\mathcal{F} \subset C(X)$ be a finite collection of functions with range in $[0, 1]$ and let ε be a positive number. We define

(i) $\mathbf{H}_1(\mathcal{F}, \varepsilon) = \log N(\mathcal{U}_{\mathcal{F}}^{\varepsilon})$,
(ii) $\mathbf{h}_1(\boldsymbol{T}, \mathcal{F}, \varepsilon) = \limsup_{n \to \infty} \frac{1}{n} \mathbf{H}_1(\mathcal{F}^n, \varepsilon)$,
(iii) $\mathbf{h}_1(\boldsymbol{T}) = \sup_{\mathcal{F}} \sup_{\varepsilon} \mathbf{h}_1(\boldsymbol{T}, \mathcal{F}, \varepsilon)$.

Definition 12.1.2 Let $\mathcal{V}$ be a cover of $[0, 1]$.

(i) $\mathbf{H}_2(\mathcal{F}, \mathcal{V}) = \log N(\mathcal{F}^{-1}(\mathcal{V}))$,
(ii) $\mathbf{h}_2(\boldsymbol{T}, \mathcal{F}, \mathcal{V}) = \limsup_{n \to \infty} \frac{1}{n} \mathbf{H}_2(\mathcal{F}^n, \mathcal{V})$,
(iii) $\mathbf{h}_2(\boldsymbol{T}) = \sup_{\mathcal{F}} \sup_{\mathcal{V}} \mathbf{h}_2(\boldsymbol{T}, \mathcal{F}, \mathcal{V})$.

Given $\mathcal{F}$ we define a pseudometric on X by

$$d_{\mathcal{F}}(x, y) = \sup_{f \in \mathcal{F}} |f(x) - f(y)|.$$

We say that a subset of X is $(d_{\mathcal{F}}, \varepsilon)$-separated if it is ε-separated in the pseudometric $d_{\mathcal{F}}$. Since the space X is compact, there exists a finite $(d_{\mathcal{F}}, \varepsilon)$-separated subset of maximal cardinality in X. We denote the number of elements of this subset by $s(d_{\mathcal{F}}, \varepsilon)$.

Definition 12.1.3 For $\mathcal{F}$ and ε we define

(i) $\mathbf{H}_3(\mathcal{F}, \varepsilon) = \log s(d_{\mathcal{F}}, \varepsilon)$,
(ii) $\mathbf{h}_3(\boldsymbol{T}, \mathcal{F}, \varepsilon) = \limsup_{n \to \infty} \frac{1}{n} \mathbf{H}_3(\mathcal{F}^n, \varepsilon)$,
(iii) $\mathbf{h}_3(\boldsymbol{T}) = \sup_{\mathcal{F}} \sup_{\varepsilon} \mathbf{h}_3(\boldsymbol{T}, \mathcal{F}, \varepsilon)$.

Theorem 12.1.4 *For every Markov operator $\boldsymbol{T}$ we have*

$$\mathbf{h}_1(\boldsymbol{T}) = \mathbf{h}_2(\boldsymbol{T}) = \mathbf{h}_3(\boldsymbol{T}).$$

Proof We begin by showing that $\mathbf{h}_1(\boldsymbol{T}) \le \mathbf{h}_2(\boldsymbol{T})$. Choose $\varepsilon > 0$ and let $\mathcal{V}$ be a finite open cover of the unit interval consisting of sets having diameters not greater than ε. We claim that the cover defined by the formula

$$\mathcal{W}_n = \left\{ U \times V : U \in (\mathcal{F}^n)^{-1}(\mathcal{V}),\ V \in \mathcal{V} \right\}$$

is inscribed in $\mathcal{U}^{\varepsilon}_{\mathcal{F}^n}$. Indeed, for each $U \times V \in \mathcal{W}_n$ we let

$$\mathcal{F}' = \{ f \in \mathcal{F}^n : \forall_{x \in U}\ f(x) \ge \inf V \},$$

and it is not hard to verify that

$$U \times V \subset \bigcap_{f \in \mathcal{F}'} U^{\varepsilon}_{<f} \cap \bigcap_{f \in \mathcal{F}^n \setminus \mathcal{F}'} U^{\varepsilon}_{>f} \in \mathcal{U}^{\varepsilon}_{\mathcal{F}^n}.$$

Thus,

$$N(\mathcal{U}^{\varepsilon}_{\mathcal{F}^n}) \le N(\mathcal{W}_n) \le N\left((\mathcal{F}^n)^{-1}(\mathcal{V}) \right) \cdot N(\mathcal{V})$$

and since $N(\mathcal{V})$ is independent of n

$$\mathbf{h}_1(\boldsymbol{T}, \mathcal{F}, \varepsilon) \le \mathbf{h}_2(\boldsymbol{T}, \mathcal{F}, \mathcal{V}).$$

The desired inequality follows by taking appropriate suprema.

Now, we prove that $\mathbf{h}_2(\boldsymbol{T}) \le \mathbf{h}_3(\boldsymbol{T})$. Let $\mathcal{V}$ be a finite open cover of the unit interval. Denote its Lebesgue number by δ and let E be a maximal $(d_{\mathcal{F}^n}, \frac{\delta}{2})$-separated set in X. It follows from the maximality of E that the collection $\{B(x, \frac{\delta}{2}) : x \in E\}$ of balls in the pseudometric $d_{\mathcal{F}^n}$ constitutes a finite open cover of X. For every $f \in \mathcal{F}^n$ and $x \in E$ the interval $(f(x) - \frac{\delta}{2}, f(x) + \frac{\delta}{2})$ is contained in some element $V_f(x)$ of $\mathcal{V}$. Hence,

$$B(x, \tfrac{\delta}{2}) = \bigcap_{f \in \mathcal{F}^n} f^{-1}\left((f(x) - \tfrac{\delta}{2}, f(x) + \tfrac{\delta}{2}) \right) \subset \bigcap_{f \in \mathcal{F}^n} f^{-1}(V_f(x)) \in (\mathcal{F}^n)^{-1}(\mathcal{V})$$

and

$$N((\mathcal{F}^n)^{-1}(\mathcal{V})) \le \#\{B(x, \tfrac{\delta}{2}) : x \in E\} = s(d_{\mathcal{F}^n}, \tfrac{\delta}{2}),$$

which implies $\mathbf{h}_2(\boldsymbol{T}) \le \mathbf{h}_3(\boldsymbol{T})$.

We end the proof showing that $\mathbf{h}_3(\boldsymbol{T}) \le \mathbf{h}_1(\boldsymbol{T})$. Let $E \subset X$ be a $(d_{\mathcal{F}}, \varepsilon)$-separated set of maximal cardinality. Put $\gamma = \varepsilon/6$ and define

$$\mathcal{F}^{\circ} = \left\{ \tfrac{1}{2} f + i\gamma : f \in \mathcal{F},\ i \in \mathbb{Z},\ 0 \le i \le \tfrac{1}{2\gamma} \right\}.$$

We will show that the cover $\mathcal{U}^{\gamma}_{\mathcal{F}^{\circ}}$ separates points of $E \times \left\{\frac{1}{2}\right\}$ in the sense that each element of the cover contains at most one point from $E \times \left\{\frac{1}{2}\right\}$.

Consider two elements x, y of E. We can choose a function $f \in \mathcal{F}$ satisfying $|f(x) - f(y)| \geq \varepsilon$ and, since both x and y play the same role in the formula, we may assume that $f(x) + \varepsilon \leq f(y)$. Since $\frac{1}{2}f(x)$ ranges between 0 and $\frac{1}{2} - 3\gamma$, there exists an integer $0 \leq i \leq \frac{1}{2\gamma}$ such that $f^\circ = \frac{1}{2}f + i\gamma \in \mathcal{F}^\circ$ satisfies

$$\tfrac{1}{2} - 2\gamma \leq f^\circ(x) \leq \tfrac{1}{2} - \gamma.$$

Since $f^\circ(y)$ is by at least $\frac{\varepsilon}{2} = 3\gamma$ larger than $f^\circ(x)$, we have

$$\tfrac{1}{2} + \gamma \leq f^\circ(y),$$

which implies that $(y, \frac{1}{2})$ belongs to $U^\gamma_{<f^\circ}$ and not to $U^\gamma_{>f^\circ}$, while, on the contrary, $(x, \frac{1}{2})$ belongs to $U^\gamma_{>f^\circ}$ and not to $U^\gamma_{<f^\circ}$. Since every element of $\mathcal{U}^\gamma_{\mathcal{F}^\circ}$ is contained either in $U^\gamma_{<f^\circ}$ or in $U^\gamma_{>f^\circ}$, this proves that the cover separates points of $E \times \{\frac{1}{2}\}$. Moreover, every subcover of $\mathcal{U}^\gamma_{\mathcal{F}^\circ}$ has the same property, so

$$s(d_{\mathcal{F}}, \varepsilon) \leq N(\mathcal{U}^\gamma_{\mathcal{F}^\circ}).$$

Recall that $\boldsymbol{T}$, as a Markov operator, is linear and preserves constants. This implies that $(\mathcal{F}^n)^\circ = (\mathcal{F}^\circ)^n$, so we can replace $\mathcal{F}$ with $\mathcal{F}^n$ obtaining

$$s(d_{\mathcal{F}^n}, \varepsilon) \leq N\Big(\mathcal{U}^\gamma_{(\mathcal{F}^\circ)^n}\Big).$$

The proof is ended by taking upper limits and suprema. □

12.2 Properties of the topological operator entropy

In the sequel we will use the symbol $\mathbf{h}(\boldsymbol{T})$ to denote the common value of $\mathbf{h}_1(\boldsymbol{T})$, $\mathbf{h}_2(\boldsymbol{T})$ and $\mathbf{h}_3(\boldsymbol{T})$. According to the next result, the coincidence with the notation for the topological entropy of continuous maps is reasonable. We prove the first basic postulate.

Theorem 12.2.1 *If $\boldsymbol{T}f = f \circ T$ is an operator generated by a continuous map $T : X \to X$, then the operator entropy $\mathbf{h}(\boldsymbol{T})$ is equal to the topological entropy $\mathbf{h}(T)$.*

The proof relies on the next lemma which is a straightforward observation.

Lemma 12.2.2 *If $\boldsymbol{T}$ is a pointwise generated Markov operator (by a continuous transformation T), $\mathcal{F}$ is a finite family of continuous functions with values in $[0, 1]$, and $\mathcal{V}$ is a finite open cover of the unit interval, then*

$$(\mathcal{F}^n)^{-1}(\mathcal{V}) = (\mathcal{F}^{-1}(\mathcal{V}))^n,$$

where the former exponent n refers to the action of the operator $\boldsymbol{T}$ on $C(X)$, and the latter, to the action of the continuous map T on X. □

Proof of Theorem 12.2.1 We will exploit Definition 12.1.2 of $\mathbf{h}_2(\boldsymbol{T})$. For every $\varepsilon > 0$, using Urysohn functions, it is easy to produce a finite family of continuous functions and a cover $\mathcal{V}$ of the unit interval, so that all cells of the cover $\mathcal{F}^{-1}(\mathcal{V})$ have diameters smaller than ε. It follows that the covers of this form refine in X and thus the topological entropy is the supremum over such covers only. The assertion now follows directly from Lemma 12.2.2, by comparing the definition of the topological entropy $\mathbf{h}(T)$ via open covers (see Section 6.1.3) with Definition 12.1.2 of the topological operator entropy. □

Next we prove that the second basic postulate is fulfilled.

Theorem 12.2.3 *If for every $f \in \mathcal{F}$ the sequence $\boldsymbol{T}^n f$ converges uniformly, (to some $g_f \in C(X)$), then $\mathbf{h}_3(\boldsymbol{T}, \mathcal{F}, \varepsilon) = 0$ for every $\varepsilon > 0$. If such convergence holds for all $f \in C(X)$, then $\mathbf{h}(\boldsymbol{T}) = 0$.*

Proof Denote $\mathcal{G} = \{g_f : f \in \mathcal{F}\}$. For every ε there exists N such that for every $f \in \mathcal{F}$, $x \in X$ and $n \geq N$ it holds that $|\boldsymbol{T}^n f(x) - g_f(x)| < \frac{\varepsilon}{3}$. Then, for every $x, y \in X$ we have

$$|\boldsymbol{T}^n f(x) - \boldsymbol{T}^n f(y)| \leq$$
$$|\boldsymbol{T}^n f(x) - g_f(x)| + |g_f(x) - g_f(y)| + |g_f(y) - \boldsymbol{T}^n f(y)| \leq$$
$$\frac{2\varepsilon}{3} + |g_f(x) - g_f(y)|,$$

implying that

$$\forall_{n \geq N} \ s(d_{\mathcal{F}^n}, \varepsilon) \leq s(d_{\mathcal{F}^N \cup \mathcal{G}}, \tfrac{\varepsilon}{3}),$$

where the right-hand side does not depend on n. The rest now is obvious. □

We can define a *topological factor* of a Markov operator the same way as it was done in the measure-theoretic Definition 11.3.2, replacing only the words "measure-preserving" by "continuous surjection." Two operators are *conjugate* if the relevant factor map is a homeomorphism. We will say that a compact set Y is *invariant* under $\boldsymbol{T}$ if for any two functions $f, f' \in C(X)$ with $f|_Y = f'|_Y$ the equality $\boldsymbol{T}f(y) = \boldsymbol{T}f'(y)$ holds at every point $y \in Y$. Then we can easily define a Markov operator on $C(Y)$, which may be treated as a restriction of $\boldsymbol{T}$. The proofs of the following statements concerning topological entropy are standard and will be omitted (see Exercise 12.1).

Theorem 12.2.4 *The following facts hold*

(i) The entropy of a factor of a Markov operator $\boldsymbol{T}$ is smaller than or equal to the entropy of $\boldsymbol{T}$.
(ii) Two conjugate Markov operators have equal entropies.
(iii) If Y is a compact invariant subset of X, then the entropy of the Markov operator $\boldsymbol{T}$ restricted to $C(Y)$ is smaller than or equal to the entropy of $\boldsymbol{T}$ on $C(X)$.
(iv) For every $n \in \mathbb{N}$, $\mathbf{h}(\boldsymbol{T}^n) = n\mathbf{h}(\boldsymbol{T})$ (the power rule*).* □

12.3 Half of the variational principle

In this section we prove the analog of the "easier" inequality in the variational principle: the topological operator entropy of a Markov operator dominates its measure-theoretic operator entropy with respect to each invariant regular Borel probability measure μ. The converse inequality completing the operator variational principle remains an open question.

Theorem 12.3.1 *Let X be a compact Hausdorff space and let $\boldsymbol{T}$ be a Markov operator acting on $C(X)$. For every invariant regular Borel probability measure μ on X we have*

$$h_\mu(\boldsymbol{T}) \le \mathbf{h}(\boldsymbol{T}).$$

Proof We will refer to h_μ as defined in Section 11.3. By (11.3.1) and regularity, for calculations of measure-theoretic entropy it suffices to consider families $\mathcal{F}$ of continuous functions with range in $[0,1]$. Let $\mathcal{F}$ be such a family of cardinality r. Choose a positive number ε such that $2r\varepsilon\log(2r\varepsilon) < 1/2^r$. For every $A \in \mathcal{A}_\mathcal{F}$ we denote by $\mathcal{F}_{\ge A}$ the set of all functions in $\mathcal{F}$, for which $f(x) \ge t$ whenever $(x,t) \in A$. Analogously, $\mathcal{F}_{<A}$ is the set of all functions from $\mathcal{F}$, such that $f(x) < t$ if $(x,t) \in A$. It is easy to see that the union of $\mathcal{F}_{<A}$ and $\mathcal{F}_{\ge A}$ is the whole $\mathcal{F}$. We define a new partition $\mathcal{B}^\varepsilon_\mathcal{F}$ of $X \times [0,1]$ consisting of compact sets

$$B_A = B^\varepsilon_{A,\mathcal{F}} = \\ \big\{(x,t) : \forall_{f\in\mathcal{F}_{\ge A}}\ t \le f(x) - \varepsilon\big\} \cap \{(x,t) : \forall_{f\in\mathcal{F}_{<A}}\ t \ge f(x) + \varepsilon\},$$

where A belongs to $\mathcal{A}_\mathcal{F}$, and the open set

$$\widetilde{B} = \widetilde{B}^\varepsilon_\mathcal{F} = \bigcap_{A\in\mathcal{A}_\mathcal{F}} {B_A}^c = \{(x,t) : \exists_{f\in\mathcal{F}}\ |f(x) - t| < \varepsilon\}.$$

Notice that $A \cap B_{A'} = B_A$ if and only if $A = A'$. Otherwise the intersection $A \cap B_{A'}$ is empty. Thus,

$$H(\mu \times \lambda, \mathcal{A}_{\mathcal{F}} | \mathcal{B}^{\varepsilon}_{\mathcal{F}}) = H(\mu \times \lambda, \mathcal{A}_{\mathcal{F}} \vee \mathcal{B}^{\varepsilon}_{\mathcal{F}}) - H(\mu \times \lambda, \mathcal{B}^{\varepsilon}_{\mathcal{F}}) = \\ - \sum_{A \in \mathcal{A}_{\mathcal{F}}} (\mu \times \lambda)(A \cap \widetilde{B}) \log((\mu \times \lambda)(A \cap \widetilde{B})).$$

Since $(\mu \times \lambda)(\widetilde{B}) < 2r\varepsilon$ and $\mathcal{A}_{\mathcal{F}}$ has at most 2^r elements, we get (by the choice of ε) that

$$H(\mu \times \lambda, \mathcal{A}_{\mathcal{F}} | \mathcal{B}^{\varepsilon}_{\mathcal{F}}) < 1.$$

Similarly, for every natural number i

$$H(\mu \times \lambda, \mathcal{A}_{\boldsymbol{T}^i \mathcal{F}} | \mathcal{B}^{\varepsilon}_{\boldsymbol{T}^i \mathcal{F}}) < 1,$$

because ε was chosen according only to the cardinality of $\mathcal{F}$. We will abbreviate $\bigvee_{i=0}^{n-1} \mathcal{B}^{\varepsilon}_{\boldsymbol{T}^i \mathcal{F}}$ by $\mathcal{B}^n$ (note that this is not equal to $\mathcal{B}^{\varepsilon}_{\mathcal{F}^n}$). Using the above estimates and the elementary Fact 1.6.11 for the entropy of partitions, we derive:

$$H(\mu \times \lambda, \mathcal{A}_{\mathcal{F}^n} | \mathcal{B}^n) < n,$$

and thus

$$H(\mu \times \lambda, \mathcal{A}_{\mathcal{F}^n}) < H(\mu \times \lambda, \mathcal{B}^n) + n.$$

Since the entropy of a partition is always less than or equal to the logarithm of its cardinality we need to estimate the number of sets in $\mathcal{B}^n$.

Let $\mathcal{U}'$ be an optimal subcover of $\mathcal{U}^{\varepsilon}_{\mathcal{F}^n}$. Obviously,

$$\#\mathcal{B}^n \leq \sum_{U \in \mathcal{U}'} \# \{B \in \mathcal{B}^n : B \cap U \neq \emptyset\}.$$

Every $U \in \mathcal{U}^{\varepsilon}_{\mathcal{F}^n}$ has the form $\bigcap_{i=0}^{n-1} U_i$, where $U_i \in \mathcal{U}^{\varepsilon}_{\boldsymbol{T}^i \mathcal{F}}$. For each U_i there exists $A \in \mathcal{A}_{\boldsymbol{T}^i \mathcal{F}}$ such that U_i is contained in the union of $B^{\varepsilon}_{A, \boldsymbol{T}^i \mathcal{F}}$ and $\widetilde{B}^{\varepsilon}_{\boldsymbol{T}^i \mathcal{F}}$. This implies that

$$\# \{B \in \mathcal{B}^n : B \cap U \neq \emptyset\} \leq 2^n$$

and hence

$$\#\mathcal{B}^n \leq 2^n \cdot N(\mathcal{U}^{\varepsilon}_{\mathcal{F}^n}).$$

Thus finally, we have

$$H(\mu \times \lambda, \mathcal{A}_{\mathcal{F}^n}) < \log N(\mathcal{U}^{\varepsilon}_{\mathcal{F}^n}) + n \log 2 + n$$

implying

$$h_\mu(\boldsymbol{T}) \leq \mathbf{h}(\boldsymbol{T}) + \log 2 + 1.$$

Repeating the argument for $\boldsymbol{T}^n$ replacing $\boldsymbol{T}$ and using Fact 11.2.16 and Theorem 12.2.4 (iv) we complete the proof by writing

$$h_\mu(\boldsymbol{T}) = \tfrac{1}{n} h_\mu(\boldsymbol{T}^n) \leq \tfrac{1}{n}(\mathbf{h}(\boldsymbol{T}^n) + \log 2 + 1) \leq \mathbf{h}(\boldsymbol{T}) + \tfrac{\log 2+1}{n}$$

for an arbitrary n. □

Exercises

12.1 Prove Theorem 12.2.4.

12.2 Let $X = \{0,1\}^{\mathbb{Z}}$. Consider the Markov operator on $C(X)$ given by transition probabilities

$$P(x,\cdot) = \tfrac{1}{2}(\boldsymbol{\delta}_{\sigma x} + \boldsymbol{\delta}_{\sigma^2 x}).$$

This operator can be thought of as the average of the shift map and its iterative square (the shift by two positions): $\boldsymbol{T}f = \frac{1}{2}(f \circ \sigma + f \circ \sigma^2)$. Notice that both the shift and its square have positive entropies. What is the topological operator entropy of $\boldsymbol{T}$? Every shift invariant measure on X is invariant under $\boldsymbol{T}$. What are entropies of $\boldsymbol{T}$ treated as a doubly stochastic operator with respect to any such measure?

12.3 Let X be a compact metric space. Any Markov operator $\boldsymbol{T}$ on $C(X)$ induces a transformation $\boldsymbol{T}^*$ on the compact space $\mathcal{M}(X)$ of all Borel probability measures on X. Simply, $\boldsymbol{T}^*$ is the dual operator restricted to probability measures. Provide an example that the topological entropy of $\boldsymbol{T}^*$ is not a good candidate to define the operator entropy of $\boldsymbol{T}$. Hint: Consider the operator $\boldsymbol{T}$ pointwise generated by the unilateral shift over the alphabet $\{0,1\}$. Prove that $\boldsymbol{T}^*$ has infinite topological entropy[1] (so it violates the first postulate of operator entropy).

12.4 Prove the product rule for Markov operators.

[1] Glasner and Weiss showed that whenever a transformation T had positive entropy, then the entropy of T^* on measures was infinite [Glasner and Weiss, 2003].

13
Open problems in operator entropy

There are clearly some gaps in the theory of entropy for operators. At least as long as we seek for similarities with the analogous theory for dynamical systems. It can be hoped that these similarities reach further than we know.

13.1 Questions on doubly stochastic operators

In the entropy theory of doubly stochastic operators, the fundamental missing issue is a relevant information theory. The notion of operator entropy is created without reference to any reasonable notion of information function. Clearly, it is most desirable that such a function depends on the family of functions $\mathcal{F}$ and is defined directly on the phase space X, however, a compromise solution with this function defined on the product $X \times [0, 1]$ seems also acceptable. In any case, the static entropy should be the integral of the information function with respect to the appropriate measure (μ or $\mu \times \lambda$, respectively). Needless to say, the notion should coincide with the classical one for a family of characteristic functions of a partition. Of course, the best justification of this notion would be an analog (generalization) of the Shannon–McMillan–Breiman Theorem. Let us verbalize the problem:

Question 13.1.1 Is there a meaningful notion of an information function with respect to a family of functions, such that the static entropy is its integral? Does a generalization of the Shannon–McMillan–Breiman Theorem hold for doubly stochastic operators?

A number of further questions can be asked:

Question 13.1.2 How can conditional operator entropy (given a factor) for doubly stochastic operators be defined? Which of the relevant facts known for measure-preserving transformations can be generalized?

Question 13.1.3 Is there an analog of the notion of a generator for a doubly stochastic operator?

Question 13.1.4 Is there an analog of the Pinsker factor of a doubly stochastic operator?

Question 13.1.5 What is the interpretation of operator entropy zero? Does it make the operator "deterministic" in any sense?

We stop here although the list can be continued.

Let us mention one interesting question which is not in the spirit of copying facts from the theory of transformations. Conversely, the question makes sense only for operators: The space of doubly stochastic operators on $L^1(\mu)$ is convex (which is not true for measure-preserving transformations). It is easy to see that operator entropy is not affine: there are easy examples of two transformations with positive entropy whose average has operator entropy zero (see Exercise 12.2). But we can still ask:

Question 13.1.6 Is the entropy of doubly stochastic operators a convex function?

13.2 Questions concerning Markov operators

The first problem concerns uniqueness of topological entropy. We do not even know this:

Question 13.2.1 Does the notion of topological operator entropy $\mathbf{h}(\boldsymbol{T})$ presented (in three ways) in Section 12.1 coincide with the notion introduced in [Langevin and Walczak, 1994]?

In the context of Exercise 12.3, a natural question arises:

Question 13.2.2 Is the entropy of an arbitrary Markov operator $\boldsymbol{T}$ equal to the entropy of the dual operator $\boldsymbol{T}^*$ restricted to the smallest compact $\boldsymbol{T}^*$-invariant subset of $\mathcal{M}(X)$ containing all Dirac measures?

The next obvious question concerns the missing inequality in the variational principle. Of course, its positive solution would provide a good reason for establishing the uniqueness of the "good" notion of topological operator entropy.

Question 13.2.3 Does the "harder" inequality of the variational principle hold for Markov operators?

The difficulties with proving this inequality arise from the lack of subadditivity of the sequence $H_\mu(\mathcal{F}^n)$. The interested reader will find more exhaustive comments in [Downarowicz and Frej, 2005].

Just like doubly stochastic operators, Markov operators form a convex set. The same examples show that topological operator entropy is not affine under convex combinations of Markov operators. We repeat the same question as was asked for doubly stochastic operator:

Question 13.2.4 Is the topological operator entropy a convex function?

Parallel to the search for a generator for a doubly stochastic operator (Question 13.1.3) should be the search for an analog of a symbolic Markov operator (or having a topological generator). Once this is done, one can ask about symbolic extensions of a Markov operator $\boldsymbol{T}$. On the other hand, the notion of an entropy structure for $\boldsymbol{T}$ can be literally copied from topological dynamics, and the function $\mathrm{E}\mathcal{H}$ can be computed without any modification in its construction. Would there be a relation between this function and the entropy of symbolic operator extensions of $\boldsymbol{T}$ (whatever they are)? Briefly speaking, the question is

Question 13.2.5 Is there a theory of symbolic extensions for Markov operators? What is the interpretation of $\mathrm{E}\mathcal{H}$?

Appendix A
Toolbox

A.1 Elementary tools

A.1.1 The rectangle rule

We begin with a simple fact concerning integrals with respect to probability measures (in particular weighted averages), a standard argument proving that in probability spaces the L^1-convergence and the convergence in measure coincide. Because we will make repeated use of it, for easy reference, we isolate this fact and call it the *rectangle rule*. The name refers to the idea of the proof (by contradiction) relying on finding a $(\delta \times \gamma)$-rectangle under the graph of f. We skip the easy details.

Fact A.1.1 *Let $(\Omega, \mathfrak{A}, \mu)$ be a probability space and let $f \geq 0$ be a measurable function on Ω such that $\int f\, d\mu < \gamma\delta$ for some $\gamma > 0$ and $\delta > 0$. Then*

$$\mu\{\omega \in \Omega : f(\omega) \geq \gamma\} < \delta. \qquad \square$$

Applying the above to the difference of two functions we get:

Fact A.1.2 *Let $(\Omega, \mathfrak{A}, \mu)$ be a probability space and let $f \leq g$ be two measurable real functions on Ω. If*

$$\int f\, d\mu > \int g\, d\mu - \gamma\delta,$$

for some $\gamma > 0$ and $\delta > 0$, then

$$\mu\{\omega \in \Omega : f(\omega) \leq g(\omega) - \gamma\} < \delta. \qquad \square$$

A.1.2 Upper semicontinuous partitions

Definition A.1.3 A partition $\varkappa$ (we do not assume it is countable) of a topological space X is *upper semicontinuous* if the cells of $\varkappa$ are closed and for every closed set F the union of all cells having nonempty intersection with F (so-called *$\varkappa$-saturation* of F) is closed.

Let $\pi : X \to Y$ be a map from one space to another. By the *fibers* we mean preimages by π of the points of Y. The collection of all fibers is a partition of X. The following elementary fact holds:

Fact A.1.4 *Let $\varkappa$ be a partition of a compact metric space X. Then $\varkappa$ is upper semicontinuous if and only if there exists a continuous map $\pi : X \to Y$ into another topological space such that $\varkappa$ coincides with the partition into fibers determined by π.*

Proof First, let $\pi : X \to Y$ be a continuous map and let $\varkappa$ be the partition of X by the fibers. Let F be a closed (hence compact) set in X. By continuity, $\pi(F)$ is also compact, hence closed in Y. By continuity again, the preimage of $\pi(F)$ is closed in X. But $\pi^{-1}(\pi(F))$ is precisely the $\varkappa$-saturation of F, where $\varkappa$ is the fiber partition.

For the converse implication assume that $\varkappa$ is an upper semicontinuous partition of X, set $Y = \varkappa$, and define the topology by the condition: $V \subset \varkappa$ is open if and only if $\bigcup V$ is open in X. The map π is defined naturally, by $\pi x = A \iff x \in A$ $(A \in \varkappa)$. We skip the elementary verification of the correctness of so defined topology and of the continuity of π. □

A.1.3 Nets

By a *net* we will understand a family (of some objects) indexed by a partially ordered set $\mathcal{K}$ satisfying the condition for being a *directed family*, as follows:

$$\forall_{\kappa_1,\kappa_2\in\mathcal{K}}\ \exists_{\kappa_3\in\mathcal{K}}\ \kappa_3 \geq \kappa_1, \kappa_3 \geq \kappa_2. \tag{A.1.5}$$

In particular the ordering can be by the natural numbers, in which case we are dealing with a usual sequence.

A *subnet* of a net is its restriction indexed by a subset $\mathcal{K}'$ of $\mathcal{K}$ satisfying

$$\forall_{\kappa\in\mathcal{K}}\ \exists_{\kappa'\in\mathcal{K}'}\ \kappa' \geq \kappa. \tag{A.1.6}$$

The family $\mathcal{K}'$ is then automatically directed. Unlike for sequences, the restriction to an infinite subfamily $\mathcal{K}' \subset \mathcal{K}$ which is directed, need not be a subnet. An example here is any net indexed by the halfline $[0, \infty)$ ordered by the usual inequality, and its subfamily indexed by the interval $[0, 1)$. Such "false subnets," obtained by restricting a net to a directed subfamily of indices which do not necessarily satisfy (A.1.6), will be called *sub-nets* (with the dash).

Recall that in compact Hausdorff spaces every net of points has a convergent subnet.

A.1.4 Semicontinuous functions

Definition A.1.7 A function $f : X \to [-\infty, \infty)$ defined on a topological space X is called *upper semicontinuous at the point* x if for every $\varepsilon > 0$ there is a neighborhood $U \ni x$ such that $f(y) < f(x) + \varepsilon$ (or $f(y) < -1/\varepsilon$ in case $f(x) = -\infty$) for every $y \in U$. This is equivalent to

$$f(x) = \inf_U \sup_{y \in U} f(y), \tag{A.1.8}$$

where U ranges over open neighborhoods of x.

The expression on the right will be written as $\limsup_{y \to x} f(y)$.

Definition A.1.9 A function $f : X \to [-\infty, \infty)$ is *upper semicontinuous* if it is upper semicontinuous at every point. An equivalent condition for f to be upper semicontinuous is that for each $t \in \mathbb{R}$ the set $\{x : f(x) \geq t\}$ is closed.

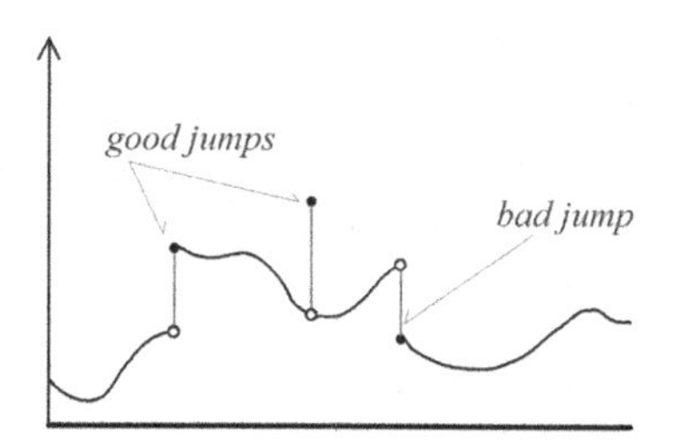

Figure A.1 Discontinuities admitted and not admitted in upper semicontinuous functions.

For example, the characteristic function of a closed set is upper semicontinuous, as well as a characteristic function of an open set U multiplied by a continuous function negative on the boundary of U.

Notice the obvious facts that the infimum of any family of upper semicontinuous functions is again an upper semicontinuous function and that both the sum and the supremum of finitely many upper semicontinuous functions are upper semicontinuous functions.

Remark A.1.10 Usually, one defines upper semicontinuity for real-valued functions. The reason why we admit $-\infty$ in the range of f is the remark above: we want the family of upper semicontinuous functions to be closed under infima (of any cardinality). We do not admit ∞, because we want all upper semicontinuous functions to be bounded from above on compact sets. Also, on compact metric spaces, so defined upper semicontinuity is maintained by the map $\mu \mapsto \int f \, d\mu$ defined on Borel probability measures on X endowed with the weak-star topology (see Section A.2.2). This fails if we do not admit $-\infty$ or admit ∞ in the range of f.

We skip the purely topological proof of the fact below. To avoid excessive and unnecessary generality we assume that X is metric.

Fact A.1.11 *A function $f : X \to [-\infty, \infty)$ is upper semicontinuous at x if and only if $f = \lim_k \downarrow f_k$, where (f_k) is a decreasing sequence of real-valued functions continuous at x. A function $f : X \to [-\infty, \infty)$ is upper semicontinuous if and only if*

$$f = \inf\{g : g \geq f, g \text{ is continuous}\}. \tag{A.1.12}$$

By taking finite infima, we also have

$$f = \lim_{\kappa} \downarrow f_\kappa,$$

where (f_κ) is a decreasing net of real-valued functions continuous at x. Since X is metric, the net can be taken a sequence. □

Since every continuous function on a compact set is bounded, we get

Corollary A.1.13 *Every upper semicontinuous function is bounded from above on every compact set.* □

The following statement will be useful:

Fact A.1.14 *If $f = \lim_\kappa \downarrow f_\kappa$ is the limit of a decreasing net of upper semicontinuous functions on a compact metric space and g is a continuous function strictly larger (at each point) than f, then g is strictly larger than f_κ for sufficiently large κ.*

Proof For each κ the function $f_\kappa - g$ is upper semicontinuous. Thus the set $\{x : f_\kappa(x) \geq g(x)\}$ is closed. These sets decrease with κ, so if they all were nonempty they would form a centered family, and thus have a nonempty intersection, on which $f(x) \geq g(x)$, a contradiction. □

Corollary A.1.15 *If f in the fact above is continuous (in particular constant), then the convergence is necessarily uniform.*

Proof Apply Fact A.1.14 to the function $g = f + \varepsilon$. □

Definition A.1.16 Let f be a function defined on a metric space X into $[-\infty, \infty)$. By $\widetilde{f}$ we will denote the *upper semicontinuous envelope* of f, i.e.,

$$\widetilde{f} = \inf\{g : g \geq f,\ g \text{ is continuous}\}$$

(compare (A.1.12)). If there are no such functions g, we set $\widetilde{f} \equiv \infty$.

The following fact is almost immediate:

Fact A.1.17 *The condition $\widetilde{f} \not\equiv \infty$ is equivalent to f being* locally bounded, *i.e., bounded from above on some neighborhood of every point x. Assuming local boundedness, we have:*

$$\widetilde{f}(x) = \limsup_{y \to x} f(y) \tag{A.1.18}$$

(which equals $\inf_U \sup_{y \in U} f(y)$, where U ranges over open neighborhoods of x; compare (A.1.8)*). Moreover, $\widetilde{f}$ is an upper semicontinuous function and any upper semicontinuous function $g \geq f$ satisfies $g \geq \widetilde{f}$. The function f is upper semicontinuous at a point x if and only if $\widetilde{f}(x) = f(x)$.* □

It is immediately seen directly from the definition that

Fact A.1.19 *For any functions f and g,*

$$\widetilde{f+g} \leq \widetilde{f} + \widetilde{g}.$$

If either f or g is continuous, then above we have equality. □

Definition A.1.20 For a real-valued function f on a metric domain X, the difference function

$$\dddot{f} = \widetilde{f} - f$$

is called the *defect of upper semicontinuity* (or, for short, the *defect*).

Like $\widetilde{f}$, the defect function $\dddot{f}$ is either finite at every point (when f is locally bounded) or it is infinite everywhere. It is always nonnegative.

Subadditivity analogous to that in Fact A.1.19 holds also for the defect functions

$$\overset{\cdots\cdots\cdots}{(f+g)} \leq \dddot{f} + \dddot{g}, \tag{A.1.21}$$

and equality holds when g (or f) is continuous (then $\dddot{g} = 0$ and $\overset{\cdots\cdots\cdots}{(f+g)} = \dddot{f}$).

It is not hard to see that for a locally bounded function f, $\dddot{f}(x) = 0$ if and only if f is upper semicontinuous at x.

We will also use lower semicontinuous functions.

Definition A.1.22 A function $f : X \to (-\infty, \infty]$ defined on a topological space X is called *lower semicontinuous* if $-f$ is upper semicontinuous.

We skip rewriting the facts about lower semicontinuous functions, which are symmetric to the facts concerning upper semicontinuous functions.

In the theory of entropy some functions are increasing limits of sequences of upper semicontinuous functions. We follow the notation of [Young, 1910]:

Definition A.1.23 We say that a function f belongs to the *Young class* LU if $f = \lim_k \uparrow f_k$, where each f_k is upper semicontinuous.

Functions in LU are, in general, neither upper nor lower semicontinuous. Yet, on compact metric spaces, they belong to the second Baire class, hence the class LU is relatively small among all Borel-measurable functions. It is known that any function which is an increasing limit of first Baire class functions is of Young class LU. On the other hand, every nonnegative LU function can be represented as a sum of a series of nonnegative upper semicontinuous functions. We refer the reader to [Natanson, 1950] for more information on this class.

A.1.5 Exchanging suprema and infima

Many arguments in this book, especially in the proofs of variational principles, rely on exchanging suprema and infima. Some of these exchanges are completely trivial, others work only under very special circumstances. We recall the trivial ones here. If $f(a, b)$ is any function of two variables $a \in A$, $b \in B$ and "term" stands for either sup, inf, lim sup, lim inf, or lim (if it exists), we always have

$$\operatorname*{term}_{a \in A} \sup_{b \in B} f(a, b) \geq \sup_{b \in B} \operatorname*{term}_{a \in A} f(a, b),$$

and

$$\operatorname*{term}_{a \in A} \inf_{b \in B} f(a, b) \leq \inf_{b \in B} \operatorname*{term}_{a \in A} f(a, b).$$

In particular, the order of two suprema (or two infima) can be exchanged. Other inequalities usually fail, unless we have the following situation:

Fact A.1.24 *Let $(f_\kappa)_\kappa$ be a decreasing net of upper semicontinuous functions on a compact domain X. Then*

$$\sup_{x\in X}\inf_\kappa f_\kappa(x) = \inf_\kappa \sup_{x\in X} f_\kappa(x).$$

Proof The inequality $\le$ is trivial. For the converse, note that the function $f = \lim_\kappa \downarrow f_\kappa$ is upper semicontinuous, and its pointwise supremum, say M is the left-hand side of the equality to be proved. Then, for every ε, $M+\varepsilon$ is a constant (hence continuous) function strictly larger than f. By Fact A.1.14, for large enough κ, the entire function f_κ (hence also its pointwise supremum) is below $M+\varepsilon$. The inequality follows by passing with ε to zero. □

A.1.6 Lifting and pushing down functions

Let $\pi : X \to Y$ be a map between two spaces. Consider functions $f : X \to [-\infty,\infty]$ and $g : Y \to [-\infty,\infty]$. We learn how to "switch" the domains of these functions using the map π.

Definition A.1.25 The *lift* of g (via π) is defined on X as the composition $g^X = g\circ\pi$. In most situations we will denote the lift by the same letter g. This usually does not lead to ambiguity (as lifting preserves all properties which we need). If π is a surjection, then the *push-down* $f^{[Y]}$ of f (via π) is defined on Y by the formula

$$f^{[Y]}(y) = \sup\{f(x) : \pi x = y\}.$$

The easy proof of the fact below is left to the reader.

Fact A.1.26 *Suppose X and Y are topological spaces and π is continuous. Then the operation "lift" preserves both upper and lower semicontinuity of g. If X is compact metric and π is onto, then the operation "push-down" preserves upper semicontinuity of f.* □

A.2 Convex analysis tools

Let $\mathbb{B}$ be a Banach space. By elementary properties of the norm, the balls in this space are convex sets, so the space is locally convex. We will consider a compact convex subset $\mathcal{K}$ of $\mathbb{B}$. We assume the reader understands the notions of an affine, convex or concave function on $\mathbb{B}$ (or on $\mathcal{K}$), even when f assumes values in the extended real line.

A.2.1 Concave upper semicontinuous functions

The fact below can be found in [Choquet, 1969, Prop. 2.18]:

Fact A.2.1 *If $f : \mathcal{K} \to [-\infty, \infty)$ is concave and upper semicontinuous, then*

$$f = \inf\{g : g \geq f, g \text{ is affine and continuous}\}. \qquad \square$$

The construction and properties of the upper semicontinuous concave envelope are analogous to those of the upper semicontinuous envelope. The only difference is that, in place of continuous functions, it uses continuous affine functions:

Definition A.2.2 Let f be a function defined on $\mathcal{K}$ into $[-\infty, \infty)$. By $\widehat{f}$ we denote the *upper semicontinuous concave envelope* of f, i.e.,

$$\widehat{f} = \inf\{g : g \geq f, g \text{ is affine and continuous}\}.$$

If there are no such functions g, then we let $\widehat{f} \equiv \infty$.

The following fact is almost trivial:

Fact A.2.3 *If $\widehat{f}$ is not infinite, then $\widehat{f}$ is upper semicontinuous concave and any upper semicontinuous concave function g satisfying $g \geq f$ satisfies also $g \geq \widehat{f}$. A function f is upper semicontinuous and concave if and only if $f = \widehat{f}$.* $\square$

It is immediately seen directly from the definition that:

Fact A.2.4 *For any functions f and g*

$$\widehat{f+g} \leq \widehat{f} + \widehat{g}.$$

If either f or g is continuous and affine, then above we have equality. $\square$

Clearly, $f \leq \widetilde{f} \leq \widehat{f}$, and

$$\sup_{x \in X} f(x) = \sup_{x \in X} \widetilde{f}(x) = \sup_{x \in X} \widehat{f}(x).$$

Fact A.2.5 *If f is affine, then $\widetilde{f}$ is concave, hence $\widetilde{f} = \widehat{f}$.*

Proof If $\widetilde{f} \equiv \infty$, then the case is trivial. Otherwise, let $x = px_1 + qx_2$ ($q = 1 - p$) and choose sequences $x_{i,n} \to x_i$, ($i = 1, 2$) such that $f(x_{i,n}) \to \widetilde{f}(x_i)$. Then

$$\widetilde{f}(x) \geq \lim_n f(px_{1,n} + qx_{2,n}) = \lim_n [pf(x_{1,n}) + qf(x_{2,n})] = p\tilde{f}(x_1) + q\tilde{f}(x_2).$$

$\square$

A.2.2 The weak-star topology

Let X be a compact metric space. The collection of all Borel probability measures on X will be denoted by $\mathcal{M}(X)$. Such measures can be identified with the nonnegative functionals of norm one on the space $C(X)$ of all continuous real functions on X with the supremum norm (the Riesz Theorem [see e.g. Rudin, 1974]). Thus $\mathcal{M}(X)$ is naturally endowed with the weak-star topology: a sequence of measures μ_i converges to μ if and only if $\int f\,d\mu_i \to \int f\,d\mu$ for every $f \in C(X)$. Because X is metric and compact, $C(X)$ is separable, hence the weak-star topology on $\mathcal{M}(X)$ is metrizable, and, by the Banach–Alaoglu Theorem [see e.g. Rudin, 1991], also compact. The standard metric compatible with the weak-star topology is given by

$$d^*(\mu,\nu) = \sum_{i=1}^{\infty} 2^{-i}\left|\int f_i\,d\mu - \int f_i\,d\nu\right|, \tag{A.2.6}$$

where $\{f_i, i \in \mathbb{N}\}$ is some (in fact arbitrary) fixed linearly dense sequence contained in the unit ball of $C(X)$. Notice that $\mathcal{M}(X)$ contains (a homeomorphic copy of) X realized as measures concentrated at one point. Every function f on X (measurable and bounded from at least one side) prolongs to a function on $\mathcal{M}(X)$ by assigning $\mu \mapsto \int f\,d\mu$.

The fact below is completely elementary and we leave the proof to the reader.

Fact A.2.7 *The function $\mu \mapsto \int f\,d\mu$ is continuous, upper semicontinuous, lower semicontinuous, or Borel-measurable, respectively, if and only if f is.*

□

It is obvious that $\mathcal{M}(X)$ is convex and that for any f the above map $\mu \mapsto \int f\,d\mu$ is affine. In most of the considerations of this book, it is either $\mathcal{M}(X)$ or the set of T-invariant measures $\mathcal{M}_T(X)$ in a topological dynamical system $(X, T, \mathbb{S})$ that plays the role of the compact convex set $\mathcal{K}$ addressed in this section of the appendix.

A.2.3 The barycenter map, harmonic functions

We continue to assume that $\mathcal{K}$ is a compact convex subset of some Banach space. Let ξ be a probability distribution on the Borel subsets of $\mathcal{K}$ ("distribution" is synonymous with "measure"; we use this word here in order to distinguish such distributions (supported by the convex set $\mathcal{K}$) from the measures defined on the space X (which, in general, lacks any convex structure).

Definition A.2.8 The *barycenter* of ξ is defined as the Pettis integral

$$\mathsf{bar}(\xi) = \int_{\mathcal{K}} x \, d\xi,$$

i.e., as the unique point $x^{\xi} \in \mathcal{K}$ such that $f(x^{\xi}) = \int f(x)\, d\xi(x)$ for every continuous and affine real function f on $\mathcal{K}$.

The existence and uniqueness of such an x follows from the general theory of Banach spaces [see e.g. Phelps, 2001]. The barycenter map is a continuous affine surjection from $\mathcal{M}(\mathcal{K})$, the set of all probability distributions on $\mathcal{K}$ endowed with the weak-star topology, onto $\mathcal{K}$.

Definition A.2.9 A Borel-measurable function $f : \mathcal{K} \to [-\infty, \infty]$ will be called *harmonic* if

$$f(\mathsf{bar}(\xi)) = \int f \, d\xi$$

for every probability distribution ξ on $\mathcal{K}$. (This statement includes the assumption that all such integrals are well defined, which is equivalent to saying that f is bounded from at least one side.) If, instead of equality, we have the inequality $\leq$ (or $\geq$), then we say that f is *subharmonic* (or *supharmonic*).

The above terminology is motivated by the classical notion of "harmonic functions" in real analysis; such functions enjoy the "mean value property" which is the same as our "barycenter property" applied to the Lebesgue measure on spheres (and also balls) in $\mathbb{R}^n$.

Since convex combinations are integrals with respect to atomic probability measures, it is immediate to see that the terms "harmonic," "subharmonic" and "supharmonic" used with respect to a function are stronger than "affine," "convex" and "concave", respectively. For some functions the converse also holds:

Fact A.2.10 *Let f be a function defined on $\mathcal{K}$, which is either continuous or upper semicontinuous, or lower semicontinuous. Then f is harmonic, subharmonic, or supharmonic if and only if it is affine, convex or concave, respectively.*

Proof Of course, the latter conditions are weaker, as they require the same as the former ones, but for finitely supported distributions only. Continuous affine functions are harmonic directly by the definition of the barycenter map. An upper semicontinuous concave function is, by Fact A.2.1, an infimum of a family of some harmonic functions, and it is immediate to see that such infima are supharmonic. We will now prove that an upper semicontinuous convex function is subharmonic. Let ξ be a probability distribution on $\mathcal{K}$ with barycenter at x. We can partition $\mathcal{K}$ into a finite number of disjoint sets K_i of diameter smaller than ε, then compute the barycenters x_i of the conditional measures ξ_{K_i}, and finally create a finitely supported measure $\xi_\varepsilon = \sum_i \xi(K_i)\boldsymbol{\delta}_{x_i}$. It is clear that ξ_ε has its barycenter at x. By convexity, $\int f\, d\xi_\varepsilon \geq f(x)$. It is also immediate to see that, as $\varepsilon \to 0$, the measures ξ_ε converge to ξ in the weak-star topology. Since f is upper semicontinuous, so is the map assigning to a measure the integral of f. Thus

$$\int f\, d\xi \geq \lim_{\varepsilon \to 0} \int f\, d\xi_\varepsilon \geq f(x),$$

and f is shown subharmonic. The remaining cases follow trivially (by changing, where necessary, the sign of f). □

A.2.4 Choquet simplices

If $\mathcal{K}$ is as in the preceding section, the set $\mathsf{ex}\mathcal{K}$ of extreme points in $\mathcal{K}$ is Borel-measurable (it is of type G_δ) and it need not be closed. The Choquet theorem states that the map $\mathsf{bar}(\cdot)$ sends the distributions supported by $\mathsf{ex}\mathcal{K}$ onto $\mathcal{K}$ [Phelps, 2001]. A special case occurs when this map is bijective.

Definition A.2.11 A *Choquet simplex* is a compact convex subset $\mathcal{K} \in \mathbb{B}$ such that every point $x \in \mathcal{K}$ is the barycenter of a *unique* probability distribution supported by $\mathsf{ex}\mathcal{K}$. This unique distribution will be denoted by ξ^x.

The following is proved by standard methods [see e.g. Phelps, 2001]:

Fact A.2.12 *The mapping from $\mathcal{K}$ to $\mathcal{M}(\mathcal{K})$ given by $x \mapsto \xi^x$ is Borel measurable.* □

Remark A.2.13 The above mapping usually fails to be continuous.

Remark A.2.14 For a more general definition of a Choquet simplex see [Phelps, 2001]. What we use as definition is a characterization of Choquet simplices valid in separable locally convex spaces (known as the *Choquet Representation Theorem* (or just *Choquet Theorem*, see [Phelps, 2001]).

Fact A.2.15 *If $f : \mathcal{K} \to [-\infty, \infty]$ is a Borel measurable function defined on a Choquet simplex, then f is harmonic if and only if, for each $x_0 \in \mathcal{K}$,*

$$f(x_0) = \int f\, d\xi^{x_0} \tag{A.2.16}$$

(this statement includes that all such integrals are well defined, i.e., that f is measurable and bounded from at least one side). Also, f is subharmonic if and only if

$$\int f\,d\xi \leq \int f\,d\xi^{x_0} \tag{A.2.17}$$

for every $x_0 \in \mathcal{K}$ and ξ with barycenter at x_0.

Proof Let ξ be any probability distribution on $\mathcal{K}$ with barycenter x_0. Consider the probability distribution ξ_0 on $\mathcal{K}$ defined by

$$\xi_0(A) = \int \xi^x(A)\,d\xi(x)$$

(which exists, by measurability of the map $x \mapsto \xi^x$). Using harmonicity of affine continuous functions one easily checks that the barycenter of ξ_0 is the same as that of ξ, i.e., x_0. Clearly, ξ_0 is supported by ex$\mathcal{K}$ (because each ξ^x is), so, by uniqueness, $\xi_0 = \xi^{x_0}$. Thus, for any integrable function we have

$$\int f\,d\xi^{x_0} = \int \left(\int f\,d\xi^x \right) d\xi.$$

If f satisfies (A.2.16), then the left-hand side above is $f(x_0)$ while the right-hand side equals $\int f(x)\,d\xi$, and f is proved harmonic (the converse implication is obvious). If f is subharmonic, then the right-hand side is not smaller than $\int f(x)\,d\xi$ and (A.2.17) is fulfilled (also here the converse implication is obvious). □

We can try to apply the integral as in (A.2.16) to an arbitrary (measurable) function f defined on the simplex $\mathcal{K}$. In fact, it suffices that f is defined on ex$\mathcal{K}$. The resulting function agrees with f on ex$\mathcal{K}$. Thus, if it happens to be well defined on the whole set $\mathcal{K}$, then, by the above fact, it is harmonic. It will be called the *harmonic prolongation* of f:

Definition A.2.18 Let f be a measurable function defined on a simplex $\mathcal{K}$ (or just on ex$\mathcal{K}$) into $[-\infty, \infty]$, bounded from at least one side. We define the *harmonic prolongation* of f as the following function on $\mathcal{K}$:

$$f^{\mathsf{har}}(x) = \int f\,d\xi^x,$$

where ξ^x is the unique probability distribution on ex$\mathcal{K}$ with barycenter at x.

By measurability of the map $x \mapsto \xi^x$, the harmonic prolongation of a measurable function is measurable. By Fact A.2.15, each harmonic function is the harmonic prolongation of its own restriction to the set of the extreme points. Thus we have the following:

Fact A.2.19 *The restriction map $f \mapsto f|_{\mathrm{ex}\mathcal{K}}$ provides a bijection (the inverse is the harmonic prolongation) between the collection of all harmonic functions on $\mathcal{K}$ and all measurable functions on $\mathrm{ex}\mathcal{K}$ bounded from at least one side.* □

Unfortunately, the harmonic prolongation of a continuous function need not be continuous or even semicountinuous (the reason is given in Remark A.2.13). We will prove, however, the preservation of upper semicontinuity for convex functions:

Fact A.2.20 *Let f be a convex upper semicontinuous function defined on a simplex $\mathcal{K}$ into $[-\infty, \infty)$. Then the harmonic prolongation f^{har} of f is an upper semicontinuous harmonic function on $\mathcal{K}$ and $f^{\mathsf{har}} \geq f$.*

Proof To begin with, take any upper semicontinuous function f on $\mathcal{K}$. Then $\xi \mapsto \int f\, d\xi$ defines an upper semicontinuous function on the set $\mathcal{M}(\mathcal{K})$ of all probability distributions on $\mathcal{K}$ (see Fact A.2.7). The barycenter map sends continuously $\mathcal{M}(\mathcal{K})$ onto $\mathcal{K}$, so the corresponding push-down function

$$f^{[\mathcal{K}]}(x) = \sup\left\{\textstyle\int f\, d\xi : \mathsf{bar}(\xi) = x\right\} \tag{A.2.21}$$

is upper semicontinuous on $\mathcal{K}$ as well (this is the second statement of Fact A.1.26). Now, if the initial function f on $\mathcal{K}$ is (in addition to being upper semicontinuous) also convex, then f is subharmonic (see Fact A.2.10) and the above supremum is achieved for $\xi = \xi^x$ (see (A.2.17)). Thus $f^{[\mathcal{K}]}(x) = f^{\mathsf{har}}(x)$. This implies that the latter function is also upper semicontinuous. As we already know, it is harmonic. It dominates f because the supremum in (A.2.21) includes the point mass at x. □

The next fact has important applications to invariant measures of dynamical systems.

Fact A.2.22 *Let $\pi\colon \mathcal{M} \to \mathcal{K}$ be an affine surjection between convex sets. Then the operation "lift" preserves affinity, convexity and concavity of any function g defined on $\mathcal{K}$. The operation "push-down" preserves concavity of any function f defined on $\mathcal{M}$. If, moreover, $\mathcal{M}$ and $\mathcal{K}$ are Choquet simplices, π is continuous and preserves the extreme points (i.e., $\pi(\mathrm{ex}\mathcal{M}) \subset \mathrm{ex}\mathcal{K}$), then the operation "push-down" preserves affinity of f.*

Proof The properties of the lifting operation are obvious. We pass to investigating the push-down.

Let $x = px_1 + qx_2$ in $\mathcal{K}$ ($p \in (0,1)$, $q = 1 - p$). Clearly

$$\pi^{-1}(x) \supset \{py_1 + qy_2 \colon y_1 \in \pi^{-1}(x_1), y_2 \in \pi^{-1}(x_2)\}, \tag{A.2.23}$$

hence concavity of $f^{[\mathcal{K}]}$ for a concave f follows.

For the last statement (with all the additional assumptions) it suffices to prove the reversed inclusion in (A.2.23). We will employ the Choquet Representation Theorem and the Radon–Nikodym Theorem. First of all, notice that since π is affine and continuous, it is harmonic, i.e., the barycenter of the image $\pi\zeta$ by π of a distribution ζ on $\mathcal{M}$ coincides with the image of the barycenter of ζ. Moreover, since π sends ex$\mathcal{M}$ to ex$\mathcal{K}$, any distribution supported by ex$\mathcal{M}$ is sent to a distribution supported by ex$\mathcal{K}$.

Let ξ_1 and ξ_2 be the unique probability distributions supported by ex$\mathcal{K}$ with barycenters at x_1 and x_2, respectively. Clearly, $\xi = p\xi_1 + q\xi_2$ is supported by ex$\mathcal{K}$ and has barycenter at x, so it is the unique such distribution. Each ξ_i ($i = 1, 2$) is absolutely continuous with respect to ξ. Let f_i denote the corresponding Radon–Nikodym derivative defined on ex$\mathcal{K}$. Note that $pf_1 + qf_2 = 1$ ξ-a.e. Fix some $y \in \pi^{-1}(x)$ and let ζ be the probability distribution on ex$\mathcal{M}$ with the barycenter at y. Let ξ' be the image of ζ via the map π. By what was said about π, the barycenter of ξ' is $\pi y = x$ and ξ' is supported by ex$\mathcal{K}$, so $\xi' = \xi$.

Now consider the distributions ζ_i on ex$\mathcal{M}$ defined by $d\zeta_i = (f_i \circ \pi)d\zeta$ (here we use again the assumption $\pi(\text{ex}\mathcal{M}) \subset \text{ex}\mathcal{K}$). We have

$$\zeta_i(\text{ex}\mathcal{M}) = \int f_i \circ \pi \, d\zeta = \int f_i \, d\xi' = \int f_i \, d\xi = 1,$$

which shows that each ζ_i is a probability distribution. Then we let y_i be the barycenter of ζ_i. The image πy_i coincides with the barycenter of $\pi\zeta_i$, and it is straightforward to verify that $\pi\zeta_i = \xi_i$, hence $\pi y_i = x_i$ ($i = 1, 2$). Finally,

$$p d\zeta_1 + q d\zeta_2 = ((pf_1 + qf_2) \circ \pi) d\zeta = d\zeta,$$

hence, passing to barycenters, we get $py_1 + qy_2 = y$, which completes the proof. □

A.2.5 Bauer simplices

Definition A.2.24 A *Bauer simplex* is a Choquet simplex $\mathcal{K}$ whose set of extreme points is closed (hence compact).

Like in every simplex, the collection of all (Borel) probability distributions on ex$\mathcal{K}$ is sent by the barycenter map bijectively, affinely and continuously onto $\mathcal{K}$, but now the domain is compact, so the map is an affine homeomorphism. In other words, the simplex can be identified with the collection of all Borel probability distributions on the compact metric space ex$\mathcal{K}$. Conversely, the set $\mathcal{M}(X)$ of all Borel probability measures on any compact metric space X is a Bauer simplex.

If now f is a continuous or upper semicontinuous function on $\mathrm{ex}\mathcal{K}$, then f^{har} is also continuous or upper semicontinuous, respectively (this is not guaranteed on general simplices). Thus, the following fact holds:

Fact A.2.25 *Let $\mathcal{K}$ be a Bauer simplex. The restriction map $f \mapsto f|_{\mathrm{ex}\mathcal{K}}$ provides a bijection (the inverse is $\cdot^{\mathrm{har}}$) between the collection of affine continuous (hence harmonic, see Fact A.2.10) functions on $\mathcal{K}$ and all continuous functions on $\mathrm{ex}\mathcal{K}$, and between all affine upper semicontinuous (hence harmonic) functions on $\mathcal{K}$ and all upper semicontinuous functions on $\mathrm{ex}\mathcal{K}$.* □

A.2.6 Separation theorems

Essential for us are various results, collected below, on separating lower semicontinuous and upper semicontinuous functions, and their variants for concave and convex functions.

Theorem A.2.26 (Sandwich Theorem) *Suppose h and f are functions defined on a compact metric space $\mathfrak{X}$, $h \leq f$, h is upper semicontinuous and f is lower semicontinuous. Then there exists a continuous function g such that $h \leq g \leq f$.*

Proof [See Tong, 1952] (The result holds for a topological space $\mathfrak{X}$ if and only if $\mathfrak{X}$ is normal). □

Theorem A.2.27 (Separation of Disjoint Epigraphs) *Suppose h and f are functions defined on a compact convex subset $\mathcal{K}$ of a locally convex linear space, $h < f$, h is upper semicontinuous and f is lower semicontinuous.*

1. *If h is concave and f is convex, then there exists an affine continuous function g such that $h < g < f$.*
2. *If h is concave, then there exists a concave continuous function g such that $h < g < f$.*

Proof 1. This is [Choquet, 1969, Theorem 21.20], which we include for context. 2. The concave upper semicontinuous function h is the pointwise infimum of the continuous affine functions g' such that $h < g'$ (see Fact A.2.1). Because f is lower semicontinuous with $h < f$, it follows that for each $x \in \mathcal{K}$ there is an open neighborhood V and an affine function g'_V such that $h < g'_V$ and $g'_V(y) < f(y)$ for $y \in V$. Then g defined as $\min g'_V$, where the minimum is over a finite cover of $\mathcal{K}$ by such sets V, satisfies $h < g < f$. □

Theorem A.2.28 (Edwards Separation Theorem) *Suppose h and f are functions defined on a Choquet simplex $\mathcal{K}$, $h \leq f$, h is convex and f is concave* *lower semicontinuous. Then there exists an affine continuous function g such that $h \le g \le f$.*

Proof [See Asimow and Ellis, 1980, Theorem 7.6]. □

Theorem A.2.29 (Combined Separation Theorem) *Suppose h and f are functions defined on a Choquet simplex, $h < f$, h is affine upper semicontinuous and f is lower semicontinuous. Then there exists an affine continuous function g such that $h < g < f$.*

Proof Choose $\varepsilon > 0$ such that $h + \varepsilon < f$. By Theorem A.2.27 item 2, there exists a concave continuous function g_1 such that $h+\varepsilon < g_1 < f$. By Theorem A.2.28, there exists an affine continuous function g_2 such that $h+\varepsilon \le g_2 \le g_1$. Then $g = g_2$ is the required function. □

A.3 Miscellanies

A class of particularly simple and useful zero-dimensional systems which are not conjugate to subshifts (because they are not expansive) are *odometers*.

Definition A.3.1 Let $(p_k)_{k\in\mathbb{N}}$ be an increasing sequence of natural numbers such that $1 < p_{k+1}/p_k = q_k \in \mathbb{N}$. Let $(X_k, T_k, \mathbb{S})$ be the periodic cycle of p_k points. For each k the system $(X_k, T_k, \mathbb{S})$ is a factor of $(X_{k+1}, T_{k+1}, \mathbb{S})$ via the obvious map π_k whose fibers are the orbits under $T_{k+1}^{p_k}$. The *odometer to base* (p_k) is defined as the inverse limit of the systems $(X_k, T_k, \mathbb{S})$ via the maps π_k.

The above odometer is conjugate to the set of all sequences $(j_k)_{k\in\mathbb{N}}$, satisfying $j_k \in \{0, 1, \dots, p_k - 1\}$ and $j_{k+1} = j_k \bmod p_k$, for every k, with the product topology, with T corresponding to adding 1 mod p_k to each coordinate j_k. In fact, this set is a compact topological group, but we will not use its algebraic properties.

Odometers have the following symbolic-array representation (one of many possibilities): Each array x has only two symbols, say 0 and 1. In row number k the symbol 1 occurs periodically every p_k positions, otherwise the row is filled with zeros. The symbol 1 in row $k+1$ is allowed to occur only under the symbols 1 in row k (one out of q_k of them; see Figure A.2).

The positioning of the symbols 1 in the array representing an element of an odometer is often used to introduce "markers" allowing an abstract array to be cut into nice rectangular blocks. We make the following definitions.

Definition A.3.2 A symbolic array x is *marked to base* (p_k) if it has the following form: each row is over an alphabet $\Lambda_k \times \{0, 1\}$, the pairs $(a, 0)$ are

$\cdots$ 10 10 10 10 10 10 10 10 10 10 10 10 10 10 10 10 10 10 $\cdots$
$\cdots$ 10 00 00 10 00 00 10 00 00 10 00 00 10 00 00 10 00 00 $\cdots$
$\cdots$ 10 00 00 00 00 00 10 00 00 00 00 00 10 00 00 00 00 00 $\cdots$
$\cdots$ 10 00 00 00 00 00 00 00 00 00 00 00 00 00 00 00 00 00 $\cdots$

Figure A.2 An element of the odometer to base (p_k) starting with $2, 6, 12, 36, \dots$

written as a ($a \in \Lambda_k$), the pairs $(a, 1)$ are written as $|a$. The symbols 1 (the vertical bars) occur in the row number k of x periodically, every p_k positions. The bars in row $k + 1$ are allowed to occur only under the bars in row k (see Figure A.3).

Definition A.3.3 Any $(k \times p_k)$-matrix occurring in an array marked to base (p_k), extending vertically through rows 1 to k and horizontally between the positions of two neighboring vertical bars in the row k, will be called a *k-rectangle*.

Every bilateral marked symbolic array has the property that for every k the initial k rows (together) are an infinite concatenation of k-rectangles. It is very easy to see that a zero-dimensional system $(X, T, \mathbb{S})$ has an odometer to base (p_k) as a topological factor if and only if $(X, T, \mathbb{S})$ admits a symbolic-array representation in which every $x \in X$ is an array marked to base (p_k).

...|0 1|1 1|**0 0**|**0 1**|**0 0**|**0 1**|**1 0**|**1 1**|1 0|1 1|0 1|1...
... 1 2 0 1|**2 0 1 1 2 0**|**0 2 1 2 1 0**|1 2 0 0 2 0|1...
... 0 1 3 0|**2 1 3 1 2 0 1 0 1 2 0 3**|2 1 3 0 3 1 2...
... 2 1 1 1 0 2 1 2 0 1 2 2 2 0 2 1|1 1 0 2 1 1 0...

$\vdots\,\vdots\,\vdots$

Figure A.3 An array marked to base (p_k) starting with $(2, 6, 12, \dots)$. The boldface numbers form a 3-rectangle.

At some point we will need to replace an odometer by its topological extension in the form of a bilateral subshift of entropy zero. Below we describe the most elementary such extension: the Toeplitz system with one hole. The reader can find more about Toeplitz systems in [Downarowicz, 2005b].

Example A.3.4 Let $(X, T, \mathbb{S})$ be the odometer to base (p_k), as described in Definition A.3.1. The *binary Toeplitz system with one hole* over the base (p_k) is the binary (i.e., over $\{0, 1\}$) bilateral symbolic system $(Y, S, \mathbb{S})$ whose elements y are characterized by the following recursive property:

- there exists $0 \leq j_1 < p_1$ such that $y(n) = 1$ for all $n \in Z$ except when $n = j_1 \bmod p_1$;
- there exists $0 \leq j_k < p_k$, $j_k = j_{k-1} \bmod p_{k-1}$ such that $y(n) = k \bmod 2$, for all $n \in Z$ such that $n = j_{k-1} \bmod p_k$ except when $n = j_k \bmod p_k$.

Practically, every element y of the system Y can be constructed as follows: we let $y(n) = 1$ for all coordinates $n \in \mathbb{Z}$ except along a p_1-periodic set containing one coordinate in every period. In the next step we fill all the unfilled places with the symbols 0 except along a p_2-periodic set containing one coordinate in every period. Then we fill all the remaining places again with 1's, except along an analogous p_3-periodic set. We continue inductively, using alternately 0's and 1's depending on the parity of the step.

$$\ldots 111\underline{0}111\underline{0}111\underline{\underline{1}}111\underline{0}111\underline{0}111\underline{\underline{1}}111\underline{0}111\underline{0}111\underline{\underline{1}}111\underline{0}111\underline{0}111\underline{\underline{\underline{0}}}111\underline{0}111\underline{0}111\underline{\underline{1}}111 \ldots$$

Figure A.4 A binary Toeplitz sequence to base $(4, 12, 48, \ldots)$. The entries filled in step k are underlined $k - 1$ times.

In the end either the entire bilateral sequence y will become completely filled or there will remain one unfilled coordinate. In the latter case we are allowed to fill this place with either 0 or 1 (both sequences belong to the Toeplitz system). The collection of y's so obtained is closed, and shift invariant. The system is minimal and it factors to the odometer via the map $y \mapsto (j_k)_{k \in \mathbb{N}}$, where each j_k is defined in the itemized recursive description above. This system has one invariant measure and is measure-theoretically isomorphic (via the just described factor map) to the odometer. In particular, it has entropy zero.

The following two lemmas are used in Section 4.5 on Ornstein Theory:

Lemma A.3.5 (Variant of the Marriage Lemma) *Let A and B be two finite sets and $R \subset A \times B$ a relation. If there exists a positive integer $K \geq 1$ such that every $a \in A$ is in relation with at least K elements of B and every $b \in B$ is in relation with at most K elements of A then the relation contains an injective function from A into B.*

Proof The assumption of the lemma will be called, for short, the K*-condition.* The characteristic function of the relation is a binary matrix (also denoted by R) with columns indexed by the elements $a \in A$ and rows indexed by the elements $b \in B$. The K-condition is expressed in terms of column and row sums. Notice that the K-condition implies $\#B \geq \#A$.

First assume $\#B = \#A$. Then the K-condition easily yields that all column and row sums equal K. So, the matrix R equals K times a doubly stochastic matrix. It is well known that every doubly stochastic matrix is a convex combination of some permutation matrices,[1] hence R is a linear combination

[1] This fact is known as the Birkhoff–von Neumann Theorem [Birkhoff, 1946].

(with positive coefficients) of some permutation matrices. Any permutation matrix P participating in this combination provides a bijection contained in R.

In other cases we proceed by a double induction. The outer induction is over K, while for each K we run an induction over the cardinality of B.

The case $K = 1$ is trivial: any way of assigning to each a a b related to a produces an injection. Suppose the lemma holds for some $K - 1 \geq 1$ and assume the K-condition. The assertion is true if $\#B = K$ (then we have a square matrix all filled with 1's). Suppose it holds for $\#B = m \geq K$. We need to check the case $\#B = m+1 > \#A$. By discarding excessive 1's from every column we can make all column sums equal to K; this can only lower the row sums, so the K-condition will be maintained. Clearly, now, at most $\#A$ rows may have sums equal to K. If there are no such rows, the $(K-1)$-condition is satisfied and the assertion follows from the inductive assumption (of the induction over K). Otherwise we consider the sub-matrix Q constituted by the rows with sum K. The transposed matrix Q^{T} has all column sums K while the row sums are at most K, so it satisfies the K-condition. The larger dimension of Q^{T} is $\#A \leq m$. By the inductive assumption (over m), Q^{T} contains an injection. By transposition, we have found within Q an injection from part of B into A. We now discard from the matrix R the 1's corresponding to this function, replacing them with 0's. By doing so we lower by a unit sums in precisely these rows where the sum was initially K, so after this step all row sums are at most $K - 1$. On the other hand, we have lowered the column sums by at most a unit, hence after this step the matrix satisfies the $(K-1)$-condition and again, by the inductive assumption over K, the assertion holds. □

Let $\mathbf{p}$ and $\mathbf{q}$ be two probability distributions on a countable space Δ. Their *coupling* is any probability $\boldsymbol{\xi}$ on $\Delta \times \Delta$ with respective marginals $\mathbf{p}$ and $\mathbf{q}$. It is clear that for any $b \in \Delta$, $\boldsymbol{\xi}(b, b)$ must not exceed $\min\{\mathbf{p}(b), \mathbf{q}(b)\}$. The coupling is called *maximal* if it assigns exactly these maximal possible values on the diagonal.

Lemma A.3.6 *For any two probability distributions* $\mathbf{p}$ *and* $\mathbf{q}$ *on a countable space* Δ *there exists at least one maximal coupling.*

Proof Let $s \leq 1$ denote the total mass of the vector $\min\{\mathbf{p}, \mathbf{q}\}$. Then the maximal coupling is defined as the sum of $\min\{\mathbf{p}, \mathbf{q}\}$ distributed along the diagonal and $\frac{1}{1-s}((\mathbf{p} - \min\{\mathbf{p}, \mathbf{q}\}) \times (\mathbf{q} - \min\{\mathbf{p}, \mathbf{q}\}))$ on the remaining part of $\Delta \times \Delta$. The verification is straightforward. □

Notice that a maximal coupling assigns to the diagonal the mass $1 - \frac{1}{2}\|\mathbf{p} - \mathbf{q}\|_1$ (which goes to 1 as $\mathbf{q}$ approaches $\mathbf{p}$).

Appendix B
Conditional S–M–B

We give the full conditional version of the Shannon–McMillan–Breiman Theorem valid for endomorphisms and subinvariant conditioning sigma-algebras. The generating partition is admitted countable and we only assume finiteness of its conditional static entropy.

Theorem B.0.1 *Let $(X, \mathcal{P}, \mu, T, \mathbb{S})$ be an ergodic process on finitely or countably many states and let $\mathfrak{B}$ be a subinvariant (or invariant) sigma-algebra. Assume that $H(\mathcal{P}|\mathfrak{B}) < \infty$. Then for μ-almost every point x we have:*

$$\lim_{n} \tfrac{1}{n} I_{\mathcal{P}^n|\mathfrak{B}}(x) = h(\mathcal{P}|\mathfrak{B}).$$

Proof The proof follows the same guidelines as the proof of the usual (unconditional) Shannon–McMillan–Breiman Theorem. However, the presence of the conditioning sigma-algebra (especially if we want it only subinvariant) adds some intricacy. First of all, unless $\mathfrak{B}$ is invariant, the functions

$$x \mapsto \liminf_{n\to\infty} \tfrac{1}{n} I_{\mathcal{P}^n|\mathfrak{B}}(x) \quad \text{and} \quad x \mapsto \limsup_{n\to\infty} \tfrac{1}{n} I_{\mathcal{P}^n|\mathfrak{B}}(x)$$

need not be subinvariant, as it was in the unconditional case. Thus, we will consider functions defined by slightly more complicated expressions

$$x \;\mapsto\; \sup_{k\geq 1} \limsup_{n\to\infty} \tfrac{1}{n} I_{\mathcal{P}^n|\mathfrak{B}}(T^k x) \quad \text{and} \quad x \;\mapsto\; \inf_{k\geq 1} \liminf_{n\to\infty} \tfrac{1}{n} I_{\mathcal{P}^n|\mathfrak{B}}(T^k x).$$

The former is clearly subinvariant and the latter is supinvariant, hence they are equal almost everywhere to some constants C and c, respectively, where $C \geq c$. We begin by showing that $c \geq h(\mathcal{P}|\mathfrak{B})$. First consider the case of a finite partition $\mathcal{P}$. Its cardinality will be denoted by l. Fix some $\varepsilon > 0$ and $\delta > 0$. There exist some k, two integers $n_0 < N_0$ both larger than k/δ, and a set Z of measure larger than $1 - \delta$, such that for every $x \in Z$

$$\tfrac{1}{n} I_{\mathcal{P}^n|\mathfrak{B}}(T^j x) < c + \varepsilon,$$

for some $j \leq k$ and some n between n_0 and N_0 (both j and n depend on x).

Recall (see (3.3.6)) that the Martingale Convergence Theorem allows one to replace the conditional information function by the almost everywhere limit over any sequence of partitions $\mathcal{Q}$ generating $\mathfrak{B}$. And so, there exists a finite $\mathfrak{B}$-measurable partition $\mathcal{Q}$ such that (making the set Z slightly smaller, but still of measure larger than $1-\delta$) the above holds on Z with $\frac{1}{n}I_{\mathcal{P}^n|\mathfrak{B}}$ replaced by $\frac{1}{n}I_{\mathcal{P}^n|\mathcal{Q}}$:

$$\tfrac{1}{n}I_{\mathcal{P}^n|\mathcal{Q}}(T^j x) < c + \varepsilon. \tag{B.0.2}$$

Like in the proof of the unconditional Shannon–McMillan–Breiman Theorem, we will need a partition into cylinders of varying lengths, such that the associated length-information function is bounded from above by $c+\varepsilon$. Finding such a partition becomes a bit complicated due to the presence of the iterates T^j. We remark that if the sigma-algebra $\mathfrak{B}$ is invariant (not only subinvariant), then the iterate T^j can be skipped in (B.0.2) (we can set $j = 0$ for every $x \in Z$) and the following portion of the proof can be simplified.

Let J be the set of points for which j can be chosen 0. Since each point of Z falls into J under at most k iterates, we have $\mu(J) > \frac{1-\delta}{k}$. For $x \in J$ there exists a length n_x, between n_0 and N_0, such that the cylinder W_x of length n_x over $\mathcal{P}$ containing x satisfies $\mu_Q(W_x) > 2^{-n_x(c+\varepsilon)}$, where Q is the cell of $\mathcal{Q}$ containing x. If we choose n_x minimal, then the sets $W_x \cap Q \cap J$ become pairwise disjoint. For a fixed $Q \in \mathcal{Q}$ we obtain a finite partition of $Q \cap J$ by the sets $W_x \cap Q \cap J$, which we denote $\{W_{Q,i}\}$ (i ranges over a finite set depending on Q, so we prefer not to mark it) together with an associated vector of lengths $(n_{Q,i})$ (where $n_{Q,i} = n_x$ for $W_{Q,i} = W_x \cap Q \cap J$). It is crucial to notice that if $x \in Q \cap J$, then every point y in $W_x \cap Q$ also belongs to J, hence $W_x \cap Q \cap J = W_x \cap Q$. This implies that the conditional measure $\mu_{Q\cap J}$ assigns to W_x a value larger than or equal to $\mu_Q(W_x)$ (i.e., larger than $2^{-n_x(c+\varepsilon)}$). Thus, the length-information function associated with the partition $\{W_{Q,i}\}$ of $Q \cap J$ is dominated by $c + \varepsilon$.

By the Ergodic Theorem, for large enough m, all points in a set $X' \subset X$ of measure larger than $1 - \delta$ visit the complement of Z no more than $m\delta$ times within the first m iterates. We can require that $N_0/m \leq \delta$. Then X' can be covered by cylinders B corresponding to two-row blocks of length m (where the first row contains the $\mathcal{P}$-name and the second row contains the $\mathcal{Q}$-name) having the following structure: in the first row we have the blocks $W_{Q,i}$ (with varying Q) preceded by portions of lengths not exceeding k (corresponding to the "waiting time" of a point in Z to enter the set J) and no more than $2m\delta$ extra entries (at most $m\delta$ visits in the complement of Z and, at the end, a possible prefix of an incomplete $W_{Q,i}$ of length at most N_0). In the second row, under the first entry of each block $W_{Q,i}$, there occurs the symbol

Q, the same as the first index of $W_{Q,i}$. (We remark that by viewing the block $W_{Q,i}$ alone we may not be able to determine neither Q nor i. Once we read Q from the second row, we can also determine i.) The structure of such a block B is shown on Figure B.1. Of course, a block B may admit many such divisions.

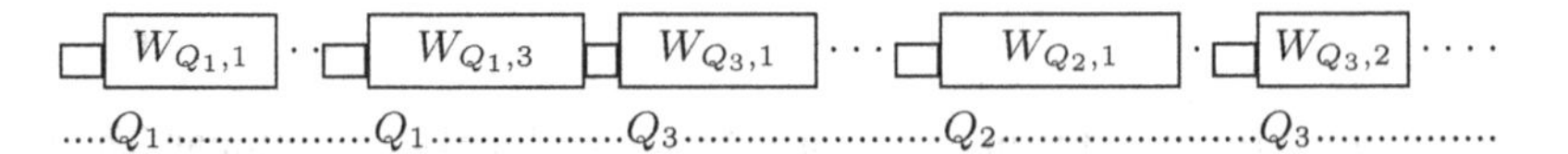

Figure B.1 The structure of the block B. The small rectangles have lengths at most k and correspond to the "waiting time" of a point in Z to enter J.

For each B we fix one such division and for each $Q \in \mathfrak{Q}$ we let $\mathbf{p}_Q = (p_{Q,i})$ be the vector of frequencies of $W_{Q,i}$ among all blocks with the first index Q in this division. By Lemma 1.1.13 and since $1/n_0 < \varepsilon$, no matter what probability vector is obtained, its entropy (not length-entropy but the usual entropy) does not exceed $\mathbf{n}_{\mathbf{p}_Q}(c+2\varepsilon)$, where $\mathbf{n}_{\mathbf{p}_Q} =$ is the average length of the blocks with the first index Q.

We want to estimate the cardinality $\mathbf{C}$ of all such blocks B with a fixed second row. At first we establish the "entry times" to the set J, i.e., the starting places of the blocks $W_{Q,i}$. Since the gaps between these times are at least n_0 (the inverse of which is at most δ), the logarithm of the cardinality of all possibilities is at most (roughly) $mH(\delta, 1-\delta)$. Fixing these entry times corresponds to "marking" some entries in the second row. For each $Q \in \mathfrak{Q}$ we let k_Q be the cardinality of the marked entries equal to Q. Next, we must fill in the blocks $W_{Q,i}$ starting at the marked positions, so that their first indices agree with the marked entries in the second row. Every such filling can be done in steps numbered by the blocks $Q \in \mathfrak{Q}$: in each step we fill only the k_Q blocks with this particular first index Q (and varying i). We must do it so that the corresponding vector of frequencies $\mathbf{p}_Q$ has entropy at most $c_Q = \mathbf{n}_{\mathbf{p}_Q}(c+2\varepsilon)$. The number of possibilities in step Q is at most $\mathbf{C}[1, k_Q, c_Q]$ and $\log(\mathbf{C}[1, k_Q, c_Q]) \leq k_Q(\mathbf{n}_{\mathbf{p}_Q}(c+2\varepsilon)+1)$ (see (2.8.5)). The filling must be such that the number of the remaining places in the first row does not exceed $3m\delta$ ($m\delta$ for the visits outside Z, $m\delta$ for the ending, and at most $k_m/n_0 \leq m\delta$ for the waiting periods to enter J.) These places can be filled in $l^{3m\delta}$ ways. Jointly, the logarithm of the number of all possibilities satisfies

$$\log(\mathbf{C}) \leq mH(\delta, 1-\delta) + \sum_{Q \in \mathfrak{Q}} k_Q(\mathbf{n}_{\mathbf{p}_Q}(c+2\varepsilon)+1) + 3m\delta \log l.$$

Note that $k_Q \mathbf{n}_{\mathbf{p}_Q}$ is the joint length of all blocks with the first index Q, so $\sum_Q k_Q \mathbf{n}_{\mathbf{p}_Q} \leq m$. Also note that $\sum_Q k_Q \leq m/n_0 \leq m\varepsilon$. So,

$$\log(\mathbf{C}) \leq m\left(H(\delta, 1-\delta) + c + 3\varepsilon + 3\delta \log l\right) \leq m(c + 4\varepsilon)$$

(for an appropriate *a priori* choice of δ).

We have proved that the conditional entropy of $\mathcal{P}^m$ on each of the cylinders of $\mathcal{Q}^m$ contained in X' does not exceed $m(c + 4\varepsilon)$. Since the conditional entropy on the remaining cylinders of $\mathcal{Q}^m$ does not exceed $m \log l$, the overall conditional entropy $H(\mathcal{P}^m|\mathcal{Q}^m)$ does not exceed $m((1-\delta)(c+4\varepsilon)+\delta \log l)$. Since ε and δ are arbitrarily small, and $\frac{1}{m}H(\mathcal{P}^m|\mathcal{Q}^m) \geq h(\mathcal{P}|\mathcal{Q})$, we have proved that $c \geq h(\mathcal{P}|\mathcal{Q})$, as desired.

Now, if $\mathcal{P}$ is infinite countable, still with finite conditional static entropy given $\mathfrak{B}$, we invoke the finite partitions $\mathcal{P}_{(m)}$. For each n the conditional information function $I_{\mathcal{P}^n_{(m)}|\mathfrak{B}}$ associated with the partition $\mathcal{P}_{(m)}$ is dominated by the analogous function associated with $\mathcal{P}$, thus it is clear that $\liminf_n \frac{1}{n} I_{\mathcal{P}^n|\mathfrak{B}}$ is not smaller than $\sup_m \liminf_n \frac{1}{n} I_{\mathcal{P}^n_{(m)}|\mathfrak{B}}$, which, by the already proved part for finite partitions and by Fact 2.4.12, is at least $h(\mathcal{P}|\mathfrak{B})$ almost everywhere.

In the second part of the proof we will need a small lemma. Recall that we are using the Martingale Convergence Theorem to establish the almost everywhere convergence $I_{\mathcal{P}|\mathcal{Q}}(x) \to I_{\mathcal{P}|\mathfrak{B}}(x)$ along any sequence $(\mathcal{Q})$ of countable partitions generating $\mathfrak{B}$ (to avoid introducing another index we will treat $(\mathcal{Q})$ as a "self-indexed" sequence). Notice that since we are assuming $H(\mathcal{P}|\mathfrak{B}) < \infty$, and (by Definition 1.4.5) $H(\mathcal{P}|\mathfrak{B}) = \inf_{\mathcal{Q}} H(\mathcal{P}|\mathcal{Q})$, we can choose the sequence $(\mathcal{Q})$ so that also the entropies $H(\mathcal{P}|\mathcal{Q})$ converge to $H(\mathcal{P}|\mathfrak{B})$. Then we have convergence of the information functions and of their integrals, which implies that the convergence holds in $L^1(\mu)$. For any fixed n the ergodic averages $S_{\mathcal{Q},n} = \frac{1}{n}\sum_{i=0}^{n-1} I_{\mathcal{P}|\mathcal{Q}}(T^i x)$ converge (with $\mathcal{Q}$) in $L^1(\mu)$ to the ergodic average $\frac{1}{n}\sum_{i=0}^{n-1} I_{\mathcal{P}|\mathfrak{B}}(T^i x)$, which, for large n, is close (on a large set) to $H(\mathcal{P}|\mathfrak{B})$. On the other hand, with increasing n (while $\mathcal{Q}$ is fixed) the ergodic averages $S_{\mathcal{Q},n}$ converge to $H(\mathcal{P}|\mathcal{Q})$, which, for fine $\mathcal{Q}$, is also close to $H(\mathcal{P}|\mathfrak{B})$. The lemma below will allow us to assume that we have convergence of the double sequence $S_{\mathcal{Q},n}$ to $H(\mathcal{P}|\mathfrak{B})$.

Lemma B.0.3 *Let f_k be a sequence of functions converging in $L^1(\mu)$ to some function f. Then there exists a subsequence k_m such that the double sequence $S_{k_m,n} = \frac{1}{n}\sum_{i=0}^{n-1} f_{k_m}(T^i x)$ converges with (m,n) almost everywhere to $\int f \, d\mu$.*

Proof The proof uses a standard Borel–Cantelli type argument. We only outline it. The subsequence is chosen so that the sequence of $L^1(\mu)$-distances $\|f - f_{k_m}\|$ is summable. Then we let $g_m = \sum_{i=m}^{\infty} |f - f_{k_i}|$. Now, we fix m_0

such that $\int g_{m_0}\,d\mu < \varepsilon$. For almost every $x \in X$ there exists n_0 such that for $n \geq n_0$ the nth ergodic average of g_{m_0} is smaller than ε. Obviously, by monotonicity of the sequence g_m, the same holds for g_m with any $m > m_0$ (and for any $n \geq n_0$). Since the ergodic average of g_m estimates from above the difference between the ergodic average of f_{k_m} and that of f, we obtain the double convergence, as declared. □

We now proceed to proving that $C \leq h(\mathcal{P}|\mathfrak{B})$ starting with the easier inequality

$$C \leq H(\mathcal{P}|\mathfrak{B}). \tag{B.0.4}$$

We shall now fix a number of constants, in a carefully chosen order (so that the way they depend on each other does not loop), according to the interplay between two convergences.

Fix some $\delta > 0$. Recall that

$$C = \sup_{k\geq 1} \limsup_{n\to\infty} \tfrac{1}{n} I_{\mathcal{P}^n|\mathfrak{B}}(T^k x)$$

almost everywhere. So there exists some k such that the above supremum taken only up to k is larger than $C - \delta$ for points x belonging to a large set, say, of measure larger than $1 - \delta$. Notice that each point of this large set falls, after at most k iterates, into the set J of points satisfying

$$\limsup_{n\to\infty} \tfrac{1}{n} I_{\mathcal{P}^n|\mathfrak{B}}(x) > C - \delta.$$

Thus, the measure of J is larger than $\frac{1-\delta}{k}$.

By Lemma B.0.3, we can find a sequence of refining $\mathfrak{B}$-measurable partitions $\mathcal{Q}$, such that the double convergence of the ergodic averages $S_{\mathcal{Q},n} = \frac{1}{n}\sum_{i=0}^{n-1} I_{\mathcal{P}|\mathcal{Q}}(T^i x)$ to $H(\mathcal{P}|\mathfrak{B})$ holds almost everywhere. Thus, there exist n' and $\mathcal{Q}_0$ such that for any $\mathcal{Q}$ (in the selected sequence) finer than $\mathcal{Q}_0$ we have $S_{\mathcal{Q},n}(x) < H(\mathcal{P}|\mathfrak{B}) + \delta$, for any $n \geq n'$ and all x in some large set X', say, of measure at least $1 - \delta'$, where δ' is negligibly small compared with $\frac{1-\delta}{k}$.

We select an integer n_0 larger than both n' and $\frac{\log 2k}{\delta}$. For each $x \in J$ there exists an $n_x \geq n_0$ for which $\frac{1}{n_x} I_{\mathcal{P}^{n_x}|\mathfrak{B}}(x) > C - \delta$. Making the set J a bit smaller (still of measure larger than $\frac{1-\delta}{k}$) we can assume that all values of n_x for $x \in J$ are bounded from above by some constant N_0.

Because the partitions $\mathcal{Q}^{N_0}$ also generate $\mathfrak{B}$, the Martingale Theorem holds along this sequence as well. So, we have

$$\tfrac{1}{n_x} I_{\mathcal{P}^{n_x}|\mathcal{Q}^{N_0}}(x) > C - \delta \tag{B.0.5}$$

for a sufficiently fine partition $\mathcal{Q}$, finer than $\mathcal{Q}_0$ and belonging to the formerly chosen sequence, for each $x \in J$ after another negligible modification (restriction) of the set J. We now enlarge J to be the set of points for which just (B.0.5) is fulfilled for some $n_x \in [n_0, N_0]$, and agree that n_x is the smallest possible choice for $x \in J$. Notice that n_x is constant on the whole cell $W_x \cap Q$ of $\mathcal{P}^{n_x} \vee \mathcal{Q}^{N_0}$ which contains x. In particular this cell is entirely contained in J and such cells form a partition of J. If, moreover, x belongs to X', then, since $n_x \geq n'$ and by the choice of $\mathcal{Q}$, we also have

$$\frac{1}{n_x} \sum_{i=0}^{n_x-1} I_{\mathcal{P}|\mathcal{Q}}(T^i x) \leq H(\mathcal{P}|\mathfrak{B}) + \delta. \tag{B.0.6}$$

This condition holds on the (containing x) cell $W_x \cap Q_x$ of $\mathcal{P}^{n_x} \vee \mathcal{Q}^{n_x}$, in particular on the smaller cell $W_x \cap Q$. At this moment we replace J by the union of all these cells $W_x \cap Q$ contained in J, on which (B.0.6) holds. The measure of J has dropped only insignificantly, by at most δ', so we can safely write that $\mu(J) \geq 1/2k$.

For every $Q \in \mathcal{Q}^{N_0}$ such that $Q \cap J \neq \emptyset$ we have obtained an at most countable partition of $Q \cap J$ by the sets $W_x \cap Q$, where $x \in Q$ and W_x is the cylinder of length n_x over $\mathcal{P}$ containing x. We denote this partition by $\{W_{Q,i}\}$ and the corresponding lengths (of the cylinders W_x) by $n_{Q,i}$.

We need to estimate the length-entropy of the partition $\{W_{Q,i}\}$ for just one (suitable) set $Q \in \mathcal{Q}^{N_0}$. There exists at least one Q such that its intersection with J occupies at least the fraction $1/2k$ of Q in measure. For this particular Q the conditional measures of the cells $W_x \cap Q$ relative to $Q \cap J$ are at most $2k$ times larger than their conditional measures relative to Q. These last measures are, by (B.0.5), at most $2^{-n_{Q,i}(C-\delta)}$. Thus, the length-entropy of $\{W_{Q,i}\}$ (relative to conditional measure $\mu_{Q\cap J}$) is at least $C - \delta - \frac{\log 2k}{n_0}$, which, by the choice of n_0, is larger than $C - 2\delta$.

Now we want to find another measure leading to exponential length-entropy for the same partition with the same lengths. Take the same Q as before (it has essential intersection with J). Notice that for each $x \in Q \cap J$ the sum on the left-hand side of (B.0.6) equals $-\log(\nu_{Q_x}(W_x))$ (recall that $W_x \cap Q_x \in \mathcal{P}^{n_x} \vee \mathcal{Q}^{n_x}$), where ν is the product (i.e., Bernoulli) measure on the formal product space $(\mathcal{P} \vee \mathcal{Q})^{\mathbb{N}_0}$, and each pair of symbols $(A, B) \in \mathcal{P} \vee \mathcal{Q}$ has measure $\mu(A \cap B)$. So,

$$\nu_{Q_x}(W_x) \geq 2^{-n_x(H(\mathcal{P}|\mathfrak{B})+\delta)}. \tag{B.0.7}$$

Since in the independent process the conditional measures do not change as we add conditions from the future, the same value will occur for the conditional measure $\nu_Q(W_x)$ (note that the block corresponding to the cylinder $Q \in \mathcal{Q}^{N_0}$

is an extension to the right of that corresponding to Q_x). Now we are dealing with the sets $W_x \cap Q$ which were denoted as $W_{Q,i}$, and (B.0.7) can be rewritten as

$$\nu_Q(W_{Q,i}) \geq 2^{-n_{Q,i}(H(\mathcal{P}|\mathfrak{B})+\delta)}. \tag{B.0.8}$$

Since the sets $W_{Q,i}$ are disjoint and contained in $Q \cap J$ (this does not depend on the measure), their measures ν_Q add over i to at most 1 and so do the numbers $2^{-n_{Q,i}(H(\mathcal{P}|\mathfrak{B})+\delta)}$. The second assertion of Lemma 1.1.13 implies that $H(\mathcal{P}|\mathfrak{B}) + \delta + 1/n_0 > C - 2\delta$. Since δ and $1/n_0$ are arbitrarily small, $C \leq H(\mathcal{P}|\mathfrak{B})$, as claimed.

The way we replace $H(\mathcal{P}|\mathfrak{B})$ by $h(\mathcal{P}|\mathfrak{B})$ is slightly more delicate than in the unconditional case. The new difficulty is that the conditional information function (at a given point, with respect to a fixed partition and given a conditioning sigma-algebra) is no longer a convex function of the measure, hence the passage to the (not necessarily ergodic) power process requires more attention.

We start the same way as in the nonconditional proof. By Definition 2.3.3, for large n_0 we have $\frac{1}{n_0}H(\mathcal{P}^{n_0}|\mathfrak{B}) \leq h(\mathcal{P}|\mathfrak{B}) + \varepsilon$, so it suffices to prove that

$$n_0 C \leq H(\mathcal{P}^{n_0}|\mathfrak{B}).$$

Consider the power process $(X, \mathcal{P}^{n_0}, \mu, T^{n_0}, \mathbb{S})$. For a point $x \in X$, the cylinder of length mn_0 containing x in the original process is (as a set) the same as the cylinder of length m containing x in the power process. We can write this as

$$I_{\mathcal{P}^{mn_0}|\mathfrak{B}}(x) = I_{(\mathcal{P}^{n_0})^m|\mathfrak{B}}(x),$$

where $(\mathcal{P}^{n_0})^m$ denotes the partition obtained through m steps in the power process. For $n = mn_0 - r$ $(0 \leq r < n_0)$, by simple inclusion of the cylinders, we have

$$\tfrac{1}{n}I_{\mathcal{P}^n|\mathfrak{B}}(x) \leq \tfrac{1}{n}I_{\mathcal{P}^{mn_0}|\mathfrak{B}}(x) = \tfrac{mn_0}{n}\tfrac{1}{mn_0}I_{\mathcal{P}^{mn_0}|\mathfrak{B}}(x),$$

so, at almost every point x, the upper limit $C = \limsup_n \frac{1}{n}I_{\mathcal{P}^n|\mathfrak{B}}(x)$ is attained along a subsequence of mn_0, and then it equals

$$\frac{1}{n_0}\limsup_{m\to\infty}\frac{1}{m}I_{(\mathcal{P}^{n_0})^m|\mathfrak{B}}(x).$$

We have proved that the upper limit analogous to C, computed for the power process is constant almost everywhere and equals $n_0 C$.

If the power process is ergodic, we simply apply the just proved inequality (B.0.4) to the power process, and obtain

$$n_0 C \leq H(\mathcal{P}^{n_0}|\mathfrak{B}),$$

which is exactly what we needed to complete the proof.

We pass to the nonergodic case. To make things easier, we can always agree that n_0 has been chosen a prime number. Then the measure μ in the power process (if it is not ergodic) has precisely n_0 ergodic components $\mu^{(i)}$ ($i = 1, 2, \ldots, n_0$) supported by disjoint sets X_i of equal measures $1/n_0$, and the power process factors to the cycle of n_0 points. We let C_i denote the constants analogous to C computed for the ergodic processes $(X, \mathcal{P}^{n_0}, \mu^{(i)}, T^{n_0}, \mathbb{S})$. We have another choice of two possibilities: either the factor system associated with the subinvariant sigma-algebra $\mathfrak{B}$ is disjoint (in the sense of Furstenberg, see Section 4.4) from the cyclic factor or not. If not, then we argue as follows: The only way for a system to not be disjoint of a cycle of a prime length is to contain it as a factor (this is an elementary fact in ergodic theory). That is, the ergodic sets X_i are measurable with respect to $\mathfrak{B}$. This implies that the conditional information function $I_{\mathcal{P}^{mn_0}|\mathfrak{B}}(x)$ computed for the measure μ is the same as the analogous function computed for $\mu^{(i_x)}$ where i_x is the index for which $x \in X_i$. As a consequence, $C_i = n_0 C$ for each i. Applying (B.0.4) to the ergodic power process we obtain

$$n_0 C = C_i \leq H(\mu^{(i)}, \mathcal{P}^{n_0}|\mathfrak{B}).$$

Now, by concavity of conditional entropy and since μ is the arithmetic average of the measures $\mu^{(i)}$, we get

$$n_0 C \leq \frac{1}{n_0} \sum H(\mu^{(i)}, \mathcal{P}^{n_0}|\mathfrak{B}) \leq H(\mu, \mathcal{P}^{n_0}|\mathfrak{B}),$$

and we are done.

What remains is the case of disjoint factors. Here, every $\mathfrak{B}$-measurable Q is independent of the partition into the sets X_i, i.e., it is cut by them into parts of equal measure μ. But this implies that the conditional measure μ_Q (μ restricted to Q and normalized) remains the arithmetic average of the conditional measures $\mu_Q^{(i)}$. In this case the conditional information function behaves (for this particular combination of measures) like a convex function, and we can apply the same argument as in the nonconditional proof:

$$n_0 C \leq \frac{1}{n_0} \sum_i C_i \leq \frac{1}{n_0} \sum_i H(\mu^{(i)}, \mathcal{P}^{n_0}|\mathfrak{B}) \leq H(\mu, \mathcal{P}^{n_0}|\mathfrak{B}). \qquad \square$$

Symbols

$\mathsf{ord}(x)$ topological order of accumulation of a point, 242
$\mathsf{ord}(\mathfrak{X})$ topological order of accumulation of a space, 242
$\mathcal{O}$ trivial family, 317
$\mathcal{O}_{\varkappa}$ auxuliary family of functions depending on $\varkappa$, 319
$\mathcal{O}_k$ oracle, 277
(p_k) base of an odometer, 362
$\mathbf{p}$ probability vector, 24
$\mathbf{p}_{(m)}$ m-dimensional restriction of a probability vector, 25
$\mathbf{p}_{n,B}$ probability vector of frequencies of blocks of length n in B, 73
$\mathbf{p}(\mu, \mathcal{P})$ distribution vector of a partition, 32
$\mathbf{p}^{\xi}$ barycenter of a distribution ξ on probability vectors, 27
PL class of all Pinsker-like systems, 192
Prob generic denotation of a probability measure, 50
$\mathbf{P}$ set of all countable probability vectors, 23
$\mathbf{P}_m$ set of all m-dimensional probability vectors, 25
$\mathcal{P}$, $\mathcal{Q}$, $\mathcal{R}$ measurable partitions, 30
$\mathcal{P}^{\mathbb{D}}$, $\mathcal{P}^{n}$, $\mathcal{P}^{-n}$, $\mathcal{P}^{+}$, $\mathcal{P}^{-}$, $\mathcal{P}^{\mathbb{S}}$ partitions and sigma-algebras generated by a partition along a set of times, 59
$\mathcal{P}_{(m)}$ m-element partition obtained from $\mathcal{P}$, 38
$\mathcal{P}_{\mathcal{U}}$ partition generated by a cover, 255
$\mathcal{P}_{\Lambda}$ zero-coordinate partition, 58
$\mathfrak{P}_m$ space of all m-element partitions with d_1 (or d_{R}), 42
$\mathfrak{P}_{\mathrm{R}}$ space of all finite entropy partitions with d_{R}, 46
$\mathfrak{P}_{\aleph_0}$ space of all countable partitions with d_1, 42
$r(n,\varepsilon)$ minimal cardinality of an (n,ε)-spanning set, 161
$\binom{R^{(1)} R^{(2)} \dots R^{(q_k)}}{B} = \binom{D}{B}$ $(k+1)$-rectangle, 277
R_B return time to a set B, 82
$\overline{\mathrm{R}}_B$ normalized return time to B, 141
$\mathrm{R}_B^{(k)}$ kth return time to B, 142
$\overline{\mathrm{R}}_B^{(k)}$ normalized kth return time to B, 142
$\mathrm{R}_n(x)$ return time to the cylinder A_x^n, 94
$\mathrm{R}_n^{(k)}(x)$ kth return time to the cylinder A_x^n, 101
$\mathsf{Rep}_n(x)$ maximal force of repelling of A_x^n, 141
$\mathsf{R}(f)$ limit of $\frac{1}{n}\mathsf{L}(f^n)$, 294
$\mathcal{R}_{k,n}$ family of all $(k \times n)$-rectangles, 223
$\mathcal{R}_k$ family of all k-rectangles, 223
$\mathcal{R}_k(X)$ family of all k-rectangles occurring in X, 277
$\mathbb{R}$ set of all real numbers, 18
$s(n,\varepsilon)$ maximal cardinality of an (n,ε)-separated set, 160
$\mathbf{S}$ set of all countable subprobability vectors, 23
$\mathbf{S}_m$ set of all m-dimensional subprobability vectors, 25
$\mathbb{S}$ $(= \mathbb{Z}$ or $\mathbb{N}_0)$ acting semigroup, 57
T, S, R main transformation, 18
T_B induced map on a set B, 82
$T \times S$ product transformation, 116, 172
$T \vee S$ topological joining within a common extension, 173
TD class of all topologically deterministic systems, 191
TEZ class of all systems with topological entropy zero, 191
TPF topological Pinsker factor, 195
$\boldsymbol{T}$ doubly stochastic operator or Markov operator, 313
$\boldsymbol{T}^*$ dual operator on measures, 314
$\mathbb{T}$ unit circle (or $[0,1]$ with $0=1$), 64
$u_{\mathcal{H}}$ smallest repair function, 234
$u_{\alpha}^{\mathcal{H}} = u_{\alpha}$ functions in the transfinite sequence, 236
$u_{\alpha}^{T} = u_{\alpha}$ elements of the transfinite sequence associated with the entropy structure, 266
$U_{<f}^{\varepsilon}$ thickening of the set below the graph, 336
$U_{>f}^{\varepsilon}$ thickening of the set above the graph, 336
$\mathcal{U}$, $\mathcal{V}$ open covers, 162
$\mathcal{U} \otimes \mathcal{V}$ product cover, 173
$\mathcal{U}_{\mathcal{F}}^{\varepsilon}$ cover of $X \times [0,1]$ associated with a family of functions, 336
$\mathbf{v}_{(\mathcal{P},\mathcal{Q})}$, $\mathbf{v}_{(\mathcal{P}_1,\dots,\mathcal{P}_k)}$ vector of static entropies associated with a collection of partitions, 48, 49
V_x^n cell of $\mathcal{V}^n$ containing x, 298
W waiting time in a signal process, 136
W_B waiting time for a set B, 139
$\overline{\mathrm{W}}_B$ normalized waiting time for B, 141
WPL class of all weakly Pinsker-like systems, 192

(x_D, t_D) point selected in the fundamental $(k+1)$-cell D, 219
$(X, \mathfrak{A}, \mu)$ standard probability space, 30
$(X, \mathfrak{A}, \mu, T, \mathbb{S})$ measure-theoretic dynamical system, 57
$(X, \mathcal{P}, \mu, T, \mathbb{S})$ process generated by a partition in a dynamical system, 58
$(X, T, \mathbb{S})$ topological dynamical system, 159
$X \times Y$ product space, 116, 172
$\mathrm{X}_t(\omega)$ stochastic process, 135
Y_c collection of arrays satisfying the column condition, 216
$\mathbb{Z}$ set of all integers, 18
$\alpha, \beta, \gamma, \delta, \varepsilon$ small positive numbers, 26
$\boldsymbol{\alpha}_0$ order of accumulation of entropy, 266
$\boldsymbol{\alpha}_0 = \boldsymbol{\alpha}_0^{\mathcal{H}}$ order of accumulation of $\mathcal{H}$, 237
$\boldsymbol{\alpha}_0(\mu)$ order of accumulation of entropy at a measure, 266
Γ_k set of all vectors of static entropies associated with k partitions, 49
$\widetilde{\Gamma}_k$ set of all vectors of dynamical entropies associated with k partitions, 50
∂A boundary of a set, 177
$\boldsymbol{\delta}_y$ measure concentrated at one point, 180
$\eta(t)$ real function $-t \log t$, 23
ϑ lattice polynomial, 323
θ_κ tails of a net, 228
Θ lattice expression, 324
$\varkappa$ generic denotation of a partition, 319
λ normalized Lebesgue measure, 64
$\boldsymbol{\lambda}$ intensity of a signal process, 136
Λ, Δ alphabets, 58
$(\Lambda^{\mathbb{S}'}, \mu, \sigma, \mathbb{S})$ measure-theoretic symbolic system, 58
μ_B conditional measure on a set B, 32
μ_y disintegration of μ, 35
ξ^x distribution on extreme points with barycenter at x, 357
ξ^μ ergodic decomposition of an invariant measure (distribution on ergodic measures), 176
$\boldsymbol{\xi}$ (maximal) coupling, 365
π, ϕ, ψ factor maps, 18
$\Pi_{\mathcal{P}}$ Pinsker sigma-algebra of a process, 85
Π_μ Pinsker sigma-algebra of a dynamical system, 110
σ shift map, 58
$\Sigma\mathcal{G}$ sum of $\mathcal{G}$, 244
ϱ parameter of circle rotation, 64
$\chi(\mu)$ Lyapunov exponent, 294
$\chi^+(\mu)$ positive Lyapunov exponent, 294
$\overline{\chi}^+(\mu)$ harmonic prolongation of χ^+ at μ, 294
Φ generic denotation of a block code, 97
$(\Omega, \mathfrak{A}, \mu)$ generic notation for a probability space, 135, 347

References

Abadi, M. 2001. Exponential approximation for hitting times in mixing processes. *Math. Phys. Electron. J.*, **7**, Paper 2, (electronic).

Abadi, M. and Galves, A. 2001. Inequalities for the occurrence times of rare events in mixing processes. The state of the art. *Markov Process. Related Fields*, **7**(1), 97–112. Inhomogeneous random systems (Cergy-Pontoise, 2000).

Abramov, L. M. 1959. The entropy of a derived automorphism. *Dokl. Akad. Nauk SSSR*, **128**, 647–650.

Abramov, L. M. and Rokhlin, V. A. 1962. Entropy of a skew product of mappings with invariant measure. *Vestnik Leningrad. Univ.*, **17**(7), 5–13.

Adler, R. L., Konheim, A. G. and McAndrew, M. H. 1965. Topological entropy. *Trans. Amer. Math. Soc.*, **114**, 309–319.

Alicki, R., Andries, J., Fannes, M. and Tuyls, P. 1996. An algebraic approach to the Kolmogorov-Sinai entropy. *Rev. Math. Phys.*, **8**(2), 167–184.

Asimow, L. and Ellis, A. J. 1980. *Convexity Theory and Its Applications in Functional Analysis*. London Mathematical Society Monographs, vol. 16. London: Academic Press Inc. [Harcourt Brace Jovanovich Publishers].

Birkhoff, G. 1946. Three observations on linear algebra. *Univ. Nac. Tucumán. Revista A.*, **5**, 147–151.

Blanchard, F. 1993. A disjointness theorem involving topological entropy. *Bull. Soc. Math. France*, **121**(4), 465–478.

Blanchard, F., Host, B. and Ruette, S. 2002. Asymptotic pairs in positive-entropy systems. *Ergodic Theory Dynam. Systems*, **22**(3), 671–686.

Blanchard, F. and Lacroix, Y. 1993. Zero entropy factors of topological flows. *Proc. Amer. Math. Soc.*, **119**(3), 985–992.

Boltzmann, L. 1877. Über die beziehung dem zweiten Haubtsatze der mechanischen Wärmetheorie und der Wahrscheinlichkeitsrechnung respektive den Sätzen über das Wärmegleichgewicht. *Wiener Berichte*, **76**, 373–435.

Bowen, R. 1971. Entropy for group endomorphisms and homogeneous spaces. *Trans. Amer. Math. Soc.*, **153**, 401–414.

Boyle, M. 1983. Lower entropy factors of sofic systems. *Ergodic Theory Dynam. Systems*, **3**(4), 541–557.

Boyle, M. 1991. Quotients of subshifts. Adler conference lecture. Unpublished.

Boyle, M. and Downarowicz, T. 2004. The entropy theory of symbolic extensions. *Invent. Math.*, **156**(1), 119–161.

Boyle, M. and Downarowicz, T. 2006. Symbolic extension entropy: C^r examples, products and flows. *Discrete Contin. Dyn. Syst.*, **16**(2), 329–341.

Boyle, M., Fiebig, D. and Fiebig, U.-R. 2002. Residual entropy, conditional entropy and subshift covers. *Forum Math.*, **14**(5), 713–757.

Breiman, L. 1957. The individual ergodic theorem of information theory. *Ann. Math. Statist.*, **28**, 809–811.

Brin, M. and Katok, A. 1983. On local entropy, in *Geometic Dynamics* (Rio de Janeiro 1981). Springer Lecture Notes in Mathematics, vol. 1007. Berlin: Springer-Verlag, pp. 30–38.

Bryant, B. F., and Walters, P. 1969. Asymptotic properties of expansive homeomorphisms. *Math. Systems Theory*, **3**, 60–66.

Burguet, D. C^2 surface diffeomorphisms have symbolic extensions. *Invent. Math.* in print.

Burguet, D. 2010. Symbolic extensions for nonuniformly entropy expanding maps. *Colloq. Math.*, **121**, 129–151.

Burguet, D. 2008. Entropie et complexité locale des systèmes dynamiques différentiables. Doctoral Dissertation. Unpublished.

Burguet, D. 2009. A direct proof of the tail variational principle and its extension to maps. *Ergodic Theory Dynam. Systems*, **29**(2), 357–369.

Burguet, D. 2010. Examples of C^r interval map with large symbolic extension entropy. *Discrete Contin. Dyn. Syst.*, **26**(3), 873–899.

Burguet, D. and McGoff, K. Orders of accumulation of entropy. *Fund. Math.* in print.

Burton, R. M., Keane, M. S. and Serafin, J. 2000. Residuality of dynamical morphisms. *Colloq. Math.*, **84/85**(, part 2), 307–317. Dedicated to the memory of Anzelm Iwanik.

Buzzi, J. 1997. Intrinsic ergodicity of smooth interval maps. *Israel J. Math.*, **100**, 125–161.

Carnot, S. 1824. *Reflections on the Motive Power of Fire and on Machines Fitted to Develop that Power.* Paris: Bachelier. French title: *Réflexions sur la puissance motrice du feu et sur les machines propres à développer cette puissance.*

Cheng, W.-C. and Newhouse, S. E. 2005. Pre-image entropy. *Ergodic Theory Dynam. Systems*, **25**(4), 1091–1113.

Choquet, G. 1969. *Lectures on Analysis. Vol. II: Representation Theory.* Edited by J. Marsden, T. Lance and S. Gelbart. New York, Amsterdam: W. A. Benjamin, Inc.

Chung, K. L. 1961. A note on the ergodic theorem of information theory. *Ann. Math. Statist.*, **32**, 612–614.

Ciesielski, K. 1997. *Set Theory for the Working Mathematician.* London Mathematical Society Student Texts, vol. 39. Cambridge: Cambridge University Press.

Clausius, R. 1850. Über die bewegende Kraft der Wärme, Part I, Part II. *Annalen der Physik*, **79**, 368–397, 500–524. English translation "On the Moving Force of Heat, and the Laws regarding the Nature of Heat itself which are deducible therefrom" Phil. Mag. (1851), 2, 1–21, 102–119.

Coelho, Z. 2000. Asymptotic laws for symbolic dynamical systems, in *Topics in Symbolic Dynamics and Applications* (Temuco, 1997). London Mathematical

Society Lecture Note Series, vol. 279. Cambridge: Cambridge University Press, pp. 123–165.

Cover, T. M. and Thomas, J. A. 1991. *Elements of Information Theory*. Wiley Series in Telecommunications. New York: John Wiley & Sons Inc. A Wiley-Interscience Publication.

del Junco, A. 1981. Finitary codes between one-sided Bernoulli shifts. *Ergodic Theory Dynam. Systems*, **1**(3), 285–301 (1982).

Diaconis, P., Holmes, S. and Montgomery, R. 2007. Dynamical bias in the coin toss. *SIAM Rev.*, **49**(2), 211–235.

Dinaburg, E. I. 1970. A correlation between topological entropy and metric entropy. *Dokl. Akad. Nauk SSSR*, **190**, 19–22.

Downarowicz, T. 2001. Entropy of a symbolic extension of a dynamical system. *Ergodic Theory Dynam. Systems*, **21**(4), 1051–1070.

Downarowicz, T. 2005a. Entropy structure. *J. Anal. Math.*, **96**, 57–116.

Downarowicz, T. 2005b. Survey of odometers and Toeplitz flows, in *Algebraic and Topological Dynamics*. Contemporary Mathematics, vol. 385. Providence, RI: American Mathematical Society, pp. 7–37.

Downarowicz, T. 2006. Minimal models for noninvertible and not uniquely ergodic systems. *Israel J. Math.*, **156**, 93–110.

Downarowicz, T. 2008. Faces of simplexes of invariant measures. *Israel J. Math.*, **165**, 189–210.

Downarowicz, T. and Frej, B. 2005. Measure-theoretic and topological entropy of operators on function spaces. *Ergodic Theory Dynam. Systems*, **25**(2), 455–481.

Downarowicz, T., Grzegorek, P. and Lacroix, Y. 2010. Attracting and repelling in homogeneous signal processes. *Nonlinearity*, **23**(11), 2793–2813.

Downarowicz, T. and Huczek, D. Zero-dimensional principal extensions. *Acta Appl. Math.* in print.

Downarowicz, T. and Lacroix, Y. 2011. The law of series. *Ergodic Theory Dynam. Systems*, **31**, 351–367.

Downarowicz, T. and Lacroix, Y. Topological entropy zero and asymptotic pairs. *Israel J. Math.* in print.

Downarowicz, T., Lacroix, Y. and Leandri, D. 2010. Spontaneous clustering in theoretical and some empirical stochastic processes. *ESAIM. Probab. Statist.*, **14**, 256–262.

Downarowicz, T. and Maass, A. 2009. Smooth interval maps have symbolic extensions: the antarctic theorem. *Invent. Math.*, **176**(3), 617–636.

Downarowicz, T. and Newhouse, S. E. 2005. Symbolic extensions and smooth dynamical systems. *Invent. Math.*, **160**(3), 453–499.

Downarowicz, T. and Serafin, J. 2002. Fiber entropy and conditional variational principle in compact non-metrizable spaces. *Fund. Math.*, **172**(3), 217–247.

Downarowicz, T. and Serafin, J. 2003. Possible entropy functions. *Israel J. Math.*, **135**, 221–250.

Downarowicz, T., and Serafin, J. A short proof of the Ornstein Theorem. *Ergodic Theory Dynam. Systems*. in print.

Durand, F. and Maass, A. 2001. Limit laws of entrance times for low-complexity Cantor minimal systems. *Nonlinearity*, **14**(4), 683–700.

Feller, W. 1968. *An Introduction to Probability Theory and Its Applications. Vol. I.* Third edn. New York: John Wiley & Sons Inc.

Fiebig, D., Fiebig, U.-R. and Nitecki, Z. 2003. Entropy and preimage sets. *Ergodic Theory Dynam. Systems*, **23**(6), 1785–1806.

Frej, B. 2006. Maličký-Riečan's entropy as a version of operator entropy. *Fund. Math.*, **189**(2), 185–193.

Frej, B., and Frej, P. An integral formula for entropy of doubly stochastic operators. preprint.

Furstenberg, H. 1967. Disjointness in ergodic theory, minimal sets, and a problem in Diophantine approximation. *Math. Systems Theory*, **1**, 1–49.

Furstenberg, H. 1981. *Recurrence in Ergodic Theory and Combinatorial Number Theory.* Princeton, NJ: Princeton University Press. M. B. Porter Lectures.

Ghys, É., Langevin, R. and Walczak, P. 1986. Entropie mesurée et partitions de l'unité. *C. R. Acad. Sci. Paris Sér. I Math.*, **303**(6), 251–254.

Glasner, S. 2003. *Ergodic Theory via Joinings*. Mathematical Surveys and Monographs, vol. 101. Providence, RI: American Mathematical Society.

Glasner, S. and Weiss, B. 2003. Quasifactors of ergodic systems with positive entropy. *Israel J. Math.*, **134**, 363–380.

Goodman, T. N. T. 1971. Relating topological entropy and measure entropy. *Bull. London Math. Soc.*, **3**, 176–180.

Goodwyn, L. W. 1971. Topological entropy bounds measure-theoretic entropy, in *Global Differentiable Dynamics*. Lecture Notes in Mathematics, vol. 235 Berlin: Springer-Verlag, pp. 69–84.

Grzegorek, P. and Kupsa, M. 2009. Return times in a process generated by a typical partition. *Nonlinearity*, **22**(2), 371–379.

Hart, Klaas Pieter, Nagata, Jun-iti and Vaughan, Jerry E. (eds). 2004. *Encyclopedia of General Topology*. Amsterdam: Elsevier Science Publishers B.V.

Haydn, N., Lacroix, Y. and Vaienti, S. 2005. Hitting and return times in ergodic dynamical systems. *Ann. Probab.*, **33**(5), 2043–2050.

Hirata, M., Saussol, B. and Vaienti, S. 1999. Statistics of return times: a general framework and new applications. *Comm. Math. Phys.*, **206**(1), 33–55.

Hirsch, M. W. 1994. *Differential Topology*. Graduate Texts in Mathematics, vol. 33. New York: Springer-Verlag. Corrected reprint of the 1976 original.

Hurley, M. 1995. On topological entropy of maps. *Ergodic Theory Dynam. Systems*, **15**(3), 557–568.

Ionescu-Tulcea, C. T. 1949. Mesures dan les espaces produits. *Atti Accad. Naz. Lincei. Rend. Cl. Sci. Fis. Mat. Nat. (8)*, **7**, 208–211 (1950).

Jung, C. G. 1977. *Jung on Synchronicity and the Paranormal: Key Readings*. Routledge.

Jung, C. G. and Pauli, W. 1955. *The Interpretation of Nature and Psyche*. New York: Pantheon Books.

Kac, M. 1947. On the notion of recurrence in discrete stochastic processes. *Bull. Amer. Math. Soc.*, **53**, 1002–1010.

Kalikow, S. A. 1982. T, T^{-1} transformation is not loosely Bernoulli. *Ann. of Math. (2)*, **115**(2), 393–409.

Kamiński, B. and de Sam Lazaro, J. 2000. A note on the entropy of a doubly stochastic operator. *Colloq. Math.*, **84/85**(, part 1), 245–254. Dedicated to the memory of Anzelm Iwanik.

Kamiński, B., Siemaszko, A. and Szymański, J. 2003. The determinism and the Kolmogorov property in topological dynamics. *Bull. Polish Acad. Sci. Math.*, **51**(4), 401–417.

Kammerer, P. 1919. *Das Gesetz der Serie. Eine Lehre von den Wiederholungen im Lebens- und Weltgeschehen.* Stuttgart/Berlin: Deutsche Verlags-Anstalt. Mit 8 Tafeln und 26 Abb.

Katok, A. 1980. Lyapunov exponents, entropy and periodic orbits for diffeomorphisms. *Inst. Hautes Études Sci. Publ. Math.*, 137–173.

Katok, A. 2007. Fifty years of entropy in dynamics: 1958–2007. *J. Mod. Dyn.*, **1**(4), 545–596.

Katok, A. and Hasselblatt, B. 1995. *Introduction to the Modern Theory of Dynamical Systems.* Encyclopedia of Mathematics and its Applications, vol. 54. Cambridge: Cambridge University Press. With a supplementary chapter by Katok and Leonardo Mendoza.

Keane, M. S. and Smorodinsky, M. 1979. Bernoulli schemes of the same entropy are finitarily isomorphic. *Ann. of Math. (2)*, **109**(2), 397–406.

Kolmogorov, A. N. 1958. A new metric invariant of transient dynamical systems and automorphisms in Lebesgue spaces. *Dokl. Akad. Nauk SSSR (N.S.)*, **119**, 861–864.

Kolmogorov, A. N. 1959. Entropy per unit time as a metric invariant of automorphisms. *Dokl. Akad. Nauk SSSR*, **124**, 754–755.

Krengel, U. 1985. *Ergodic Theorems.* de Gruyter Studies in Mathematics, vol. 6. Berlin: Walter de Gruyter & Co. With a supplement by Antoine Brunel.

Krieger, W. 1970. On entropy and generators of measure-preserving transformations. *Trans. Amer. Math. Soc.*, **149**, 453–464.

Kruskal, W. 1988. Miracles and statistics: the casual assumption of independence. *J. Amer. Statist. Assoc.*, **83**(404), 929–940.

Kryloff, N. and Bogoliouboff, N. 1937. La théorie générale de la mesure dans son application à l'étude des systèmes dynamiques de la mécanique non linéaire. *Ann. of Math. (2)*, **38**(1), 65–113.

Langevin, R. and Walczak, P. 1994. Entropy, transverse entropy and partitions of unity. *Ergodic Theory Dynam. Systems*, **14**(3), 551–563.

Ledrappier, F. 1979. A variational principle for the topological conditional entropy, in *Ergodic theory, Proc., Oberwolfach 1978.* Lect. Notes Math. 729, Springer. pp. 78–88.

Ledrappier, F. and Walters, P. 1977. A relativised variational principle for continuous transformations. *J. London Math. Soc. (2)*, **16**(3), 568–576.

Lindenstrauss, E. 1999. Mean dimension, small entropy factors and an embedding theorem. *Inst. Hautes Études Sci. Publ. Math.*, 227–262 (2000).

Lindenstrauss, E. and Weiss, B. 2000. Mean topological dimension. *Israel J. Math.*, **115**, 1–24.

Liu, P.-D. 1998. Pesin's entropy formula for endomorphisms. *Nagoya Math. J.*, **150**, 197–209.

Makarov, I. I. 2000. Dynamical entropy for Markov operators. *J. Dynam. Control Systems*, **6**(1), 1–11.

Makarychev, K., Makarychev, Y., Romashchenko, A. and Vereshchagin, N. 2002. A new class of non-Shannon-type inequalities for entropies. *Commun. Inf. Syst.*, **2**(2), 147–165.

Maličký, P., and Riečan, B. 1987. On the entropy of dynamical systems, in *Proceedings of the Conference on Ergodic Theory and Related Topics, II (Georgenthal, 1986).* Teubner-Texte Math., vol. 94, Leipzig: Teubner. pp. 135–138.

Maxwell, J. C. 1871. *Theory of Heat.* Unveränderter Nachdruck der ersten Auflage von 1932. Die Grundlehren der mathematischen Wissenschaften, Band 38. Dover Publications, Inc.

McGoff, K. Orders of accumulation of entropy on manifolds. *J. Anal. Math.* in print.

McMillan, B. 1953. The basic theorems of information theory. *Ann. Math. Statistics*, **24**, 196–219.

Mešalkin, L. D. 1959. A case of isomorphism of Bernoulli schemes. *Dokl. Akad. Nauk SSSR*, **128**, 41–44.

Misiurewicz, M. 1976. Topological conditional entropy. *Studia Math.*, **55**(2), 175–200.

Moisset, J. 2000. *La loi des séries*. Ed. JMG Editions.

Natanson, I. P. 1950. *Teoriya funkciĭ veščestvennoĭ peremennoĭ.* Gosudarstv. Izdat. Tehn.-Teor. Lit. Moscow-Leningrad.

Neveu, J. 1965. *Mathematical Foundations of the Calculus of Probability.* Translated by Amiel Feinstein. San Francisco, CA: Holden-Day Inc.

Newhouse, S. E. 1989. Continuity properties of entropy. *Ann. of Math.*, **129**(2), 215–235.

Newhouse, S. E. 1990. Corrections to: "Continuity properties of entropy" [Ann. of Math. (2) **129** (1989), no. 2, 215–235]. *Ann. of Math.*, **131**(2), 409–410.

Nitecki, Z. and Przytycki, F. 1999. Preimage entropy for mappings. *Internat. J. Bifur. Chaos Appl. Sci. Engrg.*, **9**(9), 1815–1843. Discrete dynamical systems.

Ornstein, D. S. 1970a. Bernoulli shifts with the same entropy are isomorphic. *Advances in Math.*, **4**, 337–352 (1970).

Ornstein, D. S. 1970b. Factors of Bernoulli shifts are Bernoulli shifts. *Advances in Math.*, **5**, 349–364 (1970).

Ornstein, D. S. 1970c. Two Bernoulli shifts with infinite entropy are isomorphic. *Advances in Math.*, **5**, 339–348 (1970).

Ornstein, D. S. 1973. An example of a Kolmogorov automorphism that is not a Bernoulli shift. *Advances in Math.*, **10**, 49–62.

Ornstein, D. S. and Shields, P. C. 1973. An uncountable family of K-automorphisms. *Advances in Math.*, **10**, 63–88.

Ornstein, D. S. and Weiss, B. 1975. Every transformation is bilaterally deterministic. *Israel J. Math.*, **21**(2-3), 154–158. Conference on Ergodic Theory and Topological Dynamics (Kibbutz Lavi, 1974).

Ornstein, D. S. and Weiss, B. 1993. Entropy and data compression schemes. *IEEE Trans. Inform. Theory*, **39**(1), 78–83.

Oseledets, V. I. 1968. A multiplicative ergodic theorem. Characteristic Ljapunov, exponents of dynamical systems. *Trudy Moskov. Mat. Obšč.*, **19**, 179–210.

Oxtoby, J. C. 1952. Ergodic sets. *Bull. Amer. Math. Soc.*, **58**, 116–136.

Pesin, Y. B. 1977. Characteristic Ljapunov exponents, and smooth ergodic theory. *Uspehi Mat. Nauk*, **32**(4 (196)), 55–112, 287.

Petersen, K. 1983. *Ergodic Theory*. Cambridge Studies in Advanced Mathematics, vol. 2. Cambridge: Cambridge University Press.

Phelps, R. R. 2001. *Lectures on Choquet's Theorem*. Second edn. Lecture Notes in Mathematics, vol. 1757. Berlin: Springer-Verlag.

Qian, M., Xie, J.-S. and Zhu, S. 2009. *Smooth Ergodic Theory for Endomorphisms*. Lecture Notes in Mathematics, vol. 1978. Berlin: Springer-Verlag.

Rokhlin, V. A. 1952. On the fundamental ideas of measure theory. *Amer. Math. Soc. Translation*, **1952**(71), 55.

Rokhlin, V. A. and Sinai, Y. G. 1961. The structure and properties of invariant measurable partitions. *Dokl. Akad. Nauk SSSR*, **141**, 1038–1041.

Romagnoli, P.-P. 2003. A local variational principle for the topological entropy. *Ergodic Theory Dynam. Systems*, **23**(5), 1601–1610.

Rudin, W. 1974. *Real and Complex Analysis*. Second edn. McGraw-Hill Series in Higher Mathematics. New York: McGraw-Hill.

Rudin, W. 1991. *Functional Analysis*. Second edn. International Series in Pure and Applied Mathematics. New York: McGraw-Hill.

Rudolph, D. J. 1978. If a two-point extension of a Bernoulli shift has an ergodic square, then it is Bernoulli. *Israel J. Math.*, **30**(1-2), 159–180.

Shannon, C. E. 1948. A Mathematical theory of communication. *Bell System Tech.*, 379–423, 623–656.

Shields, P. C. 1996. *The Ergodic Theory of Discrete Sample Paths*. Graduate Studies in Mathematics, vol. 13. Providence, RI: American Mathematical Society.

Sierpinski, W. 1952. *General Topology*. Mathematical Expositions, No. 7. Toronto: University of Toronto Press. Translated by C. Cecilia Krieger.

Sinai, Y. G. 1959. On the concept of entropy for a dynamic system. *Dokl. Akad. Nauk SSSR*, **124**, 768–771.

Sinai, Y. G. 1962. A weak isomorphism of transformations with invariant measure. *Dokl. Akad. Nauk SSSR*, **147**, 797–800.

Smorodinsky, M. 1972. On Ornstein's isomorphism theorem for Bernoulli shifts. *Advances in Math.*, **9**, 1–9.

Tarski, A. 1955. A lattice-theoretical fixpoint theorem and its applications. *Pacific J. Math.*, **5**, 285–309.

Tong, H. 1952. Some characterizations of normal and perfectly normal spaces. *Duke Math. J.*, **19**, 289–292.

von Mises, R. 1981. *Probability, Statistics and Truth*. English edn. New York: Dover Publications Inc.

von Neumann, J. 1968. *Mathematische Grundlagen der Quantenmechanik*. Unveränderter Nachdruck der ersten Auflage von 1932. Die Grundlehren der mathematischen Wissenschaften, Band 38. Berlin: Springer-Verlag.

Walters, P. 1982. *An Introduction to Ergodic Theory*. Graduate Texts in Mathematics, vol. 79. New York: Springer-Verlag.

Yomdin, Y. 1987. Volume growth and entropy. *Israel J. Math.*, **57**(3), 285–300.

Young, W. H. 1910. On the new theory of integration. *Proc. London Math. Soc.*, **9**, 15–50.

Zhang, Z. and Yeung, R. W. 1997. A non-Shannon-type conditional inequality of information quantities. *IEEE Trans. Inform. Theory*, **43**(6), 1982–1986.

Index

Printed in the United States
By Bookmasters